GROUP THEORY IN PHYSICS

Volume I

Techniques of Physics

Editor

N. H. MARCH

Department of Theoretical Chemistry, University of Oxford, Oxford, England

Techniques of physics find wide application in biology, medicine, engineering and technology generally. This series is devoted to techniques which have found and are finding application. The aim is to clarify the principles of each technique, to emphasize and illustrate the applications and to draw attention to new fields of possible employment.

1. D. C. Champeney: Fourier Transforms and their Physical Applications.
2. J. B. Pendry: Low Energy Electron Diffraction.
3. K. G. Beauchamp: Walsh Functions and their Applications.
4. V. Cappellini, A. G. Constantinides and P. Emiliani: Digital Filters and their applications.
5. G. Rickayzen: Green's Functions and Condensed Matter.
6. M. C. Hutley: Diffraction Gratings.
7. J. F. Cornwell: Group Theory in Physics, Volumes 1 and 2.
8. N. H. March and B. M. Deb: The Single-Particle Density in Physics and Chemistry.
9. D. B. Pearson: Quantum Scattering and Spectral Theory.
10. J. F. Cornwell: Group Theory in Physics Volume 3 (in press).
11. J. Blackledge: Quantitative Coherent Imaging (in press).

GROUP THEORY IN PHYSICS

Volume I

J. F. CORNWELL

Department of Theoretical Physics
University of St Andrews, Scotland

ACADEMIC PRESS

Harcourt Brace & Company, Publishers

London San Diego New York
Boston Sydney Tokyo

ACADEMIC PRESS LIMITED
24–28 Oval Road
London NW1 7DX

United States Edition published by
ACADEMIC PRESS INC.
San Diego, CA 92101

Second Printing 1985
Third Printing 1988
Fourth Printing 1989
Fifth Printing 1994

British Library Cataloguing in Publication Data

Cornwell, J. F.
Group theory in physics.—(Techniques of physics,
ISSN 0308-5392; 7)
Vol. 1
1. Groups, Theory of 2. Mathematical physics
I. Title II. Series
530.1'5222 QC20.7.G76 LCCCN 83-72256

ISBN 0-12-189801-6
0-12-189803-2 (Pbk)

Preface

Twenty years or so ago group theory could have been regarded merely as providing a very valuable tool for the elucidation of the symmetry aspects of physical problems, but recent developments, particularly in theoretical high-energy physics, have transformed its role, so that it now occupies a crucial and indispensable position at the centre of the stage. These developments have taken physicists increasingly deeper into the fascinating world of the pure mathematicians, and have led to an ever-growing appreciation of their achievements. That this recognition is in some respects rather belated is to a large extent due to the unfortunate fact that much of modern pure mathematics is written in a style that outsiders find difficult to penetrate. Consequently one of the main objectives of these two volumes (and particularly of the second) has been to help overcome this unnatural barrier, and to present to theoretical physicists and others the relevant mathematical developments in a form that should be easier to comprehend and appreciate.

The main aim of these two volumes has been to provide a thorough and self-contained account both of those parts of group theory that have been found to be most useful and of their major applications to physical problems. The treatment starts with the basic concepts and is carried right through to some of the most significant and recent developments. The areas of physics that appear include atomic physics, electronic energy bands in solids, vibrations in molecules and solids, and the theory of elementary particles. No prior knowledge of group theory is assumed, and for convenience various relevant algebraic concepts are summarized in Appendices A and B.

It need hardly be said that the title that has been chosen for these volumes, "Group Theory in Physics", does not imply that they contain *every* application of group theoretical ideas to physics, nor that the mathematical concepts contained within them are strictly restricted to those of group theory. Some parts of physics, such as nuclear structure theory, have had to be omitted completely. Moreover, even in those areas that have been discussed, a rigorous selection of topics has had to be made. This is particularly so in applications to elementary-particle theory, where literally thousands of papers involving group theoretical techniques have been written. On the other hand, the mathematical coverage goes outside the strict confines of group theory itself, for one is soon led to the study of Lie algebras, which, although related to Lie groups, are often developed by mathematicians as a separate subject.

No apology should be needed for combining such diverse physical applications in the same work, for it is a manifestation of the power of the theory that it has such wide applicability. However, for the benefit of those readers who may wish to concentrate on specific applications, the following list gives the relevant chapters:

(i) molecular vibrations: Chapters 1, 2, and 4 to 7,
(ii) electronic energy bands in solids: Chapters 1, 2, 3 (Section 5 only), 4, 5, 6, 8 and 9,
(iii) lattice vibrations in solids: Chapters 1, 2, and 4 to 9,
(iv) atomic physics: Chapters 1 to 6 and 10 to 12,
(v) elementary particles: Chapters 1 to 6, 10, 11, 12 (except Sections 6 to 8), and 13 to 19.

In the text the treatments of specific cases are frequently given under the heading of "Examples". The format is such that these are clearly distinguished from the main part of the text, the intention being to indicate that the detailed analysis in the Example is not essential for the general understanding of the rest of that section or succeeding sections. Nevertheless, the examples are important for two reasons. Firstly, they give concrete realizations of the concepts that have just been introduced. Secondly, they indicate how the concepts apply to certain physically important groups or algebras, thereby allowing a "parallel" treatment of a number of specific cases. For instance, many of the properties of the groups SU(2) and SU(3) are developed in a series of such Examples.

The *proofs* of theorems have been divided into three categories. First there are those that by virtue of the direct nature of their arguments assist in the appreciation of the theorem. These are included in the main text. Then there are proofs which are worth recording, if only because it is interesting to see to what extent they retain their validity when the conditions of the theorem are changed slightly, as for example when a Lie algebra is generalized to a Lie superalgebra. The arguments involved in these are usually less direct, and so they have been relegated to appendices. Finally there are proofs that are just too long, or involve ideas that have not been developed in these volumes. For these all that is given is a reference or references to works where they may be found.

In the second volume I have chosen to devote much more space to the development of mathematical techniques than to the treatment of specific physical models, largely because the mathematics is likely to be more durable. Consequently not all of the mathematical results that have been derived are actually used explicitly in the models which are discussed in detail. Indeed these models could be regarded as prototypes indicating what can be achieved by this type of argument, rather than as definitive and conclusive statements about the physical world, although the progress with them and the agreement with experimental observations are extremely encouraging. The development of semi-simple Lie algebras follows the classic approach of Cartan, which has the great advantage of being equally applicable to all cases, and treatments that are valid only for restricted types have been largely neglected. A considerable amount of useful data on semi-simple Lie algebras and groups has been presented in the Appendices, some of this having been specially obtained by computer calculation.

I would like to thank Dr A. Cant for his valuable comments on the first drafts of certain chapters, and Miss L. M. McLean, Mrs J. Kubrycht and Mrs N. Pacholek for the excellence of their typing.

J. F. CORNWELL
St Andrews
January 1984

To my wife Elizabeth and daughters
Rebecca and Jane

Contents

PART B APPLICATIONS IN MOLECULAR AND SOLID STATE PHYSICS 163

Contents of Volume II

Part A

Fundamental Concepts

1

The Basic Framework

1 The concept of a group

The aim of this chapter is to introduce the idea of a group, to give some physically important examples, and then to indicate *immediately* how this notion arises naturally in physical problems, and how the related concept of a group representation lies at the heart of the quantum mechanical formulation. With the basic framework established, the next four chapters will explore in more detail the relevant properties of groups and their representations before the application to physical problems is taken up in earnest in Chapter 6.

To mathematicians a group is an object with a very precise meaning. It is a set of elements that must obey four group axioms. On these is based a most elaborate and fascinating theory, not all of which is covered in this book. The development of the theory does not depend on the nature of the elements themselves, but in most physical applications these elements are transformations of one kind or another, which is why T will be used to denote a typical group member.

Definition *Group* $\mathcal{G}$

A set $\mathcal{G}$ of elements is called a "group" if the following four "group axioms" are satisfied:

(*a*) There exists an operation which associates with every pair of elements T and T′ of $\mathcal{G}$ another element T″ of $\mathcal{G}$. This operation is called *multiplication* and is written as T″ = TT′, T″ being described as the "product of T with T′".

(*b*) For any three elements T, T′ and T″ of $\mathscr{G}$

$$(\mathrm{TT}')\mathrm{T}'' = \mathrm{T}(\mathrm{T}'\mathrm{T}'') \tag{1.1}$$

This is known as the "associative law" for group multiplication. (The interpretation of the left-hand side of Equation (1.1) is that the product TT′ is to be evaluated first, and then multiplied by T″, whereas on the right-hand side T′ is multiplied by the product T′T″.)

(*c*) There exists an *identity element* E which is contained in $\mathscr{G}$ such that

$$\mathrm{TE} = \mathrm{ET} = \mathrm{T}$$

for every element T of $\mathscr{G}$.

(*d*) For each element T of $\mathscr{G}$ there exists an *inverse element* T^{-1} which is also contained in $\mathscr{G}$ such that

$$\mathrm{TT}^{-1} = \mathrm{T}^{-1}\mathrm{T} = \mathrm{E}.$$

This definition covers a diverse range of possibilities, as the following examples indicate.

Example I *The multiplicative group of real numbers*
The simplest example (from which the concept of a group was generalized) is the set of all real numbers (excluding zero) with ordinary multiplication as the group multiplication operation. The axioms (*a*) and (*b*) are obviously satisfied, the identity is the number 1, and each real number t ($\neq 0$) has its reciprocal $1/t$ as its inverse.

Example II *The additive group of real numbers*
To demonstrate that the group multiplication operation need not have any connection with ordinary multiplication, take $\mathscr{G}$ to be the set of all real numbers with ordinary addition as the group multiplication operation. Again axioms (*a*) and (*b*) are obviously satisfied, but in this case the identity is 0 (as $a+0=0+a=a$) and the inverse of a real number a is its negative $-a$ (as $a+(-a)=(-a)+a=0$).

Example III *A finite matrix group*
Many of the groups appearing in physical problems consist of matrices with matrix multiplication as the group multiplication operation. (A brief account of the terminology and properties of matrices is given in Appendix A.) As an example of such a group let $\mathscr{G}$ be the set

of eight matrices

$$\mathbf{M}_1=\begin{bmatrix}1 & 0\\0 & 1\end{bmatrix},\qquad \mathbf{M}_2=\begin{bmatrix}1 & 0\\0 & -1\end{bmatrix},\qquad \mathbf{M}_3=\begin{bmatrix}-1 & 0\\0 & -1\end{bmatrix},$$

$$\mathbf{M}_4=\begin{bmatrix}-1 & 0\\0 & 1\end{bmatrix},\qquad \mathbf{M}_5=\begin{bmatrix}0 & -1\\1 & 0\end{bmatrix},\qquad \mathbf{M}_6=\begin{bmatrix}0 & 1\\-1 & 0\end{bmatrix},$$

$$\mathbf{M}_7=\begin{bmatrix}0 & 1\\1 & 0\end{bmatrix},\qquad \mathbf{M}_8=\begin{bmatrix}0 & -1\\-1 & 0\end{bmatrix}.$$

By explicit calculation it can be verified that the product of any two members of $\mathcal{G}$ is also contained in $\mathcal{G}$, so that axiom (*a*) is satisfied. Axiom (*b*) is automatically true for matrix multiplication, $\mathbf{M}_1$ is the identity of axiom (*c*) as it is a unit matrix, and finally axiom (*d*) is satisfied as

$$\mathbf{M}_1^{-1}=\mathbf{M}_1,\qquad \mathbf{M}_2^{-1}=\mathbf{M}_2,\qquad \mathbf{M}_3^{-1}=\mathbf{M}_3,\qquad \mathbf{M}_4^{-1}=\mathbf{M}_4,$$

$$\mathbf{M}_5^{-1}=\mathbf{M}_6,\qquad \mathbf{M}_6^{-1}=\mathbf{M}_5,\qquad \mathbf{M}_7^{-1}=\mathbf{M}_7,\qquad \mathbf{M}_8^{-1}=\mathbf{M}_8.$$

Example IV *The groups* U(*N*) *and* SU(*N*)

U(*N*) for $N\geqslant 1$ is defined to be the set of all $N\times N$ unitary matrices $\mathbf{u}$ with matrix multiplication as the group multiplication operation. SU(*N*) for $N\geqslant 2$ is defined to be the subset of such matrices $\mathbf{u}$ for which $\det \mathbf{u}=1$, with the same group multiplication operation. (As noted in Appendix A, if $\mathbf{u}$ is unitary then $\det \mathbf{u}=\exp(i\alpha)$, where α is some real number. The "S" of SU(*N*) indicates that SU(*N*) is the "special" subset of U(*N*) for which this α is zero.)

It is easily established that these sets do form groups. Consider first the set U(*N*). As $(\mathbf{u}_1\mathbf{u}_2)^\dagger=\mathbf{u}_2^\dagger\mathbf{u}_1^\dagger$ and $(\mathbf{u}_1\mathbf{u}_2)^{-1}=\mathbf{u}_2^{-1}\mathbf{u}_1^{-1}$, if $\mathbf{u}_1$ and $\mathbf{u}_2$ are both unitary then so is $\mathbf{u}_1\mathbf{u}_2$. Again axiom (*b*) is automatically valid for matrix multiplication and, as the unit matrix $\mathbf{1}_N$ is a member of U(*N*), it provides the identity E of axiom (*c*). Finally, axiom (*d*) is satisfied, as if $\mathbf{u}$ is a member of U(*N*) then so is $\mathbf{u}^{-1}$.

For SU(*N*) the same considerations apply, but in addition if $\mathbf{u}_1$ and $\mathbf{u}_2$ both have determinant 1, Equation (A.4) shows that the same is true of $\mathbf{u}_1\mathbf{u}_2$. Moreover, $\mathbf{1}_N$ is a member of SU(*N*), so it is its identity, and $\mathbf{u}^{-1}$ is a member of SU(*N*) if that is the case for $\mathbf{u}$.

The set of groups SU(*N*) is particularly important in theoretical physics. SU(2) is intimately related to angular momentum and isotopic spin, as will be shown in Chapters 12 and 18, while SU(3) is now famous for its role in the classification of elementary particles, which will be studied in detail in Chapter 18.

Example V *The groups* O(N) *and* SO(N)
The set of all $N \times N$ real orthogonal matrices $\mathbf{R}$ (for $N \geq 2$) is denoted almost universally by O(N), although O(N, $\mathbf{R}$) would have been preferable as it indicates that only real matrices are included. The subset of such matrices $\mathbf{R}$ with $\det \mathbf{R} = 1$ is denoted by SO(N). As will be described in Section 2, O(3) and SO(3) are intimately related to rotations in a real three-dimensional Euclidean space, and so occur time and time again in physical applications.

O(N) and SO(N) are both groups with matrix multiplication as the group multiplication operation, as they can be regarded as being the subsets of U(N) and SU(N) respectively that consist only of *real* matrices. (All that has to be observed to supplement the arguments given in Example IV is that the product of any two real matrices is real, that $\mathbf{1}_N$ is real, and that the inverse of a real matrix is also real.)

If $T_1T_2 = T_2T_1$ for *every* pair of elements T_1 and T_2 of a group $\mathcal{G}$ (that is, if all T_1 and T_2 of $\mathcal{G}$ *commute*), then $\mathcal{G}$ is said to be "Abelian". It will transpire that such groups have relatively straightforward properties. However, many of the groups having physical applications are non-Abelian. Of the cases considered above the only Abelian groups are those of Examples I and II and the groups U(1) and SO(2) of Examples IV and V. (One of the non-commuting pairs of products of Example III which makes that group non-Abelian is $\mathbf{M}_5\mathbf{M}_7 = \mathbf{M}_4$, $\mathbf{M}_7\mathbf{M}_5 = \mathbf{M}_2$.)

The "order" of $\mathcal{G}$ is defined to be the number of elements in $\mathcal{G}$, which may be finite, countably infinite, or even non-countably infinite. A group with finite order is called a "finite group". The vast majority of groups that arise in physical situations are either finite groups or are "Lie groups", which are a special type of group of non-countably infinite order whose precise definition will be given in Chapter 3, Section 1. Example III is a finite group of order 8, whereas Examples I, II, IV and V are all Lie groups.

For a finite group the product of every element with every other element is conveniently displayed in a *multiplication table*, from which all information on the structure of the group can subsequently be deduced. The multiplication table of Example III is given in Table 1.1. (By convention the order of elements in a product is such that the element in the left-hand column precedes the element in the top row, so for example $\mathbf{M}_5\mathbf{M}_8 = \mathbf{M}_2$.) For groups of infinite order the construction of a multiplication table is clearly completely impractical, but fortunately for a Lie group the structure of the group is very largely determined by another *finite* set of relations, namely the commuta-

	$\mathbf{M}_1$	$\mathbf{M}_2$	$\mathbf{M}_3$	$\mathbf{M}_4$	$\mathbf{M}_5$	$\mathbf{M}_6$	$\mathbf{M}_7$	$\mathbf{M}_8$
$\mathbf{M}_1$	$\mathbf{M}_1$	$\mathbf{M}_2$	$\mathbf{M}_3$	$\mathbf{M}_4$	$\mathbf{M}_5$	$\mathbf{M}_6$	$\mathbf{M}_7$	$\mathbf{M}_8$
$\mathbf{M}_2$	$\mathbf{M}_2$	$\mathbf{M}_1$	$\mathbf{M}_4$	$\mathbf{M}_3$	$\mathbf{M}_8$	$\mathbf{M}_7$	$\mathbf{M}_6$	$\mathbf{M}_5$
$\mathbf{M}_3$	$\mathbf{M}_3$	$\mathbf{M}_4$	$\mathbf{M}_1$	$\mathbf{M}_2$	$\mathbf{M}_6$	$\mathbf{M}_5$	$\mathbf{M}_8$	$\mathbf{M}_7$
$\mathbf{M}_4$	$\mathbf{M}_4$	$\mathbf{M}_3$	$\mathbf{M}_2$	$\mathbf{M}_1$	$\mathbf{M}_7$	$\mathbf{M}_8$	$\mathbf{M}_5$	$\mathbf{M}_6$
$\mathbf{M}_5$	$\mathbf{M}_5$	$\mathbf{M}_7$	$\mathbf{M}_6$	$\mathbf{M}_8$	$\mathbf{M}_3$	$\mathbf{M}_1$	$\mathbf{M}_4$	$\mathbf{M}_2$
$\mathbf{M}_6$	$\mathbf{M}_6$	$\mathbf{M}_8$	$\mathbf{M}_5$	$\mathbf{M}_7$	$\mathbf{M}_1$	$\mathbf{M}_3$	$\mathbf{M}_2$	$\mathbf{M}_4$
$\mathbf{M}_7$	$\mathbf{M}_7$	$\mathbf{M}_5$	$\mathbf{M}_8$	$\mathbf{M}_6$	$\mathbf{M}_2$	$\mathbf{M}_4$	$\mathbf{M}_1$	$\mathbf{M}_3$
$\mathbf{M}_8$	$\mathbf{M}_8$	$\mathbf{M}_6$	$\mathbf{M}_7$	$\mathbf{M}_5$	$\mathbf{M}_4$	$\mathbf{M}_2$	$\mathbf{M}_3$	$\mathbf{M}_1$

Table 1.1 Multiplication table for the group of Example III.

tion relations between the basis elements of the corresponding *real Lie algebra*, as will be explained in detail in Chapter 10.

2 Groups of coordinate transformations

To proceed beyond an intuitive picture of the effect of symmetry operations, it is necessary to specify the operations in a precise algebraic form so that the results of successive operations can be easily deduced. Attention will be confined here to transformations in a real three-dimensional Euclidean space $\mathbb{R}^3$, as most applications in atomic, molecular and solid state physics involve only transformations of this type. (The generalization to Minkowski space–time will be introduced in Example V of Chapter 2, Section 7, and developed in more detail in Chapter 17.)

(*a*) *Rotations*

Let Ox, Oy, Oz be three mutually orthogonal Cartesian axes and let Ox', Oy', Oz' be another set of mutually orthogonal Cartesian axes with the same origin O that is obtained from the first set by a rotation T about a specified axis through O. Let (x, y, z) and (x', y', z') be the coordinates of a fixed point P in the space with respect to these two sets of axes. Then there exists a real orthogonal 3×3 matrix $\mathbf{R}(\mathrm{T})$ which depends on the rotation T, but which is independent of the position of P, such that

$$\mathbf{r}' = \mathbf{R}(\mathrm{T})\mathbf{r}, \tag{1.2}$$

where

$$\mathbf{r}' = \begin{bmatrix} x' \\ y' \\ z' \end{bmatrix} \quad \text{and} \quad \mathbf{r} = \begin{bmatrix} x \\ y \\ z \end{bmatrix}.$$

(Hereafter position vectors will always be considered as 3×1 column matrices in matrix expressions unless otherwise indicated, although for typographical reasons they will often be displayed in the text as 1×3 row matrices.) For example, if T is a rotation through an angle θ in the right-hand screw sense about the axis Ox, then, as indicated in Figures 1.1 and 1.2,

$$x' = x,$$
$$y' = y\cos\theta + z\sin\theta,$$
$$z' = -y\sin\theta + z\cos\theta,$$

so that

$$\mathbf{R}(\mathrm{T}) = \begin{bmatrix} 1 & 0 & 0 \\ 0 & \cos\theta & \sin\theta \\ 0 & -\sin\theta & \cos\theta \end{bmatrix}. \tag{1.3}$$

The matrix $\mathbf{R}(\mathrm{T})$ obeys the orthogonality condition $\tilde{\mathbf{R}}(\mathrm{T}) = \mathbf{R}(\mathrm{T})^{-1}$ because rotations leave invariant the length of every position vector and the angle between every pair of position vectors, that is, they leave invariant the scalar product $\mathbf{r}_1 . \mathbf{r}_2$ of any two position vectors. (Indeed the name "orthogonal" stems from the involvement of such matrices in the transformations being considered here between sets of orthogonal axes.) The proof that $\mathbf{R}(\mathrm{T})$ is orthogonal depends on the

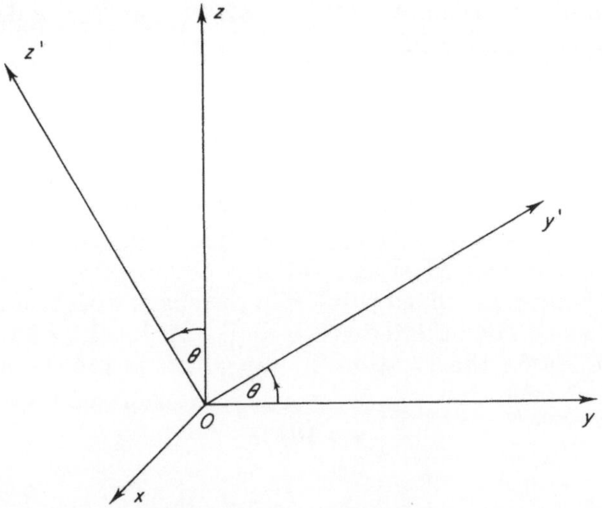

Figure 1.1 Effect of a rotation through an angle θ in the right-hand screw sense about Ox.

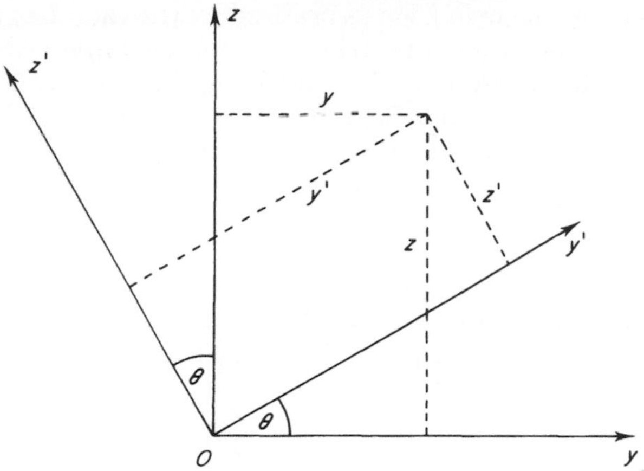

Figure 1.2 The plane containing the axes Oy, Oz, Oy' and Oz' corresponding to the rotation of Figure 1.1.

fact that $\mathbf{r}_1 \cdot \mathbf{r}_2$ can be expressed in matrix form as $\tilde{\mathbf{r}}_1\mathbf{r}_2$. Then, if $\mathbf{r}_1' = \mathbf{R}(\mathrm{T})\mathbf{r}_1$ and $\mathbf{r}_2' = \mathbf{R}(\mathrm{T})\mathbf{r}_2$, it follows that $\mathbf{r}_1' \cdot \mathbf{r}_2' = \tilde{\mathbf{r}}_1'\mathbf{r}_2' = \tilde{\mathbf{r}}_1\tilde{\mathbf{R}}(\mathrm{T})\mathbf{R}(\mathrm{T})\mathbf{r}_2$, which is equal to $\tilde{\mathbf{r}}_1\mathbf{r}_2$ for all $\mathbf{r}_1$ and $\mathbf{r}_2$ if and only if $\tilde{\mathbf{R}}(\mathrm{T})\mathbf{R}(\mathrm{T}) = \mathbf{1}$.

As noted in Appendix A, the orthogonality condition implies that $\det \mathbf{R}(\mathrm{T})$ can take only the values $+1$ or -1. If $\det \mathbf{R}(\mathrm{T}) = +1$ the rotation is said to be "proper"; otherwise it is said to be "improper". The only rotations which can be applied to a rigid body are proper rotations. The transformation of Equation (1.3) gives an example.

The simplest example of an improper rotation is the spatial inversion operation I for which $\mathbf{r}' = -\mathbf{r}$, so that

$$\mathbf{R}(\mathrm{I}) = \begin{bmatrix} -1 & 0 & 0 \\ 0 & -1 & 0 \\ 0 & 0 & -1 \end{bmatrix}.$$

Another important example is the operation of reflection in a plane. For instance, for reflection in the plane Oyz, for which $x' = -x$, $y' = y$, $z' = z$, the transformation matrix is

$$\begin{bmatrix} -1 & 0 & 0 \\ 0 & 1 & 0 \\ 0 & 0 & 1 \end{bmatrix}.$$

The "product" $\mathrm{T}_1\mathrm{T}_2$ of two rotations T_1 and T_2 may be defined to be the rotation whose transformation matrix is given by

$$\mathbf{R}(\mathrm{T}_1\mathrm{T}_2) = \mathbf{R}(\mathrm{T}_1)\mathbf{R}(\mathrm{T}_2). \tag{1.4}$$

(The validity of this definition is assured by the fact that the product of any two real orthogonal matrices is itself real and orthogonal.) In general $\mathbf{R}(T_1)\mathbf{R}(T_2) \neq \mathbf{R}(T_2)\mathbf{R}(T_1)$, in which case $T_1T_2 \neq T_2T_1$. If $\mathbf{r}' = \mathbf{R}(T_2)\mathbf{r}$ and $\mathbf{r}'' = \mathbf{R}(T_1)\mathbf{r}'$, then Equation (1.4) implies that $\mathbf{r}'' = \mathbf{R}(T_1T_2)\mathbf{r}$, so the interpretation of Equation (1.4) is that operation T_2 takes place *before* T_1. This is an example of the general convention that will be applied throughout this book that in any product of operators the operator on the right acts first.

With this definition (Equation (1.4)) every improper rotation can be considered to be the product of the spatial inversion operator I with a proper rotation. For example, for the reflection in the *Oyz* plane

$$\begin{bmatrix} -1 & 0 & 0 \\ 0 & 1 & 0 \\ 0 & 0 & 1 \end{bmatrix} = \begin{bmatrix} -1 & 0 & 0 \\ 0 & -1 & 0 \\ 0 & 0 & -1 \end{bmatrix} \begin{bmatrix} 1 & 0 & 0 \\ 0 & -1 & 0 \\ 0 & 0 & -1 \end{bmatrix},$$

and, as the second matrix on the right-hand side is the transformation of Equation (1.3) with $\theta = \pi$, it corresponds to a rotation through π about *Ox*.

If a set of matrices $\mathbf{R}(T)$ forms a group, then the corresponding set of rotations T also forms a group in which Equation (1.4) defines the group multiplication operator and for which the inverse T^{-1} of T is given by $\mathbf{R}(T^{-1}) = \mathbf{R}(T)^{-1}$. As these two groups have the same structure, they are said to be "isomorphic" (a concept which will be examined in more detail in Chapter 2, Section 6).

Example I *The group of all rotations*
The set of all rotations, both proper and improper, forms a Lie group that is isomorphic to the group O(3) that was introduced in Example V of Section 1.

Example II *The group of all proper rotations*
The set of all proper rotations forms a Lie group that is isomorphic to the group SO(3).

Example III *The crystallographic point group* D_4
A group of rotations that leave invariant a crystal lattice is called a "crystallographic point group", the "point" indicating that one point, the origin *O*, is left unmoved by the operations of the group. There are only 32 such groups, all of which are finite. A complete description is given in Appendix D. The only possible angles of rotation are $2\pi/n$, where $n = 2, 3, 4$, or 6, and it is convenient to denote a proper

rotation through $2\pi/n$ about an axis Oj by C_{nj}. The identity transformation may be denoted by E, so that $\mathbf{R}(E) = \mathbf{1}$, and improper rotations can be written in the form IC_{nj}. As an example consider the crystallographic point group D_4, the notation being that of Schönfliess (1923). D_4 consists of the eight rotations:

E: the identity
C_{2x}, C_{2y}, C_{2z}: proper rotations through π about Ox, Oy, Oz respectively;
C_{4y}, C_{4y}^{-1}: proper rotations through $\frac{1}{2}\pi$ about Oy in the right-hand and left-hand screw senses respectively;
C_{2c}, C_{2d}: proper rotations through π about Oc and Od respectively.

Here Ox, Oy, Oz are mutually orthogonal Cartesian axes, and Oc, Od are mutually orthogonal axes in the plane Oxz with Oc making an angle of $\frac{1}{4}\pi$ with both Ox and Oz, as indicated in Figure 1.3. The

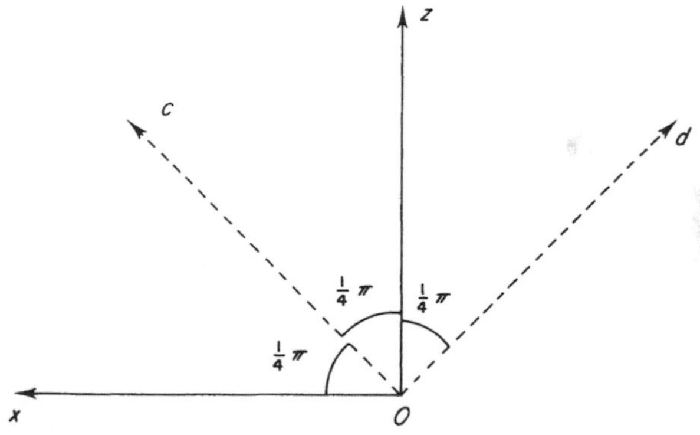

Figure 1.3 The rotation axes Ox, Oz, Oc and Od of the crystallographic point group D_4.

transformation matrices are

$$\mathbf{R}(E) = \begin{bmatrix} 1 & 0 & 0 \\ 0 & 1 & 0 \\ 0 & 0 & 1 \end{bmatrix}, \quad \mathbf{R}(C_{2x}) = \begin{bmatrix} 1 & 0 & 0 \\ 0 & -1 & 0 \\ 0 & 0 & -1 \end{bmatrix}, \quad \mathbf{R}(C_{2y}) = \begin{bmatrix} -1 & 0 & 0 \\ 0 & 1 & 0 \\ 0 & 0 & -1 \end{bmatrix},$$

$$\mathbf{R}(C_{2z}) = \begin{bmatrix} -1 & 0 & 0 \\ 0 & -1 & 0 \\ 0 & 0 & 1 \end{bmatrix}, \quad \mathbf{R}(C_{4y}) = \begin{bmatrix} 0 & 0 & -1 \\ 0 & 1 & 0 \\ 1 & 0 & 0 \end{bmatrix}, \quad \mathbf{R}(C_{4y}^{-1}) = \begin{bmatrix} 0 & 0 & 1 \\ 0 & 1 & 0 \\ -1 & 0 & 0 \end{bmatrix},$$

$$\mathbf{R}(C_{2c}) = \begin{bmatrix} 0 & 0 & 1 \\ 0 & -1 & 0 \\ 1 & 0 & 0 \end{bmatrix}, \quad \mathbf{R}(C_{2d}) = \begin{bmatrix} 0 & 0 & -1 \\ 0 & -1 & 0 \\ -1 & 0 & 0 \end{bmatrix}.$$

	E	C_{2x}	C_{2y}	C_{2z}	C_{4y}	C_{4y}^{-1}	C_{2c}	C_{2d}
E	E	C_{2x}	C_{2y}	C_{2z}	C_{4y}	C_{4y}^{-1}	C_{2c}	C_{2d}
C_{2x}	C_{2x}	E	C_{2z}	C_{2y}	C_{2d}	C_{2c}	C_{4y}^{-1}	C_{4y}
C_{2y}	C_{2y}	C_{2z}	E	C_{2x}	C_{4y}^{-1}	C_{4y}	C_{2d}	C_{2c}
C_{2z}	C_{2z}	C_{2y}	C_{2x}	E	C_{2c}	C_{2d}	C_{4y}	C_{4y}^{-1}
C_{4y}	C_{4y}	C_{2c}	C_{4y}^{-1}	C_{2d}	C_{2y}	E	C_{2z}	C_{2x}
C_{4y}^{-1}	C_{4y}^{-1}	C_{2d}	C_{4y}	C_{2c}	E	C_{2y}	C_{2x}	C_{2z}
C_{2c}	C_{2c}	C_{4y}	C_{2d}	C_{4y}^{-1}	C_{2x}	C_{2z}	E	C_{2y}
C_{2d}	C_{2d}	C_{4y}^{-1}	C_{2c}	C_{4y}	C_{2z}	C_{2x}	C_{2y}	E

Table 1.2 Multiplication table for the crystallographic point group D_4.

The multiplication table is given in Table 1.2. This example will be used to illustrate a number of concepts in Chapters 2, 4, 5 and 6.

(*b*) *Translations*

Suppose now that Ox, Oy, Oz are a set of mutually orthogonal Cartesian axes and $O'x', O'y', O'z'$ are another set, obtained by first rotating the original set about some axis through O by a rotation whose transformation matrix is $\mathbf{R}(T)$, and then translating O to O' along a vector $-\mathbf{t}(T)$ without further rotation. (In $\mathbb{R}^3$ any two sets of Cartesian axes can be related in this way.) Then Equation (1.2) generalizes to

$$\mathbf{r}' = \mathbf{R}(T)\mathbf{r} + \mathbf{t}(T) \tag{1.5}$$

It is useful to regard the rotation and translation as being two parts of a *single* coordinate transformation T, and so it is convenient to rewrite Equation (1.5) as

$$\mathbf{r}' = \{\mathbf{R}(T) \mid \mathbf{t}(T)\}\mathbf{r},$$

thereby defining the composite operator $\{\mathbf{R}(T) \mid \mathbf{t}(T)\}$. Indeed, in non-symmorphic space groups (see Chapter 9, Section 1) there exist symmetry operations in which the combined rotation and translation leave the crystal lattice invariant without this being true for the rotational and translational parts separately.

The generalization of Equation (1.4) can be deduced by considering the two successive transformations $\mathbf{r}' = \{\mathbf{R}(T_2) \mid \mathbf{t}(T_2)\}\mathbf{r} \equiv \mathbf{R}(T_2)\mathbf{r} + \mathbf{t}(T_2)$ and $\mathbf{r}'' = \{\mathbf{R}(T_1) \mid \mathbf{t}(T_1)\}\mathbf{r}' \equiv \mathbf{R}(T_1)\mathbf{r}' + \mathbf{t}(T_1)$, which give

$$\mathbf{r}'' = \mathbf{R}(T_1)\mathbf{R}(T_2)\mathbf{r} + [\mathbf{R}(T_1)\mathbf{t}(T_2) + \mathbf{t}(T_1)]. \tag{1.6}$$

Thus the natural choice of the definition of the "product" T_1T_2 of two general symmetry operations T_1 and T_2 is

$$\{\mathbf{R}(T_1T_2) \mid \mathbf{t}(T_1T_2)\} = \{\mathbf{R}(T_1)\mathbf{R}(T_2) \mid \mathbf{R}(T_1)\mathbf{t}(T_2) + \mathbf{t}(T_1)\}. \tag{1.7}$$

This product always satisfies the group associative law of Equation (1.1).

As Equation (1.5) can be inverted to give

$$\mathbf{r} = \mathbf{R}(T)^{-1}\mathbf{r}' - \mathbf{R}(T)^{-1}\mathbf{t}(T),$$

the inverse of $\{\mathbf{R}(T) \mid \mathbf{t}(T)\}$ may be defined by

$$\{\mathbf{R}(T) \mid \mathbf{t}(T)\}^{-1} = \{\mathbf{R}(T)^{-1} \mid -\mathbf{R}(T)^{-1}\mathbf{t}(T)\}. \tag{1.8}$$

It is easily verified that

$$\{\mathbf{R}(T_1T_2) \mid \mathbf{t}(T_1T_2)\}^{-1} = \{\mathbf{R}(T_2) \mid \mathbf{t}(T_2)\}^{-1}\{\mathbf{R}(T_1) \mid \mathbf{t}(T_1)\}^{-1}, \tag{1.9}$$

the order of factors being reversed on the right-hand side.

It is sometimes convenient to refer to transformations for which $\mathbf{t}(T) = \mathbf{0}$ as "pure rotations" and those for which $\mathbf{R}(T) = \mathbf{1}$ as "pure translations".

3 The group of the Schrödinger equation

(a) *The Hamiltonian operator*

The Hamiltonian operator H of a physical system plays two major roles in quantum mechanics (Schiff 1968). Firstly, its eigenvalues ε, as given by the time-independent Schrödinger equation

$$H\phi = \varepsilon\phi,$$

are the only allowed values of the energy of the system. Secondly, the time-development of the system is determined by a wave function $\psi(t)$ which satisfies the time-dependent Schrödinger equation

$$H\psi = i\hbar\, \partial\psi/\partial t.$$

Not surprisingly, a considerable amount can be learnt about the system by simply examining the set of transformations which leave the Hamiltonian invariant. Indeed the main function of group theory, as it is applied in physical problems, is to systematically extract as much information as possible from this set of transformations.

In order to present the essential features as clearly as possible, it will be assumed in the first instance that the problem involves solving a "single-particle" Schrödinger equation. That is, it will be supposed that either the system contains only one particle or, if there is more than one particle involved, then they do not interact or their inter-particle interactions have been treated in a Hartree–Fock or similar approximation in such a way that each particle experiences only the average field of all of the others. Moreover, it will be assumed that H contains no spin-dependent terms, so that the significant part of every wave function is a scalar function. For example, for an electron in this situation each wave function can be taken to be the product of an "orbital" function, which is a scalar, with one of two possible spin functions, so that the only effect of the electron's spin is to double the "orbital" degeneracy of each energy eigenvalue. (To enable spin-dependent Hamiltonians to be studied, a development of a theory of spinors along similar lines is given in Chapter 6, Section 4.)

With these assumptions a typical Hamiltonian operator for a particle of mass μ has the form

$$\mathrm{H}(\mathbf{r}) = -\frac{\hbar^2}{2\mu}\left(\frac{\partial^2}{\partial x^2}+\frac{\partial^2}{\partial y^2}+\frac{\partial^2}{\partial z^2}\right)+\mathrm{V}(\mathbf{r}), \tag{1.10}$$

where $\mathrm{V}(\mathbf{r})$ is the potential field experienced by the particle. For example, for the electron of a hydrogen atom whose nucleus is located at O,

$$\mathrm{H}(\mathbf{r}) = -\frac{\hbar^2}{2\mu}\left(\frac{\partial^2}{\partial x^2}+\frac{\partial^2}{\partial y^2}+\frac{\partial^2}{\partial z^2}\right)-e^2/\{x^2+y^2+z^2\}^{1/2}. \tag{1.11}$$

In Equations (1.10) and (1.11) the Hamiltonian is written as $\mathrm{H}(\mathbf{r})$ to emphasize its dependence on the particular coordinate system $Oxyz$.

(b) *The invariance of the Hamiltonian operator*

Let $\mathrm{H}(\{\mathbf{R}(\mathrm{T})\mid \mathbf{t}(\mathrm{T})\}\mathbf{r})$ be the operator that is obtained from $\mathrm{H}(\mathbf{r})$ by substituting the components of $\mathbf{r}'=\{\mathbf{R}(\mathrm{T})\mid \mathbf{t}(\mathrm{T})\}\mathbf{r}$ in place of the corresponding components of $\mathbf{r}$. For example, if $\mathrm{H}(\mathbf{r})$ is given by Equation (1.11), then

$$\mathrm{H}(\{\mathbf{R}(\mathrm{T})\mid \mathbf{t}(\mathrm{T})\}\mathbf{r}) = -\frac{\hbar^2}{2\mu}\left(\frac{\partial^2}{\partial x'^2}+\frac{\partial^2}{\partial y'^2}+\frac{\partial^2}{\partial z'^2}\right)-e^2/\{x'^2+y'^2+z'^2\}^{1/2}. \tag{1.12}$$

$\mathrm{H}(\{\mathbf{R}(\mathrm{T})\mid \mathbf{t}(\mathrm{T})\mathbf{r}\}$ can then be rewritten so that it depends *explicitly* on $\mathbf{r}$. For example, in Equation (1.12), if T is a pure translation $x'=x+t_1$, $y'=y+t_2$, $z'=z+t_3$, then

$$\mathrm{H}(\{\mathbf{R}(\mathrm{T})\mid \mathbf{t}(\mathrm{T})\}\mathbf{r}) = -\frac{\hbar^2}{2\mu}\left(\frac{\partial^2}{\partial x^2}+\frac{\partial^2}{\partial y^2}+\frac{\partial^2}{\partial z^2}\right) - e^2/\{(x+t_1)^2+(y+t_2)^2+(z+t_3)^2\}^{1/2},$$

so that

$$H(\{\mathbf{R}(T) \mid \mathbf{t}(T)\}\mathbf{r}) \neq H(\mathbf{r}),$$

whereas if T is a pure rotation about O, then a short algebraic calculation gives

$$H(\{\mathbf{R}(T) \mid \mathbf{t}(T)\}\mathbf{r}) = -\frac{\hbar^2}{2\mu}\left(\frac{\partial^2}{\partial x^2}+\frac{\partial^2}{\partial y^2}+\frac{\partial^2}{\partial z^2}\right)-e^2/\{x^2+y^2+z^2\}^{1/2}$$

and hence in this case

$$H(\{\mathbf{R}(T) \mid \mathbf{t}(T)\}\mathbf{r}) = H(\mathbf{r}).$$

A coordinate transformation T for which

$$H(\{\mathbf{R}(T) \mid \mathbf{t}(T)\}\mathbf{r}) = H(\mathbf{r}) \tag{1.13}$$

is said to leave the Hamiltonian "invariant". For the hydrogen atom the above analysis merely explicitly demonstrates the intuitively obvious fact that the system is invariant under pure rotations but not under pure translations.

The following key theorem shows how and why group theory plays such a significant part in quantum mechanics.

Theorem I The set of coordinate transformations that leave the Hamiltonian invariant form a group. This group is usually called "the group of the Schrödinger equation", but is sometimes referred to as "the invariance group of the Hamiltonian operator".

Proof It has only to be verified that the four group axioms are satisfied. Firstly, if the Hamiltonian is invariant under two separate coordinate transformations T_1 and T_2, then it is invariant under their product T_1T_2. (Invariance under T_1 implies that $H(\mathbf{r}'') = H(\mathbf{r}')$, where $\mathbf{r}'' = \{\mathbf{R}(T_1) \mid \mathbf{t}(T_1)\}\mathbf{r}'$, and invariance under T_2 implies that $H(\mathbf{r}') = H(\mathbf{r})$, where $\mathbf{r}' = \{\mathbf{R}(T_2) \mid \mathbf{t}(T_2)\}\mathbf{r}$, so that $H(\mathbf{r}'') = H(\mathbf{r})$, where, by Equation (1.7), $\mathbf{r}'' = \{\mathbf{R}(T_1T_2) \mid \mathbf{t}(T_1T_2)\}\mathbf{r}$). Secondly, as noted in Section 2(b), the associative law is valid for *all* coordinate transformations. Thirdly, the identity transformation obviously leaves the Hamiltonian invariant, and finally, as Equation (1.13) can be rewritten as $H(\mathbf{r}') = H(\{\mathbf{R}(T) \mid \mathbf{t}(T)\}^{-1}\mathbf{r}')$, where $\mathbf{r}' = \{\mathbf{R}(T) \mid \mathbf{t}(T)\}\mathbf{r}$, if T leaves the Hamiltonian invariant then so does T^{-1}.

For the case of the hydrogen atom, or any other spherically symmetric system in which $V(\mathbf{r})$ is a function of $|\mathbf{r}|$ alone, the group of the Schrödinger equation is the group of all pure rotations in $\mathbb{R}^3$.

(c) *The scalar transformation operators* P(T)

A "scalar field" is *defined* to be a quantity that takes a value at each point in the space $\mathbb{R}^3$ (in general taking different values at different points), the value at a point being independent of the choice of cooordinate system that is used to designate the point. One of the simplest examples to visualize is the density of particles. The concept is relevant to the present consideration because the "orbital" part of an electron's wave function is a scalar field.

Suppose that the scalar field is specified by a function $\psi(\mathbf{r})$ when the coordinates of points of $\mathbb{R}^3$ are defined by a coordinate system Ox, Oy, Oz, and that the same scalar field is specified by a function $\psi'(\mathbf{r}')$ when another coordinate system $O'x'$, $O'y'$, $O'z'$ is used instead. If $\mathbf{r}$ and $\mathbf{r}'$ are the position vectors of the *same* point referred to the two coordinate systems, then the definition of the scalar field implies that

$$\psi'(\mathbf{r}') = \psi(\mathbf{r}). \tag{1.14}$$

Now suppose that $O'x'$, $O'y'$, $O'z'$ are obtained from Ox, Oy, Oz by a coordinate transformation T, so that $\mathbf{r}' = \{\mathbf{R}(\mathrm{T}) \mid \mathbf{t}(\mathrm{T})\}\mathbf{r}$.

Then Equation (1.14) can be written as

$$\psi'(\mathbf{r}') = \psi(\{\mathbf{R}(\mathrm{T}) \mid \mathbf{t}(\mathrm{T})\}^{-1}\mathbf{r}'), \tag{1.15}$$

which provides a concrete prescription for determining the function ψ' from the function ψ, namely that $\psi'(\mathbf{r}')$ is the function obtained by replacing the components of $\mathbf{r}$ in $\psi(\mathbf{r})$ by the corresponding components of $\{\mathbf{R}(\mathrm{T}) \mid \mathbf{t}(\mathrm{T})\}^{-1}\mathbf{r}'$. For example, if $\psi(\mathbf{r}) = x^2y^3$ and T is the pure rotation of Equation (1.3), as

$$\{\mathbf{R}(\mathrm{T}) \mid \mathbf{t}(\mathrm{T})\}^{-1}\mathbf{r}' = \mathbf{R}(\mathrm{T})^{-1}\mathbf{r}' = \tilde{\mathbf{R}}(\mathrm{T})\mathbf{r}'$$
$$= (x', y' \cos\theta - z' \sin\theta, y' \sin\theta + z' \cos\theta),$$

then

$$\psi'(\mathbf{r}') = x'^2(y' \cos\theta - z' \sin\theta)^3.$$

It is very convenient in the following analysis to replace the argument $\mathbf{r}'$ of ψ' by $\mathbf{r}$ (without changing the functional form of ψ'). Thus in the above example

$$\psi'(\mathbf{r}) = x^2(y \cos\theta - z \sin\theta)^3$$

and Equation (1.15) can be rewritten as

$$\psi'(\mathbf{r}) = \psi(\{\mathbf{R}(\mathrm{T}) \mid \mathbf{t}(\mathrm{T})\}^{-1}\mathbf{r}). \tag{1.16}$$

As $\psi'(\mathbf{r})$ is uniquely determined from $\psi(\mathbf{r})$ for the coordinate transformation T, ψ' can be regarded as being obtained from ψ by the action of an *operator* P(T), which is therefore *defined* by $\psi' = P(T)\psi$, or, equivalently, from Equation (1.16) by

$$(P(T)\psi)(\mathbf{r}) = \psi(\{\mathbf{R}(T) \mid \mathbf{t}(T)\}^{-1}\mathbf{r}).$$

The typography can be simplified without causing confusion by removing one of the sets of brackets on the left-hand side, giving

$$P(T)\psi(\mathbf{r}) = \psi(\{\mathbf{R}(T) \mid \mathbf{t}(T)\}^{-1}\mathbf{r}). \tag{1.17}$$

These scalar transformation operators perform a particularly important role in the application of group theory to quantum mechanics. Their properties will now be established.

Clearly $P(T_1) = P(T_2)$ only if $T_1 = T_2$. (Here $P(T_1) = P(T_2)$ means that $P(T_1)\psi(\mathbf{r}) = P(T_2)\psi(\mathbf{r})$ for *every* function $\psi(\mathbf{r})$.) Moreover, each operator P(T) is *linear*, that is

$$P(T)\{a\phi(\mathbf{r}) + b\psi(\mathbf{r})\} = aP(T)\phi(\mathbf{r}) + bP(T)\psi(\mathbf{r}) \tag{1.18}$$

for any two functions $\phi(\mathbf{r})$ and $\psi(\mathbf{r})$ and any two complex numbers a and b, as can be verified directly from Equation (1.17); (see Appendix B, Section 4). The other major properties of the operators P(T) are most succinctly stated in the following four theorems.

Theorem II Each operator P(T) is a *unitary operator* in the Hilbert space L^2 with inner product (ϕ, ψ) defined by

$$(\phi, \psi) = \int \phi^*(\mathbf{r})\psi(\mathbf{r})\, dx\, dy\, dz, \tag{1.19}$$

where the integral is over the whole of the space $\mathbb{R}^3$, that is,

$$(P(T)\phi, P(T)\psi) = (\phi, \psi) \tag{1.20}$$

for any two functions ϕ and ψ of L^2; (see Appendix B, Sections 3 and 4).

Proof With $\mathbf{r}''$ defined by $\mathbf{r}'' = \{\mathbf{R}(T) \mid \mathbf{t}(T)\}^{-1}\mathbf{r}$, from Equations (1.17) and (1.19)

$$(P(T)\phi, P(T)\psi) = \int_{-\infty}^{\infty}\int_{-\infty}^{\infty}\int_{-\infty}^{\infty} \phi^*(\mathbf{r}'')\psi(\mathbf{r}'')\, dx\, dy\, dz. \tag{1.21}$$

However, $dx\, dy\, dz = J\, dx''\, dy''\, dz''$, where the Jacobian J is defined by

$$J = \det \begin{bmatrix} \partial x/\partial x'' & \partial x/\partial y'' & \partial x/\partial z'' \\ \partial y/\partial x'' & \partial y/\partial y'' & \partial y/\partial z'' \\ \partial z/\partial x'' & \partial z/\partial y'' & \partial z/\partial z'' \end{bmatrix}.$$

As $\mathbf{r} = \mathbf{R}(\mathrm{T})\mathbf{r}'' + \mathbf{t}(\mathrm{T})$, it follows that $\partial x/\partial x'' = \mathrm{R}(\mathrm{T})_{11}$, $\partial x/\partial y'' = \mathrm{R}(\mathrm{T})_{12}$ etc., so that $J = \det \mathbf{R}(\mathrm{T}) = \pm 1$. In converting the right-hand side of Equation (1.21) to a triple integral with respect to x'', y'', z'', there appears an odd number of interchanges of upper and lower limits for an improper rotation, whereas for a proper rotation there is an even number of such interchanges. (For example, for spatial inversion I, $x'' = -x$, $y'' = -y$, $z'' = -z$, so the upper and lower limits are interchanged three times, while for a rotation through π about Oz the limits are interchanged twice.) Thus in all cases Equation (1.21) can be written as

$$(\mathrm{P}(\mathrm{T})\phi, \mathrm{P}(\mathrm{T})\psi) = \int_{-\infty}^{\infty} \int_{-\infty}^{\infty} \int_{-\infty}^{\infty} \phi^*(\mathbf{r}'')\psi(\mathbf{r}'')\, dx''\, dy''\, dz'',$$

from which Equation (1.20) follows immediately.

Theorem III For any two coordinate transformations T_1 and T_2,

$$\mathrm{P}(\mathrm{T}_1\mathrm{T}_2) = \mathrm{P}(\mathrm{T}_1)\mathrm{P}(\mathrm{T}_2). \tag{1.22}$$

Proof It is required to show that for *any* function $\psi(\mathbf{r})$, $\mathrm{P}(\mathrm{T}_1\mathrm{T}_2)\psi(\mathbf{r}) = \mathrm{P}(\mathrm{T}_1)\mathrm{P}(\mathrm{T}_2)\psi(\mathbf{r})$, where in the right-hand side $\mathrm{P}(\mathrm{T}_2)$ acts first on $\psi(\mathbf{r})$ and $\mathrm{P}(\mathrm{T}_1)$ acts on the resulting expression. Let $\phi(\mathbf{r}) = \mathrm{P}(\mathrm{T}_2)\psi(\mathbf{r})$, so that $\phi(\mathbf{r}) = \psi(\{\mathbf{R}(\mathrm{T}_2) \mid \mathbf{t}(\mathrm{T}_2)\}^{-1}\mathbf{r})$. Then

$$\mathrm{P}(\mathrm{T}_1)\phi(\mathbf{r}) = \phi(\{\mathbf{R}(\mathrm{T}_1) \mid \mathbf{t}(\mathrm{T}_1)\}^{-1}\mathbf{r}) = \psi(\{\mathbf{R}(\mathrm{T}_2) \mid \mathbf{t}(\mathrm{T}_2)\}^{-1}\{\mathbf{R}(\mathrm{T}_1) \mid \mathbf{t}(\mathrm{T}_1)\}^{-1}\mathbf{r}),$$

the last equality being a consequence of the fact that $\phi(\{\mathbf{R}(\mathrm{T}_1) \mid \mathbf{t}(\mathrm{T}_1)\}^{-1}\mathbf{r})$ is by definition the function obtained from $\phi(\mathbf{r})$ by simply replacing the components of $\mathbf{r}$ by the components of $\{\mathbf{R}(\mathrm{T}_1) \mid \mathbf{t}(\mathrm{T}_1)\}^{-1}\mathbf{r}$. Thus, on using Equation (1.9),

$$\mathrm{P}(\mathrm{T}_1)\mathrm{P}(\mathrm{T}_2)\psi(\mathbf{r}) = \psi(\{\mathbf{R}(\mathrm{T}_1\mathrm{T}_2) \mid \mathbf{t}(\mathrm{T}_1\mathrm{T}_2)\}^{-1}\mathbf{r}) = \mathrm{P}(\mathrm{T}_1\mathrm{T}_2)\psi(\mathbf{r}).$$

Theorem IV The set of operators P(T) that correspond to the coordinate transformations T of the group of the Schrödinger equation forms a group that is isomorphic to the group of the Schrödinger equation.

Proof The product $\mathrm{P}(\mathrm{T}_1)\mathrm{P}(\mathrm{T}_2)$, as defined in the proof of the previous theorem, may be taken to specify the group multiplication operation, so that the associative law of axiom (*b*) is satisfied. The previous theorem then implies that group axiom (*a*) is fulfilled, and with P(E) being the identity operator it also implies that the inverse operator $\mathrm{P}(\mathrm{T})^{-1}$ may be defined by $\mathrm{P}(\mathrm{T})^{-1} = \mathrm{P}(\mathrm{T}^{-1})$. Finally, it also indicates that the two groups are isomorphic.

Theorem V For every coordinate transformation T of the group of the Schrödinger equation

$$P(T)H(\mathbf{r}) = H(\mathbf{r})P(T). \tag{1.23}$$

Proof It has to be established that for any $\psi(\mathbf{r})$ and any T of $\mathscr{G}$

$$P(T)\{H(\mathbf{r})\psi(\mathbf{r})\} = H(\mathbf{r})\{P(T)\psi(\mathbf{r})\}. \tag{1.24}$$

Let $\phi(\mathbf{r}) = H(\mathbf{r})\psi(\mathbf{r})$. Then, by Equation (1.17)

$$\begin{aligned} P(T)\phi(\mathbf{r}) &= \phi(\{\mathbf{R}(T) \mid \mathbf{t}(T)\}^{-1}\mathbf{r}) \\ &= H(\{\mathbf{R}(T) \mid \mathbf{t}(T)\}^{-1}\mathbf{r})\psi(\{\mathbf{R}(T) \mid \mathbf{t}(T)\}^{-1}\mathbf{r}) \\ &= H(\{\mathbf{R}(T) \mid \mathbf{t}(T)\}^{-1}\mathbf{r})\{P(T)\psi(\mathbf{r})\}, \end{aligned}$$

from which Equation (1.24) follows by Equation (1.13).

4 The role of matrix representations

Having shown how groups arise naturally in quantum mechanics, in this preliminary survey it remains only to introduce the concept of a group representation and to demonstrate that it too has a fundamental role to play.

Definition *Representation of a group*
If each element T of a group $\mathscr{G}$ can be assigned a non-singular $d \times d$ matrix $\boldsymbol{\Gamma}(T)$ contained in a group of matrices having matrix multiplication as its group multiplication operation in such a way that

$$\boldsymbol{\Gamma}(T_1T_2) = \boldsymbol{\Gamma}(T_1)\boldsymbol{\Gamma}(T_2) \tag{1.25}$$

for every pair of elements T_1 and T_2 of $\mathscr{G}$, then this set of matrices is said to provide a d-dimensional "representation" $\boldsymbol{\Gamma}$ of $\mathscr{G}$.

Example I *A representation of the crystallographic point group* D_4
The group D_4 introduced in Example III of Section 2 has the following two-dimensional representation:

$$\boldsymbol{\Gamma}(E) = \mathbf{M}_1, \quad \boldsymbol{\Gamma}(C_{2x}) = \mathbf{M}_2, \quad \boldsymbol{\Gamma}(C_{2y}) = \mathbf{M}_3, \quad \boldsymbol{\Gamma}(C_{2z}) = \mathbf{M}_4,$$

$$\boldsymbol{\Gamma}(C_{4y}) = \mathbf{M}_5, \quad \boldsymbol{\Gamma}(C_{4y}^{-1}) = \mathbf{M}_6, \quad \boldsymbol{\Gamma}(C_{2c}) = \mathbf{M}_7, \quad \boldsymbol{\Gamma}(C_{2d}) = \mathbf{M}_8,$$

where $\mathbf{M}_1, \mathbf{M}_2, \ldots$ are the 2×2 matrices defined in Example III of Section 1. That Equation (1.25) is satisfied can be verified simply by comparing Tables 1.1 and 1.2.

It will be shown in Chapter 4 that every group has an infinite number of different representations, but they are derivable from a smaller number of basic representations, the so-called "irreducible representations". A finite group has only a finite number of such irreducible representations that are essentially different.

The representations of the group of the Schrödinger equation are of particular interest. The intimate connection between them and the eigenfunctions of the time-independent Schrödinger equation is provided by the notion of "basis functions" of the representations.

Definition *Basis functions of a group of coordinate transformations* $\mathcal{G}$

A set of d linearly independent functions $\psi_1(\mathbf{r}), \psi_2(\mathbf{r}), \ldots, \psi_d(\mathbf{r})$ forms a basis for a d-dimensional representation Γ of $\mathcal{G}$ if, for every coordinate transformation T of $\mathcal{G}$,

$$P(T)\psi_n(\mathbf{r}) = \sum_{m=1}^{d} \Gamma(T)_{mn}\psi_m(\mathbf{r}), \qquad n = 1, 2, \ldots, d. \tag{1.26}$$

The function $\psi_n(\mathbf{r})$ is then said to "transform as the nth row" of the representation Γ.

The definition implies that not only is each function $P(T)\psi_n(\mathbf{r})$ required to be a linear combination of $\psi_1(\mathbf{r}), \psi_2(\mathbf{r}), \ldots, \psi_d(\mathbf{r})$, but the coefficients are required to be equal to specified matrix elements of $\Gamma(T)$. The rather unusual ordering of row and column indices on the right-hand side of Equation (1.26) ensures the consistency of the definition for every product T_1T_2, for, according to Equations (1.18), (1.22), (1.25) and (1.26),

$$\begin{aligned} P(T_1T_2)\psi_n(\mathbf{r}) &= P(T_1)P(T_2)\psi_n(\mathbf{r}) \\ &= P(T_1)\left\{\sum_{m=1}^{d} \Gamma(T_2)_{mn}\psi_m(\mathbf{r})\right\} \\ &= \sum_{m=1}^{d} \Gamma(T_2)_{mn}P(T_1)\psi_m(\mathbf{r}) \\ &= \sum_{m=1}^{d}\sum_{p=1}^{d} \Gamma(T_2)_{mn}\Gamma(T_1)_{pm}\psi_p(\mathbf{r}) \\ &= \sum_{p=1}^{d} \Gamma(T_1T_2)_{pn}\psi_p(\mathbf{r}). \end{aligned}$$

Example II *Some basis functions of the crystallographic point group* D_4

The functions $\psi_1(\mathbf{r}) = x$, $\psi_2(\mathbf{r}) = z$ provide a basis for the representation Γ of D_4 that has been constructed in Example I above, as can be verified by inspection. (This set has been deduced by a method that will be described in detail in Chapter 5, Section 1.)

Theorem I The eigenfunctions of a d-fold degenerate eigenvalue ε of the time-independent Schrödinger equation

$$H(\mathbf{r})\psi(\mathbf{r}) = \varepsilon\psi(\mathbf{r})$$

form a basis for a d-dimensional representation of the group of the Schrödinger equation $\mathcal{G}$.

Proof Let $\psi_1(\mathbf{r}), \psi_2(\mathbf{r}), \ldots, \psi_d(\mathbf{r})$ be a set of linearly independent eigenfunctions of $H(\mathbf{r})$ with eigenvalue ε, so that

$$H(\mathbf{r})\psi_n(\mathbf{r}) = \varepsilon\psi_n(\mathbf{r}), \qquad n = 1, 2, \ldots, d,$$

and any other eigenfunction of $H(\mathbf{r})$ with eigenvalue ε is a linear combination of $\psi_1(\mathbf{r}), \psi_2(\mathbf{r}), \ldots, \psi_d(\mathbf{r})$. For any transformation T of the group of the Schrödinger equation, Equation (1.23) implies that

$$H(\mathbf{r})\{P(T)\psi_n(\mathbf{r})\} = P(T)\{H(\mathbf{r})\psi_n(\mathbf{r})\} = \varepsilon\{P(T)\psi_n(\mathbf{r})\},$$

demonstrating that $P(T)\psi_n(\mathbf{r})$ is also an eigenfunction of $H(\mathbf{r})$ with eigenvalue ε, so that $P(T)\psi_n(\mathbf{r})$ may be written in the form

$$P(T)\psi_n(\mathbf{r}) = \sum_{m=1}^{d} \Gamma(T)_{mn}\psi_m(\mathbf{r}), \qquad n = 1, 2, \ldots, d. \tag{1.27}$$

At this stage the $\Gamma(T)_{mn}$ are merely a set of coefficients with the m, n and T dependence explicitly displayed. For each T the set $\Gamma(T)_{mn}$ can be arranged to form a $d \times d$ matrix $\boldsymbol{\Gamma}(T)$. It will now be shown that

$$\Gamma(T_1T_2)_{mn} = \sum_{p=1}^{d} \Gamma(T_1)_{mp}\Gamma(T_2)_{pn} \tag{1.28}$$

for any two transformations T_1 and T_2 of $\mathcal{G}$, thereby demonstrating that the matrices $\boldsymbol{\Gamma}(T)$ do actually form a representation of $\mathcal{G}$. Equation (1.27) then implies that the eigenfunctions $\psi_1(\mathbf{r}), \psi_2(\mathbf{r}), \ldots, \psi_d(\mathbf{r})$ form a basis for this representation.

From Equation (1.27), with T replaced by T_1, T_2 and T_1T_2 in turn,

$$P(T_1)\psi_p(\mathbf{r}) = \sum_{m=1}^{d} \Gamma(T_1)_{mp}\psi_m(\mathbf{r}), \tag{1.29}$$

$$P(T_2)\psi_n(\mathbf{r}) = \sum_{p=1}^{d} \Gamma(T_2)_{pn}\psi_p(\mathbf{r}), \tag{1.30}$$

$$P(T_1T_2)\psi_n(\mathbf{r}) = \sum_{m=1}^{d} \Gamma(T_1T_2)_{mn}\psi_m(\mathbf{r}). \tag{1.31}$$

From Equations (1.29) and (1.30)

$$P(T_1)P(T_2)\psi_n(\mathbf{r}) = \sum_{m=1}^{d} \sum_{p=1}^{d} \Gamma(T_1)_{mp}\Gamma(T_2)_{pn}\psi_m(\mathbf{r}), \tag{1.32}$$

and, as $P(T_1)P(T_2)\psi_n(\mathbf{r}) = P(T_1T_2)\psi_n(\mathbf{r})$ by Equation (1.22), the right-hand sides of Equations (1.31) and (1.32) must be equal. As the functions $\psi_1(\mathbf{r}), \psi_2(\mathbf{r}), \ldots$ have been assumed to be linearly independent, Equation (1.28) follows on equating coefficients of each $\psi_m(\mathbf{r})$.

This theorem implies that each energy eigenvalue can be labelled by a representation of the group of the Schrödinger equation. In Chapter 12 it will be shown that the familiar categorization of electronic states of an atom into s states, p states, d states etc. is actually just a special case of this type of description. More precisely, every s-state eigenfunction is a basis function of particular representation of the group of rotations in three dimensions, the p-state eigenfunctions are basis functions of another representation of that group, and so on.

Having established a prima facie case that groups and their representations play a significant role in the quantum mechanical study of physical systems, the next chapters will be devoted to a detailed examination of the structure of groups and the theory of their representations. So far only a brief indication has been given of what can be achieved, but the ensuing chapters will show that the group theoretical approach is capable of dealing with a very wide range of profound and detailed questions.

2

The Structure of Groups

1 Some elementary considerations

This section will be devoted to some immediate consequences of the definition of a group that was given in Chapter 1, Section 1. As many statements will be made about the contents of various sets, it is convenient to introduce an abbreviated notation in which "$T \in \mathcal{S}$" means "the element T is a member of the set $\mathcal{S}$" and "$T \notin \mathcal{S}$" means "the element T is not a member of the set $\mathcal{S}$".

The associative law of Equation (1.1) implies that in any product of three or more elements no ambiguity arises if the brackets are removed completely. Moreover, they can be inserted freely around any chosen subset or subsets of elements in the product, provided of course that the order of elements is unchanged. The proof that

$$(T_1T_2)^{-1} = T_2^{-1}T_1^{-1} \tag{2.1}$$

for any $T_1, T_2 \in \mathcal{G}$ provides some examples of this, for

$$(T_2^{-1}T_1^{-1})(T_1T_2) = T_2^{-1}(T_1^{-1}T_1)T_2 = T_2^{-1}ET_2 = T_2^{-1}(ET_2) = T_2^{-1}T_2 = E,$$

there being a similar argument for $(T_1T_2)(T_2^{-1}T_1^{-1})$.

Definition *Subgroup*
A subset $\mathcal{S}$ of a group $\mathcal{G}$ that is itself a group with the same multiplication operation as $\mathcal{G}$ is called a "subgroup" of $\mathcal{G}$.

By convention, a set may be considered to be a subset of itself, so $\mathcal{G}$ can be regarded as being a subgroup of itself. All other subgroups of

$\mathcal{G}$ are called ***proper subgroups***. Obviously the identity E must be a member of every subgroup of $\mathcal{G}$. Indeed one subgroup of $\mathcal{G}$ is the set {E} consisting only of E. It will be shown in Section 4 that if g and s are the orders of $\mathcal{G}$ and $\mathcal{S}$ respectively, then g/s must be an integer.

A concise criterion for a subset of a group to be a subgroup is provided by the following theorem.

Theorem I If $\mathcal{S}$ is a subset of a group $\mathcal{G}$ such that $S'S^{-1} \in \mathcal{S}$ for any two elements S and S′ of $\mathcal{S}$, then $\mathcal{S}$ is a subgroup of $\mathcal{G}$.

Proof It has only to be verified that the group axioms (*a*), (*c*) and (*d*) are satisfied by $\mathcal{S}$, axiom (*b*) being automatically obeyed for any subset of $\mathcal{G}$. Putting $S' = S$ gives $S'S^{-1} = E$, so $E \in \mathcal{S}$ and hence axiom (*c*) is satisfied. Putting $S' = E$ gives $S'S^{-1} = S^{-1}$, so $S^{-1} \in \mathcal{S}$, thereby fulfilling axiom (*d*). Finally, as $S^{-1} \in \mathcal{S}$, $S'(S^{-1})^{-1} = S'S \in \mathcal{S}$, so (*a*) is also true.

Example I *Subgroups of the crystallographic point group* D_4
The group D_4 defined in Chapter 1, Section 2 has the following subgroups:

(i) s = 1 (i.e. g/s = 8): {E};
(ii) s = 2 (i.e. g/s = 4): $\{E, C_{2x}\}$, $\{E, C_{2y}\}$, $\{E, C_{2z}\}$, $\{E, C_{2c}\}$, $\{E, C_{2d}\}$;
(iii) s = 4 (i.e. g/s = 2): $\{E, C_{2x}, C_{2y}, C_{2z}\}$, $\{E, C_{2y}, C_{4y}, C_{4y}^{-1}\}$, $\{E, C_{2y}, C_{2c}, C_{2d}\}$;
(iv) s = 8 (i.e. g/s = 1): $\{E, C_{2x}, C_{2y}, C_{2z}, C_{4y}, C_{4y}^{-1}, C_{2c}, C_{2d}\}$.

The following theorem displays an interesting property of multiplication in a group.

Theorem II For any fixed element T′ of a group $\mathcal{G}$, the sets $\{T'T; T \in \mathcal{G}\}$ and $\{TT'; T \in \mathcal{G}\}$ both contain every element of $\mathcal{G}$ once and only once.

(Here $\{T'T; T \in \mathcal{G}\}$ denotes the set of elements T′T where T varies over the whole of $\mathcal{G}$. For example, in the special case in which $\mathcal{G}$ is a finite group of order g with elements $T_1, T_2, \ldots, T_g$ and $T' = T_n$, this set consists of $T_nT_1, T_nT_2, \ldots, T_nT_g$. The interpretation of $\{TT'; T \in \mathcal{G}\}$ is similar. The theorem is often called the "Rearrangement Theorem", as it asserts that each of the two sets $\{T'T; T \in \mathcal{G}\}$ and $\{TT'; T \in \mathcal{G}\}$ merely consists of the elements of $\mathcal{G}$ rearranged in order.)

Proof An explicit proof will be given for the set $\{T'T; T \in \mathcal{G}\}$, the proof for the other set being similar.

If T'' is any element of $\mathcal{G}$, then with T defined by $T=(T')^{-1}T''$ it follows that $T'T=T''$. Thus $\{T'T; T\in\mathcal{G}\}$ certainly contains every element of $\mathcal{G}$ at least once. Now suppose that $\{T'T; T\in\mathcal{G}\}$ contains some element of $\mathcal{G}$ twice (or more), i.e. for some $T_1, T_2\in\mathcal{G}$ $T'T_1=T'T_2$ but $T_1\neq T_2$. However, these statements are inconsistent, for pre-multiplying the first by $(T')^{-1}$ gives $T_1=T_2$, so no element of $\mathcal{G}$ appears more than once in $\{T'T; T\in\mathcal{G}\}$.

The Rearrangement Theorem implies that in the multiplication table of a finite group every element of the group appears *once and only once* in every *row*, and *once and only once* in every *column*. This provides a useful check on the computation of the multiplication table. Tables 1.1 and 1.2 exemplify these properties.

2 Classes

Whereas in ordinary everyday language the word "class" is often synonymous with the word "set", in the context of group theory a class is defined to be a special type of set. In fact it is a subset of a group having a certain property which causes it to play an important role in representation theory, as will be shown in Chapter 4.

As a preliminary it is necessary to introduce the idea of "conjugate elements" of a group.

Definition *Conjugate elements*
An element T' of a group $\mathcal{G}$ is said to be "conjugate" to another element T of $\mathcal{G}$ if there exists an element X of $\mathcal{G}$ such that

$$T'=XTX^{-1}. \tag{2.2}$$

If T' is conjugate to T, then T is conjugate to T', as Equation (2.2) can be rewritten as $T=X^{-1}T'(X^{-1})^{-1}$. Moreover, if T, T' and T'' are three elements of $\mathcal{G}$ such that T' and T'' are both conjugate to T, then T' is conjugate to T''. This follows because there exist elements X and Y of $\mathcal{G}$ such that $T'=XTX^{-1}$ and $T''=YTY^{-1}$, so that $T'=X(Y^{-1}T''Y)X^{-1}=(XY^{-1})T''(XY^{-1})^{-1}$ (by Equation (2.1)), which has the form of Equation (2.2) as $XY^{-1}\in\mathcal{G}$. It is therefore permissible to talk of a set of *mutually* conjugate elements.

Definition *Class*
A class of a group $\mathcal{G}$ is a set of mutually conjugate elements of $\mathcal{G}$. (For extra precision this is sometimes called a "conjugacy class".)

A class can be constructed from any $T \in \mathcal{G}$ by forming the set of products XTX^{-1} for every $X \in \mathcal{G}$, retaining only the distinct elements. This class contains T itself as $T = ETE^{-1}$.

Example I *Classes of the crystallographic point group* D_4
For the group D_4 this procedure when applied to C_{2x} gives (on using Table 1.2):

$$XC_{2x}X^{-1} = C_{2x} \quad \text{for} \quad X = E, C_{2x}, C_{2y}, C_{2z},$$
$$XC_{2x}X^{-1} = C_{2z} \quad \text{for} \quad X = C_{4y}, C_{4y}^{-1}, C_{2c}, C_{2d}.$$

Thus $\{C_{2x}, C_{2z}\}$ is one of the classes of D_4. The same class would have been found if the procedure had been applied to C_{2z}. D_4 has four other classes, namely $\{E\}$, $\{C_{2y}\}$, $\{C_{4y}, C_{4y}^{-1}\}$ and $\{C_{2c}, C_{2d}\}$, which may be deduced in a similar way.

The properties of classes are conveniently summarized in the following three theorems.

Theorem I

(*a*) Every element of a group $\mathcal{G}$ is a member of some class of $\mathcal{G}$.
(*b*) No element of $\mathcal{G}$ can be a member of two different classes of $\mathcal{G}$.
(*c*) The identity E of $\mathcal{G}$ always forms a class on its own.

Proof

(*a*) As noted above, for any $T \in \mathcal{G}$, $ETE^{-1} = T$, so that T is in the class constructed from itself.
(*b*) Suppose that $T \in \mathcal{G}$ is a member of a class containing T′ and is also a member of a class containing T″. Then T is conjugate to T′ and T″, so T′ and T″ must be conjugate and must therefore be in the same class.
(*c*) For any $X \in \mathcal{G}$, $XEX^{-1} = XX^{-1} = E$, so E forms a class on its own.

Theorem II If $\mathcal{G}$ is an *Abelian* group, *every* element of $\mathcal{G}$ forms a class on its own.

Proof For any T and X of an Abelian group $\mathcal{G}$

$$XTX^{-1} = XX^{-1}T = ET = T,$$

so T forms a class on its own.

Theorem III If $\mathcal{G}$ is a group consisting entirely of pure rotations, no class of $\mathcal{G}$ contains both proper and improper rotations. Moreover, in each class of proper rotations all the rotations are through the same angle. Similarly, in each class of improper rotations the proper parts are all through the same angle.

Proof If T and T′ are two pure rotations in the same class, Equations (1.4) and (2.2) imply that $\mathbf{R}(T') = \mathbf{R}(X)\mathbf{R}(T)\mathbf{R}(X^{-1}$, so that

$$\det \mathbf{R}(T') = \det \mathbf{R}(T) \tag{2.3}$$

and

$$\operatorname{tr} \mathbf{R}(T') = \operatorname{tr} \mathbf{R}(T) \tag{2.4}$$

(see Appendix A). Equation (2.3) shows that T and T′ are either both proper or are both improper. Moreover, for any proper rotation T through an angle θ (in the right- or left-hand screw sense) $\operatorname{tr} \mathbf{R}(T) = 1 + 2\cos\theta$, as will be proved in Chapter 12, Section 2. Equation (2.4) then implies that all proper rotations in a class are through the same angle θ. Finally, by expressing any improper rotation T as the product of the spatial inversion operator I with a proper rotation through an angle θ, it follows that $\operatorname{tr} \mathbf{R}(T) = -\{1 + 2\cos\theta\}$, so all proper parts involved in a class are through the same angle θ.

It should be noted that the converse of the last theorem is not necessarily true, in that there is no requirement for all rotations of the same type to be in the same class. Indeed, in the above example of the point group D_4 the proper rotations C_{2x} and C_{2y} are in different classes, even though they are rotations through the same angle π.

3 Invariant subgroups

The main object of this and the following section is to introduce two concepts that are involved in the construction of factor groups. These groups play an essential part in the treatment of electron spin in atomic, molecular and solid state physics, in the determination of representations of non-symmorphic space groups, such as that of the diamond structure, and in many aspects of elementary particle symmetries.

Definition *Invariant subgroup*
A subgroup $\mathcal{S}$ of a group $\mathcal{G}$ is said to be an "invariant" subgroup if

$$XSX^{-1} \in \mathcal{S} \tag{2.5}$$

for every $S \in \mathcal{S}$ and every $X \in \mathcal{G}$.

Invariant subgroups are sometimes called "normal subgroups" or "normal divisors". Because of the occurrence of the same forms in Equation (2.2) and Condition (2.5), there is a close connection between invariant subgroups and classes.

Theorem I A subgroup $\mathcal{S}$ of a group $\mathcal{G}$ is an ***invariant*** subgroup if and only if $\mathcal{S}$ consists entirely of ***complete*** classes of $\mathcal{G}$.

Proof Suppose first that $\mathcal{S}$ is an invariant subgroup of $\mathcal{G}$. Then if S is any member of $\mathcal{S}$ and T is any member of the same class of $\mathcal{G}$ as S, by Equation (2.2) there exists an element X of $\mathcal{G}$ such that $T = XSX^{-1}$. Condition (2.5) then implies that $T \in \mathcal{S}$, so the whole class of $\mathcal{G}$ containing S is contained in $\mathcal{S}$.

Now suppose that $\mathcal{S}$ consists entirely of complete classes, and let S be any member of $\mathcal{S}$. Then the set of products XSX^{-1} for all $X \in \mathcal{G}$ form the class containing S, which by assumption is contained in $\mathcal{S}$. Thus $XSX^{-1} \in \mathcal{S}$ for all $S \in \mathcal{S}$ and $X \in \mathcal{G}$, so $\mathcal{S}$ is an invariant subgroup of $\mathcal{G}$.

This theorem provides a very easy method of determining which of the subgroups of a group are invariant when the classes have been previously calculated.

Example I ***Invariant subgroups of the crystallographic point group D_4***

For the group D_4 it follows immediately from the lists of subgroups and classes given in Sections 1 and 2 that the invariant subgroups are $\{E\}$, $\{E, C_{2y}\}$, $\{E, C_{2x}, C_{2y}, C_{2z}\}$, $\{E, C_{2y}, C_{4y}, C_{4y}^{-1}\}$, $\{E, C_{2y}, C_{2c}, C_{2d}\}$, and D_4 itself. (The subgroup $\{E, C_{2x}\}$ is not an invariant subgroup as C_{2x} is part of a class $\{C_{2x}, C_{2z}\}$ that is not wholly contained in the subgroup. The same is true of $\{E, C_{2c}\}$ and $\{E, C_{2d}\}$.)

For every $\mathcal{G}$ the trivial subgroups $\{E\}$ and $\mathcal{G}$ are both invariant subgroups.

4 Cosets

Definition *Coset*

Let $\mathcal{S}$ be a subgroup of a group $\mathcal{G}$. Then for any fixed $T \in \mathcal{G}$ (which may or may not be a member of $\mathcal{S}$) the set of elements ST, where S varies over the whole of $\mathcal{S}$, is called the "right coset" of $\mathcal{S}$ with respect to T, and is denoted by $\mathcal{S}$T. Similarly, the set TS is called the "left coset" of $\mathcal{S}$ with respect to T and is denoted by T$\mathcal{S}$.

In particular, if $\mathcal{S}$ is a finite subgroup of order s with elements $S_1, S_2, \ldots, S_s$, then $\mathcal{S}$T is the set of s elements $S_1T, S_2T, \ldots, S_sT$, and T$\mathcal{S}$ is the set of s elements $TS_1, TS_2, \ldots, TS_s$. In the following discussions two sets will be said to be *identical* if they merely contain the same elements, the ordering of the elements within the sets being immaterial.

Example I *Some cosets of the crystallographic point group* D_4
Let $\mathcal{G}$ be D_4 and let $\mathcal{S} = \{E, C_{2x}\}$. Then from Table 1.2 the right cosets are

$$\begin{aligned}
\mathcal{S}E &= \mathcal{S}C_{2x} = \{E, C_{2x}\},\\
\mathcal{S}C_{2y} &= \mathcal{S}C_{2z} = \{C_{2y}, C_{2z}\},\\
\mathcal{S}C_{4y} &= \mathcal{S}C_{2d} = \{C_{4y}, C_{2d}\},\\
\mathcal{S}C_{4y}^{-1} &= \mathcal{S}C_{2c} = \{C_{4y}^{-1}, C_{2c}\},
\end{aligned}$$

and the left cosets are

$$\begin{aligned}
E\mathcal{S} &= C_{2x}\mathcal{S} = \{E, C_{2x}\},\\
C_{2y}\mathcal{S} &= C_{2z}\mathcal{S} = \{C_{2y}, C_{2z}\},\\
C_{4y}\mathcal{S} &= C_{2c}\mathcal{S} = \{C_{4y}, C_{2c}\},\\
C_{4y}^{-1}\mathcal{S} &= C_{2d}\mathcal{S} = \{C_{4y}^{-1}, C_{2d}\}.
\end{aligned}$$

It should be noted that $C_{4y}\mathcal{S} \neq \mathcal{S}C_{4y}$ and $C_{4y}^{-1}\mathcal{S} \neq \mathcal{S}C_{4y}^{-1}$.

This example shows that the right and left cosets $\mathcal{S}$T and T$\mathcal{S}$ formed from the same element $T \in \mathcal{G}$ are not necessarily identical. The properties of cosets are summarized in the following two theorems. The first theorem is stated for right cosets, but every statement applies equally to left cosets. It is worthwhile checking that the above example of the point group D_4 does satisfy all the assertions of this theorem.

Theorem I

(*a*) If $T \in \mathcal{S}$, then $\mathcal{S}T = \mathcal{S}$.
(*b*) If $T \notin \mathcal{S}$, then $\mathcal{S}$T is not a subgroup of $\mathcal{G}$.
(*c*) Every element of $\mathcal{G}$ is a member of some right coset.
(*d*) Any two elements ST and S′T of $\mathcal{S}$T are different, provided that $S \neq S'$. In particular, if $\mathcal{S}$ is a finite subgroup of order s, $\mathcal{S}$T contains s different elements.

(*e*) Two right cosets of $\mathcal{S}$ are either identical or have no elements in common.
(*f*) If $T' \in \mathcal{S}T$, then $\mathcal{S}T' = \mathcal{S}T$.
(*g*) If $\mathcal{G}$ is a finite group of order g and $\mathcal{S}$ has order s, then the number of distinct right cosets is g/s.

Proof

(*a*) If $T \in \mathcal{S}$, the Rearrangement Theorem of Section 1 applied to $\mathcal{S}$ considered as a group shows that $\mathcal{S}T$ is merely a rearrangement of $\mathcal{S}$.
(*b*) If $\mathcal{S}T$ is a subgroup of $\mathcal{G}$, it must contain the identity E, so there must exist an element $S \in \mathcal{S}$ such that $ST = E$. This implies $T = S^{-1}$, so $T \in \mathcal{S}$. Thus if $T \notin \mathcal{S}$, $\mathcal{S}T$ cannot be a subgroup of $\mathcal{G}$.
(*c*) For any $T \in \mathcal{G}$, as $T = ET$ and $E \in \mathcal{S}$, it follows that $T \in \mathcal{S}T$.
(*d*) Suppose that $ST = S'T$ and $S \neq S'$. Post-multiplying by T^{-1} gives $S = S'$, a contradiction.
(*e*) Suppose that $\mathcal{S}T$ and $\mathcal{S}T'$ are two right cosets with a common element. It will be shown that $\mathcal{S}T = \mathcal{S}T'$. Let $ST = S'T'$ be the common element of $\mathcal{S}T$ and $\mathcal{S}T'$. Here $S, S' \in \mathcal{S}$. Then $T'T^{-1} = (S')^{-1}S$, so $T'T^{-1} \in \mathcal{S}$, and hence by (*a*) $\mathcal{S}(T'T^{-1}) = \mathcal{S}$. As $\mathcal{S}(T'T^{-1})$ is the set of elements of the form $ST'T^{-1}$, the set obtained from this by post-multiplying each member by T consists of the elements ST', that is, it is the coset $\mathcal{S}T'$. Thus $\mathcal{S}T' = \mathcal{S}T$.
(*f*) As in (*c*), $T' \in \mathcal{S}T'$. If $T \in \mathcal{S}T'$ then $\mathcal{S}T'$ and $\mathcal{S}T$ have a common element and must therefore be identical by (*e*).
(*g*) Suppose that there are *M* distinct right cosets of $\mathcal{S}$. By (*d*) each contains s different elements, so the collection of distinct cosets contains *Ms* different elements of $\mathcal{G}$. But by (*c*) every element of $\mathcal{G}$ is in this collection of distinct cosets, so $Ms = g$.

The property (*f*) is particularly important. It shows that the same coset is formed starting from *any* member of the coset. All members of a coset therefore appear on an equal footing, so that *any* member of the coset can be taken as the "coset representative" that labels the coset and from which the coset can be constructed. For example, for the right coset $\{C_{4y}, C_{2d}\}$ of the point group D_4, the coset representatives could equally well be chosen to be C_{4y} or C_{2d}.

As the number of distinct right cosets is necessarily a positive integer, property (g) demonstrates that s must divide g, as was mentioned in Section 1.

Theorem II The right and left cosets of a subgroup $\mathcal{S}$ of a group $\mathcal{G}$

are *identical* (i.e. $\mathcal{S}T = T\mathcal{S}$ for all $T \in \mathcal{G}$) if and only if $\mathcal{S}$ is an *invariant* subgroup.

Proof Suppose that $\mathcal{S}$ is an invariant subgroup. It will be shown that if $T' \in \mathcal{S}T$ then $T' \in T\mathcal{S}$. (A similar argument proves that if $T' \in T\mathcal{S}$ then $T' \in \mathcal{S}T$, so, on combining the two, it follows that $\mathcal{S}T = T\mathcal{S}$.) If $T' \in \mathcal{S}T$ there exists an element S of $\mathcal{S}$ such that $T' = ST$. Then $T^{-1}T' = T^{-1}ST$, which is a member of $\mathcal{S}$ as $\mathcal{S}$ is an invariant subgroup. Thus $T^{-1}T' \in \mathcal{S}$, so $T' = T(T^{-1}T')$ must be a member of $T\mathcal{S}$.

Now suppose that $T\mathcal{S} = \mathcal{S}T$ for every $T \in \mathcal{G}$. This implies that for any $S \in \mathcal{S}$ and any $T \in \mathcal{G}$ there exists an $S' \in \mathcal{S}$ such that $TS = S'T$, so $TST^{-1} = S'$ and hence $TST^{-1} \in \mathcal{S}$. Thus $\mathcal{S}$ is an invariant subgroup of $\mathcal{G}$.

Of course in the above example concerning the point group D_4 the subgroup $\mathcal{S} = \{E, C_{2x}\}$ was carefully chosen so as *not* to be an invariant subgroup, in order to demonstrate that right and left cosets are not always identical.

5 Factor groups

Let $\mathcal{S}$ be an *invariant* subgroup of a group $\mathcal{G}$. Each right coset of $\mathcal{S}$ can be considered to be an "element" of the set of distinct right cosets of $\mathcal{S}$, the internal structure of each coset now being disregarded. With the following definition of the product of two right cosets, the set of cosets then forms a group called a "factor group".

Definition *Product of right cosets*
The product of two right cosets $\mathcal{S}T_1$ and $\mathcal{S}T_2$ of an *invariant* subgroup $\mathcal{S}$ is defined by

$$\mathcal{S}T_1 \,.\, \mathcal{S}T_2 = \mathcal{S}(T_1T_2). \tag{2.6}$$

Proof of consistency It will be shown that Equation (2.6) provides a meaningful definition in that if alternative coset representatives are chosen for the cosets on the left-hand side of the equation, the coset on the right-hand side remains unchanged.

Suppose that T_1' and T_2' are alternative coset representatives for $\mathcal{S}T_1$ and $\mathcal{S}T_2$ respectively, so that $T_1' \in \mathcal{S}T_1$ and $T_2' \in \mathcal{S}T_2$. It has to be proved that $\mathcal{S}(T_1'T_2') = \mathcal{S}(T_1T_2)$.

As $T_1' \in \mathcal{S}T_1$ and $T_2' \in \mathcal{S}T_2$, there exist $S, S' \in \mathcal{S}$ such that $T_1' = ST_1$ and $T_2' = S'T_2$. Then $T_1'T_2' = ST_1S'T_2$. But $T_1S' \in T_1\mathcal{S}$, so, as $\mathcal{S}$ is an invariant subgroup, $T_1S' \in \mathcal{S}T_1$. Consequently there exists an $S'' \in \mathcal{S}$ such that $T_1S' = S''T_1$. Then $T_1'T_2' = (SS'')(T_1T_2)$, so that $T_1'T_2' \in \mathcal{S}(T_1T_2)$ and hence, by property (*f*) of the first theorem of Section 4, $\mathcal{S}(T_1'T_2') = \mathcal{S}(T_1T_2)$.

Theorem I The set of right cosets of an *invariant* subgroup $\mathcal{S}$ of a group $\mathcal{G}$ forms a group, with Equation (2.6) defining the group multiplication operation. This group is called a "factor group" and is denoted by $\mathcal{G}/\mathcal{S}$.

Proof It has only to be verified that the four group axioms are satisfied.

(*a*) By Equation (2.6), the product of any two right cosets of $\mathcal{S}$ is itself a right coset of $\mathcal{S}$ and is therefore a member of $\mathcal{G}/\mathcal{S}$.

(*b*) The associative law is valid for coset multiplication because, if $\mathcal{S}T$, $\mathcal{S}T'$ and $\mathcal{S}T''$ are any three right cosets,

$$(\mathcal{S}T \,.\, \mathcal{S}T') \,.\, \mathcal{S}T'' = \mathcal{S}(TT') \,.\, \mathcal{S}T'' = \mathcal{S}((TT')T'')$$

and

$$\mathcal{S}T \,.\, (\mathcal{S}T' \,.\, \mathcal{S}T'') = \mathcal{S}T \,.\, \mathcal{S}(T'T'') = \mathcal{S}(T(T(T'T''))),$$

where the two cosets on the right-hand sides are equal by virtue of the associative law $(TT')T'' = T(T'T'')$ for $\mathcal{G}$.

(*c*) The identity element of $\mathcal{G}/\mathcal{S}$ is $\mathcal{S}E$ $(=\mathcal{S})$, as for any right coset

$$\mathcal{S}E \,.\, \mathcal{S}T = \mathcal{S}(ET) = \mathcal{S}T = \mathcal{S}(TE) = \mathcal{S}T \,.\, \mathcal{S}E.$$

(*d*) The inverse of $\mathcal{S}T$ is $\mathcal{S}(T^{-1})$, as

$$\mathcal{S}T \,.\, \mathcal{S}(T^{-1}) = \mathcal{S}(TT^{-1}) = \mathcal{S}E = \mathcal{S}(T^{-1}T) = \mathcal{S}(T^{-1}) \,.\, \mathcal{S}T.$$

The coset $\mathcal{S}(T^{-1})$ is a member of $\mathcal{G}/\mathcal{S}$ as $T^{-1} \in \mathcal{G}$.

If $\mathcal{G}$ is a *finite* group of order g and $\mathcal{S}$ has order s, part (g) of the first theorem of Section 4 shows that there are g/s distinct right cosets. Thus $\mathcal{G}/\mathcal{S}$ is a group of order g/s with elements $\mathcal{S}T_1, \mathcal{S}T_2, \ldots,$ ($T_1, T_2, \ldots$ being a set of coset representatives). As $\mathcal{S}$ itself is one of the cosets, one can take $T_1 = E$.

Example I *A factor group formed from the crystallographic point group* D_4

Let $\mathcal{G}$ be D_4 and let $\mathcal{S} = \{E, C_{2y}\}$, which is an invariant subgroup of $\mathcal{G}$ (see Example I of Section 3). Then $\mathcal{G}/\mathcal{S}$ is a group of order 4 with

	$\mathscr{S}$E	$\mathscr{S}C_{2x}$	$\mathscr{S}C_{4y}$	$\mathscr{S}C_{2c}$
$\mathscr{S}$E	$\mathscr{S}$E	$\mathscr{S}C_{2x}$	$\mathscr{S}C_{4y}$	$\mathscr{S}C_{2c}$
$\mathscr{S}C_{2x}$	$\mathscr{S}C_{2x}$	$\mathscr{S}$E	$\mathscr{S}C_{2c}$	$\mathscr{S}C_{4y}$
$\mathscr{S}C_{4y}$	$\mathscr{S}C_{4y}$	$\mathscr{S}C_{2c}$	$\mathscr{S}$E	$\mathscr{S}C_{2x}$
$\mathscr{S}C_{2c}$	$\mathscr{S}C_{2c}$	$\mathscr{S}C_{4y}$	$\mathscr{S}C_{2x}$	$\mathscr{S}$E

Table 2.1 Multiplication table for the factor group $\mathscr{G}/\mathscr{S}$, where $\mathscr{G}$ is the crystallographic point group D_4 and $\mathscr{S}=\{E, C_{2y}\}$.

elements

$$\mathscr{S}E=\mathscr{S}C_{2y}=\{E, C_{2y}\}, \qquad \mathscr{S}C_{2x}=\mathscr{S}C_{2z}=\{C_{2x}, C_{2z}\},$$
$$\mathscr{S}C_{4y}=\mathscr{S}C_{4y}^{-1}=\{C_{4y}, C_{4y}^{-1}\}, \qquad \mathscr{S}C_{2c}=\mathscr{S}C_{2d}=\{C_{2c}, C_{2d}\},$$

whose multiplication table is given in Table 2.1.

6 Homomorphic and isomorphic mappings

Let $\mathscr{G}$ and $\mathscr{G}'$ be two groups. A "mapping" ϕ of $\mathscr{G}$ onto $\mathscr{G}'$ is simply a rule by which each element T of $\mathscr{G}$ is assigned to some element $T=\phi(T)$ of $\mathscr{G}'$, with every element of $\mathscr{G}'$ being the "image" of at least one element of $\mathscr{G}$. If ϕ is a *one-to-one* mapping, that is, if each element T′ of $\mathscr{G}'$ is the image of *only one* element T of $\mathscr{G}$, then the inverse mapping ϕ^{-1} of $\mathscr{G}'$ onto $\mathscr{G}$ may be defined by $\phi^{-1}(T')=T$ if and only if $T'=\phi(T)$.

Definition *Homomorphic mapping of a group $\mathscr{G}$ onto a group $\mathscr{G}'$*
If ϕ is a mapping of a group $\mathscr{G}$ onto a group $\mathscr{G}'$ such that

$$\phi(T_1)\phi(T_2)=\phi(T_1T_2) \tag{2.7}$$

for all $T_1, T_2 \in \mathscr{G}$, then ϕ is said to be a "homomorphic" mapping.

On the right-hand side of Equation (2.7) the product of T_1 with T_2 is evaluated using the group multiplication operation for $\mathscr{G}$, whereas on the left-hand side the product of $\phi(T_1)$ with $\phi(T_2)$ is obtained from the group multiplication operation for $\mathscr{G}'$. Although these operations may be different, there is no need to introduce any special notations to distinguish between them, because the relevant operation can always be deduced from the context and there is really no possibility of confusion.

Example I *A homomorphic mapping of the point group* D_4
Let $\mathscr{G}$ be D_4 and let $\mathscr{G}'$ be the group of order 2 with elements +1 and −1, with ordinary multiplication as the group multiplication opera-

tion. Then

$$\left.\begin{aligned}\phi(E)=\phi(C_{2y})=\phi(C_{2x})=\phi(C_{2z})=+1,\\ \phi(C_{4y})=\phi(C_{4y}^{-1})=\phi(C_{2c})=\phi(C_{2d})=-1,\end{aligned}\right\}$$

is a homomorphic mapping of $\mathscr{G}$ onto $\mathscr{G}'$, as may be confirmed by examination of Table 1.2. For example, $\phi(C_{2x})\phi(C_{2c})=(+1)(-1)=-1$, while Table 1.2 gives $\phi(C_{2x}C_{2c})=\phi(C_{4y}^{-1})=-1$.

Clearly, if g and g' are the orders of $\mathscr{G}$ and $\mathscr{G}'$ respectively, then $g \geqslant g'$. Actually, the First Homomorphism Theorem, which will be proved shortly, implies that if g and g′ are both finite, then g/g' must be an integer.

One major example of a homomorphic mapping has already been encountered in the concept of a representation of a group. Indeed, the definition in Chapter 1, Section 4 can now be rephrased as follows:

Definition *Representation of a group* $\mathscr{G}$

If there exists a *homomorphic* mapping of a group $\mathscr{G}$ onto a group of non-singular $d \times d$ matrices $\mathbf{\Gamma}(T)$ with matrix multiplication as the group multiplication operation, then the group of matrices $\mathbf{\Gamma}(T)$ forms a d-dimensional representation Γ of $\mathscr{G}$.

There is no requirement in the definition of a homomorphic mapping that the mapping should be one-to-one. However, as such mappings are particularly important, they are given a special name:

Definition *Isomorphic mapping of a group* $\mathscr{G}$ *onto a group* $\mathscr{G}'$

If ϕ is a *one-to-one* mapping of a group $\mathscr{G}$ onto a group $\mathscr{G}'$ of the same order such that

$$\phi(T_1)\phi(T_2)=\phi(T_1T_2)$$

for all $T_1, T_2 \in \mathscr{G}$, then ϕ is said to be an "isomorphic" mapping.

In the case of representations, if the homomorphic mapping is actually isomorphic, then the representation is said to be "faithful".

Clearly, if ϕ is an isomorphic mapping of $\mathscr{G}$ onto $\mathscr{G}'$, then the inverse mapping ϕ^{-1} is an isomorphic mapping of $\mathscr{G}'$ onto $\mathscr{G}$. (There is *no* analogous result for general homomorphic mappings, as ϕ^{-1} is only well defined when ϕ is a one-to-one mapping.)

Although two isomorphic groups may differ in the nature of their elements, they have the same structure of subgroups, cosets, classes, and so on. Most important of all, isomorphic groups necessarily have identical representations.

The following theorem clarifies various aspects of homomorphic mappings. As it is the first of a series of such theorems it is often called the "First Homomorphism Theorem", but the others in the series will not be needed in this book.

Definition *Kernel $\mathscr{K}$ of a homomorphic mapping*
Let ϕ be a homomorphic mapping of a group $\mathscr{G}$ onto a group $\mathscr{G}'$. Then the set of elements $T \in \mathscr{G}$ such that $\phi(T) = E'$, the identity of $\mathscr{G}'$, is said to form the "kernel" $\mathscr{K}$ of the mapping.

Theorem I Let ϕ be a homomorphic mapping of $\mathscr{G}$ onto $\mathscr{G}'$ and let $\mathscr{K}$ be the kernel of this mapping. Then

(*a*) $\mathscr{K}$ is an invariant subgroup of $\mathscr{G}$;
(*b*) every element of the right coset $\mathscr{K}T$ maps onto the *same* element $\phi(T)$ of $\mathscr{G}'$ and the mapping θ thereby defined by

$$\theta(\mathscr{K}T) = \phi(T) \tag{2.8}$$

is a one-to-one mapping of the factor group $\mathscr{G}/\mathscr{K}$ onto $\mathscr{G}'$; and
(*c*) θ is an *isomorphic* mapping of $\mathscr{G}/\mathscr{K}$ onto $\mathscr{G}'$.

Proof

(*a*) It has first to be established that $\mathscr{K}$ is a subgroup of $\mathscr{G}$. Clearly, if $T_1, T_2 \in \mathscr{K}$, then $\phi(T_1T_2) = \phi(T_1)\phi(T_2) = E'E' = E'$, so $T_1T_2 \in \mathscr{K}$. Also, as $\phi(T) = \phi(ET) = \phi(E)\phi(T)$ for any $T \in \mathscr{G}$, it follows that $\phi(E) = E'$ and so $E \in \mathscr{K}$. Finally, for any $T \in \mathscr{K}$, as $E' = \phi(E) = \phi(TT^{-1}) = \phi(T)\phi(T^{-1}) = E'\phi(T^{-1}) = \phi(T^{-1})$, then $T^{-1} \in \mathscr{K}$.
Now let $X \in \mathscr{G}$ and $T \in \mathscr{K}$. As $\phi(XTX^{-1}) = \phi(X)\phi(T)\phi(X^{-1}) = \phi(X)\phi(X^{-1}) = \phi(E) = E'$, it follows that $XTX^{-1} \in \mathscr{K}$ and hence $\mathscr{K}$ is an *invariant* subgroup of $\mathscr{G}$.
(*b*) A typical element of $\mathscr{K}T$ is KT, where $K \in \mathscr{K}$. Then $\phi(KT) = \phi(K)\phi(T) = E'\phi(T) = \phi(T)$, so every element of $\mathscr{K}T$ maps into the same element $\phi(T)$ of $\mathscr{G}'$. The mapping θ of Equation (2.8) is therefore well defined, and it remains only to demonstrate that it is one-to-one. This merely requires showing that if $\theta(\mathscr{K}T_1) = \theta(\mathscr{K}T_2)$ then $\mathscr{K}T_1 = \mathscr{K}T_2$. However, if $\theta(\mathscr{K}T_1) = \theta(\mathscr{K}T_2)$, then $\phi(T_1) = \phi(T_2)$, so if T_3 is the element of $\mathscr{G}$ such that $T_1 = T_3T_2$, then $\phi(T_3) = E'$, so $T_3 \in \mathscr{K}$ and hence $T_1 \in \mathscr{K}T_2$. By part (*f*) of the first theorem of Section 4, it then follows that $\mathscr{K}T_1 = \mathscr{K}T_2$.
(*c*) Let $\mathscr{K}T_1$ and $\mathscr{K}T_2$ be any two right cosets of $\mathscr{G}/\mathscr{K}$. Then, by Equations (2.6) and (2.8),

$$\theta(\mathscr{K}T_1)\theta(\mathscr{K}T_2) = \phi(T_1)\phi(T_2) = \phi(T_1T_2) = \theta(\mathscr{K}(T_1T_2)) = \theta(\mathscr{K}T_1 \,.\, \mathscr{K}T_2),$$

so θ is an isomorphic mapping of $\mathscr{G}/\mathscr{K}$ onto $\mathscr{G}'$.

Example II *The First Homomorphism Theorem applied to the point group* D_4

Let $\mathcal{G}$, $\mathcal{G}'$ and the homomorphic mapping ϕ be as in Example I above. Then $\mathcal{K}=\{E, C_{2y}, C_{2x}, C_{2z}\}$ which, as shown in Example I of Section 3, is an invariant subgroup of $\mathcal{G}=D_4$) and the other distinct right coset is $\mathcal{K}C_{4y}=\{C_{4y}, C_{4y}^{-1}, C_{2c}, C_{2d}\}$. The group multiplication tables for $\mathcal{G}/\mathcal{K}$ and $\mathcal{G}'$ are given in Tables 2.2 and 2.3 respectively, from which it is obvious that the mapping θ of $\mathcal{G}/\mathcal{K}$ onto $\mathcal{G}'$ here defined by $\theta(\mathcal{K}E)=1$, $\theta(\mathcal{K}C_{4y})=-1$ is an isomorphic mapping.

	$\mathcal{K}E$	$\mathcal{K}C_{4y}$
$\mathcal{K}E$	$\mathcal{K}E$	$\mathcal{K}C_{4y}$
$\mathcal{K}C_{4y}$	$\mathcal{K}C_{4y}$	$\mathcal{K}E$

Table 2.2 Multiplication table for the factor group $\mathcal{G}/\mathcal{K}$ of Example II.

	1	−1
1	1	−1
−1	−1	1

Table 2.3 Multiplication table for the group $\mathcal{G}'$ of Example II.

One consequence of the theorem is that every element of $\mathcal{G}'$ is the image of the *same* number of elements of $\mathcal{G}$. This has the further implication that the mapping is an isomorphism if and only if $\mathcal{K}$ consists only of the identity E of $\mathcal{G}$.

In the special case in which $\mathcal{G}'$ is identical to $\mathcal{G}$ (so that ϕ is a mapping of $\mathcal{G}$ onto itself) an isomorphic mapping is known as an "automorphism." For each $X\in\mathcal{G}$ the mapping ϕ_X of $\mathcal{G}$ onto itself defined by

$$\phi_X(T)=XTX^{-1}$$

is an automorphism, as it is certainly one-to-one and

$$\phi_X(T_1)\phi_X(T_2)=(XT_1X^{-1})(XT_2X^{-1})=X(T_1T_2)X^{-1}=\phi_X(T_1T_2)$$

for all $T_1, T_2\in\mathcal{G}$. Such a mapping is called an "inner automorphism", and any automorphism that is not of this form is known as an "outer automorphism".

Example III *Outer automorphism of* SU(N) *for* $N\geqslant 3$

It is obvious that the mapping ϕ defined by

$$\phi(\mathbf{u})=\mathbf{u}^*$$

for all $\mathbf{u} \in SU(N)$ is an automorphism of SU(N). (This is not an inner automorphism because $\mathbf{u}$ and $\mathbf{u}^*$ provide two non-equivalent representations of SU(N) for $N \geq 3$) (see Chapter 15, Section 4).)

Many groups do not possess any outer automorphisms. If $\mathcal{S}$ is a subgroup of $\mathcal{G}$ then an automorphic mapping maps $\mathcal{S}$ onto a subgroup $\mathcal{S}'$ that is isomorphic to $\mathcal{S}$. Two subgroups $\mathcal{S}$ and $\mathcal{S}'$ of $\mathcal{G}$ are said to be "conjugate" if they are related this way, but it is possible to have isomorphic subgroups that are not conjugate.

7 Direct product groups and semi-direct product groups

Although the abstract construction of direct product groups appears at first sight rather artificial, a number of examples of groups having this structure occur naturally in physical problems.

Let $\mathcal{G}_1$ and $\mathcal{G}_2$ be *any* two groups, and suppose that E_1 and E_2 are the identities of $\mathcal{G}_1$ and $\mathcal{G}_2$ respectively. Consider the set of pairs (T_1, T_2), where $T_1 \in \mathcal{G}_1$ and $T_2 \in \mathcal{G}_2$, and *define* the product of two such pairs (T_1, T_2) and (T_1', T_2') by

$$(T_1, T_2)(T_1', T_2') = (T_1T_1', T_2T_2') \tag{2.9}$$

for all $T_1, T_1' \in \mathcal{G}_1$ and $T_2, T_2' \in \mathcal{G}_2$.

Theorem I The set of pairs (T_1, T_2) (for $T_1 \in \mathcal{G}_1$, $T_2 \in \mathcal{G}_2$) form a group with Equation (2.9) as the group multiplication operation. This group is denoted by $\mathcal{G}_1 \otimes \mathcal{G}_2$, and is called the "direct product of $\mathcal{G}_1$ with $\mathcal{G}_2$".

Proof All that has to be verified is that the four group axioms of Chapter 1, Section 1, are satisfied. By Equation (2.9), the product of any two pairs of $\mathcal{G}_1 \otimes \mathcal{G}_2$ is also a member of $\mathcal{G}_1 \otimes \mathcal{G}_2$, so axiom ($a$) is fulfilled. Axiom ($b$) is observed, as

$$\{(T_1, T_2)(T_1', T_2')\}(T_1'', T_2'') = ((T_1T_1')T_1'', (T_2T_2')T_2'')$$

and

$$(T_1, T_2)\{(T_1', T_2')(T_1'', T_2'')\} = (T_1(T_1'T_1''), T_2(T_2'T_2'')),$$

the pairs on the right-hand sides being equal because the associative law applies to $\mathcal{G}_1$ and $\mathcal{G}_2$ separately. The identity of $\mathcal{G}_1 \otimes \mathcal{G}_2$ is (E_1, E_2), as for all $T_1 \in \mathcal{G}_1$ and $T_2 \in \mathcal{G}_2$

$$(T_1, T_2)(E_1, E_2) = (E_1, E_2)(T_1, T_2) = (T_1, T_2).$$

Finally, the inverse of (T_1, T_2) is (T_1^{-1}, T_2^{-1}), which is also a member of $\mathcal{G}_1 \otimes \mathcal{G}_2$.

If $\mathcal{G}_1$ and $\mathcal{G}_2$ are finite groups of orders g_1 and g_2 respectively, then $\mathcal{G}_1 \otimes \mathcal{G}_2$ has order $g_1 g_2$.

The properties of $\mathcal{G}_1 \otimes \mathcal{G}_2$ are best presented in the form of a theorem (all the assertions of which have trivial proofs).

Theorem II

(*a*) $\mathcal{G}_1 \otimes \mathcal{G}_2$ contains a subgroup consisting of the elements (T_1, E_2), $T_1 \in \mathcal{G}_1$, that is isomorphic to $\mathcal{G}_1$, the isomorphic mapping being $\phi((T_1, E_2)) = T_1$.

(*b*) $\mathcal{G}_1 \otimes \mathcal{G}_2$ contains a subgroup consisting of the elements (E_1, T_2), $T_2 \in \mathcal{G}_2$, that is isomorphic to $\mathcal{G}_2$, the isomorphic mapping being $\psi((E_1, T_2)) = T_2$.

(*c*) The elements of these two subgroups commute with each other, that is

$$(T_1, E_2)(E_1, T_2) = (E_1, T_2)(T_1, E_2) \quad (= (T_1, T_2))$$

for all $T_1 \in \mathcal{G}_1$ and $T_2 \in \mathcal{G}_2$.

(*d*) These two subgroups have only one element in common, namely the identity (E_1, E_2).

(*e*) Every element of $\mathcal{G}_1 \otimes \mathcal{G}_2$ is the product of an element of the first subgroup with an element of the second subgroup. That is, for all $T_1 \in \mathcal{G}_1$ and $T_2 \in \mathcal{G}_2$

$$(T_1, T_2) = (T_1, E_2)(E_1, T_2).$$

As isomorphic groups have identical structures, it is natural to now *extend* the definition of a direct product.

Enlarged definition *Direct product group*
A group $\mathcal{G}'$ is said to be a "direct product group" if it is *isomorphic* to a group $\mathcal{G}_1 \otimes \mathcal{G}_2$ constructed as in the first theorem above.

With this extension the elements of a direct product group need no longer be in the form of pairs. Such a group can be identified by the following theorem, which is essentially the converse of that immediately above.

Theorem III If a group $\mathcal{G}'$ possesses two subgroups $\mathcal{G}_1'$ and $\mathcal{G}_2'$ such that

(*a*) the elements of $\mathcal{G}_1'$ commute with the elements of $\mathcal{G}_2'$,

(*b*) $\mathcal{G}_1'$ and $\mathcal{G}_2'$ have only the identity element in common, and (*c*) every element of $\mathscr{G}'$ can be written as a product of an element of $\mathscr{G}'_1$ with an element of $\mathscr{G}'_2$,

then $\mathscr{G}'$ is a direct product group that is isomorphic to $\mathscr{G}'_1 \otimes \mathscr{G}'_2$.

Proof It follows from (*b*) and (*c*) that every element T′ of $\mathscr{G}'$ can be written *uniquely* in the form $T' = T'_1T'_2$, ($T'_1 \in \mathscr{G}'_1$ and $T'_2 \in \mathscr{G}'_2$). (Suppose to the contrary that $T' = T'_1T'_2 = T''_1T''_2$ where $T'_1, T''_1 \in \mathscr{G}'_1$ and $T'_2, T''_2 \in \mathscr{G}'_2$. Then $(T''_1)^{-1}T'_1 = T''_2(T'_2)^{-1}$, the left-hand and right-hand sides of which are members of $\mathscr{G}'_1$ and $\mathscr{G}'_2$ respectively. Thus by (*b*) $(T''_1)^{-1}T'_1 = T''_2(T'_2)^{-1} = E$, the common identity of $\mathscr{G}'$, $\mathscr{G}'_1$ and $\mathscr{G}'_2$, so $T'_1 = T''_1$ and $T'_2 = T''_2$.)

Now let $\mathscr{G}'_1 \otimes \mathscr{G}'_2$ be the direct product group whose elements are the pairs (T'_1, T'_2), where $T'_1 \in \mathscr{G}'_1$ and $T'_2 \in \mathscr{G}'_2$. Then with $T' = T'_1T'_2$, the mapping θ defined by

$$\theta(T') = (T'_1, T'_2) \tag{2.10}$$

is a well defined mapping of $\mathscr{G}'$ onto $\mathscr{G}'_1 \otimes \mathscr{G}'_2$. It is obviously one-to-one and is isomorphic because, if $T' = T'_1T'_2$ and $T'' = T''_1T''_2$, then from Equations (2.9) and (2.10)

$$\theta(T')\theta(T'') = (T'_1, T'_2)(T''_1, T''_2) = (T'_1T''_1, T'_2T''_2),$$

while by (*a*) $T'T'' = (T'_1T'_2)(T''_1T''_2) = (T'_1T''_1)(T'_2T''_2)$, so that

$$\theta(T'T'') = (T'_1T''_1, T'_2T''_2).$$

Thus $\mathscr{G}'$ is isomorphic to $\mathscr{G}'_1 \otimes \mathscr{G}'_2$.

Example I *The group* O(3) *as a direct product group*
The group O(3) is isomorphic to $SO(3) \otimes \mathscr{G}'_2$, where $\mathscr{G}'_2$ is the matrix group of order 2 consisting of the matrices $\mathbf{1}_3$ and $-\mathbf{1}_3$, as the properties (*a*), (*b*) and (*c*) of the preceding theorem are obviously satisfied.

As O(3) is isomorphic to the group of all rotations in three dimensions and SO(3) is isomorphic to the subgroup of proper rotations (see Chapter 1, Section 2), this implies that the group of *all* rotations is isomorphic to the direct product of the group of *proper* rotations and the group {E, I} consisting of the identity transformation E and the spatial inversion operator I.

Example II *The crystallographic point group* O_h *as a direct product group*
As shown in Appendix D, the point group O_h consists of 48 rotations, 24 being proper rotations and the remaining 24 being products of these proper rotations with the spatial inversion operator I. The

proper rotations of O_h form the point group O. Thus O_h is isomorphic to $O \otimes \{E, I\}$. A number of other crystallographic point groups have a similar direct product structure.

A special case of interest, particularly in elementary particle physics, is that in which $\mathscr{G}_1$ and $\mathscr{G}_2$ are both isomorphic to the *same* group $\mathscr{G}$, giving $\mathscr{G} \otimes \mathscr{G}$. $\mathscr{G} \otimes \mathscr{G}$ possesses a "diagonal subgroup", consisting of the pairs (T, T) (for all $T \in \mathscr{G}$), which is isomorphic to $\mathscr{G}$. This diagonal subgroup is not conjugate to either of the two other subgroups isomorphic to $\mathscr{G}$ consisting of the pairs (T, E) and (E, T) that were mentioned earlier.

It should be observed that condition (*a*) of the last theorem can be replaced by an equivalent condition (*a'*), which reads:

(*a'*) $\mathscr{G}'_1$ and $\mathscr{G}'_2$ are both invariant subgroups of $\mathscr{G}'$.

(Obviously (*a*) implies (*a'*). Conversely, if (*a'*) is true, then for any $T'_1 \in \mathscr{G}'_1$ and $T'_2 \in \mathscr{G}'_2$, $T'_1 T'_2 (T'_1)^{-1} = T''_2 \in \mathscr{G}'_2$. Similarly $(T'_2)^{-1} T'_1 T'_2 = T''_1 \in \mathscr{G}'_1$, so that $(T'_2)^{-1} T'_1 T'_2 (T'_1)^{-1} = T''_1 (T'_1)^{-1} = (T'_2)^{-1} T''_2$. As $T''_1 (T'_1)^{-1} \in \mathscr{G}'_1$ and $(T'_2)^{-1} T''_2 \in \mathscr{G}'_2$, (*b*) implies that $T'_1 = T''_1$ and $T'_2 = T''_2$. Thus $(T'_2)^{-1} T'_1 T'_2 = T'_1$ for all $T'_1 \in \mathscr{G}'_1$ and $T'_2 \in \mathscr{G}'_2$, so that $\mathscr{G}'_1$ and $\mathscr{G}'_2$ commute.)

The notion of a semi-direct product group $\mathscr{G}'$ is essentially a generalization of that of a direct product group in which conditions (*b*) and (*c*) of the last theorem are retained intact but condition (*a'*) is weakened to the requirement that only $\mathscr{G}'_1$ must be an invariant subgroup, but $\mathscr{G}'_2$, although remaining a subgroup of $\mathscr{G}'$, need not be invariant.

Definition *Semi-direct product group*

A group $\mathscr{G}'$ is said to be a "semi-direct product group" if it possesses two subgroups $\mathscr{G}'_1$ and $\mathscr{G}'_2$ such that

(*a*) $\mathscr{G}'_1$ is an invariant subgroup of $\mathscr{G}$;
(*b*) $\mathscr{G}'_1$ and $\mathscr{G}'_2$ have only the identity element in common; and
(*c*) every element of $\mathscr{G}'$ can be written as a product of an element of $\mathscr{G}'_1$ with an element of $\mathscr{G}'_2$.

$\mathscr{G}'$ may then be said to be isomorphic to $\mathscr{G}'_1 \circledS \mathscr{G}'_2$.

As in the special case of a direct product group, the requirement (*b*) always implies that the decomposition (*c*) is *unique*.

Example III *The crystallographic point group D_4 as a semi-direct product group*

Let $\mathscr{G}' = D_4$, $\mathscr{G}'_1 = \{E, C_{2y}, C_{4y}, C_{4y}^{-1}\}$ and $\mathscr{G}'_2 = \{E, C_{2x}\}$. Then the conditions of the definition of a semi-direct product group are satisfied for

D_4. Firstly, as noted in the Example of Section 3, $\mathcal{G}'_1$ is an invariant subgroup of D_4. Secondly, $\mathcal{G}'_1$ and $\mathcal{G}'_2$ have in common only E, and finally, from Table 1.2,

$$\left.\begin{array}{llll} E = EE, & C_{2y} = C_{2y}E, & C_{4y} = C_{4y}E, & C_{4y}^{-1} = C_{4y}^{-1}E, \\ C_{2x} = EC_{2x}, & C_{2z} = C_{2y}C_{2x}, & C_{2c} = C_{4y}C_{2x}, & C_{2d} = C_{4y}^{-1}C_{2x}. \end{array}\right\}$$

Example IV *The Euclidean group of* $\mathbb{R}^3$ *as a semi-direct product group*

The Euclidean group $\mathcal{G}'$ of $\mathbb{R}^3$ is defined to be the group of all linear coordinate transformations T, with Equation (1.7) giving the group multiplication operation. Let $\mathcal{G}'_1$ be the subgroup of pure translations and $\mathcal{G}'_2$ the subgroup of pure rotations. Then for any $T_1 \in \mathcal{G}'_1$ and any $T \in \mathcal{G}'$, from Equations (1.7) and (1.8),

$$\{\mathbf{R}(T) \mid \mathbf{t}(T)\}\{\mathbf{1} \mid \mathbf{t}(T_1)\}\{\mathbf{R}(T) \mid \mathbf{t}(T)\}^{-1} = \{\mathbf{1} \mid \mathbf{R}(T)\mathbf{t}(T_1)\},$$

so that $\mathcal{G}'_1$ is an invariant subgroup of $\mathcal{G}'$. Moreover, for any $T \in \mathcal{G}'$,

$$\{\mathbf{R}(t) \mid \mathbf{t}(T)\} = \{\mathbf{1} \mid \mathbf{t}(T)\}\{\mathbf{R}(T) \mid \mathbf{0}\}$$

so that requirement (*c*) is also satisfied, while (*b*) is obvious. Thus $\mathcal{G}'$ is isomorphic to $\mathcal{G}'_1 Ⓢ \mathcal{G}'_2$.

Example V *The Poincaré group of Minkowski space–time as a semi-direct product group*

The Poincaré group $\mathcal{G}'$ may be similarly defined as the group of all coordinate transformations in Minkowski space–time of the form

$$\begin{bmatrix} x' \\ y' \\ z' \\ ct' \end{bmatrix} = \mathbf{\Lambda}(T) \begin{bmatrix} x \\ y \\ z \\ ct \end{bmatrix} + \mathbf{t}(T), \tag{2.11}$$

(c being the speed of light), such that the 4×4 real matrices $\mathbf{\Lambda}(T)$ satisfy $\tilde{\mathbf{\Lambda}}(T)\mathbf{g}\mathbf{\Lambda}(T) = \mathbf{g}$, where the matrix $\mathbf{g}$ is given by

$$\mathbf{g} = \begin{bmatrix} -1 & 0 & 0 & 0 \\ 0 & -1 & 0 & 0 \\ 0 & 0 & -1 & 0 \\ 0 & 0 & 0 & +1 \end{bmatrix} \tag{2.12}$$

and where $\mathbf{t}(T)$ is a 4×1 real matrix. For such a transformation

$$c^2t'^2 - x'^2 - y'^2 - z'^2 = c^2t^2 - x^2 - y^2 - z^2$$

and, conversely, any linear transformation for which this holds must be of the above form (Pauli 1958). By analogy with the notation of Chapter 1, Section 2, such a transformation may be denoted by the composite symbol $\{\mathbf{\Lambda}(\mathrm{T}) \mid \mathbf{t}(\mathrm{T})\}$. Then the group transformation law is a generalization of Equation (1.7), namely

$$\{\mathbf{\Lambda}(\mathrm{T}_1\mathrm{T}_2) \mid \mathbf{t}(\mathrm{T}_1\mathrm{T}_2)\} = \{\mathbf{\Lambda}(\mathrm{T}_1)\mathbf{\Lambda}(\mathrm{T}_2) \mid \mathbf{\Lambda}(\mathrm{T}_1)\mathbf{t}(\mathrm{T}_2) + \mathbf{t}(\mathrm{T}_1)\}. \tag{2.13}$$

The subgroup of all such transformations for which $\mathbf{t}(\mathrm{T}) = \mathbf{0}$, that is, those involving no translations, is called the "homogeneous Lorentz group". (Sometimes the Poincaré group is called the "inhomogeneous Lorentz group".)

Let $\mathcal{G}_1'$ be the subgroup of all pure space–time translations (that is, transformations such that $\mathbf{\Lambda}(\mathrm{T}) = \mathbf{1}_4$) and let $\mathcal{G}_2'$ be the homogeneous Lorentz group. Then, exactly as in Example IV, $\mathcal{G}'$ is isomorphic to $\mathcal{G}_1' \circledS \mathcal{G}_2'$.

Example VI *The linear infinite point groups* $\mathrm{D}_{\infty h}$ *and* $\mathrm{C}_{\infty h}$
$\mathrm{D}_{\infty h}$ consists of the rotations:

E: the identity;
C_θ: rotation through an angle θ about the axis Ox in the right-hand screw sense for $0 < \theta \leq \pi$ and in the left-hand screw sense for $-\pi < \theta < 0$;
$\mathrm{C}_{2\Phi}$: rotation through π about the axis $O\Phi$ lying in the plane Oyz and making an angle $\frac{1}{2}\phi$ with Oz, $-\pi < \phi \leq \pi$, as in Figure 2.1; (in particular $\mathrm{C}_{2\Phi} = \mathrm{C}_{2z}$ when $\phi = 0$);
I: the spatial inversion operation;
IC_θ: $-\pi < \theta \leq \pi$, $\theta \neq 0$;
$\mathrm{IC}_{2\Phi}$: $-\pi < \phi \leq \pi$.

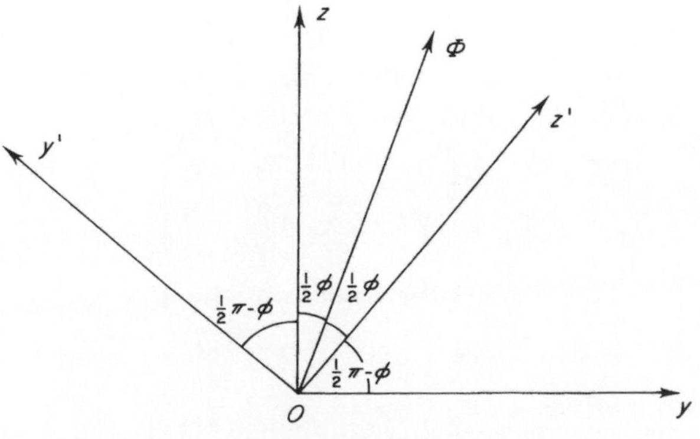

Figure 2.1 The axis $O\Phi$ of the rotation $\mathrm{C}_{2\Phi}$.

The matrices **R**(T) for $T = C_\theta$ are given in Equation (1.3). For $T = C_{2\Phi}$

$$\mathbf{R}(C_{2\Phi}) = \begin{bmatrix} -1 & 0 & 0 \\ 0 & -\cos\phi & \sin\phi \\ 0 & \sin\phi & \cos\phi \end{bmatrix}.$$

All the rotations $C_{2\Phi}$ lie in the *same* class, and all the rotations $IC_{2\Phi}$ lie in another class. For each θ such that $0 < \theta < \pi$, C_θ and $C_{-\theta}$ together form one class and IC_θ and $IC_{-\theta}$ together form another class. Otherwise every element lies in a class of its own. The subgroup $\mathscr{G}_1'$ consisting of E together with all C_θ ($-\pi < \theta \leq \pi$, $\theta \neq 0$) is therefore an Abelian invariant subgroup of $D_{\infty h}$. Then $\mathscr{G}_1'$ and $\mathscr{G}_2' = \{E, C_{2z}, I, IC_{2z}\}$ obviously satisfy condition (*b*), and as $C_{2\Phi} = C_\phi C_{2z}$ for all $-\pi < \phi \leq \pi$, they also satisfy (*c*) as well. Thus $D_{\infty h}$ is isomorphic to $\mathscr{G}_1' \circledS \mathscr{G}_2'$.

$C_{\infty h}$ is a subgroup of $D_{\infty h}$ consisting only of E, C_θ (for $-\pi < \theta \leq \pi$, $\theta \neq 0$), and $IC_{2\Phi}$ ($-\pi < \phi \leq \pi$). Again all the reflections lie in the same class, and for each θ such that $0 < \theta < \pi$, C_θ and $C_{-\theta}$ together form one class, while E and C_π each form classes of their own. With $\mathscr{G}_1'$ as for $D_{\infty h}$ and $\mathscr{G}_2' = \{E, IC_{2z}\}$, $C_{\infty h}$ is isomorphic to $\mathscr{G}_1' \circledS \mathscr{G}_2'$.

$D_{\infty h}$ and $C_{\infty h}$ appear in the theory of vibrations of linear molecules (see Chapter 7, Section 3, Example II).

A further important set of examples is provided by the symmorphic crystallographic space groups. These will be discussed in detail in Chapter 9.

Although it is possible to give an abstract construction of a semi-direct product of certain groups in terms of pairs of elements from the two groups, the procedure is much more elaborate than for the direct product (Lomont 1959, page 29). Fortunately, all the physically important examples of groups having a semi-direct product structure occur naturally, so this abstract construction will be omitted.

3

Lie Groups

It is now time to formulate a definition of a Lie group and to describe some of the major properties of such groups. Readers whose interests lie only in the applications to solid state physics (where only finite groups appear) may safely omit this chapter, except for Section 5.

1 Definition of a linear Lie group

A Lie group embodies three different forms of mathematical structure. Firstly, it satisfies the group axioms of Chapter 1 and so has the group structure described in Chapter 2. Secondly, the elements of the group also form a "topological space", so that it may be described as being a special case of a "topological group". Finally, the elements also constitute an "analytic manifold". Consequently a Lie group can be defined in several different (but equivalent) ways, depending on the degree of emphasis that is being accorded to the various aspects. In particular, it can be defined as a topological group with certain additional analytic properties (Pontrjagin 1946) or, alternatively, as an analytic manifold with additional group properties (Chevalley 1946, Adams 1969, Varadarajan 1974, Warner 1971). Both of these formulations involve the introduction of a series of ancillary concepts of a rather abstract nature.

Very fortunately, every Lie group that is important in physical problems is of a type, known as a "linear Lie group", for which a relatively straightforward definition can be given. As will be seen, this definition is both precise and simple, in that it involves only familiar concrete objects such as matrices and contains no mention of topological spaces or analytic manifolds. (Readers who are interested in the *general* definition of a Lie group in terms of analytic manifolds will find this formulation in Appendix J.)

The basic feature of any Lie group is that it has a non-countable number of elements lying in a region "near" its identity and that the structure of this region both very largely determines the structure of the whole group and is itself determined by its corresponding real Lie algebra. To ensure that this is so, the elements in this region must be parametrized in a particular analytic way. Of course, to say that certain elements are "near" the identity means that a notion of "distance" has to be composed, and it is here that the complications of the general treatment start. However, all the Lie groups of physical interest are "linear", in the sense that they at least one faithful finite-dimensional representation. This representation can be used to provide the necessary precise formulation of distance and to ensure that all the other topological requirements are automatically observed.

Definition *Linear Lie group of dimension n*

A group $\mathcal{G}$ is a linear Lie group of dimension n if it satisfies the following conditions (A), (B), (C) and (D):

(A) *$\mathcal{G}$ must possess at least one faithful finite-dimensional representation* Γ.

Suppose that this representation has dimension m. Then the "distance" $d(\mathrm{T},\mathrm{T}')$ between two elements T and T′ of $\mathcal{G}$ may be *defined* by

$$d(\mathrm{T},\mathrm{T}') = +\left\{\sum_{j=1}^{m}\sum_{k=1}^{m}\left|\Gamma(\mathrm{T})_{jk}-\Gamma(\mathrm{T}')_{jk}\right|^2\right\}^{1/2}. \tag{3.1}$$

This distance function $d(\mathrm{T},\mathrm{T}')$ will be called the "metric". Then

(i) $d(\mathrm{T}',\mathrm{T}) = d(\mathrm{T},\mathrm{T}')$;
(ii) $d(\mathrm{T},\mathrm{T}) = 0$;
(iii) $d(\mathrm{T},\mathrm{T}') > 0$ if $\mathrm{T} \neq \mathrm{T}'$;
(iv) if T,T' and T'' are any three elements of $\mathcal{G}$, $d(\mathrm{T},\mathrm{T}') \leq d(\mathrm{T},\mathrm{T}') + d(\mathrm{T}',\mathrm{T}'')$,

all of which are essential for the interpretation of $d(\mathrm{T},\mathrm{T}')$ as a distance. (The choice of this metric implies that the group is being endowed with the topology of the m^2-dimensional complex Euclidean space $\mathbb{C}^{m^2}$ (see Example II of Appendix B, Section 2).) The set of element T of $\mathcal{G}$ such that

$$d(\mathrm{T},\mathrm{E}) < \delta,$$

where δ is positive real number, are then said to "lie in a sphere of radius δ centred on the identity E", which will be denoted by M_δ. Such a sphere will be sometimes referred to as a "small neighbourhood" of E.

(B) *There must exist a* $\delta > 0$ *such that every element* T *of* $\mathscr{G}$ *lying in the sphere* M_δ *of radius* δ *centred on the identity can be parametrized by* n *real parameters* $x_1, x_2, \ldots, x_n$, *(no two such sets of parameters corresponding to the same element T of* $\mathscr{G}$*), the identity* E *being parametrized by* $x_1 = x_2 = \ldots = x_n = 0$.

Thus every element of M_δ corresponds to one and only one point in a n-dimensional real real Euclidean space $\mathbb{R}^n$, the identity E corresponding to the origin $(0, 0, \ldots, 0)$ of $\mathbb{R}^n$. Moreover, no point in $\mathbb{R}^n$ corresponds to more than one element T in M_δ.

(C) *There must exist a* $\eta > 0$ *such that every point in* $\mathbb{R}^n$ *for which*

$$\sum_{j=1}^{n} x_j^2 < \eta^2 \tag{3.2}$$

corresponds to some element T *in* M_δ.

The set of point elements T so obtained will be denoted by R_η. Thus R_η is a subset of M_δ, and there is a *one-to-one* correspondence between elements of $\mathscr{G}$ in R_η and points in $\mathbb{R}^n$ satisfying Condition (3.2).

The final set of conditions ensures that in terms of this parametrization the group multiplication operation is expressible in terms of *analytic* functions. Let $T(x_1, x_2, \ldots, x_n)$ denote the element of $\mathscr{G}$ corresponding to a point satisfying Condition (3.2) and define $\mathbf{\Gamma}(x_1, x_2, \ldots, x_n)$ by $\mathbf{\Gamma}(x_1, x_2, \ldots, x_n) = \mathbf{\Gamma}(T(x_1, x_2, \ldots, x_n))$ for all $(x_1, x_2, \ldots, x_n)$ satisfying Condition (3.2).

(D) *Each of the matrix elements of* $\mathbf{\Gamma}(x_1, x_2, \ldots, x_n)$ *must be an analytic function of* $x_1, x_2, \ldots, x_n$ *for all* $(x_1, x_2, \ldots, x_n)$ *satisfying Condition* (3.2).

The term "analytic" here means that each of the matrix elements Γ_{jk} must be expressible as a power series in $x_1 - x_1^0, x_2 - x_2^0, \ldots, x_n - x_n^0$ for all $(x_1^0, x_2^0, \ldots, x_n^0)$ satisfying Condition (3.2). This implies that *all* the derivatives $\partial\Gamma_{jk}/\partial x^p$, $\partial^2\Gamma_{jk}/\partial x^p\,\partial x^q$ etc. must exist for all $j, k = 1, 2, \ldots, m$ at all points satisfying Condition (3.2), including in particular the point $(0, 0, \ldots, 0)$ (Fleming 1977).

Define the n $m \times m$ matrices $\mathbf{a}_1, \mathbf{a}_2, \ldots, \mathbf{a}_n$ by

$$(\mathbf{a}_p)_{jk} = (\partial\Gamma_{jk}/\partial x_p)_{x_1 = x_2 = \ldots = x_n = 0}. \tag{3.3}$$

These conditions together imply the following very important theorem.

Theorem I The matrices $\mathbf{a}_1, \mathbf{a}_2, \ldots, \mathbf{a}_n$ defined by Equation (3.3) form the basis for a n-dimensional *real* vector space.

Proof It is required to show that the only solution of the equation $\sum_{p=1}^{n} \lambda_p \mathbf{a}_p = \mathbf{0}$ with $\lambda_1, \lambda_2, \ldots, \lambda_n$ all *real* is $\lambda_1 = \lambda_2 = \ldots = \lambda_n = 0$.

Let $\Gamma_{jk} = A_{jk} + iB_{jk}$ for all $j, k = 1, 2, \ldots, m$. Then there must exist a subset $\mathscr{S}$ of the set

$$\mathscr{S}' = \{A_{11}, A_{12}, \ldots, A_{mm}, B_{11}, B_{12}, \ldots, B_{mm}\}$$

consisting of n members, such that every member of $\mathscr{S}'$ that is not in $\mathscr{S}$ is an analytic function of members of $\mathscr{S}$. Let the members of $\mathscr{S}$ be denoted by $C_1, C_2, \ldots, C_n$. By condition (D) each C_j is a real-valued analytic function of $(x_1, x_2, \ldots, x_n)$ for $(x_1, x_2, \ldots, x_n)$ satisfying Condition (3.2). As the mapping between elements of $\mathscr{G}$ in R_η and points in $\mathbb{R}^n$ satisfying Condition (3.2) is *one-to-one*, it follows that

$$\det\,(\partial C_j/\partial x_p)_{x_1=x_2=\ldots=x_n=0} \neq 0,$$

the left-hand side here being the Jacobian of the functions $C_1, C_2, \ldots, C_n$ (Fleming 1977). This implies that the only solution of the n simultaneous linear algebraic equations

$$\sum_{p=1}^{n} \lambda_p (\partial C_j/\partial x_p)_{x_1=x_2=\ldots=\lambda_n=0} = 0.$$

(for $j = 1, 2, \ldots, n$) is $\lambda_1 = \lambda_2 = \ldots = \lambda_n = 0$. Consequently the only solution of the equations

$$\sum_{p=1}^{n} \lambda_p (\partial \Gamma_{jk}/\partial x_p)_{x_1=x_2=\ldots=x_n=0} = 0$$

$(j, k = 1, 2, \ldots, m)$ with $\lambda_1, \lambda_1, \ldots, \lambda_n$ real is $\lambda_1 = \lambda_2 = \ldots = \lambda_n = 0$, thereby giving the desired result.

It should be noted that, although $\mathbf{a}_1, \mathbf{a}_2, \ldots, \mathbf{a}_n$ form the basis of a *real* vector space, there is *no* requirement that the matrix elements of these matrices need be real. (This point is demonstrated explicitly in Example III.) It will be shown in Chapter 10 that the matrices $\mathbf{a}_1, \mathbf{a}_2, \ldots, \mathbf{a}_n$ actually form the basis of a "real Lie algebra", a vital observation on which most of the subsequent theory is founded. However, the rest of the present chapter will be devoted to "group theoretical" aspects of linear Lie groups.

The next two theorems show the group multiplication operation and the operation of taking inverses are both "analytic", in a sense that will now be defined. First suppose that U is a subset of R_η such that if T and T′ are in U then T″ = TT′ is in R_η. Let $(x_1, x_2, \ldots, x_n)$, $(x'_1, x'_2, \ldots, x'_n)$ and $(x''_1, x''_2, \ldots, x''_n)$ be the parameters of T, T′ and T″ respectively. Then $(x''_1, x''_2, \ldots, x''_n)$ depends on both $(x_1, x_2, \ldots, x_n)$ and $(x'_n, x'_2, \ldots, x'_n)$. This may be indicated by introducting n functions f_j such that

$$x''_j = f_j(x_1, x_2, \ldots, x_n, x'_1, x'_2, \ldots, x'_n), \qquad j = 1, 2, \ldots, n. \tag{3.4}$$

As TE = T for all $T \in R_\eta$,

$$f_j(x_1, x_2, \ldots, x_n, 0, 0, \ldots, 0) = x_j, \qquad j = 1, 2, \ldots, n. \tag{3.5a}$$

Similarly, as ET′ = T′ for all $T' \in R_\eta$,

$$f_j(0, 0, \ldots, 0, x'_1, x'_2, \ldots, x'_n) = x'_j, \qquad j = 1, 2, \ldots, n \tag{3.5b}$$

Theorem II The functions $f_j(x_1, x_2, \ldots, x_n, x'_1, x'_2, \ldots, x'_n)$ are *analytic* functions for all T and T′ in U.

Proof With the notation of the previous theorem, as the Jacobian $\det(\partial C_j/\partial x_p)_{x_1=x_2=\ldots=x_n=0}$ is non-zero and as the mapping from $(x_1, x_2, \ldots, x_n)$ to $(C_1, C_2, \ldots, C_n)$ is one-to-one, the "inverse function theorem" (Fleming 1977) implies that one can write $x_j = \phi_j(C_1, C_2, \ldots, C_n)$, where the functions $\phi_1, \phi_2, \ldots, \phi_n$ are *analytic* functions of $C_1, C_2, \ldots, C_n$. In particular, for the product T″ = TT′, $x''_j = \phi_j(C''_1, C''_2, \ldots, C''_n)$, where $C''_1, C''_2, \ldots, C''_n$, being directly related to the real and imaginary parts of $\Gamma(T'')$, are analytic functions of the real and imaginary parts of $\Gamma(T)$ and $\Gamma(T')$ (as $\Gamma(T'') = \Gamma(T)\Gamma(T')$) and hence, by (D), of $x_1, x_2, \ldots, x_n$ and $x'_1, x'_2, \ldots, x'_n$.

It is clear that if an element $T \in \mathscr{G}$ is close to the identity E of $\mathscr{G}$, then its inverse T^{-1} is also close to E.

Theorem III The coordinates $(x'_1, x'_2, \ldots, x'_n)$ of T^{-1} are given in terms of the coordinates $(x_1, x_2, \ldots, x_n)$ of T by

$$x'_j = g_j(x_1, x_2, \ldots, x_n), \qquad j = 1, 2, \ldots, n,$$

where $g_1, g_2, \ldots, g_n$ are *analytic* functions, and T and T^{-1} are both in R_η.

Proof By Equation (3.4), as T″ = E has coordinates $(0, 0, 0, \ldots, 0)$,

$$f_j(x_1, x_2, \ldots, x_n, x'_1, x'_2, \ldots, x'_n) = 0$$

for $j = 1, 2, \ldots, n$. The stated result then follows immediately from the "Implicit Function Theorem" (Fleming 1977).

The above definition requires a parametrization *only* of the group elements belonging to a small neighbourhood of the identity element. In some cases this parametrization by a *single* set of n parameters $x_1, x_2, \ldots, x_n$ is valid over a large part of the group or even over the whole group, but this is not essential. In Section 2 it will be shown that the *whole* of the "connected" subgroup of a linear Lie group of dimension n can be given a parametrization in terms of single set of n real numbers which will be denoted by $y_1, y_2, \ldots, y_n$. However, this latter parametrization is *not* required to satisfy all the conditions of the above definition, and so need bear little relation to the parametrization by $x_1, x_2, \ldots, x_n$.

The first three of the following examples have been chosen because they illustrate all the essential points of the definition without involving any heavy algebra.

Example I *The multiplicative group of real numbers*
As in Example I of Chapter 1, Section 1, let $\mathcal{G}$ be the group of real numbers t ($\neq 0$) with ordinary multiplication as the group multiplication operation, the identity E being the number 1. $\mathcal{G}$ has the obvious one-dimensional faithful representation $\Gamma(t) = [t]$, so condition (A) is satisfied and the metric d of Equation (3.1) is given by $d(t, t') = |t - t'|$. In particular, $d(t, 1) = |t - 1|$. Let $\delta = \frac{1}{2}$ so that $\frac{1}{2} < t < \frac{3}{2}$ for all t in M_δ. A convenient parametrization for $t \in M_\delta$ is then

$$t = \exp x_1. \tag{3.6}$$

As required in (B), the identity 1 corresponds to $x_1 = 0$. Condition (C) is obeyed with $\eta = \log \frac{3}{2}$, as $x_1^2 < (\log \frac{3}{2})^2$ implies $\frac{2}{3} < \exp x_1 < \frac{3}{2}$. By Equation (3.6) $\Gamma(x_1)_{11} = \exp x_1$, which is certainly analytic, so that condition (D) is satisfied. Thus $\mathcal{G}$ is a linear Lie group of dimension 1. It should be noted that Equation (3.3) implies that $\mathbf{a}_1 = [1]$, thereby confirming first theorem above.

It is significant that the parametrization in Equation (3.6) extends to *all* $t > 0$ (with $-\infty < x_1 < +\infty$) and that this set forms a subgroup of $\mathcal{G}$. Moreover, every group element t such that $t < 0$ can be written in the form $t = (-1) \exp x_1$ for some x_1.

Example II *The groups* O(2) *and* SO(2)
O(2) is the group of all real orthogonal 2×2 matrices $\mathbf{A}$, SO(2) being the subgroup for which $\det \mathbf{A} = +1$.

If $\mathbf{A} \in \mathrm{O}(2), \Gamma(\mathbf{A}) = \mathbf{A}$ provides a faithful finite-dimensional representation. The orthogonality conditions $\tilde{\mathbf{A}}\mathbf{A} = \mathbf{A}\tilde{\mathbf{A}} = \mathbf{1}$ require that

$$A_{11}^2 + A_{12}^2 = A_{11}^2 + A_{21}^2 = A_{21}^2 + A_{22}^2 = A_{12}^2 + A_{22}^2 = 1 \tag{3.7}$$

and

$$A_{11}A_{21}+A_{22}A_{12}=A_{11}A_{12}+A_{22}A_{21}=0. \tag{3.8}$$

Equations (3.7) imply that $A_{11}^2=A_{22}^2$ and $A_{12}^2=A_{21}^2$, so that there are only two sets of solutions of Equations (3.8), namely:

(i) $A_{11}=A_{22}$ and $A_{12}=-A_{21}$.
In this case Equations (3.7) imply that $\det \mathbf{A}=1$, i.e. $\mathbf{A}\in SO(2)$. Moreover, from Equations (3.7), $d(\mathbf{A},\mathbf{1})=2(1-A_{11})^{1/2}$.

(ii) $A_{11}=-A_{22}$ and $A_{12}=A_{21}$.
In this case $\det \mathbf{A}=-1$ and $d(\mathbf{A},\mathbf{1})=2$.

With the choice $\delta=\sqrt{2}$, condition (B) requires the parametrization of part of set (i) but it is *not* necessary to include set (ii), as it is completely outside M_δ. A convenient parametrization is

$$\mathbf{A}=\boldsymbol{\Gamma}(\mathbf{A})=\begin{bmatrix}\cos x_1 & \sin x_1\\ -\sin x_1 & \cos x_1\end{bmatrix}. \tag{3.9}$$

Clearly $x_1=0$ corresponds to the group identity $\mathbf{1}$ and the dimension $n=1$.

Every point of $\mathbb{R}^1$ such that $x_1^2<(\frac{1}{3}\pi)^2$ gives a matrix $\mathbf{A}$ in M_δ, so condition (C) is satisfied. In fact the parametrization of Equation (3.9) extends to the whole of the set (i) with $-\pi\leqslant x_1\leqslant\pi$, that is, to the whole of SO(2). Condition (D) is obviously obeyed, so O(2) and SO(2) are both linear Lie groups of dimension 1. Further, Equation (3.3) gives

$$\mathbf{a}_1=\begin{bmatrix}0 & 1\\ -1 & 0\end{bmatrix},$$

again confirming the first theorem above.

Although the parametrization of Equation (3.9) extends to the whole of SO(2), it cannot apply to the set (ii). However, every $\mathbf{A}$ of set (ii) can be written as

$$\mathbf{A}=\begin{bmatrix}0 & 1\\ 1 & 0\end{bmatrix}\begin{bmatrix}\cos x_1 & \sin x_1\\ -\sin x_1 & \cos x_1\end{bmatrix}=\begin{bmatrix}-\sin x_1 & \cos x_1\\ \cos x_1 & \sin x_1\end{bmatrix} \tag{3.10}$$

for some x_1 such that $-\pi\leqslant x_1\leqslant\pi$.

Example III *The group* SU(2)
SU(2) is the group of 2×2 unitary matrices $\mathbf{u}$ with $\det\mathbf{u}=1$. A faithful finite-dimensional representation is provided by $\boldsymbol{\Gamma}(\mathbf{u})=\mathbf{u}$. The defining conditions imply that every $\mathbf{u}\in SU(2)$ has the form

$$\mathbf{u}=\begin{bmatrix}\alpha & \beta\\ -\beta^* & \alpha^*\end{bmatrix}, \tag{3.11}$$

where α and β are two complex numbers such that $|\alpha|^2+|\beta|^2=1$. With $\alpha=\alpha_1+i\alpha_2$, $\beta=\beta_1+i\beta_2$ (α_1, α_2, β_1, β_2 being real), this latter condition becomes $\alpha_1^2+\alpha_2^2+\beta_1^2+\beta_2^2=1$. An appropriate parametrization is then

$$\alpha_2=\tfrac{1}{2}x_3, \qquad \beta_1=\tfrac{1}{2}x_2, \qquad \beta_2=\tfrac{1}{2}x_1, \qquad \alpha_1=+\{1-\tfrac{1}{4}(x_1^2+x_2^2+x_3^2)\}^{1/2},$$

for then $x_1=x_2=x_3=0$ corresponds to the identity $\mathbf{1}$, and

$$d(\mathbf{u},\mathbf{1})=2[1-\{1-\tfrac{1}{4}(x_1^2+x_2^2+x_3^2)\}^{1/2}]^{1/2},$$

so that $d(\mathbf{u},\mathbf{1})<\delta$ if and only if $x_1^2+x_2^2+x_3^2<2\delta^2-\frac{1}{4}\delta^4$. Thus with $\delta<2\sqrt{2}$ and $\eta=\{2\delta^2-\frac{1}{4}\delta^4\}^{1/2}$ conditions (B) and (C) are satisfied and M_δ and R_η coincide. Condition (D) is clearly true, so SU(2) is a linear Lie group of dimension 3. Incidentally, Equation (3.3) gives

$$\mathbf{a}_1=\frac{1}{2}\begin{bmatrix}0 & i\\ i & 0\end{bmatrix}, \qquad \mathbf{a}_2=\frac{1}{2}\begin{bmatrix}0 & 1\\ -1 & 0\end{bmatrix}, \qquad \mathbf{a}_3=\frac{1}{2}\begin{bmatrix}i & 0\\ 0 & -i\end{bmatrix},$$

so that the first theorem above is yet again confirmed.

Although this parametrization is the most convenient for establishing that SU(2) is a linear Lie group, it is not the most useful for some practical calculations. Indeed only the matrices $\mathbf{u}$ with $\alpha_1\geqslant 0$ can be parametrized this way, whereas it will be shown in Example III of Section 2 that there exist parametrizations of the whole of SU(2).

There is no difficulty in principle in generalizing the arguments used in Examples II and III to show that for all $N\geqslant 2$, O(N), SO(N), U(N) and SU(N) are linear Lie groups of dimensions $\frac{1}{2}N(N-1)$, $\frac{1}{2}N(N-1)$, N^2 and N^2-1 respectively, but the detailed algebra is rather more lengthy. (U(1) is a special case that is very easy to treat along the lines of Example I, because $\mathbf{u}=[\exp ix_1]$, $-\pi\leqslant x_1\leqslant\pi$, is a parametrization.)

All the examples given so far have been essentially groups of matrices, so that the condition (A) has been trivially satisfied. The final example is rather different in this respect.

Example IV *The Euclidean group of* $\mathbb{R}^3$

The Euclidean group of $\mathbb{R}^3$ is defined to be the group of all coordinate transformations T with Equation (1.7) giving the group multiplication operation. As Equation (1.5) can be written as

$$\begin{bmatrix}x'\\ y'\\ z'\\ 1\end{bmatrix}=\begin{bmatrix}R(T)_{11} & R(T)_{12} & R(T)_{13} & t(T)_1\\ R(T)_{21} & R(T)_{22} & R(T)_{23} & t(T)_2\\ R(T)_{31} & R(T)_{32} & R(T)_{33} & t(T)_3\\ 0 & 0 & 0 & 1\end{bmatrix}\begin{bmatrix}x\\ y\\ z\\ 1\end{bmatrix},$$

it follows that

$$\boldsymbol{\Gamma}(\{\mathbf{R}(\mathrm{T}) \mid \mathbf{t}(\mathrm{T})\}) = \left[\begin{array}{c:c} \mathbf{R}(\mathrm{T}) & \mathbf{t}(\mathrm{T}) \\ \hdashline \mathbf{0} & 1 \end{array}\right] \tag{3.12}$$

provides a faithful four-dimensional representation of the group. (For a brief review of partitioned matrices see Appendix A, Section 1.) It is then easy to show that the Euclidean group of $\mathbb{R}^3$ is a linear Lie group of dimension 6, three parameters specifying the proper rotations and three parameters specifying the translations.

Finally, a Lie subgroup can be defined in the obvious way.

Definition *Lie subgroup of a linear Lie group*
A subgroup $\mathcal{G}'$ of a linear Lie group $\mathcal{G}$ that is itself a linear Lie group is called a "Lie subgroup" of $\mathcal{G}$.

2 The connected components of a linear Lie group

Definition *Connected component of a linear Lie group* $\mathcal{G}$
A maximal set of elements T of $\mathcal{G}$ that can be obtained from each other by *continuously* varying one or more of the matrix elements $\Gamma(\mathrm{T})_{jk}$ of the faithful finite dimensional representation $\boldsymbol{\Gamma}$ is said to form a "connected component" of $\mathcal{G}$.

(It can be shown that the concept of connectedness as defined for a general topological space (Simmons 1963) is equivalent, for the type of space being considered here, to that implied by the above definition.)

Example I *The multiplicative group of real numbers*
This group was considered in Example I of Section 1. The set $t>0$ forms one connected component (which actually constitutes a subgroup) and the set $t<0$ forms another connected component. As $t=0$ is excluded from the group, one set cannot be obtained continuously from the other.

Example II *The groups* O(2) *and* SO(2)
In the group O(2) that was examined in Example II of Section 1, every matrix $\mathbf{A}$ of SO(2) can be parametrized by Equation (3.9) with $-\pi \leqslant x_1 \leqslant \pi$, whereas if $\mathbf{A}$ is a member of the set (ii) (i.e. if $\det \mathbf{A} = -1$), $\mathbf{A}$ can be written in the form of Equation (3.10). Thus

SO(2) constitutes one connected component and the set (ii) is another connected component. It is obvious that these two sets cannot be connected to each other, because in a connected component $\det \boldsymbol{\Gamma}(T)$ must vary *continuously* with T (if it varies at all), but $\det \mathbf{A}$ cannot take any values between $+1$ and -1 for $\mathbf{A} \in O(2)$.

These examples suggest the following general theorem.

Theorem I The connected component of a linear Lie group $\mathscr{G}$ that contains the identity E is an ***invariant subgroup*** of $\mathscr{G}$. This component is often referred to as "the connected subgroup of $\mathscr{G}$".

Proof Let $\mathscr{S}$ be the connected component of $\mathscr{G}$ that contains E. It will first be shown that $\mathscr{S}$ is a subgroup of $\mathscr{G}$. Let S be any element of $\mathscr{S}$. Then it is obvious that the set of elements $S'S^{-1}$, where S' varies over $\mathscr{S}$, forms a connected component of $\mathscr{G}$. But with $S' = S$, $S'S^{-1} = E$, so this component contains E and must therefore be $\mathscr{S}$ itself. Thus for any S' and S of $\mathscr{S}$, $S'S^{-1} \in \mathscr{S}$. The first theorem of Chapter 2, Section 1 then shows that $\mathscr{S}$ must be a subgroup of $\mathscr{G}$.

It will now be demonstrated that $\mathscr{S}$ is an ***invariant*** subgroup. Let X be any element of $\mathscr{G}$, and consider the set of elements XSX^{-1}, where S varies over $\mathscr{S}$. This set is a connected component of $\mathscr{G}$ and, as it contains E, it must be $\mathscr{S}$ itself. Thus for any $X \in \mathscr{G}$ and any $S \in \mathscr{S}$, $XSX^{-1} \in \mathscr{S}$, so $\mathscr{S}$ is an invariant subgroup.

Corollary Each connected component of a linear Lie group $\mathscr{G}$ is a right coset of the connected subgroup $\mathscr{S}$.

Proof Let $\mathscr{C}$ be a connected component containing an element $T \in \mathscr{G}$. Then the set of elements CT^{-1}, $C \in \mathscr{C}$, also forms a connected component. However, $CT^{-1} = E$ with $C = T$, so this set is $\mathscr{S}$ and consequently $\mathscr{C}$ is the right coset $\mathscr{S}T$.

In principle $\mathscr{G}$ may have a countably infinite number of connected components, but in all cases of physical interest this number is finite. The axioms imply that the connected subgroup is always a linear Lie group. In the case in which there is only one component $\mathscr{G}$ has a special name:

Definition *Connected linear Lie group*
A linear Lie group is said to be "connected" if it possesses only one connected component.

The whole of a connected linear Lie group of dimension n can be parametrized by n real numbers $y_1, y_2, \ldots, y_n$, which form a connected set in $\mathbb{R}^n$, in such a way that all the matrix elements $\Gamma(T)_{jk}$ are *continuous* functions of the parameters. There is *no* requirement that these functions be analytic or that they provide a one-to-one mapping. Consequently this parametrization does not necessarily satisfy all the conditions appearing in the definition of a linear Lie group. As the sets $x_1, x_2, \ldots, x_n$ and $y_1, y_2, \ldots, y_n$ are required for different purposes, they need not be interchangeable. The parametrizations do coincide in Examples I and II above, but Example III below reflects the general situation.

Example III *The group* SU(2)
Every pair of complex numbers α and β of Equation (3.11) that satisfy $|\alpha|^2+|\beta|^2=1$ can be written as

$$\alpha=\cos y_1 \exp(iy_2), \qquad \beta=\sin y_1 \exp(iy_3)$$

where

$$0\leqslant y_1\leqslant\tfrac{1}{2}\pi, 0\leqslant y_2\leqslant 2\pi, 0\leqslant y_3\leqslant 2\pi. \tag{3.13}$$

Thus

$$\mathbf{u}=\boldsymbol{\Gamma}(\mathbf{u})=\begin{bmatrix} \cos y_1 \exp(iy_2) & \sin y_1 \exp(iy_3) \\ -\sin y_1 \exp(-iy_3) & \cos y_1 \exp(-iy_2)\end{bmatrix}, \tag{3.14}$$

whose matrix elements are obviously continuous functions of y_1, y_2 and y_3. This is therefore a parametrization of the whole of SU(2). (This parametrization fails to satisfy the conditions involved in the definition of a linear Lie group because it does not provide a *one-to-one* mapping of the appropriate regions, for the identity corresponds to the whole set of points $y_1=0$, $y_2=0$, $0\leqslant y_3\leqslant 2\pi$. Consequently $\partial\boldsymbol{\Gamma}/\partial y_3=\mathbf{0}$ at $y_1=y_2=y_3=0$.)

A very similar parametrization of the whole of SU(2), whose significance will become apparent in Section 5, is

$$\mathbf{u}=\begin{bmatrix} \cos\tfrac{1}{2}\theta \exp\{\tfrac{1}{2}i(\psi+\phi)\} & \sin\tfrac{1}{2}\theta \exp\{\tfrac{1}{2}i(\psi-\phi)\} \\ -\sin\tfrac{1}{2}\theta \exp\{-\tfrac{1}{2}i(\psi-\phi)\} & \cos\tfrac{1}{2}\theta \exp\{-\tfrac{1}{2}i(\psi+\phi)\}\end{bmatrix}, \tag{3.15}$$

where

$$0\leqslant\theta\leqslant\pi, \qquad 0\leqslant\psi\leqslant 4\pi, \qquad 0\leqslant\phi\leqslant 2\pi. \tag{3.16}$$

Example IV *The groups* O(3) *and* SO(3)
As noted in Chapter 1, Section 2, O(3) is isomorphic to the group of *all* rotations in $\mathbb{R}^3$ and SO(3) is isomorphic to the subgroup of *proper* rotations.

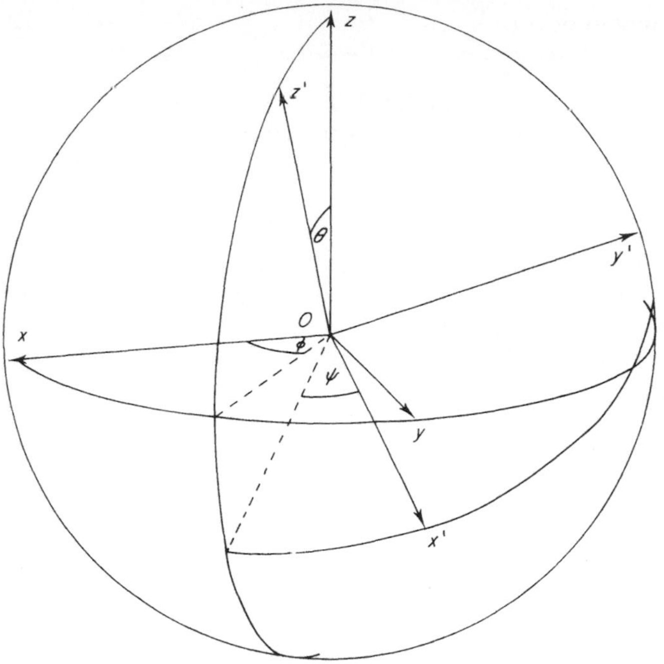

Figure 3.1 The Eulerian angles θ, ψ and ϕ.

A similar argument to that already given in Example II for O(2) shows that the group of all rotations has two connected components, namely the group of proper rotations (which forms the connected subgroup) and the set of improper rotations.

A very useful parametrization of the whole group of proper rotations is given by $y_1 = \theta, y_2 = \psi, y_2 = \phi$, where θ, ψ and ϕ are the *Eulerian angles*, as defined in Figure 3.1. Then, by inspection,

$$\left.\begin{aligned}
R(T)_{11} &= -\sin\phi\sin\psi + \cos\theta\cos\phi\cos\psi,\\
R(T)_{12} &= \cos\phi\sin\psi + \cos\theta\sin\phi\cos\psi,\\
R(T)_{13} &= -\sin\theta\cos\psi,\\
R(T)_{21} &= -\sin\phi\cos\psi - \cos\theta\cos\phi\sin\psi,\\
R(T)_{22} &= \cos\phi\cos\psi - \cos\theta\sin\phi\sin\psi,\\
R(T)_{23} &= \sin\theta\sin\psi,\\
R(T)_{31} &= \sin\theta\cos\phi,\\
R(T)_{32} &= \sin\theta\sin\phi,\\
R(T)_{33} &= \cos\theta,
\end{aligned}\right\} \qquad (3.17)$$

where

$$0 \leqslant \theta \leqslant \pi, \qquad 0 \leqslant \psi \leqslant 2\pi, \qquad 0 \leqslant \phi \leqslant 2\pi. \tag{3.18}$$

The parametrization of Equations (3.17) does *not* extend to improper rotations but, as observed previously, every improper rotation is a product of the spatial inversion operator I with a proper rotation.

Similar arguments show that SO(N) and SU(N) are connected linear Lie groups for all $N \geqslant 2$, as is U(N) for all $N \geqslant 1$. The relationship between a connected linear Lie group and its corresponding real Lie algebra will be studied in some detail in Chapter 10, where it will be shown that the Lie algebra very largely determines the structure of the group. Indeed, it is for this purpose that the parametrization in terms of $x_1, x_2, \ldots, x_n$ is required. However, the rest of this chapter is devoted to certain "global" properties of linear Lie groups, and for these it is the parametrization in terms of $y_1, y_2, \ldots, y_n$ that is relevant.

3 Compact and non-compact linear Lie groups

Although the concept of a "compact" set in a general topological space has a curiously elusive quality, the following theorem, often referred to as the "Heine–Borel Theorem", provides a very straightforward characterization of such sets in finite-dimensional real and complex Euclidean spaces. As this will suffice to distinguish a compact linear Lie group from a non-compact linear Lie group, no attempt will be made to give a detailed account of compactness, nor even a definition of the notion. (A lucid account of this and other general topological ideas may be found in the book of Simmons (1963).)

Theorem I A subset of points of a real or complex finite-dimensional Euclidean space is "compact" if and only if it is *closed* and *bounded*.

As mentioned in Section 1, by introducing the faithful m-dimensional representation Γ, the Lie group has been endowed with the topology of $\mathbb{C}^{m^2}$. However, it is often helpful to invoke the continuous parametrization of the connected subgroup by $y_1, y_2, \ldots, y_n$ introduced in Section 2. As the continuous image of a compact set is always another compact set (Simmons 1963), it follows that if the linear Lie group has only a finite number of connected components and the parameters $y_1, y_2, \ldots, y_n$ range over a closed and bounded set in $\mathbb{R}^n$, then the group is compact.

A "bounded" set of a real or complex Euclidean space is merely a set that can be contained in a *finite* "sphere" of the space. The term "closed" implies something more involved, so perhaps a few words of explanation may be needed. Although the specification of a general closed set can be fairly difficult, the only subsets of $\mathbb{R}^n$ that are relevant here are connected, and for these the characterization is straightforward. Indeed, in $\mathbb{R}^1$ every connected closed set is of the form $a_1 \leq y_1 \leq b_1$. Similarly, the set $a_j \leq y_j \leq b_j$, $j = 1, 2, \ldots, n$ of $\mathbb{R}^n$ is closed, but if any of the end points a_j and b_j are not attained the set is not closed. This set is bounded if and only if all the a_j and b_j are finite.

These considerations imply the following identification.

Characterization *Compact linear Lie group of dimension n*
A linear Lie group of dimension n with a finite number of connected components is compact if the parameters $y_1, y_2, \ldots, y_n$ range over the closed finite intervals $a_j \leq y_j \leq b_j$, $j = 1, 2, \ldots, n$.

The Lie groups of physical interest that are non-compact usually fail to be compact by virtue of giving an unbounded set of matrix elements in $\mathbb{C}^{m^2}$. As the sets of matrix elements $\Gamma(T)_{jk}$ of a linear Lie group $\mathcal{G}$ are bounded if and only if there exists a finite real number M such that $d(T, E) < M$ for all $T \in \mathcal{G}$, such groups are very easy to recognize in practice.

If a Lie group $\mathcal{G}$ is compact, then every Lie subgroup $\mathcal{S}$ of $\mathcal{G}$ must also be compact (except in the rare case when $\mathcal{S}$ has a "non-closed" parametrization). On the other hand, if $\mathcal{G}$ is non-compact, $\mathcal{G}$ may easily possess compact Lie subgroups.

For semi-simple Lie groups there exists a criterion for compactness that is expressed purely in Lie algebraic terms, as will be shown in Chapter 14, Section 2.

The importance of the distinction between compact and non-compact groups lies in the fact that the representation theory of compact Lie groups is very largely the same as that for finite groups, whereas for non-compact groups the theory is entirely different.

Example I *The multiplicative group of real numbers*
As noted in Example I of Section 1, a faithful one-dimensional representation of this group is provided by $\boldsymbol{\Gamma}(t) = [t]$. Obviously this set is unbounded in $\mathbb{C}^1$, so the group is non-compact.

Example II *The groups* O(*N*) *and* SO(*N*)
For O(2) and SO(2), Examples II of Sections 1 and 2 imply that the range of the only parameter y_1 $(=x_1)$ is $-\pi \leq y_1 \leq \pi$. For O(3) and

SO(3) the three parameters involved have ranges given in Conditions (3.18), which are closed and finite. Similar statements are true for O(N) and SO(N) for $N=4, 5, \ldots$. Thus O(N) and SO(N) are compact for all $N \geqslant 2$.

Example III *The groups* U(N) and SU(N)
As all the intervals in Conditions (3.13) are closed and finite, SU(2) is compact. The same is true of SU(N) for all $N \geqslant 2$, and of U(N) for all $N \geqslant 1$.

Example IV *The group of "standard" Lorentz transformations in one space and one time dimension*
The "standard" Lorentz transformation T between a frame $Oxyzt$ and a frame $Ox'y'z't'$ moving with velocity v along Ox is

$$\left.\begin{aligned} x' &= \gamma(v)\{x-(v/c)ct\}, \\ ct' &= \gamma(v)\{ct-(v/c)x\}, \end{aligned}\right\} \qquad \left.\begin{aligned} y' &= y, \\ z' &= z, \end{aligned}\right\}$$

where $\gamma(v)=\{1-(v^2/c^2)\}^{-1/2}$ and $-c<v<c$ (Pauli 1958).

Putting $x_1=\sinh^{-1}\{v\gamma(v)/c\}$, the transformation T can be expressed in the form of Equation (2.11) with $\mathbf{t}(\mathrm{T})=\mathbf{0}$ and

$$\boldsymbol{\Lambda}(\mathrm{T}) = \begin{bmatrix} \cosh x_1 & 0 & 0 & -\sinh x_1 \\ 0 & 1 & 0 & 0 \\ 0 & 0 & 1 & 0 \\ -\sinh x_1 & 0 & 0 & \cosh x_1 \end{bmatrix},$$

with $-\infty<x_1<+\infty$. As the matrices $\boldsymbol{\Lambda}(\mathrm{T})$ form a four-dimensional faithful representation of the group, it is easily shown that x_1 provides a parametrization that satisfies all the requirements for the group to be a linear Lie group of dimension 1. Moreover, this group is connected and $y_1=x_1$ is a parametrization of the whole group. However, $d(\mathrm{E},\mathrm{T})=\{2(1-\cosh y_1)^2+(\sinh y_1)^2\}^{1/2}\to\infty$ as $y_1\to\pm\infty$, so the set of matrix elements is unbounded in $\mathbb{C}^4$ and so this group is non-compact.

The same result is true for the homogeneous and inhomogeneous Lorentz groups in three space dimensions and one time dimension (see Example V of Chapter 2, Section 7), as these groups contain the group just considered as a subgroup.

4 Invariant integration

If to each element T of a group $\mathscr{G}$ a complex number $f(\mathrm{T})$ is assigned, then $f(\mathrm{T})$ is said to be a "complex-valued function defined on $\mathscr{G}$". One example that has been met already is the set of matrix elements $\Gamma(\mathrm{T})_{jk}$ (for j, k fixed) of a matrix representation Γ of $\mathcal{G}$.

For a *finite* group sums of the form $\sum_{\mathrm{T}\in\mathcal{G}} f(\mathrm{T})$ are frequently encountered, particularly in representation theory. Because the Rearrangement Theorem shows that the set $\{\mathrm{T'T}; \mathrm{T}\in\mathcal{G}\}$ has exactly the same members as $\mathcal{G}$, it follows that for any $\mathrm{T'}\in\mathcal{G}$

$$\sum_{\mathrm{T}\in\mathcal{G}} f(\mathrm{T'T}) = \sum_{\mathrm{T}\in\mathcal{G}} f(\mathrm{T}),$$

and the sum is said to be "left-invariant". Similarly

$$\sum_{\mathrm{T}\in\mathcal{G}} f(\mathrm{TT'}) = \sum_{\mathrm{T}\in\mathcal{G}} f(\mathrm{T}),$$

so such sums are also "right-invariant". Moreover, with $f(\mathrm{T}) = 1$ for all $\mathrm{T}\in\mathcal{G}$, the sum is finite in the sense that $\sum_{\mathrm{T}\in\mathcal{G}} 1 = g$, the order of $\mathcal{G}$.

In generalizing to a connected linear Lie group, it is natural to make the hypothesis that the sum can be replaced by an integral with respect to the parameters $y_1, y_2, \ldots, y_n$. However, questions immediately arise about the left-invariance, right-invariance and finiteness of such integrals. For general topological groups these become problems in measure theory. Using this theory Haar (1933) showed that for a very large class of topological groups, which includes the linear Lie groups, there always exists a left-invariant integral and there always exists a right-invariant integral. (Accounts of these developments, including proofs of the theorems that follow, may be found in the books of Halmos (1950), Loomis (1953) and Hewitt and Ross (1963).)

Let

$$\int_{\mathcal{G}} f(\mathrm{T})\,\mathrm{d}_l\mathrm{T} \equiv \int_{\alpha_1}^{b_1} \mathrm{d}y_1 \int_{\alpha_2}^{b_2} \mathrm{d}y_2 \ldots \int_{\alpha_n}^{b_n} \mathrm{d}y_n f(\mathrm{T}(y_1, y_2, \ldots, y_n))\sigma_l(y_1, y_2, \ldots, y_n) \tag{3.19}$$

and

$$\int_{\mathcal{G}} f(\mathrm{T})\,\mathrm{d}_r\mathrm{T} \equiv \int_{\alpha_1}^{b_1} \mathrm{d}y_1 \int_{\alpha_2}^{b_2} \mathrm{d}y_2 \ldots \int_{\alpha_n}^{b_n} \mathrm{d}y_n f(\mathrm{T}(y_1, y_2, \ldots, y_n))\sigma_r(y_1, y_2, \ldots, y_n) \tag{3.20}$$

be the left- and right-invariant integrals of a linear Lie group $\mathcal{G}$, so that

$$\int_{\mathcal{G}} f(\mathrm{T'T})\,\mathrm{d}_l\mathrm{T} = \int_{\mathcal{G}} f(\mathrm{T})\,\mathrm{d}_l\mathrm{T} \tag{3.21}$$

and

$$\int_{\mathcal{G}} f(\mathrm{TT}')\,\mathrm{d}_r\mathrm{T} = \int_{\mathcal{G}} f(\mathrm{T})\,\mathrm{d}_r\mathrm{T} \tag{3.22}$$

for any $\mathrm{T}' \in \mathcal{G}$ and any functions $f(\mathrm{T})$ for which the integrals are well defined. Here $\sigma_l(y_1, \ldots, y_n)$ and $\sigma_r(y_1, \ldots, y_n)$ are left- and right-invariant "weight functions", which are each unique up to multiplication by arbitrary constants. The left- and right-invariant integrals may be said to be *finite* if

$$\int_{\mathcal{G}} \mathrm{d}_l\mathrm{T} \equiv \int_{a_1}^{b_1} \mathrm{d}y_1 \int_{a_2}^{b_2} \mathrm{d}y_2 \ldots \int_{a_n}^{b_n} \mathrm{d}y_n \sigma_l(y_1, y_2, \ldots, y_n)$$

and

$$\int_{\mathcal{G}} \mathrm{d}_r\mathrm{T} \equiv \int_{a_1}^{b_1} \mathrm{d}y_1 \int_{a_2}^{b_2} \mathrm{d}y_2 \ldots \int_{a_n}^{b_n} \mathrm{d}y_n \sigma_r(y_1, y_2, \ldots, y_n)$$

are finite. If the multiplicative constants can be chosen so that $\sigma_l(y_1, y_2, \ldots, y_n)$ and $\sigma_r(y_1, y_2, \ldots, y_n)$ are equal, so that the integrals are both left- and right-invariant, then $\mathcal{G}$ is said to be "unimodular", and one may write

$$\mathrm{d}_l\mathrm{T} = \mathrm{d}_r\mathrm{T} = \mathrm{dT}$$

and

$$\sigma_l(y_1, y_2, \ldots, y_n) = \sigma_r(y_1, y_2, \ldots, y_n) = \sigma(y_1, y_2, \ldots, y_n).$$

If $\mathcal{G}$ has more than one connected component, the integrals in Equations (3.19) and (3.20) can be generalized in the obvious way to include a sum over the components (see Example V below).

The significance of the distinction between compact and non-compact Lie groups lies in the first two of the following theorems, the first of which was originally proved by Peter and Weyl (1927). They imply that compact Lie groups have many of the properties of finite groups, summation over a finite group merely being replaced by an invariant integral over the compact Lie groups, whereas for non-compact groups the situation is completely different.

Theorem I If $\mathcal{G}$ is a *compact* Lie group, then $\mathcal{G}$ is *unimodular* and the invariant integral

$$\int_{\mathcal{G}} f(\mathrm{T})\,\mathrm{dT} \equiv \int_{a_1}^{b_1} \mathrm{d}y_1 \int_{a_2}^{b_1} \mathrm{d}y_2 \ldots \int_{a_n}^{b_n} \mathrm{d}y_n f(\mathrm{T}(y_1, y_2, \ldots, y_n))\sigma(y_1\, y_2, \ldots, y_n)$$

exists and is *finite* for every continuous function $f(\mathrm{T})$. Consequently

$\sigma(y_1, y_2, \ldots, y_n)$ can be chosen so that

$$\int_{\mathscr{G}} dT = \int_{a_1}^{b_1} dy_1 \int_{a_2}^{b_2} dy_2 \ldots \int_{a_n}^{b_n} dy_n \sigma(y_1, y_2, \ldots, y_n) = 1.$$

(A function $f(T)$ is continuous if and only if $f(T(y_1, y_2, \ldots, y_n))$ is a continuous function of $y_1, y_2, \ldots, y_n$.)

Theorem II If $\mathscr{G}$ is a *non-compact* Lie group then the left- and right-invariant integrals are *both infinite.*

For non-compact groups the question of when $\mathscr{G}$ is unimodular is partially answered by the following theorem.

Theorem III If $\mathscr{G}$ is Abelian or semi-simple then $\mathscr{G}$ is unimodular.

The definition of a semi-simple Lie group is given in Chapter 13, Section 2.

The other non-compact linear Lie groups have to be investigated individually. The Euclidean group of $\mathbb{R}^3$ can be shown to be unimodular, whereas in Example VI below a group is examined for which this is not the case.

Example I *The multiplicative group of positive real numbers*
This is a connected non-compact linear Lie group but, as it is Abelian, it is unimodular. With parametrization $\Gamma(t) = [t] = [\exp y_1]$, $-\infty < y_1 < \infty$, the weight function $\sigma(y_1)$ can be chosen so that $\sigma(y_1) = 1$. Then with $F(y_1)$ defined by $F(y_1) = f(\exp y_1)$ and with $t' = \exp y_1'$ the invariance conditions, Equations (3.21) and (3.22), reduce to

$$\int_{-\infty}^{\infty} F(y_1 + y_1')\, dy_1 = \int_{-\infty}^{\infty} F(y_1)\, dy_1,$$

which is obviously satisfied. As expected,

$$\int_{\mathscr{G}} dT = \int_{-\infty}^{\infty} dy_1$$

is not finite.

Example II *The group* SO(2)
As this is a connected Abelian compact Lie group it is also unimodular. With $y_1 = x_1$ in the parametrization of Equation (3.9), the appropriate choice of weight function is $\sigma(y_1) = (1/2\pi)$. If T' and $T'' = T'T = TT'$ are parametrized by y_1' and y_1'' respectively, then

$$y_1'' = \begin{cases} y_1 + y_1', & -\pi \leqslant y_1 + y_1' \leqslant \pi, \\ y_1 + y_1' - 2\pi, & \pi < y_1 + y_1', \\ y_1 + y_1' + 2\pi, & y_1 + y_1' < -\pi, \end{cases}$$

and the invariance conditions in Equations (3.21) and (3.22) reduce to

$$\int_{-\pi}^{\pi} F(y_1'')\,dy_1 = \int_{-\pi}^{\pi} F(y_1)\,dy_1,$$

where $F(y_1) = f(T(y_1))$, which is obviously satisfied. Also

$$\int_{\mathscr{G}} dT \equiv \int_{-\pi}^{\pi} \sigma(y_1)\,dy_1 = 1.$$

Example III *The group* SU(2)
SU(2) is compact and hence is unimodular. With the parametrization of Equation (3.14) it can be shown (Naimark 1957) that the weight function is $\sigma(y_1, y_2, y_3) = (1/4\pi^2)\sin 2y_1$.

Example IV *The group* SO(3)
This group is compact and hence is unimodular. It can be shown (Gel'fand and Shapiro 1956) that with the parametrization of Equations (3.17) the weight function is $\sigma(\theta, \psi, \phi) = (1/8\pi^2)\sin\theta$.

Example V *The linear infinite point group* $D_{\infty h}$
$D_{\infty h}$ was defined in Example VI of Chapter 2, Section 7. It is a linear Lie group of dimension 1 with four connected components $\mathscr{S}E$, $\mathscr{S}C_{2x}$, $\mathscr{S}I$ and $\mathscr{S}(IC_{2z})$, where the connected subgroup $\mathscr{S}$ consists of E and all rotations C_θ $(-\pi \leq \theta \leq \pi, \theta \neq 0)$. As $\mathscr{S}$ is isomorphic to SO(2), $D_{\infty h}$ is a compact Lie group. It follows from Example II above that the invariant integral is given by

$$\int_{\mathscr{G}} f(T)\,dT = \frac{1}{8\pi}\sum_{j=1}^{4}\int_{-\pi}^{\pi} f(C_\theta T_j)\,d\theta,$$

where $T_1 = E$, $T_2 = C_{2z}$, $T_3 = I$, and $T_4 = IC_{2z}$.

Example VI *A non-unimodular group*
The set of matrices

$$\begin{bmatrix} y_1 & y_2 \\ 0 & 1 \end{bmatrix}$$

with $0 < y_1 < \infty$, $-\infty < y_2 < \infty$ form a connected linear Lie group of dimension 2. This group is clearly non-Abelian and non-compact and, as the matrices

$$\begin{bmatrix} 1 & y_2 \\ 0 & 1 \end{bmatrix}$$

form an Abelian invariant subgroup, it is also non-semi-simple. It is easily verified that the left- and right-invariant weight functions are $\sigma_l(y_1, y_2) = 1/y_1^2$ and $\sigma_r(y_1, y_2) = 1/y_1$ respectively, so this group is not unimodular.

The invariant weight functions for the groups $O(N)(N \geqslant 2)$, $U(N)$ $(N \geqslant 1)$ and the unitary symplectic groups (see Chapter 10, Section 5(*b*)) have been derived by Weyl (1939) and Toyama (1948). In practice, explicit expressions for weight functions are seldom needed. Indeed, in dealing with the compact Lie groups all that is usually required is the knowledge (embodied in the first theorem above) that finite left- and right-invariant integrals always exist. Nevertheless, a prescription for calculating the weight functions will be given in Chapter 11, Section 6.

5 The homomorphic mapping of SU(2) onto SO(3)

The whole theory of spin for electrons and other elementary particles in non-relativistic quantum mechanics is based on the following theorem.

Theorem I There exists a two-to-one homomorphic mapping of the group SU(2) onto the group SO(3). If $\mathbf{u} \in SU(2)$ maps onto $\mathbf{R}(\mathbf{u}) \in SO(3)$, then $\mathbf{R}(\mathbf{u}) = \mathbf{R}(-\mathbf{u})$ and the mapping may be chosen so that

$$R(\mathbf{u})_{jk} = \tfrac{1}{2}\mathrm{tr}\,\{\boldsymbol{\sigma}_j \mathbf{u} \boldsymbol{\sigma}_k \mathbf{u}^{-1}\}, \qquad j, k = 1, 2, 3, \tag{3.23}$$

where

$$\boldsymbol{\sigma}_1 = \begin{bmatrix} 0 & 1 \\ 1 & 0 \end{bmatrix}, \quad \boldsymbol{\sigma}_2 = \begin{bmatrix} 0 & -i \\ i & 0 \end{bmatrix}, \quad \boldsymbol{\sigma}_3 = \begin{bmatrix} 1 & 0 \\ 0 & -1 \end{bmatrix} \tag{3.24}$$

are the Pauli spin matrices. This implies that if $\mathbf{u}$ is parametrized as in Equation (3.15) then the components of $\mathbf{R}(\mathbf{u})$ are given by Equations (3.17).

Proof It is easiest to establish the homomorphism indirectly by considering a traceless 2×2 Hermitian matrix $\mathbf{m}(\mathbf{r})$, the components of which are defined in terms of the position vector $\mathbf{r} = (x, y, z)$ by

$$\mathbf{m}(\mathbf{r}) = \begin{bmatrix} z & x - iy \\ x + iy & -z \end{bmatrix}. \tag{3.25}$$

For any $\mathbf{u} \in SU(2)$, $\mathbf{u}\mathbf{m}(\mathbf{r})\mathbf{u}^{-1}$ is also a traceless Hermitian matrix, which can be written as

$$\mathbf{u}\mathbf{m}(\mathbf{r})\mathbf{u}^{-1} = \mathbf{m}(\mathbf{r}') = \begin{bmatrix} z' & x' - iy' \\ x' + iy' & -z' \end{bmatrix}, \tag{3.26}$$

thereby defining another position vector $\mathbf{r}' = (x', y', z')$. On evaluating the left-hand side of Equation (3.26) and equating corresponding matrix elements, a real *linear* transformation from $\mathbf{r}$ to $\mathbf{r}'$ is generated which can be written as

$$\mathbf{r}' = \mathbf{R}(\mathbf{u})\mathbf{r}. \tag{3.27}$$

This transformation defines a real 3×3 matrix $\mathbf{R}(\mathbf{u})$ which depends on $\mathbf{u}$ but not on $\mathbf{r}$.

The transformation of Equation (3.26) preserves scalar products of position vectors, so Equation (3.27) must represent a *rotation* in $\mathbb{R}^3$. This follows because if $\mathbf{r}_1$ and $\mathbf{r}_2$ are any two position vectors and $\mathbf{r}'_1 = \mathbf{R}(\mathbf{u})\mathbf{r}_1$ and $\mathbf{r}'_2 = \mathbf{R}(\mathbf{u})\mathbf{r}_2$, then

$$\mathbf{r}'_1 \cdot \mathbf{r}'_2 = \tfrac{1}{2}\mathrm{tr}\,\{\mathbf{m}(\mathbf{r}'_1)\mathbf{m}(\mathbf{r}'_2)\} = \tfrac{1}{2}\mathrm{tr}\,\{\mathbf{m}(\mathbf{r}_1)\mathbf{m}(\mathbf{r}_2)\} = \mathbf{r}_1 \cdot \mathbf{r}_2.$$

Clearly $\mathbf{u} = \mathbf{1}_2$ corresponds to the identity transformation for which $\mathbf{R}(\mathbf{u}) = \mathbf{1}_3$ and $\det \mathbf{R}(\mathbf{u}) = 1$. However, SU(2) is a *connected* Lie group, so as $\mathbf{u}$ is varied continuously in Equation (3.26) the components of $\mathbf{R}(\mathbf{u})$ must vary continuously and consequently $\det \mathbf{R}(\mathbf{u})$ cannot experience a discontinuous change. Thus $\det \mathbf{R}(\mathbf{u}) = 1$ for *all* $\mathbf{u} \in \mathrm{SU}(2)$, so $\mathbf{R}(\mathbf{u}) \in \mathrm{SO}(3)$.

It will now be shown that $\mathbf{R}(\mathbf{u}_1)\mathbf{R}(\mathbf{u}_2) = \mathbf{R}(\mathbf{u}_1\mathbf{u}_2)$ for any $\mathbf{u}_1, \mathbf{u}_2 \in \mathrm{SU}(2)$, thereby establishing that the mapping of $\mathbf{u}$ to $\mathbf{R}(\mathbf{u})$ is a homomorphic mapping. Let

$$\mathbf{u}_2\mathbf{m}(\mathbf{r})\mathbf{u}_2^{-1} = \mathbf{m}(\mathbf{r}'), \qquad \mathbf{u}_1\mathbf{m}(\mathbf{r})\mathbf{u}_1^{-1} = \mathbf{m}(\mathbf{r}''), \tag{3.28}$$

so that $(\mathbf{u}_1\mathbf{u}_2)\mathbf{m}(\mathbf{r})(\mathbf{u}_1\mathbf{u}_2)^{-1} = \mathbf{m}(\mathbf{r}'')$, which implies that

$$\mathbf{r}'' = \mathbf{R}(\mathbf{u}_1\mathbf{u}_2)\mathbf{r}. \tag{3.29}$$

However, Equations (3.28) also imply that $\mathbf{r}' = \mathbf{R}(\mathbf{u}_2)\mathbf{r}$ and $\mathbf{r}'' = \mathbf{R}(\mathbf{u}_1)\mathbf{r}'$, so that

$$\mathbf{r}'' = \mathbf{R}(\mathbf{u}_1)\mathbf{R}(\mathbf{u}_2)\mathbf{r}. \tag{3.30}$$

The desired result follows from comparing Equations (3.29) and (3.30).

If $\mathbf{u} \in \mathcal{K}$, the kernel of the homomorphic mapping (see Chapter 2, Section 6), then $\mathbf{R}(\mathbf{u}) = \mathbf{1}_3$. This implies that $\mathbf{u}\mathbf{m}(\mathbf{r})\mathbf{u}^{-1} = \mathbf{m}(\mathbf{r})$ for all $\mathbf{r}$, which can only be satisfied if $\mathbf{u} = \mu\mathbf{1}_2$, where μ is a constant. As $\det \mathbf{u} = 1$ for $\mathbf{u} \in \mathrm{SU}(2)$, it follows that $\mu = \pm 1$, so that $\mathcal{K}$ consists only of $\mathbf{1}_2$ and $-\mathbf{1}_2$. The homomorphic mapping is therefore a two-to-one mapping. Indeed it is obvious from Equation (3.26) that $\mathbf{u}$ and $-\mathbf{u}$ both correspond to the same rotation, that is $\mathbf{R}(\mathbf{u})=\mathbf{R}(-\mathbf{u})$.

As $\mathbf{m}(\mathbf{r})=x\boldsymbol{\sigma}_1+y\boldsymbol{\sigma}_2+z\boldsymbol{\sigma}_3$ and $\mathbf{m}(\mathbf{r}')=x'\boldsymbol{\sigma}_1+y'\boldsymbol{\sigma}_2+z'\boldsymbol{\sigma}_3$, Equation (3.26) can be written as

$$x'\boldsymbol{\sigma}_1+y'\boldsymbol{\sigma}_2+z'\boldsymbol{\sigma}_3=x\mathbf{u}\boldsymbol{\sigma}_1\mathbf{u}^{-1}+y\mathbf{u}\boldsymbol{\sigma}_2\mathbf{u}^{-1}+z\mathbf{u}\boldsymbol{\sigma}_3\mathbf{u}^{-1}. \qquad (3.31)$$

However, it is easily verified that

$$\operatorname{tr}\{\boldsymbol{\sigma}_j\boldsymbol{\sigma}_k\}=2\delta_{jk}, \qquad j,k=1,2,3. \qquad (3.32)$$

Thus, multiplying Equation (3.31) by $\boldsymbol{\sigma}_1$, $\boldsymbol{\sigma}_2$ and $\boldsymbol{\sigma}_3$ in turn, taking the trace and using Equation (3.32) gives expressions for x', y' and z' in terms of x, y and z. Equations (3.23) are then obtained by comparison with Equation (3.27).

Direct substitution in Equation (3.22) of the expression for $\mathbf{u}$ in Equation (3.15) leads to Equations (3.17) for the elements of $\mathbf{R}(\mathbf{u})$. As *every* proper rotation can be parametrized by Equations (3.17), *every* element of SO(3) is the image of some $\mathbf{u}\in$ SU(2).

In the application to the theory of spin it is convenient to invert the relationship between $\mathbf{u}$ and $\mathbf{R}$ and regard each member of SU(2) as depending on some member of SO(3). Then to each $\mathbf{R}$ of SO(3) there correspond *two* members of SU(2) which may be denoted by $\mathbf{u}(\mathbf{R})$ and $-\mathbf{u}(\mathbf{R})$. As

$$\mathbf{u}(\mathbf{R}_1)\mathbf{u}(\mathbf{R}_2)=\pm\mathbf{u}(\mathbf{R}_1\mathbf{R}_2)$$

for any $\mathbf{R}_1, \mathbf{R}_2\in$ SO(3), the matrices $\mathbf{u}(\mathbf{R})$ and $-\mathbf{u}(\mathbf{R})$ are said to constitute a "two-valued" two-dimensional representation of SO(3). The sign of $\mathbf{u}(\mathbf{R})$ is entirely a matter of convention. Whether a + or a − appears in the right-hand side depends entirely on the conventions chosen for $\mathbf{u}(\mathbf{R}_1)$, $\mathbf{u}(\mathbf{R}_2)$ and $\mathbf{u}(\mathbf{R}_1\mathbf{R}_2)$.

6 The homomorphic mapping of the group SL(2,*C*) onto the proper orthochronous Lorentz group $L_+^\uparrow$

The considerations of the previous section generalize in a very straightforward manner to the relativistic situation.

The analogue of the group of rotations in $\mathbb{R}^3$ is the group of homogeneous Lorentz transformations $\boldsymbol{\Lambda}(\mathrm{T})$ in Minkowski space–time (see Example V of Chapter 2, Section 7). This latter group contains as a subgroup the set of "proper" homogeneous Lorentz transformations, namely those $\boldsymbol{\Lambda}(\mathrm{T})$ for which $\det\boldsymbol{\Lambda}(\mathrm{T})=1$. However, in contrast to the case of rotations in $\mathbb{R}^3$, this "proper homogeneous Lorentz group" L_+

is not connected. In fact its connected component is the subgroup of L_+ for which $\Lambda(T)_{44}>0$. As these transformations preserve the direction of time, this subgroup is called the "proper orthochronous homogeneous Lorentz group", and is denoted by $L_+^\uparrow$. (Further discussion of these points may be found in the books of Gel'fand *et al.* (1963) and Naimark (1964).)

In the notation that will be developed in Chapter 10 (see 10.1), the group of 4×4 real matrices $\boldsymbol{\Lambda}$ satisfying $\tilde{\boldsymbol{\Lambda}}\mathbf{g}\boldsymbol{\Lambda}=\mathbf{g}$, with $\mathbf{g}$ defined by Equation (2.12), is denoted by O(3, 1), while SO(3, 1) denotes the subgroup for which $\det \boldsymbol{\Lambda}=1$. The connected component of this latter group is sometimes indicated as $SO_0(3, 1)$. Thus the homogeneous Lorentz group, the proper homogeneous Lorentz group L_+ and the proper orthochronous Lorentz group $L_+^\uparrow$ are isomorphic to O(3, 1), SO(3, 1) and $SO_0(3, 1)$ respectively.

With the final definition that $SL(2, C)$ is the group of all 2×2 matrices $\mathbf{a}$ with real or complex entries and matrix multiplication as the group multiplication operation such that $\det \mathbf{a}=1$, it is now possible to give the generalization of the theorem of the previous section.

Theorem I There exists a two-to-one homomorphic mapping of the group $SL(2, C)$ onto the proper orthochronous homogeneous Lorentz group $L_+^\uparrow$ (or, equivalently, onto $SO_0(3, 1)$. If $\mathbf{a}\in SL(2, C)$ maps onto $\boldsymbol{\Lambda}(\mathbf{a})\in L_+^\uparrow$ (or $SO_0(3, 1)$), then $\boldsymbol{\Lambda}(\mathbf{a})=\boldsymbol{\Lambda}(-\mathbf{a})$ and the mapping may be chosen so that

$$\boldsymbol{\Lambda}(\mathbf{a})_{jk}=\tfrac{1}{2}\mathrm{tr}\{\boldsymbol{\sigma}_j\mathbf{a}\boldsymbol{\sigma}_k\mathbf{a}^\dagger\}, \qquad j, k=1, 2, 3, 4, \tag{3.33}$$

where $\boldsymbol{\sigma}_1, \boldsymbol{\sigma}_2, \boldsymbol{\sigma}_3$ are the Pauli spin matrices defined in Equations (3.24) and $\boldsymbol{\sigma}_4=\mathbf{1}_2$.

Proof The method of proof is very similar to that of the theorem of the previous section, so attention will be confined to those points that are different. Consider the 2×2 Hermitian matrix $\mathbf{m}((x, y, z, t))$ defined in terms of the event (x, y, z, t) by

$$\mathbf{m}((x, y, z, t))=\begin{bmatrix} z+ct & x-iy \\ x+iy & -z+ct \end{bmatrix},$$

so that

$$\mathbf{m}((x, y, z, t))=x\boldsymbol{\sigma}_1+y\boldsymbol{\sigma}_2+z\boldsymbol{\sigma}_3+ct\boldsymbol{\sigma}_4.$$

(Here c denotes the speed of light.) Then, for any $\mathbf{a}\in SL(2, C)$,

$\mathbf{am}((x, y, z, t))\mathbf{a}^\dagger$ is also a Hermitian matrix, which may be written as

$$\mathbf{am}((x, y, z, t))\mathbf{a}^\dagger = \mathbf{m}((x', y', z', t')) \equiv \begin{bmatrix} z' + ct' & x' - iy' \\ x' + iy' & -z' + ct' \end{bmatrix}, \quad (3.34)$$

thereby defining another set of space–time coordinates (x', y', z', t') for the event. On evaluating the left-hand side of Equation (3.34) and equating corresponding matrix elements, a real linear transformation from (x, y, z, t) to (x', y', z', t') is generated which can be written in the form

$$\begin{bmatrix} x' \\ y' \\ z' \\ ct' \end{bmatrix} = \Lambda(\mathbf{a}) \begin{bmatrix} x \\ y \\ z \\ ct \end{bmatrix}$$

This transformation defines a real 4×4 matrix $\Lambda(\mathbf{a})$, which depends on $\mathbf{a}$ but not on (x, y, z, t).

This transformation is a Lorentz transformation because

$$c^2t'^2 - x'^2 - y'^2 - z'^2 = c^2t^2 - x^2 - y^2 - z^2,$$

as the left- and right-hand sides here are $\det \mathbf{m}((x', y', z', t'))$ and $\det \mathbf{m}((x, y, z, t))$ respectively, and these two determinants are equal by Equation (3.34), as $\det \mathbf{a} = \det \mathbf{a}^\dagger = 1$.

As SL(2, C) is connected, it follows that all the transformations belong to the connected component of the homogeneous Lorentz group, so that they must belong to $L_+^\uparrow$. Further detailed investigation shows that *every* element of $L_+^\uparrow$ can be obtained in this way (Gel'fand *et al.* 1963 and Naimark 1964).

That the mapping is homomorphic follows by essentially the same argument as in the proof of the previous section, as does the deduction that the kernel $\mathcal{K}$ again consists only of $\mathbf{1}_2$ and $-\mathbf{1}_2$. The deduction of Equation (3.33) also follows the same course as the argument leading to Equation (3.23), because Equation (3.32) is valid for $j, k = 1, 2, 3$ *and* 4.

It should be noted that if $\mathbf{u} \in \mathrm{SU}(2)$ (that is $\mathbf{u} \in \mathrm{SL}(2, C)$ and $\mathbf{u}^\dagger = \mathbf{u}^{-1}$) then Equations (3.23) and (3.33) imply that

$$\Lambda(\mathbf{u}) = \begin{bmatrix} \mathbf{R}(\mathbf{u}) & \mathbf{0} \\ \mathbf{0} & \mathbf{1}_1 \end{bmatrix}.$$

4

Representations of Groups —Principal Ideas

1 Definitions

The concept of the representation of a group was introduced in Chapter 1, Section 4, where it was shown that representations occur in a natural and significant way in quantum mechanics. It is worthwhile starting the detailed study of representations by repeating the definition as rephrased in Chapter 2, Section 6:

Definition *Representation of a group $\mathscr{G}$*
If there exists a homomorphic mapping of a group $\mathscr{G}$ onto a group of non-singular $d \times d$ matrices $\boldsymbol{\Gamma}(\mathrm{T})$, with matrix multiplication as the group multiplication operation, then the group of matrices $\boldsymbol{\Gamma}(\mathrm{T})$ forms a d-dimensional representation $\boldsymbol{\Gamma}$ of $\mathscr{G}$.

It will be recalled that the representation is described as being "faithful" if the mapping is one-to-one. It is fairly obvious that $\boldsymbol{\Gamma}(\mathrm{E}) = \mathbf{1}$ for every representation $\boldsymbol{\Gamma}$. (The precise proof is as follows. As $\mathrm{E}^2 = \mathrm{E}$ then $\boldsymbol{\Gamma}(\mathrm{E})\{\boldsymbol{\Gamma}(\mathrm{E}) - \mathbf{1}\} = \mathbf{0}$. Suppose first that this is the minimal equation for $\boldsymbol{\Gamma}(\mathrm{E})$ (see Appendix A, Section 2). This implies that $\boldsymbol{\Gamma}(\mathrm{E})$ is diagonalizable and has at least one eigenvalue equal to zero, which in turn implies that $\det \boldsymbol{\Gamma}(\mathrm{E}) = 0$. As this is not permitted, the minimal equation must be of degree less than two and so must be of the form $\boldsymbol{\Gamma}(\mathrm{E}) - \gamma\mathbf{1} = \mathbf{0}$, and clearly the only allowed value of γ is 1.) It follows that $\boldsymbol{\Gamma}(\mathrm{T}^{-1}) = \boldsymbol{\Gamma}(\mathrm{T})^{-1}$ for all $\mathrm{T} \in \mathscr{G}$. Every group $\mathscr{G}$ possesses an "identity" representation, which is a one-dimensional representation for which $\Gamma(T)=[1]$ for all $T \in \mathcal{G}$. Although mathematically extremely trivial, physically this representation can be very important.

Example I *Some representations of the crystallographic point group* D_4

Several representations of the group D_4 have already been encountered either explicitly or implicitly, and it is worthwhile gathering them together for future reference. As the subsequent developments will show, this list is far from being exhaustive.

(i) Equation (1.4) implies that the matrices $\mathbf{R}(T)$ listed in Example III of Chapter 1, Section 2, form a faithful three-dimensional representation of D_4.

(ii) A faithful two-dimensional representation of D_4 was explicitly noted in Example I of Chapter 1, Section 4.

(iii) A non-faithful but non-trivial one-dimensional representation of D_4 is given implicitly in Example I of Chapter 2, Section 6. In this representation

$$\mathbf{\Gamma}(E)=\mathbf{\Gamma}(C_{2y})=\mathbf{\Gamma}(C_{2x})=\mathbf{\Gamma}(C_{2z})=[1],$$
$$\mathbf{\Gamma}(C_{4y})=\mathbf{\Gamma}(C_{4y}^{-1})=\mathbf{\Gamma}(C_{2c})=\mathbf{\Gamma}(C_{2d})=[-1].$$

(iv) Finally there is the identity representation for which $\mathbf{\Gamma}(T)=[1]$ for all $T \in \mathcal{G}$.

For a Lie group it is necessary to supplement the definition by the requirement that the homomorphic mapping must be *continuous*. For a connected linear Lie group this implies that the matrix elements of the representation must be continuous functions of the parameters $y_1, y_2, \ldots, y_n$ of Chapter 2, Section 2. (The extension to *analytic* representations and the relationship between the two concepts will be considered in Chapter 11, Section 4.)

For groups of coordinate transformations in three-dimensional Euclidean space $\mathbb{R}^3$ it has already been demonstrated how useful are the operators $P(T)$ and the basis functions $\psi_n(\mathbf{r})$ that were defined in Chapter 1, Sections 2 and 4 respectively. It is profitable to partially generalize these concepts to make them available for *any* group $\mathcal{G}$. To this end, consider a d-dimensional representation $\mathbf{\Gamma}$ of $\mathcal{G}$, let $\psi_1, \psi_2, \ldots, \psi_d$ be the basis of a d-dimensional abstract complex inner product space (see Appendix B, Section 2) called the "carrier space" V, and for each $T \in \mathcal{G}$ *define* the operator $\Phi(T)$ acting on the basis by

$$\Phi(T)\psi_n = \sum_{m=1}^{d} \Gamma(T)_{mn}\psi_m, \qquad n=1,2,\ldots,d. \tag{4.1}$$

With the further definition that

$$\Phi(T)\left\{\sum_{j=1}^{d} b_j\psi_j\right\} = \sum_{j=1}^{d} b_j\{\Phi(T)\psi_j\}$$

for any set of complex numbers $b_1, b_2, \ldots, b_d$, such an operator is a linear operator. Moreover, Equation (4.1) implies the operator equalities

$$\Phi(T_1T_2) = \Phi(T_1)\Phi(T_2) \tag{4.2}$$

for all $T_1, T_2 \in \mathcal{G}$, so that the operators form a group and there is a homomorphic mapping of $\mathcal{G}$ onto this group. The operators $\Phi(T)$ and the carrier space V are sometimes said to collectively form a "module". However, there is *no* guarantee that for a given representation the operators are unitary, that is, in general

$$(\Phi(T)\phi, \Phi(T)\psi) \neq (\phi, \psi)$$

(see Section 3 for further discussion of this point). Finally, if the basis is chosen to be an ortho-normal set, then Equation (4.1) implies that

$$\Gamma(T)_{mn} = (\psi_m, \Phi(T)\psi_n) \tag{4.3}$$

for any $T \in \mathcal{G}$. Conversely, any set of operators acting on a d-dimensional inner product space and satisfying Equation (4.2) will produce, by Equation (4.3), a d-dimensional matrix representation. (This provides the best way of introducing infinite-dimensional representations, the finite-dimensional inner product space merely being replaced by an infinite-dimensional Hilbert space, but apart from a brief mention in Chapter 17, Section 8, these will not be discussed in this book.)

It is entirely a matter of taste and convenience whether one works with an explicit matrix representation or with the corresponding module consisting of the operators $\Phi(T)$ and the carrier space V on which they act. Theoretical physicists normally prefer to deal with the more concrete matrix representations, whereas pure mathematicians tend to prefer the module formulation.

It should be noted that for groups of coordinate transformations in $\mathbb{R}^3$, for which both the operators $\Phi(T)$ and $P(T)$ are defined, there are two major differences between these sets of operators. Firstly, the $\Phi(T)$ depend on the representation Γ under consideration, whereas the $P(T)$ are independent of the representation. Secondly, the operators $\Phi(T)$ act in a finite-dimensional space, whereas the $P(T)$ act in the infinite-dimensional Hilbert space L^2.

As the theory is developed in this chapter it will become apparent that every group has an infinite number of different representations, but these can be formed out of certain basic representations, the so-called "irreducible representations". For a finite group there is essentially only a finite number of these.

It will have become evident already that vector spaces and inner product spaces play an important part in representation theory. Readers who are not very familiar with them are advised to study Appendix B before proceeding further.

In order that the most significant ideas on group representations should be developed as simply and as concisely as possible, several of the longer proofs of theorems have been relegated to Appendix C.

2 Equivalent representations

Theorem I Let $\boldsymbol{\Gamma}$ be a d-dimensional representation of a group $\mathcal{G}$, and let $\mathbf{S}$ be any $d \times d$ non-singular matrix. Define for each $\mathrm{T} \in \mathcal{G}$ a $d \times d$ matrix $\boldsymbol{\Gamma}'(\mathrm{T})$ by

$$\boldsymbol{\Gamma}'(\mathrm{T}) = \mathbf{S}^{-1}\boldsymbol{\Gamma}(\mathrm{T})\mathbf{S}. \tag{4.4}$$

Then this set of matrices also forms a d-dimensional representation of $\mathcal{G}$. The representations $\boldsymbol{\Gamma}$ and $\boldsymbol{\Gamma}'$ are said to be "equivalent", and the transformation, Equation (4.4), is called a "similarity transformation".

Proof For any $\mathrm{T}_1, \mathrm{T}_2 \in \mathcal{G}$, by Equations (1.25) and (4.4),

$$\begin{aligned}\boldsymbol{\Gamma}'(\mathrm{T}_1)\boldsymbol{\Gamma}'(\mathrm{T}_2) &= \mathbf{S}^{-1}\boldsymbol{\Gamma}(\mathrm{T}_1)\mathbf{S}\mathbf{S}^{-1}\boldsymbol{\Gamma}(\mathrm{T}_2)\mathbf{S} = \mathbf{S}^{-1}\boldsymbol{\Gamma}(\mathrm{T}_1)\boldsymbol{\Gamma}(\mathrm{T}_2)\mathbf{S}\\ &= \mathbf{S}^{-1}\boldsymbol{\Gamma}(\mathrm{T}_1\mathrm{T}_2)\mathbf{S} = \boldsymbol{\Gamma}'(\mathrm{T}_1\mathrm{T}_2).\end{aligned}$$

In Section 6 there will be given a simple direct test for the equivalence of two representations which does *not* require actually finding the matrix $\mathbf{S}$ that induces the similarity transformation.

As all 1×1 matrices commute, if $d = 1$ then $\boldsymbol{\Gamma}'(\mathrm{T}) = \boldsymbol{\Gamma}(\mathrm{T})$ for all $\mathrm{T} \in \mathcal{G}$ and for *every* 1×1 non-singular matrix $\mathbf{S}$. Thus two one-dimensional representations of $\mathcal{G}$ are either identical or are not equivalent.

For $d \geqslant 2$ the situation is not so simple. In general a similarity transformation will produce an equivalent representation whose matrices $\boldsymbol{\Gamma}'(\mathrm{T})$ are different from those of $\boldsymbol{\Gamma}(\mathrm{T})$. However, these differences are in a sense superficial, for it will become clear that to a very large extent equivalent representations have essentially the same content. The following theorem on basis functions provides the first indication of this.

Theorem II Let Γ be a d-dimensional representation of a group of coordinate transformations in $\mathbb{R}^3$, let $\psi_1(\mathbf{r}), \psi_2(\mathbf{r}), \ldots, \psi_d(\mathbf{r})$ be a set of basis functions of Γ and let $\mathbf{S}$ be any $d \times d$ non-singular matrix. Then the set of d linearly independent functions $\psi'_1(\mathbf{r}), \psi'_2(\mathbf{r}), \ldots, \psi'_d(\mathbf{r})$ defined by

$$\psi'_n(\mathbf{r}) = \sum_{m=1}^{d} S_{mn}\psi_m(\mathbf{r}), \qquad n = 1, 2, \ldots, d \tag{4.5}$$

form a set of basis functions for the equivalent representation Γ', where for all $T \in \mathcal{G}$

$$\Gamma'(T) = \mathbf{S}^{-1}\Gamma(T)\mathbf{S}. \tag{4.6}$$

Proof For any $T \in \mathcal{G}$, from Equations (1.18), (1.26) and (4.5),

$$P(T)\psi'_n(\mathbf{r}) = \sum_{m=1}^{d} S_{mn}\{P(T)\psi_m(\mathbf{r})\} = \sum_{m,p=1}^{d} S_{mn}\Gamma(T)_{pm}\psi_p(\mathbf{r}).$$

However, inverting Equation (4.5) gives

$$\psi_p(\mathbf{r}) = \sum_{q=1}^{d} (S^{-1})_{qp}\psi'_q(\mathbf{r}).$$

Thus, from Equation (4.6),

$$P(T)\psi'_n(\mathbf{r}) = \sum_{m,p,q=1}^{d} (S^{-1})_{qp}\Gamma(T)_{pm}S_{mn}\psi'_q(\mathbf{r}) = \sum_{q=1}^{d} \Gamma'(T)_{qn}\psi'_q(\mathbf{r}).$$

The functions $\psi_1(\mathbf{r}), \psi_2(\mathbf{r}), \ldots, \psi_d(\mathbf{r})$ form a basis for a complex d-dimensional inner product space. With the definition in Equation (4.5) the functions $\psi'_1(\mathbf{r}), \psi'_2(\mathbf{r}), \ldots, \psi'_d(\mathbf{r})$ form an *alternative* basis for the *same* space. ***Thus the effect of a similarity transformation is merely to rearrange the basis of this space without changing the space itself.***

This result has particular significance for the solutions of the time-independent Schrödinger equation. It was shown in Chapter 1, Section 4, that the eigenfunctions of a d-fold degenerate energy eigenvalue form a basis for a d-dimensional representation Γ of the group of the Schrödinger equation. However, *any* d linearly independent linear combinations of these eigenfunctions also form a set of eigenfunctions belonging to the *same* eigenvalue, and there is no reason to prefer the original set to this new set, or vice versa. As the new set form a basis for a representation equivalent to Γ, ***the representation of the group of the Schrödinger equation that corresponds to an energy eigenvalue is determined only up to equivalence.***

This section will be concluded by stating the analogous theorem which is valid for the carrier space of any representation of *any* group.

Theorem III Let $\mathbf{\Gamma}$ be a d-dimensional representation of a group $\mathscr{G}$, let $\psi_1, \psi_2, \ldots, \psi_d$ be a basis of its carrier space and define the operators $\Phi(\mathrm{T})$ for all $\mathrm{T} \in \mathscr{G}$ by Equation (4.1) and its extension. Let $\mathbf{S}$ be any $d \times d$ non-singular matrix. Then the set of d linearly independent vectors $\psi'_1, \psi'_2, \ldots, \psi'_d$ defined by

$$\psi'_n = \sum_{m=1}^{d} \mathrm{S}_{mn}\psi_m, \qquad n = 1, 2, \ldots, d,$$

form a basis for the equivalent representation $\mathbf{\Gamma}'$, where for all $\mathrm{T} \in \mathscr{G}$

$$\mathbf{\Gamma}'(\mathrm{T}) = \mathbf{S}^{-1}\mathbf{\Gamma}(\mathrm{T})\mathbf{S},$$

in the sense that

$$\Phi(\mathrm{T})\psi'_n = \sum_{m=1}^{d} \Gamma'(\mathrm{T})_{mn}\psi'_m \tag{4.7}$$

for all $\mathrm{T} \in \mathscr{G}$ and $n = 1, 2, \ldots, d$.

Proof This is essentially identical in content to that given above.

Again $\psi_1, \psi_2, \ldots, \psi_d$ and $\psi'_1, \psi'_2, \ldots, \psi'_d$ are merely two different bases for the *same* carrier space.

3 Unitary representations

Definition *Unitary representation of a group*
A "unitary" representation of a group $\mathscr{G}$ is a representation $\mathbf{\Gamma}$ in which the matrices $\mathbf{\Gamma}(\mathrm{T})$ are unitary for every $\mathrm{T} \in \mathscr{G}$.

The following theorems show the profound difference between compact and non-compact Lie groups and the affinity between compact Lie groups and finite groups.

Theorem I If $\mathscr{G}$ is a *finite* group or a *compact* Lie group then *every* representation of $\mathscr{G}$ is *equivalent* to a *unitary* representation.

Proof This is given in Appendix C. The main theoretical interest lies in the use of the right-invariant integral for the case of a compact Lie group.

It will be recalled that all the point groups and space groups of solid state physics are finite. Likewise, the rotation groups in three dimensions and the internal symmetry groups of elementary particles are compact Lie groups. Thus, in all these situations, advantage may be taken of the considerable simplifications that result from using representations that are unitary.

Although the technical definition of "simple" and "semi-simple" Lie groups must be deferred until Chapter 13, Section 2, this is the appropriate place to mention some relevant properties of their representations.

Theorem II If $\mathcal{G}$ is a non-compact simple Lie group then $\mathcal{G}$ possesses *no* finite-dimensional unitary representations apart from the trivial representations in which $\mathbf{\Gamma}(T)=\mathbf{1}$ for all $T\in\mathcal{G}$.

Proof This will be given in Chapter 15, Section 2.

As the connected component of the homogeneous Lorentz group is simple but not compact (see Chapter 14, Section 5), none of its finite-dimensional representations can be transformed to a unitary representation by a similarity transformation (except for the trivial representations for which $\mathbf{\Gamma}(T)=\mathbf{1}$ for all T). All the non-trivial unitary representations of this group are infinite-dimensional and are outside the scope of this book. (Very good accounts of them have been given by Gel'fand *et al.* (1963) and Naimark (1957).)

A non-compact Lie group that is not simple may possess both unitary representations and representations that are not equivalent to unitary representations, as the following examples show.

Example I *The multiplicative group of positive real numbers*
This group was considered previously in Examples I of Chapter 3, Sections 1, 2 and 3. A typical element is $\exp y_1$, $-\infty<y_1<\infty$. It has a set of one-dimensional unitary representations defined by

$$\mathbf{\Gamma}(\exp y_1)=[\exp(i\alpha y_1)],$$

where α is any fixed real number. It has also a set of one-dimensional non-unitary representations given by

$$\mathbf{\Gamma}(\exp y_1)=[\exp(\beta y_1)], \tag{4.8}$$

where β is any fixed real number. These latter representations, being one-dimensional, cannot be transformed by any similarity transformation into unitary representations. (Incidentally, for the representation (4.8),

$$\int_{\mathscr{G}} \boldsymbol{\Gamma}(\mathrm{T})^{\dagger}\boldsymbol{\Gamma}(\mathrm{T})\, d\mathrm{T} = \left[\int_{-\infty}^{\infty} \exp(2\beta y_1)\, dy_1\right],$$

which is not finite, so the matrix $\mathbf{H}$ that appears in Equation (C.1) of the proof of the first theorem above is not well defined, so demonstrating the failure of that line of argument for this particular group.)

Example II *The Euclidean group of* $\mathbb{R}^3$
This was examined previously in Example IV of Chapter 3, Section 1. It is neither compact nor simple. A set of finite-dimensional unitary representations is given by

$$\boldsymbol{\Gamma}(\{\mathbf{R}(\mathrm{T}) \mid \mathbf{t}(\mathrm{T})\}) = \boldsymbol{\Gamma}'(\mathbf{R}(\mathrm{T})),$$

where $\boldsymbol{\Gamma}'(\mathbf{R}(\mathrm{T}))$ is a representation of O(3) which can be taken to be unitary as O(3) is compact. However, *no* finite-dimensional representation in which the translations are represented non-trivially (that is, for which $\boldsymbol{\Gamma}(\{\mathbf{1} \mid \mathbf{t}(\mathrm{T})\}) \neq \mathbf{1}$ for some $\mathbf{t}(\mathrm{T}) \neq \mathbf{0}$) is equivalent to a unitary representation (Miller 1964). In particular, the four-dimensional faithful representation of Equation (3.12) is not equivalent to a unitary representation.

Theorem III If $\mathscr{G}$ is a group of coordinate transformations in $\mathbb{R}^3$ and if the representation $\boldsymbol{\Gamma}$ of $\mathscr{G}$ possesses a set of basis functions, then $\boldsymbol{\Gamma}$ is unitary if the basis functions form an ortho-normal set.

Proof Suppose that the basis functions $\psi_1(\mathbf{r}), \psi_2(\mathbf{r}), \ldots, \psi_d(\mathbf{r})$ of $\boldsymbol{\Gamma}$ form an ortho-normal set, i.e. $(\psi_m, \psi_n) = \delta_{mn}$ for $m, n = 1, 2, \ldots, d$. As the operators P(T) are unitary, it follows from Equations (1.19), (1.20) and (1.26) that for each $\mathrm{T} \in \mathscr{G}$

$$\begin{aligned}\delta_{mn} = (\psi_m, \psi_n) &= (P(\mathrm{T})\psi_m, P(\mathrm{T})\psi_n)\\ &= \sum_{p,q=1}^{d} \Gamma(\mathrm{T})^*_{pm}\Gamma(\mathrm{T})_{qn}(\psi_p, \psi_q)\\ &= \sum_{p=1}^{d} \Gamma(\mathrm{T})^*_{pm}\Gamma(\mathrm{T})_{pn},\end{aligned}$$

so that $\boldsymbol{\Gamma}(\mathrm{T})^{\dagger}\boldsymbol{\Gamma}(\mathrm{T}) = \mathbf{1}$ and hence $\boldsymbol{\Gamma}(\mathrm{T})$ is unitary.

From a set of basis functions $\psi_1(\mathbf{r}), \psi_2(\mathbf{r}), \ldots, \psi_d(\mathbf{r})$ of a non-unitary representation $\mathbf{\Gamma}$ an ortho-normal set $\psi'_1(\mathbf{r}), \psi'_2(\mathbf{r}), \ldots, \psi'_d(\mathbf{r})$ can always be constructed by the Schmidt orthogonalization process (see Appendix B, Section 2). As each $\psi'_j(\mathbf{r})$ is a linear combination of the $\psi_k(\mathbf{r})$, the set $\psi'_1(\mathbf{r}), \psi'_2(\mathbf{r}), \ldots, \psi'_d(\mathbf{r})$ must be basis functions for a unitary representation $\mathbf{\Gamma}'$ that is equivalent to $\mathbf{\Gamma}$. Indeed, on defining the coefficients S_{mn} by $\psi'_n(\mathbf{r}) = \sum_{m=1}^{d} S_{mn}\psi_m(\mathbf{r})$, the matrix $\mathbf{S}$ having these coefficients as elements is precisely the matrix that induces the similarity transformation from $\mathbf{\Gamma}$ to $\mathbf{\Gamma}'$.

However, there exist groups of coordinate transformations in $\mathbb{R}^3$ that have at least some representations that do *not* possess basis functions, so this argument does *not* imply that every representation of every group of coordinate transformations is equivalent to a unitary representation. (In fact a specific instance where this is not the case was mentioned in Example II above.)

For any abstract group $\mathcal{G}$ there exists a generalization of the last theorem. If an ortho-normal basis is used in the construction of the operators $\Phi(T)$ of Equations (4.1) and (4.2), it follows by an argument similar to that given in the above proof that, for each $T \in \mathcal{G}$, $\Phi(T)$ is a unitary operator if and only if $\mathbf{\Gamma}(T)$ is a unitary matrix.

The amount of attention that has just been devoted to non-unitary representations should not be allowed to obscure the main point, which is that in most cases of physical interest all the representations can be chosen to be unitary.

This section will be concluded with an important theorem that demonstrates the special role played in similarity transformations by matrices that are unitary. (As noted in Appendix B, Section 2, such transformations transform ortho-normal bases into ortho-normal bases.)

Theorem IV If $\mathbf{\Gamma}$ and $\mathbf{\Gamma}'$ are two equivalent representations of a group $\mathcal{G}$ related by the similarity transformation

$$\mathbf{\Gamma}'(T) = \mathbf{S}^{-1}\mathbf{\Gamma}(T)\mathbf{S}$$

for all $T \in \mathcal{G}$, and if $\mathbf{\Gamma}$ is a unitary representation and $\mathbf{S}$ is a unitary matrix, then $\mathbf{\Gamma}'$ is also a unitary representation. Conversely, if $\mathbf{\Gamma}$ and $\mathbf{\Gamma}'$ are equivalent representations that are both unitary, then the matrix $\mathbf{S}$ in the similarity transformation relating them can always be chosen to be unitary.

Proof The first proposition is almost obvious, but the converse requires a rather lengthy proof, so the whole argument is relegated to Appendix C.

4 Reducible and irreducible representations

Suppose that the d-dimensional representation $\mathbf{\Gamma}$ of a group $\mathscr{G}$ can be partitioned so that it has the form

$$\mathbf{\Gamma}(\mathrm{T}) = \begin{bmatrix} \mathbf{\Gamma}_{11}(\mathrm{T}) & \mathbf{\Gamma}_{12}(\mathrm{T}) \\ \mathbf{0} & \mathbf{\Gamma}_{22}(\mathrm{T}) \end{bmatrix} \tag{4.9}$$

for every $\mathrm{T} \in \mathscr{G}$, where $\mathbf{\Gamma}_{11}(\mathrm{T})$, $\mathbf{\Gamma}_{12}(\mathrm{T})$, $\mathbf{\Gamma}_{22}(\mathrm{T})$ and the zero matrix $\mathbf{0}$ have dimensions $s_1 \times s_1$, $s_1 \times s_2$, $s_2 \times s_2$ and $s_2 \times s_1$ respectively. (Here $s_1 + s_2 = d$, $s_1 \geq 1$, $s_2 \geq 1$, and s_1 and s_2 are the same for all $\mathrm{T} \in \mathscr{G}$.) Then (cf. Equation (A.7)) for any $\mathrm{T}_1, \mathrm{T}_2 \in \mathscr{G}$

$$\mathbf{\Gamma}(\mathrm{T}_1)\mathbf{\Gamma}(\mathrm{T}_2) = \begin{bmatrix} \mathbf{\Gamma}_{11}(\mathrm{T}_1)\mathbf{\Gamma}_{11}(\mathrm{T}_2) & \mathbf{\Gamma}_{11}(\mathrm{T}_1)\mathbf{\Gamma}_{12}(\mathrm{T}_2) + \mathbf{\Gamma}_{12}(\mathrm{T}_1)\mathbf{\Gamma}_{22}(\mathrm{T}_2) \\ \mathbf{0} & \mathbf{\Gamma}_{22}(\mathrm{T}_1)\mathbf{\Gamma}_{22}(\mathrm{T}_2) \end{bmatrix}$$

so that, as the matrices $\mathbf{\Gamma}(\mathrm{T})$ form a representation of $\mathscr{G}$,

$$\mathbf{\Gamma}_{11}(\mathrm{T}_1\mathrm{T}_2) = \mathbf{\Gamma}_{11}(\mathrm{T}_1)\mathbf{\Gamma}_{11}(\mathrm{T}_2) \tag{4.10}$$

and

$$\mathbf{\Gamma}_{22}(\mathrm{T}_1\mathrm{T}_2) = \mathbf{\Gamma}_{22}(\mathrm{T}_1)\mathbf{\Gamma}_{22}(\mathrm{T}_2). \tag{4.11}$$

Equations (4.10) and (4.11) imply that the matrices $\mathbf{\Gamma}_{11}(\mathrm{T})$ and the matrices $\mathbf{\Gamma}_{22}(\mathrm{T})$ both form representations of $\mathscr{G}$. Thus the representation $\mathbf{\Gamma}$ of $\mathscr{G}$ is made up of two other representations of smaller dimensions, so it is natural to describe such a representation as being "reducible". In order that this description should apply equally to all equivalent representations, the formal definition can be stated as follows:

Definition *Reducible representation of a group* $\mathscr{G}$
A representation of a group $\mathscr{G}$ is said to be "reducible" if it is *equivalent* to a representation $\mathbf{\Gamma}$ of $\mathscr{G}$ that has the form of Equation (4.9) for all $\mathrm{T} \in \mathscr{G}$.

It follows from Equations (4.1) and (4.9) that for $n = 1, 2, \ldots, s_1$ and all $\mathrm{T} \in \mathscr{G}$

$$\Phi(\mathrm{T})\psi_n = \sum_{m=1}^{s_1} \Gamma_{11}(\mathrm{T})_{mn}\psi_m.$$

Thus the s_1-dimensional subspace of the carrier space V having basis $\psi_1, \psi_2, \ldots, \psi_{s_1}$ is *invariant* under all the operations of $\mathscr{G}$ in the sense that if ψ is any vector of this subspace then $\Phi(\mathrm{T})\psi$ is also a member of this subspace for all $\mathrm{T} \in \mathscr{G}$. (It should be noted that in general the s_2-dimensional subspace with basis $\psi_{s_1+1}, \psi_{s_1+2}, \ldots, \psi_d$ is *not* invariant.)

Definition *Irreducible representation of a group $\mathscr{G}$*
A representation of a group $\mathscr{G}$ is said to be "irreducible" if it is not reducible.

This definition implies that an irreducible representation *cannot* be transformed by a similarity transformation to the form of Equation (4.9). Consequently the carrier space V of an irreducible representation has no invariant subspace of smaller dimension. Some simple tests for irreducibility will be developed in Sections 5 and 6.

Returning to the reducible representation $\boldsymbol{\Gamma}$ of Equation (4.9), the question arises as to whether $\boldsymbol{\Gamma}_{11}$ and $\boldsymbol{\Gamma}_{22}$ are also reducible or not. If $\boldsymbol{\Gamma}_{11}$ is reducible, then by a similarity transformation it too can be put in the form of Equation (4.9) with submatrices of some dimensions. The same is true of $\boldsymbol{\Gamma}_{22}$. Obviously this process can be continued until all the representations involved are irreducible. Thus every reducible representation $\boldsymbol{\Gamma}$ by an appropriate similarity transformation $\mathbf{S}$ can be put into the form

$$\boldsymbol{\Gamma}'(\mathrm{T}) = \mathbf{S}^{-1}\boldsymbol{\Gamma}(\mathrm{T})\mathbf{S} = \begin{bmatrix} \boldsymbol{\Gamma}'_{11}(\mathrm{T}) & \boldsymbol{\Gamma}'_{12}(\mathrm{T}) & \boldsymbol{\Gamma}'_{13}(\mathrm{T}) & \dots & \boldsymbol{\Gamma}'_{1r}(\mathrm{T}) \\ \mathbf{0} & \boldsymbol{\Gamma}'_{22}(\mathrm{T}) & \boldsymbol{\Gamma}'_{23}(\mathrm{T}) & \dots & \boldsymbol{\Gamma}'_{2r}(\mathrm{T}) \\ \mathbf{0} & \mathbf{0} & \boldsymbol{\Gamma}'_{33}(\mathrm{T}) & \dots & \boldsymbol{\Gamma}'_{3r}(\mathrm{T}) \\ \vdots & \vdots & \vdots & & \vdots \\ \mathbf{0} & \mathbf{0} & \mathbf{0} & \dots & \boldsymbol{\Gamma}'_{rr}(\mathrm{T}) \end{bmatrix}$$

where *all* the matrices $\boldsymbol{\Gamma}'_{jj}(\mathrm{T})$ form *irreducible* representations, $j = 1, 2, \dots, r$. (Here $\boldsymbol{\Gamma}'_{jk}(\mathrm{T})$ is a $s'_j \times s'_k$ matrix, $\sum_{j=1}^{r} s'_j = d$, $s'_j \geqslant 1$ for each $j = 1, 2, \dots, r$ and $s'_1, s'_2, \dots, s'_r$ are the same for all $\mathrm{T} \in \mathscr{G}$.) ***It is now apparent that the irreducible representations are the basic building blocks from which all reducible representations can be constructed.***

The final question is whether all the upper off-diagonal submatrices $\boldsymbol{\Gamma}'_{jk}(\mathrm{T})$ $(k > j)$ can be transformed into zero matrices by a further similarity transformation, leaving only the diagonal submatrices non-zero. If so, $\boldsymbol{\Gamma}$ is equivalent to a representation of the form

$$\boldsymbol{\Gamma}''(\mathrm{T}) = \begin{bmatrix} \boldsymbol{\Gamma}''_{11}(\mathrm{T}) & \mathbf{0} & \mathbf{0} & \dots & \mathbf{0} \\ \mathbf{0} & \boldsymbol{\Gamma}''_{22}(\mathrm{T}) & \mathbf{0} & \dots & \mathbf{0} \\ \mathbf{0} & \mathbf{0} & \boldsymbol{\Gamma}''_{33}(\mathrm{T}) & \dots & \mathbf{0} \\ \vdots & \vdots & \vdots & & \vdots \\ \mathbf{0} & \mathbf{0} & \mathbf{0} & \dots & \boldsymbol{\Gamma}''_{rr}(\mathrm{T}) \end{bmatrix} \quad (4.12)$$

in which the $\boldsymbol{\Gamma}''_{jj}$ are all *irreducible* representations of $\mathscr{G}$.

Definition *Completely reducible representations of a group $\mathscr{G}$*
A representation $\boldsymbol{\Gamma}$ of a group $\mathscr{G}$ is said to be "completely reducible" if

it is equivalent to a representation $\boldsymbol{\Gamma}''$ that has the form in Equation (4.12) for all $T \in \mathcal{G}$.

A completely reducible representation is sometimes referred to as a "decomposable" representation.

Theorem I If $\mathcal{G}$ is a finite group or a compact Lie group then *every* reducible representation of $\mathcal{G}$ is *completely reducible.* The same is true of every reducible representation of a connected, non-compact, semi-simple Lie group and of any *unitary* reducible representation of any other group.

Proof It will be established in a lemma in Appendix C, Section 2, that every *unitary* reducible representation of any group $\mathcal{G}$ is completely reducible. As it has already been pointed out in Section 3 above that every representation of a finite group or a compact Lie group is equivalent to a unitary representation, this conclusion applies to every reducible representation of such groups.

As non-compact, semi-simple Lie groups have no non-trivial unitary representations, a totally different line of argument is needed for these groups. This is best given in terms of Lie algebras, and so is deferred until Chapter 15, Section 1.

Suppose that $\phi_1, \phi_2, \ldots, \phi_d$ form a basis for the carrier space of the completely reducible representation $\boldsymbol{\Gamma}''$ of Equation (4.12) and that the irreducible representation $\boldsymbol{\Gamma}''_{jj}$ has dimension d_j, $j = 1, 2, \ldots, r$, so that $\sum_{j=1}^{r} d_j = d$. Then it follows from Equations (4.1) and (4.12) that $\phi_1, \phi_2, \ldots, \phi_{d_1}$ form a basis for the carrier space of $\boldsymbol{\Gamma}''_{11}$, $\phi_{d_1+1}, \phi_{d_1+2}, \ldots, \phi_{d_1+d_2}$ form a basis for the carrier space of $\boldsymbol{\Gamma}''_{22}$ and so on. The carrier space of $\boldsymbol{\Gamma}''$ is therefore a direct sum of carrier spaces belonging to each of the irreducible representations $\boldsymbol{\Gamma}''_{11}, \boldsymbol{\Gamma}''_{22}, \ldots, \boldsymbol{\Gamma}''_{rr}$ (see Appendix B, Section 1). Correspondingly, the completely reducible representation $\boldsymbol{\Gamma}''$ is said to be the "direct sum" of the irreducible representations $\boldsymbol{\Gamma}''_{11}, \boldsymbol{\Gamma}''_{22}, \ldots, \boldsymbol{\Gamma}''_{rr}$, this statement being expressed concisely by

$$\boldsymbol{\Gamma}'' = \boldsymbol{\Gamma}''_{11} \oplus \boldsymbol{\Gamma}''_{22} \oplus \ldots \oplus \boldsymbol{\Gamma}''_{rr}.$$

(The symbol $\oplus$ here indicates that the sum involved is *not* that of ordinary matrix addition.) Similarly, the equivalence of a representation $\boldsymbol{\Gamma}$ to a direct sum of irreducible representations $\boldsymbol{\Gamma}''_{11}, \boldsymbol{\Gamma}''_{22}, \ldots, \boldsymbol{\Gamma}''_{rr}$ can be written as

$$\boldsymbol{\Gamma} \approx \boldsymbol{\Gamma}''_{11} \oplus \boldsymbol{\Gamma}''_{22} \oplus \ldots \oplus \boldsymbol{\Gamma}''_{rr}.$$

In the case in which $\boldsymbol{\Gamma}$ is equivalent to a unitary representation, all the irreducible representations in the direct sum are themselves equivalent to unitary representations.

The theorem shows that nearly all representations of physical interest are either completely reducible or are irreducible. However, just for academic interest, a simple example will be given which demonstrates that groups do exist with reducible representations that are not completely reducible.

Example I *The multiplicative group of positive real numbers*
As shown in Example I of Chapter 3, Section 3, this is a non-compact Abelian group, a typical element being $\exp y_1$, $-\infty<y_1<\infty$. It has a faithful two-dimensional reducible representation

$$\boldsymbol{\Gamma}(\exp y_1)=\begin{bmatrix}1 & y_1\\ 0 & 1\end{bmatrix}.$$

Suppose that this is completely reducible, so that there exists a 2×2 non-singular matrix $\mathbf{S}$ such that

$$\boldsymbol{\Gamma}'(\exp y_1)=\mathbf{S}^{-1}\begin{bmatrix}1 & y_1\\ 0 & 1\end{bmatrix}\mathbf{S}=\begin{bmatrix}\alpha(y_1) & 0\\ 0 & \beta(y_1)\end{bmatrix}.$$

The invariance of the determinant and trace under similarity transformations gives $\alpha(y_1)\beta(y_1)=1$ and $\alpha(y_1)+\beta(y_1)=2$, which imply that $\alpha(y_1)=\beta(y_1)=1$. Thus $\boldsymbol{\Gamma}'(\exp y_1)=\mathbf{1}$ and hence $\boldsymbol{\Gamma}(\exp y_1)=\mathbf{1}$, which is not true. Thus $\boldsymbol{\Gamma}$ is *not* completely reducible.

5 Schur's Lemmas and the orthogonality theorem for matrix representations

The name "Schur's Lemma" is often attached to one or other (or sometimes both) of the following theorems.

Theorem I Let $\boldsymbol{\Gamma}$ and $\boldsymbol{\Gamma}'$ be two *irreducible* representations of a group $\mathcal{G}$, of dimensions d and d' respectively, and suppose that there exists a $d\times d'$ matrix $\mathbf{A}$ such that

$$\boldsymbol{\Gamma}(T)\mathbf{A}=\mathbf{A}\boldsymbol{\Gamma}'(T)$$

for all $T\in\mathcal{G}$. Then either $\mathbf{A}=\mathbf{0}$, or $d=d'$ and $\det\mathbf{A}\neq0$.

Proof This is given in Appendix C, Section 3.

Theorem II If $\boldsymbol{\Gamma}$ is a d-dimensional *irreducible* representation of a group $\mathcal{G}$ and $\mathbf{B}$ is a $d\times d$ matrix such that $\boldsymbol{\Gamma}(T)\mathbf{B}=\mathbf{B}\boldsymbol{\Gamma}(T)$ for every $T\in\mathcal{G}$, then $\mathbf{B}$ must be a multiple of the unit matrix.

Proof Let $\mathbf{A}=\mathbf{B}-\beta\mathbf{1}$, where the complex number β is chosen so that $\det \mathbf{A}=0$. Then $\mathbf{\Gamma}(T)\mathbf{A}=\mathbf{A}\mathbf{\Gamma}(T)$ for all $T\in\mathcal{G}$, so by the previous theorem the only alternative not excluded is $\mathbf{A}=\mathbf{0}$, that is, $\mathbf{B}=\beta\mathbf{1}$.

The following corollary shows how very straightforward are the irreducible representations of Abelian groups.

Theorem III Every irreducible representation of an *Abelian* group is *one*-dimensional.

Proof Let $\mathbf{\Gamma}$ be an irreducible representation of an Abelian group $\mathcal{G}$. As $\mathbf{\Gamma}(T)\mathbf{\Gamma}(T')=\mathbf{\Gamma}(T')\mathbf{\Gamma}(T)$ for all T and T′ of $\mathcal{G}$, it follows from the preceding theorem that, for each $T'\in\mathcal{G}$, $\mathbf{\Gamma}(T')=\gamma(T')\mathbf{1}$, where $\gamma(T')$ is some complex number that depends on T′. Clearly, such a representation is irreducible if and only if it is one-dimensional.

The "orthogonality theorem for matrix representations" is a second corollary which will be used time and time again. As will be seen, it applies both to finite groups and compact Lie groups.

Theorem IV Suppose that $\mathbf{\Gamma}^p$ and $\mathbf{\Gamma}^q$ are two *unitary irreducible* representations of a finite group $\mathcal{G}$ which are not equivalent if $p\neq q$ (but which are identical if $p=q$). Then

$$(1/g)\sum_{T\in\mathcal{G}}\Gamma^p(T)^*_{jk}\Gamma^q(T)_{st}=(1/d_p)\,\delta_{pq}\,\delta_{js}\,\delta_{kt},$$

where g is the order of $\mathcal{G}$ and d_p is the dimension of $\mathbf{\Gamma}^p$.

Similarly, if $\mathcal{G}$ is a compact Lie group the summation can be replaced by an invariant integration, giving

$$\int_{\mathcal{G}}\Gamma^p(T)^*_{jk}\Gamma^q(T)_{st}\,dT=(1/d_p)\,\delta_{pq}\,\delta_{js}\,\delta_{kt}.$$

Proof This is given in Appendix C, Section 3.

It is in the application of this theorem that the main practical advantage of working with unitary representations lies. For example, one immediate consequence is the following partial converse to Theorem III of Section 3.

Theorem V If $\phi_1^p(\mathbf{r}),\phi_2^p(\mathbf{r}),\ldots$ and $\psi_1^q(\mathbf{r}),\psi_2^q(\mathbf{r}),\ldots$ are respectively basis functions for the unitary *irreducible* representations $\mathbf{\Gamma}^p$ and $\mathbf{\Gamma}^q$ of a group of coordinate transformations $\mathcal{G}$ that is either a finite group or a compact Lie group, and $\mathbf{\Gamma}^p$ and $\mathbf{\Gamma}^q$ are not equivalent if $p \neq q$ (but are identical if $p = q$), then

$$(\phi^p_m, \psi^q_n) = 0$$

unless $p = q$ and $m = n$. If $p = q$ and $m = n$, then (ϕ^p_m, ψ^p_m) is a constant independent of m.

Proof From Equations (1.20) and (1.26), for any $T \in \mathcal{G}$,

$$(\phi^p_m, \psi^q_n) = (P(T)\phi^p_m, P(T)\psi^q_n)$$

$$= \sum_{j=1}^{d_p} \sum_{k=1}^{d_q} \Gamma^p(T)^*_{jm} \Gamma^q(T)_{kn} (\phi^p_j, \psi^q_k),$$

d_p and d_q being the dimensions of Γ^p and Γ^q respectively. Summing or integrating over all the transformations T of $\mathcal{G}$, the orthogonality theorem for matrix representations gives

$$(\phi^p_m, \psi^q_n) = (1/d_p)\, \delta_{pq}\, \delta_{mn} \sum_{j=1}^{d_p} (\phi^p_j, \psi^p_j).$$

Thus when $p \neq q$, or when $p = q$ but $m \neq n$, it follows immediately that $(\phi^p_m, \psi^q_n) = 0$. When $p = q$ and $m = n$ the right-hand side of this last equation is independent of m, so (ϕ^p_m, ψ^p_m) must be independent of m.

One immediate implication of this theorem is that

$$(\psi^p_m, \psi^p_n) = \delta_{mn} (\psi^p_1, \psi^p_1)$$

for all $m, n = 1, 2, \ldots, d_p$. Thus if $\psi^p_1(\mathbf{r})$ is normalized then so too are $\psi^p_2(\mathbf{r}), \psi^p_3(\mathbf{r}) \ldots$. Henceforth it will usually be assumed that every set of basis functions of an irreducible representation is a mutually ortho-normal set, that is,

$$(\psi^p_m, \psi^p_n) = \delta_{mn}$$

for all $m, n = 1, 2, \ldots, d_p$.

It should be noted that if *every* member of such a set of basis functions is multiplied by the *same* phase factor $\exp(i\omega)$, where ω is any real number, then the resulting functions are basis functions for the same representation and they too form an ortho-normal set Consequently the choice of such an overall constant phase factor is entirely a matter of convention.

If Γ is a *reducible* unitary representation of a finite or compact Lie group of coordinate transformations, then the situation is more complicated. Although any given set of basis functions $\psi_1(\mathbf{r}), \psi_2(\mathbf{r}), \ldots$ of Γ *need not* even form an orthogonal set, it is always possible to *construct* a set of basis functions for Γ that do form an ortho-normal set. This will be demonstrated by considering the simplest possible case in which

$$\mathbf{S}^{-1}\mathbf{\Gamma}(\mathrm{T})\mathbf{S} = \begin{bmatrix} \mathbf{\Gamma}^1(\mathrm{T}) & \mathbf{0} \\ \mathbf{0} & \mathbf{\Gamma}^2(\mathrm{T}) \end{bmatrix}$$

for all $\mathrm{T} \in \mathcal{G}$, where $\mathbf{\Gamma}^1$ and $\mathbf{\Gamma}^2$ are two unitary irreducible representations of $\mathcal{G}$ of dimensions d_1 and d_2 respectively. Theorem IV of Section 3 shows that $\mathbf{S}$ may be chosen to be unitary. Thus if $\psi_j^1(\mathbf{r})$, $j = 1, 2, \ldots, d_1$, and $\psi_j^2(\mathbf{r})$, $j = 1, 2, \ldots, d_2$, are ortho-normal sets of basis functions of $\mathbf{\Gamma}^1$ and $\mathbf{\Gamma}^2$ respectively, so that $(\psi_m^p, \psi_n^q) = \delta_{pq}\,\delta_{mn}$, the basis functions

$$\psi_m(\mathbf{r}) = \sum_{n=1}^{d_1} (\mathbf{S}^{-1})_{nm}\psi_n^1(\mathbf{r}) + \sum_{n=d_1+1}^{d_1+d_2} (\mathbf{S}^{-1})_{nm}\psi_{n-d_1}^2(\mathbf{r})$$

of $\mathbf{\Gamma}$ also form an ortho-normal set (see Appendix B, Section 2). However, the functions $\phi_j^1(\mathbf{r}) = \alpha\psi_j^1(\mathbf{r})$, $j = 1, 2, \ldots, d_1$, and $\phi_j^2(\mathbf{r}) = \beta\psi_j^2(\mathbf{r})$, $j = 1, 2, \ldots, d_2$, are also basis functions of $\mathbf{\Gamma}^1$ and $\mathbf{\Gamma}^2$, for any constants α and β that are assumed independent of j. Consequently the functions

$$\phi_m(\mathbf{r}) = \sum_{n=1}^{d_1} (\mathbf{S}^{-1})_{nm}\phi_n^1(\mathbf{r}) + \sum_{n=d_1+1}^{d_1+d_2} (\mathbf{S}^{-1})_{nm}\phi_{n-d_1}^2(\mathbf{r})$$

are also basis functions of $\mathbf{\Gamma}$, but they do *not* form a mutually orthogonal set if $\alpha \neq \beta$.

6 Characters

Although equivalent representations have essentially the same content, there is a large degree of arbitrariness in the explicit forms of their matrices. However, the characters provide a set of quantities which are the same for *all* equivalent representations. Indeed, for finite groups and compact Lie groups the characters *uniquely* determine the representations up to equivalence.

The characters have a number of other very useful properties which, for the most part, are valid for finite groups or compact Lie groups but not for non-compact Lie groups.

Definition *Characters of a representation*
Suppose that $\mathbf{\Gamma}$ is a d-dimensional representation of a group $\mathcal{G}$. Then

$$\chi(\mathrm{T}) = \operatorname{tr} \mathbf{\Gamma}(\mathrm{T}) \qquad \left(= \sum_{j=1}^{d} \Gamma(\mathrm{T})_{jj}\right)$$

is defined to be the "character" of the group element T in this representation. The set of characters corresponding to a representation is called the "character system" of the representation.

As $\mathbf{\Gamma}(\mathrm{E}) = \mathbf{1}_d$ for the identity E of $\mathcal{G}$, then $\chi(\mathrm{E}) = d$.

Theorem I A necessary condition for two representations of a group to be equivalent is that they must have identical character systems.

Proof Let $\mathbf{\Gamma}$ and $\mathbf{\Gamma}'$ be two equivalent representations of a group $\mathcal{G}$, both of dimension d, so that there exists a $d \times d$ non-singular matrix $\mathbf{S}$ such that $\mathbf{\Gamma}'(\mathrm{T}) = \mathbf{S}^{-1}\mathbf{\Gamma}(\mathrm{T})\mathbf{S}$ for all $\mathrm{T} \in \mathcal{G}$. Then, as noted in Appendix A, $\mathrm{tr}\,\mathbf{\Gamma}'(\mathrm{T}) = \mathrm{tr}\,\mathbf{\Gamma}(\mathrm{T})$. Thus, if $\chi(\mathrm{T})$ and $\chi'(\mathrm{T})$ are the characters of T in $\mathbf{\Gamma}$ and $\mathbf{\Gamma}'$ respectively, then $\chi'(\mathrm{T}) = \chi(\mathrm{T})$ for all $\mathrm{T} \in \mathcal{G}$.

The characters therefore provide a set of quantities that are unchanged by similarity transformations. The converse proposition will be considered shortly. The invariance property of the trace also provides another simple result:

Theorem II In a given representation of a group $\mathcal{G}$ all the elements in the same class have the same character.

Proof Suppose that the elements T′ and T of $\mathcal{G}$ are in the same class. Then (see Chapter 2, Section 2) there exists a group element X such that $\mathrm{T}' = \mathrm{XTX}^{-1}$, so that $\mathbf{\Gamma}(\mathrm{T}') = \mathbf{\Gamma}(\mathrm{X})\mathbf{\Gamma}(\mathrm{T})\mathbf{\Gamma}(\mathrm{X})^{-1}$. Consequently $\mathrm{tr}\,\mathbf{\Gamma}(\mathrm{T}') = \mathrm{tr}\,\mathbf{\Gamma}(\mathrm{T})$ and hence $\chi(\mathrm{T}') = \chi(\mathrm{T})$.

There are two orthogonality theorems for characters. The first is as follows:

Theorem III Let $\chi^{\mathrm{p}}(\mathrm{T})$ and $\chi^{\mathrm{q}}(\mathrm{T})$ be the characters of two irreducible representations of a finite group $\mathcal{G}$ of order g, these representations being assumed to be inequivalent if $p \neq q$. Then

$$(1/g) \sum_{\mathrm{T} \in \mathcal{G}} \chi^{\mathrm{p}}(\mathrm{T})^* \chi^{\mathrm{q}}(\mathrm{T}) = \delta_{\mathrm{pq}}.$$

Similarly, if $\mathcal{G}$ is a compact Lie group the summation can be replaced by an invariant integration, giving

$$\int_{\mathcal{G}} \chi^{\mathrm{p}}(\mathrm{T})^* \chi^{\mathrm{q}}(\mathrm{T})\, d\mathrm{T} = \delta_{\mathrm{pq}}.$$

Proof Theorem I of Section 3 shows that for the groups under consideration similarity transformations may be applied to the two irreducible representations to produce unitary representations. The result then follows immediately from the orthogonality theorem for matrix representations (Theorem IV of Section 5) on putting $j = k$ and $s = t$ and summing over j and s.

The converse theorem referred to previously can now be proved fairly easily.

Theorem IV If $\mathcal{G}$ is a finite group or a compact Lie group then a sufficient condition for two representations to be equivalent is provided by the equality of their character systems.

Proof The proof may be found in Appendix C, Section 4.

It is interesting to note that this sufficient condition does not extend to all *non*-compact Lie groups, for in Example I of Section 4, which involves a non-compact Abelian group, the representation Γ is not equivalent to $\Gamma''(\exp y_1) = \mathbf{1}_2$, even though they have identical character systems.

The characters provide a complete specification (up to equivalence) of the irreducible representations that appear in a reducible representation. This knowledge can prove very useful, as will be seen later. The details are as given by the following theorem.

Theorem V The number of times n_p that an irreducible representation Γ^p (or a representation equivalent to Γ^p) appears in a reducible representation Γ is given for a finite group $\mathcal{G}$ by

$$n_p = (1/g) \sum_{T \in \mathcal{G}} \chi(T)\chi^p(T)^*,$$

where $\chi^p(T)$ and $\chi(T)$ are the characters of Γ^p and Γ respectively and g is the order of $\mathcal{G}$. For a compact Lie group this generalizes to

$$n_p = \int_{\mathcal{G}} \chi(T)\chi^p(T)^* \, dT.$$

Proof The proof will be given for a finite group, the generalization to a compact Lie group being obvious. As noted in Section 5, Γ is completely reducible. Suppose that in its direct sum decomposition the irreducible representation Γ^1 appears n_1 times, the irreducible representation Γ^2 appears n_2 times and so on. This can be concisely expressed in the form

$$\Gamma \approx n_1\Gamma^1 \oplus n_2\Gamma^2 \oplus \dots \qquad (4.13)$$

and implies that

$$\chi(\mathrm{T}) = \sum_{\mathrm{q}} n_{\mathrm{q}} \chi^{\mathrm{q}}(\mathrm{T}) \tag{4.14}$$

for all $\mathrm{T} \in \mathcal{G}$. The desired result follows on multiplying both sides of Equation (4.14) by $\chi^{\mathrm{p}}(\mathrm{T})^*$ and summing over all $\mathrm{T} \in \mathcal{G}$.

Similar considerations give the following convenient criterion for irreducibility expressed solely in terms of characters.

Theorem VI A necessary and sufficient condition for a representation Γ of a finite group $\mathcal{G}$ to be irreducible is that

$$(1/g) \sum_{\mathrm{T} \in \mathcal{G}} |\chi(\mathrm{T})|^2 = 1,$$

where $\chi(\mathrm{T})$ is the character of the group element T in Γ and g is the order of $\mathcal{G}$. The corresponding condition for a compact Lie group is

$$\int_{\mathcal{G}} |\chi(\mathrm{T})|^2 \, d\mathrm{T} = 1.$$

Proof Suppose that the representation Γ is equivalent to the direct sum decomposition of Equation (4.13), so that Equation (4.14) holds for every $\mathrm{T} \in \mathcal{G}$. Then, for a finite group $\mathcal{G}$,

$$\begin{aligned}(1/g) \sum_{\mathrm{T} \in \mathcal{G}} |\chi(\mathrm{T})|^2 &= (1/g) \sum_{\mathrm{p}} \sum_{\mathrm{q}} \sum_{\mathrm{T} \in \mathcal{G}} n_{\mathrm{p}} n_{\mathrm{q}} \chi^{\mathrm{p}}(\mathrm{T})^* \chi^{\mathrm{q}}(\mathrm{T}) \\ &= \sum_{\mathrm{p}} n_{\mathrm{p}}^2,\end{aligned}$$

from the second theorem above. However, it is obvious that Γ is an irreducible representation if and only if $\sum_{\mathrm{p}} n_{\mathrm{p}}^2 = 1$, from which the stated result follows. Again the generalization to a compact Lie group is trivial.

This criterion provides a very simple test for irreducibility, particularly for finite groups.

Characters may also be used to prove a theorem on the number of inequivalent irreducible representations of a finite group $\mathcal{G}$, as well as a useful result on their dimensions. Also involved in the argument are the concepts of the "regular representation" of $\mathcal{G}$ and "class multiplication". The details are given in Appendix C, Section 4.

Theorem VII For a finite group $\mathcal{G}$, the sum of the squares of the dimensions of the inequivalent irreducible representations is equal to the order of $\mathcal{G}$.

Theorem VIII For a finite group $\mathcal{G}$, the number of inequivalent irreducible representations is equal to the number of classes of $\mathcal{G}$.

These two theorems taken together are often sufficient to uniquely specify the dimensions of the inequivalent irreducible representations.

Example I *Dimensions of the inequivalent irreducible representations of the crystallographic point group* D_4

As noted in Chapter 2, Section 2, D_4 is of order 8 and has five classes. Thus it has five inequivalent irreducible representations. Let d_j, $j=1,2,\ldots,5$, be their dimensions, so that $\sum_{j=1}^{5} d_j^2=8$, which has the solution $d_1=d_2=d_3=d_4=1$ and $d_5=2$. This solution is unique up to a relabelling of representations.

The second orthogonality theorem for characters appears as part of the argument leading to the proof of the preceding theorem. Its precise statement is as follows.

Theorem IX If $\chi^p(\mathscr{C}_j)$ is the character of the class $\mathscr{C}_j$ of a finite group $\mathcal{G}$ for the irreducible representation Γ^p of $\mathcal{G}$, then

$$\sum_{p} \chi^p(\mathscr{C}_j)^* \chi^p(\mathscr{C}_k) N_j = g\delta_{jk},$$

where the sum is over all the inequivalent irreducible representations of $\mathcal{G}$, g is the order of $\mathcal{G}$, and N_j is the number of elements in the class $\mathscr{C}_j$.

The character systems of the irreducible representations of a finite group are conveniently displayed in the form of a "character table". The classes of the group are usually listed along the top of the table and the inequivalent irreducible representations are listed down the left-hand side. As a consequence of the Theorem VIII above, this table is always square.

For groups of low order it is quite easy to completely determine the character table *directly* from the theorems that have just been stated, without first obtaining explicit forms for the matrices, as the following example will show. The construction of character tables and irreducible representations for other groups will be considered further in Section 7.

Example II *Character table for the crystallographic point group* D_4

The classes of D_4 (see Chapter 2, Section 2) are $\mathscr{C}_1=\{E\}$, $\mathscr{C}_2=\{C_{2x}, C_{2z}\}$, $\mathscr{C}_3=\{C_{2y}\}$, $\mathscr{C}_4=\{C_{4y}, C_{4y}^{-1}\}$, $\mathscr{C}_5=\{C_{2c}, C_{2d}\}$.

Consider first the four one-dimensional representations Γ^1, Γ^2, Γ^3 and Γ^4. As $C_{2x}^2 = C_{2c}^2 = E$ then $\chi^p(C_{2x})^2 = \chi^p(C_{2c})^2 = 1$, $p = 1, 2, 3, 4$. Moreover, from Table 1.2, $C_{4y} = C_{2c}C_{2x}$, so that $\chi^p(\mathscr{C}_4) = \chi^p(\mathscr{C}_2)\chi^p(\mathscr{C}_5)$. Finally, $C_{2y} = C_{4y}^2$, so $\chi^p(C_{2y}) = 1$ for $p = 1, 2, 3, 4$. Thus the four one-dimensional irreducible representations of D_4 may be chosen to be such that: $\chi^1(\mathscr{C}_2) = 1$, $\chi^1(\mathscr{C}_5) = 1$; $\chi^2(\mathscr{C}_2) = 1$, $\chi^2(\mathscr{C}_5) = -1$; $\chi^3(\mathscr{C}_2) = -1$, $\chi^3(\mathscr{C}_5) = -1$; $\chi^4(\mathscr{C}_2) = -1$, $\chi^4(\mathscr{C}_5) = 1$.

From the first orthogonality theorem for characters (Theorem III) the two-dimensional representation Γ^5 must satisfy the conditions:

$$\left.\begin{aligned}
\chi^5(\mathscr{C}_1)+2\chi^5(\mathscr{C}_2)+\chi^5(\mathscr{C}_3)+2\chi^5(\mathscr{C}_4)+2\chi^5(\mathscr{C}_5)=0,\\
\chi^5(\mathscr{C}_1)+2\chi^5(\mathscr{C}_2)+\chi^5(\mathscr{C}_3)-2\chi^5(\mathscr{C}_4)-2\chi^5(\mathscr{C}_5)=0,\\
\chi^5(\mathscr{C}_1)-2\chi^5(\mathscr{C}_2)+\chi^5(\mathscr{C}_3)+2\chi^5(\mathscr{C}_4)-2\chi^5(\mathscr{C}_5)=0,\\
\chi^5(\mathscr{C}_1)-2\chi^5(\mathscr{C}_2)+\chi^5(\mathscr{C}_3)-2\chi^5(\mathscr{C}_4)+2\chi^5(\mathscr{C}_5)=0.
\end{aligned}\right\}$$

Adding these equations gives $\chi^5(\mathscr{C}_1)+\chi^5(\mathscr{C}_3)=0$ and, as $\chi^5(\mathscr{C}_1) = \chi^5(E) = 2$, this implies $\chi^5(\mathscr{C}_3) = -2$. Moreover, Theorem VI above gives $\sum_{T\in\mathscr{G}} |\chi^5(T)|^2 = 8$, while $|\chi^5(E)|^2 + |\chi^5(C_{2y})|^2 = 8$, so that $\chi^5(\mathscr{C}_2) = \chi^5(\mathscr{C}_4) = \chi^5(\mathscr{C}_5) = 0$.

The complete character table for D_4 is given in Table 4.1. It is interesting to relate these irreducible representations to the representations of D_4 discussed in Example I of Section 1. Γ^1 is clearly the "identity" representation (iv), Γ^2 is the one-dimensional representation (iii), Γ^5 is the two-dimensional representation (ii), and the three-dimensional representation (i) is reducible, being given by the direct sum $\Gamma^3 \oplus \Gamma^5$.

	E	$C_{2x,2z}$	C_{2y}	C_{4y}, C_{4y}^{-1}	C_{2c}, C_{2d}
Γ^1	1	1	1	1	1
Γ^2	1	1	1	−1	−1
Γ^3	1	−1	1	1	−1
Γ^4	1	−1	1	−1	1
Γ^5	2	0	−2	0	0

Table 4.1 Character table for the crystallographic point group D_4.

Although a number of results of physical significance follow immediately from a knowledge of the characters, it is often necessary to obtain explicit expressions for the matrices of the representations. A method for constructing such explicit expressions from the characters is described in Chapter 5, Section 1. Of course, for one-dimensional representations the characters themselves *are* the matrix elements.

Hitherto all results on finite groups have had an immediate generalization for compact Lie groups. For Theorems VII and VIII above this generalization is more far-reaching and is embodied in the following theorem due to Peter and Weyl (1927).

Theorem X For a compact Lie group $\mathcal{G}$, the number of inequivalent irreducible representations is infinite but *countable.*

This theorem implies that the irreducible representations of a compact Lie group can be specified by a parameter that only takes *integral* values (or, if more convenient, by a set of parameters taking integral values). This result has been anticipated in some of the notations already employed (but not in any of the proofs).

7 Methods for finding the irreducible representations of the physically important groups

It is now appropriate to look ahead and survey the various methods that are used to construct the irreducible representations of the physically most important groups, so that the reader can identify those sections that are most relevant to his particular interests.

(a) *Finite groups*

(i) *Finite Abelian groups*
It will be shown in Chapter 5, Section 6, that there exists a simple method for finding all the irreducible representations of every finite Abelian group.

(ii) *Crystallographic point groups*
In Appendix D a complete description is given of the irreducible representations of all 32 crystallographic point groups (and of their "double" groups (see Chapter 6, Section 4)). There are no particular difficulties involved in their construction. Indeed, of the 32 "single" point groups, one consists of the identity alone, 15 are Abelian (so that the method of (i) applies), and a number of the others are direct products of smaller crystallographic point groups (see Example II of Chapter 2, Section 7) (so that their irreducible representations are direct products of the irreducible representations of the smaller groups involved (see Chapter 5, Section 5)). The method illustrated in Example II of Section 6 can easily be applied to those groups that remain.

(iii) *Crystallographic space groups*
These groups have arbitrarily large order, so the direct application of the method of Section 6 is inappropriate. However, the induced representation method of Chapter 5, Section 7, produces the complete set of irreducible representations. A detailed account will be given in Chapter 9.

(*b*) *Lie groups*

(i) *Abelian Lie groups*
Every Abelian Lie group is either isomorphic to one of the following two groups or is isomorphic to a set of direct products involving only these groups (together possibly with finite Abelian groups).

(*a*) U(1).
Consider a typical element $[\exp ix_1]$ of U(1), where $-\pi \leq x_1 \leq \pi$. Every irreducible representation of U(1) is one-dimensional (see Theorem III of Section 5) and, as U(1) is compact, these representations must be unitary. Thus each irreducible representation of U(1) has the form

$$\Gamma^p([\exp(ix_1)]) = [\exp(ipx_1)],$$

where p is real number. Moreover, as $\exp(i\pi) = \exp(-i\pi)$, p must be an integer. Clearly, Γ^p is equivalent to Γ^q if and only if $p = q$, so the infinite set $p = 0, \pm 1, \pm 2, \ldots$ specifies the complete set of irreducible representations of U(1).

(*b*) *The multiplicative group of real numbers.*
A typical element is $\exp y_1$, where $-\infty < y_1 < +\infty$. Every irreducible representation is again one-dimensional, but as this group is non-compact (see Example I of Chapter 3, Section 3) these representations need not be unitary. The most general form is

$$\Gamma^\mu(\exp y_1) = [\exp(\mu y_1)],$$

where μ is some complex number. This is unitary if and only if μ is purely imaginary.

(ii) SU(2), SO(3) *and* O(3)
The irreducible representations of these groups are described in detail in Chapter 12, Section 4.

(iii) *The homogeneous Lorentz groups*
The detailed investigation of the irreducible representations of these groups appears in Chapter 17, Sections 2 and 4.

(iv) *Other connected semi-simple Lie groups*
The finite-dimensional irreducible representations of these groups are best found by the Lie algebraic method described in Chapters 15 and 16.

(v) *The Euclidean group of* $\mathbb{R}^3$, *the Poincaré group and the Galilei group*
The representations of most interest are infinite-dimensional. References may be found in Chapter 5, Section 7. The case of Poincaré group is treated explicitly in Chapter 17, Section 8.

5

Representations of Groups —Developments

Having laid the foundations of the theory of group representations in the previous chapter, attention will now be concentrated on certain developments that are particularly significant in the applications to quantum mechanics.

1 Projection operators

For any *finite* group $\mathscr{G}$ of coordinate transformations in $\mathbb{R}^3$, in particular for any crystallographic point group or space group, the basis functions of unitary irreducible representations are easily determined by a purely automatic process involving certain "projection operators". Before defining these it is necessary to state a theorem which has many applications.

Theorem I Any function $\phi(\mathbf{r})$ of L^2 can be written as a linear combination of basis functions of the unitary irreducible representations of a group $\mathscr{G}$ of coordinate transformations in $\mathbb{R}^3$. That is

$$\phi(\mathbf{r}) = \sum_{p} \sum_{j=1}^{d_p} a_j^p \phi_j^p(\mathbf{r}), \tag{5.1}$$

where $\phi_j^p(\mathbf{r})$ is a normalized basis function transforming as the jth row of the d_p-dimensional unitary irreducible representation Γ^p of $\mathscr{G}$, a_j^p are a set of complex numbers and the sum over p is over all the inequivalent unitary irreducible representations of $\mathscr{G}$.

L^2 is the space of square-integrable functions, as defined in Appendix B, Section 3. The basis functions $\phi^p_j(\mathbf{r})$ and coefficients a^p_j depend on $\phi(\mathbf{r})$ and some of the coefficients a^p_j may be zero. For example, it will be demonstrated shortly for the crystallographic point group D_4 that with $\phi(\mathbf{r})=(x+z)e^{-r}$ (where $r=\{x^2+y^2+z^2\}^{1/2}$) then $\phi(\mathbf{r})=A^{-1}\{\phi^5_1(\mathbf{r})+\phi^5_2(\mathbf{r})\}$ with $\phi^5_1(\mathbf{r})=Axe^{-r}$ and $\phi^5_2(\mathbf{r})=Aze^{-r}$, whereas with $\phi(\mathbf{r})=y(x+z)e^{-r}$ then $\phi(\mathbf{r})=B^{-1}\{\phi^5_1(\mathbf{r})-\phi^5_2(\mathbf{r})\}$ with $\phi^5_1(\mathbf{r})=Byze^{-r}$ and $\phi^5_2(\mathbf{r})=-Bxye^{-r}$. There is no suggestion in the theorem that the functions $\phi^p_j(\mathbf{r})$ on the right-hand side of Equation (5.1) form a *fixed* basis for the space L^2. Indeed this would be impossible for a finite group as L^2 is infinite-dimensional, whereas there is only a finite number of functions on the right-hand side of Equation (5.1) when $\mathscr{G}$ is a finite group. On the other hand, the theorem can be applied to every member of a complete basis for the space L^2 in turn, thereby producing a basis for L^2, all of whose members are basis functions of unitary irreducible representations of $\mathscr{G}$. A situation where this proves very useful is examined in Chapter 6, Section 1.

Proof The theorem is proved in Appendix C, Section 5.

Definition *Projection operators*
Let Γ^p be a unitary irreducible representation of dimension d_p of a finite group of coordinate transformations $\mathscr{G}$ of order g. Then the projection operators are defined by

$$\mathscr{P}^p_{mn}=(d_p/g)\sum_{T\in\mathscr{G}}\Gamma^p(T)^*_{mn}P(T). \tag{5.2}$$

If the group of coordinate transformations form a compact Lie group, the definition may be generalized to

$$\mathscr{P}^p_{mn}=d_p\int_{\mathscr{G}} dT\,\Gamma^p(T)^*_{mn}P(T). \tag{5.3}$$

Theorem II The projection operators $\mathscr{P}^p_{mn}$ have the following properties:

(*a*) For any two functions $\phi(\mathbf{r})$ and $\psi(\mathbf{r})$ of L^2

$$(\mathscr{P}^p_{mn}\phi, \psi)=(\phi, \mathscr{P}^p_{nm}\psi). \tag{5.4}$$

In particular

$$(\mathscr{P}^p_{nn}\phi, \psi)=(\phi, \mathscr{P}^p_{nn}\psi), \tag{5.5}$$

so that $\mathscr{P}^p_{nn}$ is a self-adjoint operator.

(*b*) For any two projection operators $\mathscr{P}^{p}_{mn}$ and $\mathscr{P}^{q}_{jk}$

$$\mathscr{P}^{p}_{mn}\mathscr{P}^{q}_{jk} = \delta_{pq}\delta_{nj}\mathscr{P}^{q}_{mk}. \tag{5.6}$$

In particular

$$(\mathscr{P}^{p}_{nn})^2 = \mathscr{P}^{p}_{nn}. \tag{5.7}$$

(*c*) If $\psi^q_1(\mathbf{r}), \psi^q_2(\mathbf{r}), \ldots$ are basis functions transforming as the unitary irreducible representation Γ^q of $\mathscr{G}$, then

$$\mathscr{P}^{p}_{mn}\psi^q_j(\mathbf{r}) = \delta_{pq}\delta_{nj}\psi^p_m(\mathbf{r}). \tag{5.8}$$

(*d*) For any function $\phi(\mathbf{r})$ of L^2

$$\mathscr{P}^{p}_{nn}\phi(\mathbf{r}) = a^p_n\phi^p_n(\mathbf{r}), \tag{5.9}$$

where a^p_n and $\phi^p_n(\mathbf{r})$ are the coefficients and basis functions of the expansion of $\phi(\mathbf{r})$ (Equation (5.1)) that relate to the nth row of Γ^p.

Proof As only routine manipulations are involved, the proof is relegated to Appendix C, Section 5.

The properties in Equations (5.5) and (5.7) are characteristic of any projection operators. The nature of the projection associated with the operator $\mathscr{P}^{p}_{nn}$ is apparent from Equation (5.9), which shows that $\mathscr{P}^{p}_{nn}$ projects into the subspace of L^2 consisting of functions transforming as the nth row of Γ^p. As operators with the property in Equation (5.7) are known as "idempotent" operators, the projection operator technique is sometimes called the "idempotent method".

For a finite group this theorem provides a simple automatic method for the construction of basis functions. (For a compact Lie group it is preferable to use other methods, as for example in Chapter 12, Section 4.) A set of ortho-normal basis functions transforming as the rows of Γ^p can be found by first selecting a function $\phi(\mathbf{r})$ such that $\mathscr{P}^{p}_{nn}\phi(\mathbf{r})$ is not identically zero for some arbitrarily chosen $n = 1, 2, \ldots, d_p$. With $c^p_n = (\mathscr{P}^{p}_{nn}\phi, \mathscr{P}^{p}_{nn}\phi)^{1/2}$, the function $\phi^p_n(\mathbf{r})$ defined by

$$\phi^p_n(\mathbf{r}) = (1/c^p_n)\mathscr{P}^{p}_{nn}\phi(\mathbf{r})$$

is normalized and transforms as the nth row of Γ^p. Should its ortho-normal partners $\phi^p_m(\mathbf{r})$ ($m = 1, 2, \ldots, d_p$; $m \neq n$) be required, they can be found by operating on $\phi^p_n(\mathbf{r})$ with $\mathscr{P}^{p}_{mn}$. It will be seen later (Chapter 6, Section 1) that in physical problems it is usually only necessary to work with basis functions belonging to *one* arbitrarily chosen row of each irreducible representation.

Example I *Construction of basis functions of irreducible representations of the crystallographic point group* D_4

First let $\phi(\mathbf{r}) = (x+z)e^{-r}$, where $r = \{x^2+y^2+z^2\}^{1/2}$. Then, from Equations (1.8) and (1.17), for any pure rotation T,

$$P(T)\phi(\mathbf{r}) = \phi(\mathbf{R}(T)^{-1}\mathbf{r}) = \phi(\tilde{\mathbf{R}}(T)\mathbf{r}),$$

the orthogonal matrices $\mathbf{R}(T)$ for D_4 being given in Example III of Chapter 1, Section 2. For example, for $T = C_{4y}$,

$$\tilde{\mathbf{R}}(C_{4y})\mathbf{r} = \begin{bmatrix} 0 & 0 & 1 \\ 0 & 1 & 0 \\ -1 & 0 & 0 \end{bmatrix} \begin{bmatrix} x \\ y \\ z \end{bmatrix} = \begin{bmatrix} z \\ y \\ -x \end{bmatrix}$$

and as $\phi(\tilde{\mathbf{R}}(C_{4y})\mathbf{r})$ is defined to be the function in which the x, y and z in $\phi(\mathbf{r})$ are replaced by the 11, 21 and 31 components of $\tilde{\mathbf{R}}(C_{4y})\mathbf{r}$ respectively, and here $\phi(\mathbf{r}) = (x+z)e^{-r}$, then

$$P(C_{4y})\phi(\mathbf{r}) = (z-x)e^{-r}.$$

The following is a complete list of functions $P(T)\phi(\mathbf{r})$ obtained in this way:

$$\left.\begin{aligned} P(E)\phi(\mathbf{r}) &= (x+z)e^{-r}, \\ P(C_{2x})\phi(\mathbf{r}) &= (x-z)e^{-r}, \\ P(C_{2y})\phi(\mathbf{r}) &= (-x-z)e^{-r}, \\ P(C_{2z})\phi(\mathbf{r}) &= (-x+z)e^{-r}, \\ P(C_{4y})\phi(\mathbf{r}) &= (-x+z)e^{-r}, \\ P(C_{4y}^{-1})\phi(\mathbf{r}) &= (x-z)e^{-r}, \\ P(C_{2c})\phi(\mathbf{r}) &= (x+z)e^{-r}, \\ P(C_{2d})\phi(\mathbf{r}) &= (-x-z)e^{-r}. \end{aligned}\right\} \qquad (5.10)$$

Being one-dimensional, the matrices of the irreducible representations Γ^1, Γ^2, Γ^3 and Γ^4 are given directly in terms of the characters of Table 4.1 by $\Gamma^j(T) = [\chi^j(T)]$ for all T of D_4 and $j = 1, 2, 3, 4$. As noted in Example II of Chapter 4, Section 6, the matrices of the two-dimensional irreducible representation Γ^5 may be taken to be:

$$\Gamma^5(E) = \begin{bmatrix} 1 & 0 \\ 0 & 1 \end{bmatrix}, \quad \Gamma^5(C_{2x}) = \begin{bmatrix} 1 & 0 \\ 0 & -1 \end{bmatrix}, \quad \Gamma^5(C_{2y}) = \begin{bmatrix} -1 & 0 \\ 0 & -1 \end{bmatrix},$$

$$\Gamma^5(C_{2z}) = \begin{bmatrix} -1 & 0 \\ 0 & 1 \end{bmatrix}, \quad \Gamma^5(C_{4y}) = \begin{bmatrix} 0 & -1 \\ 1 & 0 \end{bmatrix}, \quad \Gamma^5(C_{4y}^{-1}) = \begin{bmatrix} 0 & 1 \\ -1 & 0 \end{bmatrix},$$

$$\Gamma^5(C_{2c}) = \begin{bmatrix} 0 & 1 \\ 1 & 0 \end{bmatrix}, \quad \Gamma^5(C_{2d}) = \begin{bmatrix} 0 & -1 \\ -1 & 0 \end{bmatrix}.$$

Then, from Equation (5.2), $\mathscr{P}^{p}_{11}\phi(\mathbf{r})=0$ for $p=1,2,3,4$, whereas $\mathscr{P}^{5}_{11}\phi(\mathbf{r})=x\mathrm{e}^{-r}$. Thus $Ax\mathrm{e}^{-r}$ (where $A=(1/c^5_1)=(x\mathrm{e}^{-r},x\mathrm{e}^{-r})^{-1/2}$) is a normalized basis function transforming as the first row of $\boldsymbol{\Gamma}^5$, its partner transforming as the second row of $\boldsymbol{\Gamma}^5$ is $\mathscr{P}^{5}_{21}\{Ax\mathrm{e}^{-r}\}$, which is equal to $Az\mathrm{e}^{-r}$.

It will be seen that, as $\mathrm{P(T)e}^{-r}=\mathrm{e}^{-r}$ for all $\mathrm{T}\in\mathscr{G}$, the factor e^{-r} plays no role in the construction apart from ensuring that the basis functions can be normalized. Consequently, if e^{-r} is replaced by *any* function $F(r)$ such that $(xF(r),xF(r))$ is finite, then $A'xF(r)$ and $A'zF(r)$ (where $A'=(xF(r),xF(r))^{-1/2}$) are ortho-normal basis functions of $\boldsymbol{\Gamma}^5$ transforming as the first and second rows respectively. Clearly no harm comes from temporarily being less precise than usual and saying that "x and z transform as the first and second rows of $\boldsymbol{\Gamma}^5$". Such statements about basis functions of irreducible representations of groups of pure rotations appear quite commonly in the literature.

A similar analysis applied to $\phi(\mathbf{r})=(xy+yz)\mathrm{e}^{-r}$ shows that $\mathscr{P}^{p}_{11}\phi(\mathbf{r})=0$ for $p=1,2,3,4$, but $\mathscr{P}^{5}_{11}\phi(\mathbf{r})=yz\mathrm{e}^{-r}$. Thus $Byz\mathrm{e}^{-r}$ (where $B=(1/c^5_1)=(yz\mathrm{e}^{-r},yz\mathrm{e}^{-r})^{-1/2}$) is a normalized basis function transforming as the first row of $\boldsymbol{\Gamma}^5$. Its partner transforming as the second row of $\boldsymbol{\Gamma}^5$ is $\mathscr{P}^{5}_{21}\{Byz\mathrm{e}^{-r}\}$, which is equal to $-Bxy\mathrm{e}^{-r}$. Again, loosely one could say that "yz and $-xy$ transform as the first and second rows of $\boldsymbol{\Gamma}^5$".

The procedure for constructing basis functions that has just been described requires an explicit knowledge of the matrix elements of the representations, and not merely a knowledge of the character system alone, which is usually the only information which is given in the published literature. Of course, for one-dimensional representations the characters give the matrix elements immediately, but for the other representations some further analysis is needed. A method involving "character projection operators" which can be used in such cases will now be described.

Definition *Character projection operator*
Let $\boldsymbol{\Gamma}^p$ be an irreducible representation of dimension d_p of a finite group of coordinate transformations $\mathscr{G}$ of order g, $\chi^p(\mathrm{T})$ being the character of $\mathrm{T}\in\mathscr{G}$ in $\boldsymbol{\Gamma}^p$. Then the character projection operator for $\boldsymbol{\Gamma}^p$ is defined by

$$\mathscr{P}^p=(d_p/g)\sum_{\mathrm{T}\in\mathscr{G}}\chi^p(\mathrm{T})^*\mathrm{P(T)}. \qquad (5.11)$$

Obviously $\mathscr{P}^p$ can be constructed from the character table alone

and

$$\mathcal{P}^p = \sum_{n=1}^{d_p} \mathcal{P}^p_{nn},$$

so that $\mathcal{P}^p$ has the property of projecting out of a function $\phi(\mathbf{r})$ the sum of *all* the parts transforming according to the rows of Γ^p. This implies that if $\mathcal{P}^p\phi(\mathbf{r})$ is not identically zero, it is a linear combination of basis functions of Γ^p (which are as yet undetermined). However, as noted in Chapter 4, Section 2, linear combinations of basis functions are themselves basis functions in an equivalent representation, so $\mathcal{P}^p\phi(\mathbf{r})$ may be taken to transform as the first row of some form of the pth irreducible representation. This particular form will henceforth be denoted by Γ^p. (Up to this stage Γ^p was completely unspecified up to a similarity transformation.) The procedure to be described then generates explicit matrix elements for this form of Γ^p, which is, of course, as good as any other equivalent form.

Having chosen a normalizable $\phi(\mathbf{r})$ such that $\mathcal{P}^p\phi(\mathbf{r})$ is not identically zero, construct $P(T)\{\mathcal{P}^p\phi(\mathbf{r})\}$ for each $T \in \mathcal{G}$. (Each of these must be linear combinations of the d_p basis functions of Γ^p.) From these functions abstract d_p linearly independent functions, taking one of these to be $\mathcal{P}^p\phi(\mathbf{r})$ itself. Apply the Schmidt orthogonalization process (see Appendix B, Section 2) to these functions to produce d_p orthonormal functions $\psi^p_n(\mathbf{r})$, $n = 1, 2, \ldots, d_p$, $\psi^p_1(\mathbf{r})$ being a multiple of $\mathcal{P}^p\phi(\mathbf{r})$. These functions can be taken as the basis functions of a unitary representation of Γ^p. The matrix elements can then be found from Equation (1.26), that is from

$$P(T)\psi^p_n(\mathbf{r}) = \sum_{m=1}^{d_p} \Gamma^p(T)_{mn}\psi^p_m(\mathbf{r}), \tag{5.12}$$

as $P(T)\psi^p_n(\mathbf{r})$ can be found for each $T \in \mathcal{G}$ using Equation (1.17).

This method will be illustrated by using it to obtain a set of matrices for the irreducible representation Γ^5 of D_4, which is a purely academic exercise here, as a set is already known.

Example II *Determination of matrix elements of the two-dimensional irreducible representation Γ^5 of the crystallographic point group D_4 from its character system*

Take $\phi(\mathbf{r}) = zF(r)$, where $F(r)$ is any function of r such that $\phi(\mathbf{r})$ is normalized. Then, from Table 4.1 and Equation (5.11), $\mathcal{P}^p\phi(\mathbf{r}) = zF(r)$. As $P(C_{4y}^{-1})\{\mathcal{P}^p\phi(\mathbf{r})\} = xF(r)$, and as $d_5 = 2$, $zF(r)$ and $xF(r)$ together give the totality of linearly independent functions $P(T)$ $\{\mathcal{P}^p\phi(\mathbf{r})\}$. It happens here that $zF(r)$ and $xF(r)$ are orthogonal, so the Schmidt process is not needed. Then, as $xF(r)$ is also normalized, one may take $\psi_1^5(r)=zF(r)$ and $\psi_2^5(r)=xF(r)$. Then, for example,

$$\left.\begin{aligned}P(C_{4y})\psi_1^5(\mathbf{r}) &= -xF(r) = -\psi_2^5(\mathbf{r}),\\ P(C_{4y})\psi_2^5(\mathbf{r}) &= zF(r) = \psi_1^5(\mathbf{r}),\end{aligned}\right\}$$

which, in comparison with Equation (5.12), gives as the matrix representing C_{4y}

$$\begin{bmatrix} 0 & 1 \\ -1 & 0 \end{bmatrix}.$$

The matrices representing the other elements of D_4 may be found in the same way. They are not identical to those quoted in Example I above, but could be obtained from them by a similarity transformation (Equation (4.4)) with

$$\mathbf{S}=\begin{bmatrix} 0 & 1 \\ 1 & 0 \end{bmatrix}.$$

2 Direct product representations

In Appendix A, Section 1, the definition is given of the *direct product* $\mathbf{A}\otimes\mathbf{B}$ of an $m\times m$ matrix $\mathbf{A}$ and an $n\times n$ matrix $\mathbf{B}$ in which $\mathbf{A}\otimes\mathbf{B}$ is an $mn\times mn$ matrix whose rows and columns are each labelled by a *pair* of indices in such a way that (cf. Equation (A.8))

$$(\mathbf{A}\otimes\mathbf{B})_{js,kt}=A_{jk}B_{st}$$

$(1\leq j, k\leq m; 1\leq s, t\leq n)$. Certain properties of direct product matrices are also deduced in that section.

The following theorem shows that this definition allows the construction of "direct product representations", which, through the Wigner–Eckart Theorem, play a major role in the applications of group theory in quantum mechanical problems. (They are sometimes called "Kronecker product representations" or "tensor product representations".)

Theorem I If Γ^p and Γ^q are two unitary irreducible representations of a group $\mathcal{G}$ of dimensions d_p and d_q respectively, then the set of matrices defined by

$$\Gamma(T)=\Gamma^p(T)\otimes\Gamma^q(T) \tag{5.13}$$

for all $T\in\mathcal{G}$ form a unitary representation of $\mathcal{G}$ of dimension d_pd_q. The character $\chi(T)$ of $T\in\mathcal{G}$ in this representation is given by

$$\chi(T)=\chi^p(T)\chi^q(T). \tag{5.14}$$

Proof For any T_1 and T_2 of $\mathscr{G}$, by Equation (5.13),

$$\boldsymbol{\Gamma}(T_1)\boldsymbol{\Gamma}(T_2) = \{\boldsymbol{\Gamma}^p(T_1)\otimes\boldsymbol{\Gamma}^q(T_1)\}\{\boldsymbol{\Gamma}^p(T_2)\otimes\boldsymbol{\Gamma}^q(T_2)\}$$
$$= \{\boldsymbol{\Gamma}^p(T_1)\boldsymbol{\Gamma}^p(T_2)\}\otimes\{\boldsymbol{\Gamma}^q(T_1)\boldsymbol{\Gamma}^q(T_2)\}$$

(on using Equation (A.9))

$$= \boldsymbol{\Gamma}^p(T_1T_2)\otimes\boldsymbol{\Gamma}^q(T_1T_2)$$

(as $\boldsymbol{\Gamma}^p$ and $\boldsymbol{\Gamma}^q$ are themselves representations)

$$= \boldsymbol{\Gamma}(T_1T_2).$$

Thus the matrices $\boldsymbol{\Gamma}(T)$ of Equation (5.13) certainly form a representation of $\mathscr{G}$ and its dimension is obviously d_pd_q.

As the direct product of any two unitary matrices is itself unitary (see Appendix A, Section 1), each matrix $\boldsymbol{\Gamma}(T)$ of Equation (5.13) must be unitary.

Finally, as the diagonal elements of $\boldsymbol{\Gamma}^p(T)\otimes\boldsymbol{\Gamma}^q(T)$ are labelled by the pairs (j, s) and (k, t) with $j = k$ and $s = t$, for any $T\in\mathscr{G}$

$$\chi(T) = \sum_{j=1}^{d_p}\sum_{s=1}^{d_q}(\boldsymbol{\Gamma}^p(T)\otimes\boldsymbol{\Gamma}^q(T))_{js,js}$$
$$= \sum_{j=1}^{d_p}\sum_{s=1}^{d_q}\Gamma^p(T)_{jj}\Gamma^q(T)_{ss}$$
$$= \chi^p(T)\chi^q(T).$$

The direct product representation defined by Equation (5.13) will be denoted by $\boldsymbol{\Gamma}^p\otimes\boldsymbol{\Gamma}^q$. Although its definition in terms of matrix elements of $\boldsymbol{\Gamma}^p$ and $\boldsymbol{\Gamma}^q$ may appear complicated, in terms of bases the definition is completely natural, as will be demonstrated in the next two sections.

In general the representation $\boldsymbol{\Gamma}^p\otimes\boldsymbol{\Gamma}^q$ is *reducible*, even if $\boldsymbol{\Gamma}^p$ and $\boldsymbol{\Gamma}^q$ are themselves irreducible. For example, for the crystallographic point group D_4, as $\boldsymbol{\Gamma}^5$ is two-dimensional, $\boldsymbol{\Gamma}^5\otimes\boldsymbol{\Gamma}^5$ must be four-dimensional. However, D_4 has no irreducible representations of dimension greater than two, so $\boldsymbol{\Gamma}^5\otimes\boldsymbol{\Gamma}^5$ must be reducible.

Henceforth it will be assumed that $\mathscr{G}$ is a *finite* group or a *compact* Lie group. Then all the irreducible representations of $\mathscr{G}$ may be assumed to be unitary and every direct product $\boldsymbol{\Gamma}^p\otimes\boldsymbol{\Gamma}^q$ is either irreducible or is completely reducible. Suppose that a similarity transformation with a $d_pd_q\times d_pd_q$ non-singular matrix $\mathbf{C}$ is applied to $\boldsymbol{\Gamma}^p\otimes\boldsymbol{\Gamma}^q$ to give an equivalent representation that is a direct sum of unitary irreducible representations, and the unitary irreducible representation $\boldsymbol{\Gamma}^r$ of $\mathscr{G}$ appears n_{pq}^r times in this sum. This can be written formally as

$$\Gamma^{p} \otimes \Gamma^{q} \approx \sum_{r} \oplus n_{pq}^{r} \Gamma^{r}, \tag{5.15}$$

or more precisely as

$$C^{-1}(\Gamma^{p} \otimes \Gamma^{q})C = \sum_{r} \oplus n_{pq}^{r} \Gamma^{r}, \tag{5.16}$$

where the right-hand side is called the "Clebsch–Gordan series for $\Gamma^{p} \otimes \Gamma^{q}$". For the case in which $\mathcal{G}$ is a finite group of order g, Theorem V of Chapter 4, Section 6 gives

$$n_{pq}^{r} = (1/g) \sum_{T \in \mathcal{G}} \chi^{p}(T)\chi^{q}(T)\chi^{r}(T)^{*}, \tag{5.17}$$

the corresponding expression when $\mathcal{G}$ is a compact Lie group being

$$n_{pq}^{r} = \int_{\mathcal{G}} \chi^{p}(T)\chi^{q}(T)\chi^{r}(T)^{*}\, dT. \tag{5.18}$$

Thus in these cases the Clebsch–Gordan series is determined solely by the characters. Obviously, as $\Gamma^{p} \otimes \Gamma^{q}$ is of dimension $d_p d_q$,

$$d_p d_q = \sum_{r} n_{pq}^{r} d_r,$$

where d_r is the dimension of the irreducible representation Γ^{r}.

Example I *Clebsch–Gordan series for the crystallographic point group* D_4

Table 4.1 and Equation (5.17) together imply that

$$\Gamma^{5} \otimes \Gamma^{4} \approx \Gamma^{5}$$

(i.e. $n_{54}^{1} = n_{54}^{2} = n_{54}^{3} = n_{54}^{4} = 0,\ n_{54}^{5} = 1$),

$$\Gamma^{5} \otimes \Gamma^{3} \approx \Gamma^{5},$$

and

$$\Gamma^{5} \otimes \Gamma^{5} \approx \Gamma^{1} \oplus \Gamma^{2} \oplus \Gamma^{3} \oplus \Gamma^{4}.$$

These particular Clebsch–Gordan series will be needed in later examples. The other series may be found in the same way.

Two useful symmetry properties follow immediately from Equations (5.17) and (5.18), namely

$$n_{qp}^{r} = n_{pq}^{r} \tag{5.19}$$

and

$$n_{p^{*}r}^{q} = n_{pq}^{r}, \tag{5.20}$$

where in Equation (5.20) Γ^{p^*} is the irreducible representation of $\mathcal{G}$ defined by $\Gamma^{p^*}(T) = \{\Gamma^p(T)\}^*$ for all $T \in \mathcal{G}$, the identity (Equation (5.20)) being a consequence of the fact that n^r_{pq} must be an integer and so must be real.

3 The Wigner–Eckart Theorem for groups of coordinate transformations in $\mathbb{R}^3$

It is possibly easiest to introduce the notions of Clebsch–Gordan coefficients and irreducible tensor operators for the case in which $\mathcal{G}$ is a group of coordinate transformations in $\mathbb{R}^3$, and to first state the Wigner–Eckart Theorem for this case. The theory will then be developed in the next section in more general terms, employing the more abstract concept of carrier spaces, thereby substantially extending the range of applicability of the results.

Theorem I Suppose that $\mathcal{G}$ is a group of coordinate transformations having irreducible representations Γ^p and Γ^q of dimensions d_p and d_q respectively, and $\phi^p_j(\mathbf{r})$, $j = 1, 2, \ldots, d_p$, and $\psi^q_s(\mathbf{r})$, $s = 1, 2, \ldots, d_q$, are basis functions of Γ^p and Γ^q respectively. Then the set of $d_p d_q$ functions $\phi^p_j(\mathbf{r})\psi^q_s(\mathbf{r})$ (where $j = 1, 2, \ldots, d_p$ and $s = 1, 2, \ldots, d_q$) form a basis for the direct product representation $\Gamma^p \otimes \Gamma^q$, provided that they form a linearly independent set.

Proof For any $T \in \mathcal{G}$, from Equation (1.17),

$$\begin{aligned}
P(T)\{\phi^p_j(\mathbf{r})\psi^q_s(\mathbf{r})\} &= \phi^p_j(\{\mathbf{R}(T) \mid \mathbf{t}(T)\}^{-1}\mathbf{r})\psi^q_s(\{\mathbf{R}(T) \mid \mathbf{t}(T)\}^{-1}\mathbf{r}) \\
&= \{P(T)\phi^p_j(\mathbf{r})\}\{P(T)\psi^q_s(\mathbf{r})\} \\
&= \left\{\sum_{k=1}^{d_p} \Gamma^p(T)_{kj}\phi^p_k(\mathbf{r})\right\}\left\{\sum_{t=1}^{d_q} \Gamma^q(T)_{ts}\psi^p_t(\mathbf{r})\right\} \\
&= \sum_{k=1}^{d_p}\sum_{t=1}^{d_q} (\Gamma^p(T) \otimes \Gamma^q(T))_{kt,js}\{\phi^p_k(\mathbf{r})\psi^q_t(\mathbf{r})\},
\end{aligned}$$

which is of the form of Equation (1.26).

As noted in the previous section, in general $\Gamma^p \otimes \Gamma^q$ is reducible, even when Γ^p and Γ^q are both irreducible. A minor implication (see Chapter 4, Section 5) is that the basis functions $\phi^p_j(\mathbf{r})\psi^q_s(\mathbf{r})$ need *not* form an ortho-normal set, nor even a mutually orthogonal set, even if the functions $\phi^p_j(\mathbf{r})$ and $\psi^q_s(\mathbf{r})$ each form ortho-normal sets. Indeed, for D_4 the functions $\phi^5_1(\mathbf{r}) = xF(r)$, $\phi^5_2(\mathbf{r}) = zF(r)$ and $\psi^5_1(\mathbf{r}) = x^3G(r)$,

$\psi_2^5(\mathbf{r}) = z^3 G(r)$ are two linearly independent sets of basis functions for Γ^5 such that $(\phi_j^5, \phi_k^5) = \delta_{jk}$ and $(\psi_j^5, \psi_k^5) = \delta_{jk}$ (provided that $F(r)$ and $G(r)$ are chosen so that $(xF(r), xF(r)) = 1$ and $(x^3G(r), x^3G(r)) = 1$. However,

$$(\phi_1^5\psi_2^5, \phi_2^5\psi_1^5) = \iiint x^4 z^4 \, |F(r)G(r)|^2 \, dx \, dy \, dz,$$

which is definitely *non*-zero as the integrand is always positive (except when $x = 0$ or $z = 0$).

Now suppose that $\mathcal{G}$ is a *finite* group or a *compact* Lie group and n_{pq}^r is the number of times that the irreducible representation Γ^r appears in $\Gamma^p \otimes \Gamma^q$, Γ^p, Γ^q and Γ^r all being assumed to be unitary. If $n_{pq}^r \neq 0$ there must be n_{pq}^r linearly independent sets of basis functions for Γ^r formed from linear combinations of the products $\phi_j^p(\mathbf{r})\psi_s^p(\mathbf{r})$. Let these be denoted by $\theta_l^{r,\alpha}(\mathbf{r})$, where $\alpha = 1, 2, \ldots, n_{pq}^r$, and $l = 1, 2, \ldots, d_r$, d_r being the dimension of Γ^r, so that

$$P(T)\theta_l^{r,\alpha}(\mathbf{r}) = \sum_{u=1}^{d_r} \Gamma^r(T)_{ul}\theta_u^{r,\alpha}(\mathbf{r}) \tag{5.21}$$

for all $T \in \mathcal{G}$, $l = 1, 2, \ldots, d_r$, and $\alpha = 1, 2, \ldots, n_{pq}^r$. These may be written in the form

$$\theta_l^{r,\alpha}(\mathbf{r}) = \sum_{j=1}^{d_p} \sum_{k=1}^{d_q} \left(\begin{smallmatrix} p & q \\ j & k \end{smallmatrix}\middle|\begin{smallmatrix} r, & \alpha \\ l & \end{smallmatrix}\right) \phi_j^p(\mathbf{r})\psi_k^q(\mathbf{r}). \tag{5.22}$$

The coefficients $\left(\begin{smallmatrix} p & q \\ j & k \end{smallmatrix}\middle|\begin{smallmatrix} r, & \alpha \\ l & \end{smallmatrix}\right)$ are called "Clebsch–Gordan coefficients". They can be regarded as forming a $d_p d_q \times d_p d_q$ non-singular matrix, the rows being labelled by the pairs (j, k), where $j = 1, 2, \ldots, d_p$ and $k = 1, 2, \ldots, d_q$, and the columns being labelled by the triples (r, α, l), where r appears only if $n_{pq}^r \neq 0$, in which case $\alpha = 1, 2, \ldots, n_{pq}^r$ and $l = 1, 2, \ldots, d_r$. In fact this is the matrix $\mathbf{C}$ of Equation (5.16).

The inverse of Equation (5.22) can be written as

$$\phi_j^p(\mathbf{r})\psi_k^q(\mathbf{r}) = \sum_r \sum_{\alpha=1}^{n_{pq}^r} \sum_{l=1}^{d_r} \left(\begin{smallmatrix} r, & \alpha \\ l & \end{smallmatrix}\middle|\begin{smallmatrix} p & q \\ j & k \end{smallmatrix}\right) \theta_l^{r,\alpha}(\mathbf{r}) \tag{5.23}$$

for all $j = 1, 2, \ldots, d_p$ and $k = 1, 2, \ldots, d_q$, where the sum over r is over all those irreducible representations Γ^r for which $n_{pq}^r \neq 0$. The coefficients $\left(\begin{smallmatrix} r, & \alpha \\ l & \end{smallmatrix}\middle|\begin{smallmatrix} p & q \\ j & k \end{smallmatrix}\right)$ again form a $d_p d_q \times d_p d_q$ non-singular matrix, but this time the rows are labelled by the triples (r, α, l) and the columns by the pairs (j, k). Clearly this is the matrix $\mathbf{C}^{-1}$ of Equation (5.16).

As $\Gamma^p \otimes \Gamma^q$ is unitary and the direct sum on the right-hand side of Equation (5.16) is also unitary, Theorem IV of Chapter 4, Section 3 shows that **C** may be *chosen* to be *unitary*, which implies that

$$\left(\begin{smallmatrix} r, & \alpha \\ l & \end{smallmatrix}\middle|\begin{smallmatrix} p & q \\ j & k \end{smallmatrix}\right)=\left(\begin{smallmatrix} p & q \\ j & k \end{smallmatrix}\middle|\begin{smallmatrix} r, & \alpha \\ l & \end{smallmatrix}\right)^*. \tag{5.24}$$

It will now be shown that if $n_{pq}^r \leqslant 1$ for every r, then it is easy to construct a simple formula for the Clebsch–Gordan coefficients that corresponds to this choice. When there exists an n_{pq}^r such that $n_{pq}^r > 1$ the construction of a unitary matrix of Clebsch–Gordan coefficients is more difficult and for finite groups there is little advantage in making such a choice. (A detailed discussion may be found in the work of van den Broek and Cornwell (1978).)

Applying the projection operator $\mathscr{P}_{ul}^r$ to Equation (5.23) gives

$$\mathscr{P}_{ul}^r\{\phi_j^p(\mathbf{r})\psi_k^q(\mathbf{r})\}=\sum_{\alpha=1}^{n_{pq}^r}\left(\begin{smallmatrix} r, & \alpha \\ l & \end{smallmatrix}\middle|\begin{smallmatrix} p & q \\ j & k \end{smallmatrix}\right)\theta_u^{r,\alpha}(\mathbf{r}),$$

so, from Equation (5.22),

$$\mathscr{P}_{ul}^r\{\phi_j^p(\mathbf{r})\psi_k^q(\mathbf{r})\}=\sum_{s=1}^{d_p}\sum_{t=1}^{d_q}\sum_{\alpha=1}^{n_{pq}^r}\left(\begin{smallmatrix} r, & \alpha \\ l & \end{smallmatrix}\middle|\begin{smallmatrix} p & q \\ j & k \end{smallmatrix}\right)\left(\begin{smallmatrix} p & q \\ s & t \end{smallmatrix}\middle|\begin{smallmatrix} r, & \alpha \\ u & \end{smallmatrix}\right)\phi_s^p(\mathbf{r})\psi_t^q(\mathbf{r}). \tag{5.25}$$

However, if $\mathscr{G}$ is a finite group of order g, from Equation (5.2),

$$\mathscr{P}_{ul}^r\{\phi_j^p(\mathbf{r})\psi_k^q(\mathbf{r})\}=(d_r/g)\sum_{s=1}^{d_p}\sum_{t=1}^{d_q}\sum_{T\in\mathscr{G}}\Gamma^p(T)_{sj}\Gamma^q(T)_{tk}\Gamma^r(T)_{ul}^*\phi_s^p(\mathbf{r})\psi_t^q(\mathbf{r}), \tag{5.26}$$

so, on equating the right-hand sides of Equations (5.25) and (5.26), as the products $\phi_j^p(\mathbf{r})\psi_k^q(\mathbf{r})$ are assumed to be linearly independent,

$$\sum_{\alpha=1}^{n_{pq}^r}\left(\begin{smallmatrix} p & q \\ s & t \end{smallmatrix}\middle|\begin{smallmatrix} r, & \alpha \\ u & \end{smallmatrix}\right)\left(\begin{smallmatrix} r, & \alpha \\ l & \end{smallmatrix}\middle|\begin{smallmatrix} p & q \\ j & k \end{smallmatrix}\right)=(d_r/g)\sum_{T\in\mathscr{G}}\Gamma^p(T)_{sj}\Gamma^q(T)_{tk}\Gamma^r(T)_{ul}^*. \tag{5.27}$$

If the matrix of Clebsch–Gordan coefficients is assumed to be unitary, then Equation (5.27) reduces, by Equation (5.24), to

$$\sum_{\alpha=1}^{n_{pq}^r}\left(\begin{smallmatrix} p & q \\ s & t \end{smallmatrix}\middle|\begin{smallmatrix} r, & \alpha \\ u & \end{smallmatrix}\right)\left(\begin{smallmatrix} p & q \\ j & k \end{smallmatrix}\middle|\begin{smallmatrix} r, & \alpha \\ l & \end{smallmatrix}\right)^*=(d_r/g)\sum_{T\in\mathscr{G}}\Gamma^p(T)_{sj}\Gamma^q(T)_{tk}\Gamma^r(T)_{ul}^*. \tag{5.28}$$

If $\mathscr{G}$ is a compact Lie group, obviously Equations (5.27) and (5.28) generalize to

$$\sum_{\alpha=1}^{n_{pq}^r}\left(\begin{smallmatrix} p & q \\ s & t \end{smallmatrix}\middle|\begin{smallmatrix} r, & \alpha \\ u & \end{smallmatrix}\right)\left(\begin{smallmatrix} r, & \alpha \\ l & \end{smallmatrix}\middle|\begin{smallmatrix} p & q \\ j & k \end{smallmatrix}\right)=d_r\int_{\mathscr{G}}\Gamma^p(T)_{sj}\Gamma^q(T)_{tk}\Gamma^r(T)_{ul}^*\,dT \tag{5.29}$$

and

$$\sum_{\alpha=1}^{n_{pq}^r}\left(\begin{smallmatrix} p & q \\ s & t \end{smallmatrix}\middle|\begin{smallmatrix} r, & \alpha \\ u & \end{smallmatrix}\right)\left(\begin{smallmatrix} p & q \\ j & k \end{smallmatrix}\middle|\begin{smallmatrix} r, & \alpha \\ l & \end{smallmatrix}\right)^*=d_r\int_{\mathscr{G}}\Gamma^p(T)_{sj}\Gamma^q(T)_{tk}\Gamma^r(T)_{ul}^*\,dT \tag{5.30}$$

respectively.

Before proceeding further it should be noted that there is a certain degree of arbitrariness in the Clebsch–Gordan coefficients, even if the assumption is made that they form a unitary matrix. Consider first the case $n^r_{pq}=1$. If the Clebsch–Gordan coefficients $\left(\begin{smallmatrix} p & q \\ j & k \end{smallmatrix}\middle|\begin{smallmatrix} r, & 1 \\ l & \end{smallmatrix}\right)$ satisfy Equation (5.28), then, for any real number ω independent of j, k and l, the coefficients $\left(\begin{smallmatrix} p & q \\ j & k \end{smallmatrix}\middle|\begin{smallmatrix} r, & 1 \\ l & \end{smallmatrix}\right)'$ defined by

$$\left(\begin{smallmatrix} p & q \\ j & k \end{smallmatrix}\middle|\begin{smallmatrix} r, & 1 \\ l & \end{smallmatrix}\right)'=\exp(i\omega)\left(\begin{smallmatrix} p & q \\ j & k \end{smallmatrix}\middle|\begin{smallmatrix} r, & 1 \\ l & \end{smallmatrix}\right)$$

also satisfy Equation (5.28), an identical result being true in the compact Lie group case. That is, the Clebsch–Gordan coefficients $\left(\begin{smallmatrix} p & q \\ j & k \end{smallmatrix}\middle|\begin{smallmatrix} r, & 1 \\ l & \end{smallmatrix}\right)$ contain an arbitrary phase factor whose choice is entirely a matter of convention. (Much confusion has been caused in the past by the use of different conventions, particularly for the groups SU(2) and SO(3). This matter will be examined in detail for these groups in Chapter 12, Section 5.) If $n^r_{pq}>1$ the situation is even more complicated. In this case if the coefficients $\left(\begin{smallmatrix} p & q \\ q & k \end{smallmatrix}\middle|\begin{smallmatrix} r, & \alpha \\ l & \end{smallmatrix}\right)'$ are defined by

$$\left(\begin{smallmatrix} p & q \\ j & k \end{smallmatrix}\middle|\begin{smallmatrix} r, & \alpha \\ l & \end{smallmatrix}\right)'=\sum_{\beta=1}^{n^r_{pq}} \mathrm{s}_{\alpha\beta}\left(\begin{smallmatrix} p & q \\ j & k \end{smallmatrix}\middle|\begin{smallmatrix} r, & \beta \\ l & \end{smallmatrix}\right),$$

where s is any $n^r_{pq}\times n^r_{pq}$ unitary matrix, then if $\left(\begin{smallmatrix} p & q \\ j & k \end{smallmatrix}\middle|\begin{smallmatrix} r, & \alpha \\ l & \end{smallmatrix}\right)$ satisfy Equation (5.28), so do $\left(\begin{smallmatrix} p & q \\ j & k \end{smallmatrix}\middle|\begin{smallmatrix} r, & \alpha \\ l & \end{smallmatrix}\right)'$, the corresponding result again also being true for compact Lie groups. Consequently the arbitrariness is extended from a single phase factor to a $n^r_{pq}\times n^r_{pq}$ unitary matrix.

Both when $n^r_{pq}=1$ and $n^r_{pq}>1$ this element of choice in determining the Clebsch–Gordan coefficients is *additional* to the element of choice that lies in the construction of any representations involved that have dimension greater than one. By contrast the quantities n^r_{pq} are *uniquely* determined by Equation (5.17) or Equation (5.18).

When $\mathscr{G}$ is a finite group Equations (5.27) provide the most straightforward way of evaluating the Clebsch–Gordan coefficients. The discussion here will be confined to the case in which $n^r_{pq}<2$ for all r. (Details of the general case may be found in the work of van den Broek and Cornwell (1978). See also Dirl (1979a).) First choose some set j, k, l, such that

$$(d_r/g)\sum_{\mathrm{T}\in\mathscr{G}}\Gamma^p(\mathrm{T})_{jj}\Gamma^q(\mathrm{T})_{kk}\Gamma^r(\mathrm{T})^*_{ll}$$

is non-zero. From Equation (5.28) it follows that this quantity must then be real and positive. Adopting the phase convention that $\left(\begin{smallmatrix} p & q \\ j & k \end{smallmatrix}\middle|\begin{smallmatrix} r, & 1 \\ l & \end{smallmatrix}\right)$ is real and positive, Equation (5.28) with $j=s$, $k=t$ and $l=u$ implies that

$$\left(\begin{smallmatrix} p & q \\ j & k \end{smallmatrix}\middle|\begin{smallmatrix} r, & 1 \\ l & \end{smallmatrix}\right)=\left\{(d_r/g)\sum_{\mathrm{T}\in\mathscr{G}}\Gamma^p(\mathrm{T})_{jj}\Gamma^q(\mathrm{T})_{kk}\Gamma^r(\mathrm{T})^*_{ll}\right\}^{1/2}.$$

Then for *all* $s=1,2,\ldots,d_p$, $t=1,2,\ldots,d_q$, and $u=1,2,\ldots,d_r$, Equation (5.29) gives

$$\left(\begin{array}{cc|cc} p & q & r, & 1 \\ s & t & u & \end{array}\right)=\frac{(d_r/g)^{1/2}\sum_{T\in\mathcal{G}}\Gamma^p(T)_{sj}\Gamma^q(T)_{tk}\Gamma^r(T)^*_{ul}}{\{\sum_{T\in\mathcal{G}}\Gamma^p(T)_{jj}\Gamma^q(T)_{kk}\Gamma^r(T)^*_{ll}\}^{1/2}}. \tag{5.31}$$

Although this formula generalizes in the obvious way when $\mathcal{G}$ is a compact Lie group, for such a group it is much easier to use Lie algebraic methods to calculate the Clebsch–Gordan coefficients. These will be described for SO(3) and SU(2) in Chapter 12, Section 5, and for other simple compact Lie groups in Chapter 16, Section 6.

The case in which $\Gamma^q=\Gamma^1$, where Γ^1 is the one-dimensional identity representation defined by $\Gamma^1(T)=[1]$ for all $T\in\mathcal{G}$, provides a simple but important example. In this situation

$$\Gamma^p\otimes\Gamma^1=\Gamma^p \tag{5.32}$$

for every irreducible representation Γ^p of $\mathcal{G}$, while Equation (5.31) and its generalization for compact Lie groups, taken with the orthogonality theorem for matrix representations (Theorem IV of Chapter 4, Section 5) shows that

$$\left(\begin{array}{cc|cc} p & 1 & p, & 1 \\ s & 1 & u & \end{array}\right)=\delta_{su}. \tag{5.33}$$

Example I *Clebsch–Gordan coefficients for the crystallographic point group* D_4

Using the matrices of the two-dimensional irreducible representation Γ^5 specified in Example I of Section 1, the coefficients corresponding to the series $\Gamma^5\otimes\Gamma^4\approx\Gamma^5$ are given by Equation (5.31) (with $j=1$, $k=1$, $l=2$) as

$$\left.\begin{array}{l}
\left(\begin{array}{cc|cc} 5 & 4 & 5, & 1 \\ 1 & 1 & 2 & \end{array}\right)=\left(\begin{array}{cc|cc} 5 & 4 & 5, & 1 \\ 2 & 1 & 1 & \end{array}\right)=1, \\
\left(\begin{array}{cc|cc} 5 & 4 & 5, & 1 \\ 1 & 1 & 1 & \end{array}\right)=\left(\begin{array}{cc|cc} 5 & 4 & 5, & 1 \\ 2 & 1 & 2 & \end{array}\right)=0.
\end{array}\right\}$$

Similarly for $\Gamma^5\otimes\Gamma^3\approx\Gamma^5$, (5.31) (with $j=1$, $k=1$, $l=2$) gives

$$\left.\begin{array}{ll}
\left(\begin{array}{cc|cc} 5 & 3 & 5, & 1 \\ 1 & 1 & 2 & \end{array}\right)=1, & \left(\begin{array}{cc|cc} 5 & 3 & 5, & 1 \\ 2 & 1 & 1 & \end{array}\right)=-1, \\
\left(\begin{array}{cc|cc} 5 & 3 & 5, & 1 \\ 1 & 1 & 1 & \end{array}\right)=\left(\begin{array}{cc|cc} 5 & 3 & 5, & 1 \\ 2 & 1 & 2 & \end{array}\right)=0. &
\end{array}\right\}$$

Likewise, for $\Gamma^5\otimes\Gamma^5\approx\Gamma^1\oplus\Gamma^2\oplus\Gamma^3\oplus\Gamma^4$,

$$\left.\begin{array}{ll}
\left(\begin{array}{cc|cc} 5 & 1 & 1, & 1 \\ 1 & 1 & 1 & \end{array}\right)=\left(\begin{array}{cc|cc} 5 & 5 & 1, & 1 \\ 2 & 2 & 1 & \end{array}\right)=2^{-1/2}, & \\
\left(\begin{array}{cc|cc} 5 & 5 & 2, & 1 \\ 1 & 1 & 1 & \end{array}\right)=2^{-1/2}, & \left(\begin{array}{cc|cc} 5 & 5 & 2, & 1 \\ 2 & 2 & 1 & \end{array}\right)=-2^{-1/2}, \\
\left(\begin{array}{cc|cc} 5 & 5 & 3, & 1 \\ 2 & 1 & 1 & \end{array}\right)=2^{-1/2}, & \left(\begin{array}{cc|cc} 5 & 5 & 3, & 1 \\ 1 & 2 & 1 & \end{array}\right)=-2^{-1/2}, \\
\left(\begin{array}{cc|cc} 5 & 5 & 4, & 1 \\ 2 & 1 & 1 & \end{array}\right)=\left(\begin{array}{cc|cc} 5 & 5 & 4, & 1 \\ 1 & 2 & 1 & \end{array}\right)=2^{-1/2}, &
\end{array}\right\}$$

all other Clebsch–Gordan coefficients involved being zero.

The Wigner–Eckart Theorem depends on one further concept, that of "irreducible tensor operators".

Definition *Irreducible tensor operators for a group of coordinate transformations in* $\mathbb{R}^3$

Let $Q_1^q, Q_2^q, \ldots$ be a set of d_q linear operators that act on functions belonging to the Hilbert space L^2 and which satisfy the equations

$$P(T)Q_j^q P(T)^{-1} = \sum_{k=1}^{d_q} \Gamma^q(T)_{kj} Q_k^q \tag{5.34}$$

for every $j = 1, 2, \ldots, d_q$ and every T of a group of coordinate transformations $\mathcal{G}$, where $\boldsymbol{\Gamma}^q$ is an irreducible representation of $\mathcal{G}$ of dimension d_q. Then $Q_1^q, Q_2^q, \ldots$ are said to be a set of "irreducible tensor operators" of the irreducible representation $\boldsymbol{\Gamma}^q$ of $\mathcal{G}$.

Equations (5.34) are to be interpreted as operator equations, that is, both sides must produce the same result when acting on any function of the common domain in L^2 of the operators Q_j^q. Moreover, on the left-hand side of Equations (5.34) each operator acts on everything to its right.

Example II *The Hamiltonian operator as an irreducible tensor operator*

Let $\mathcal{G}$ be the group of the Schrödinger equation for some system. Then, from Equation (1.23), $P(T)H(\mathbf{r})P(T)^{-1} = H(\mathbf{r})$ for all $T \in \mathcal{G}$. Comparison with Equations (5.34) shows that $H(\mathbf{r})$ is an irreducible tensor operator for the one-dimensional identity representation of the group of the Schrödinger equation.

Example III *Differential operators as irreducible tensor operators of the crystallographic point group* D_4

For any rotation T,

$$\left.\begin{aligned}
P(T)\,\partial/\partial x P(T)^{-1} &= R(T)_{11}\,\partial/\partial x + R(T)_{21}\,\partial/\partial y + R(T)_{31}\,\partial/\partial z,\\
P(T)\,\partial/\partial y P(T)^{-1} &= R(T)_{12}\,\partial/\partial x + R(T)_{22}\,\partial/\partial y + R(T)_{32}\,\partial/\partial z,\\
P(T)\,\partial/\partial z P(T)^{-1} &= R(T)_{13}\,\partial/\partial x + R(T)_{23}\,\partial/\partial y + R(T)_{33}\,\partial/\partial z,
\end{aligned}\right\} \tag{5.35}$$

where $\mathbf{R}(T)$ is the 3×3 orthogonal matrix specifying T.

The first of Equations (5.35) will now be proved in some detail to illustrate the type of manipulation that is usually involved. For any differential function $f(\mathbf{r})$ of L^2, Equation (1.17) gives

$$P(T)^{-1}f(\mathbf{r}) = P(T^{-1})f(\mathbf{r}) = f(\mathbf{r}')$$

where $\mathbf{r}' = \mathbf{R}(T)\mathbf{r}$. Thus

$$\frac{\partial}{\partial x}\{P(T)^{-1}f(\mathbf{r})\} = \frac{\partial x'}{\partial x}\frac{\partial f(\mathbf{r}')}{\partial x'} + \frac{\partial y'}{\partial x}\frac{\partial f(\mathbf{r}')}{\partial y'} + \frac{\partial z'}{\partial x}\frac{\partial f(\mathbf{r}')}{\partial z'}$$

$$= R(T)_{11}\frac{\partial f(\mathbf{r}')}{\partial x'} + R(T)_{21}\frac{\partial f(\mathbf{r}')}{\partial y'} + R(T)_{31}\frac{\partial f(\mathbf{r}')}{\partial z'}, \tag{5.36}$$

on using Equation (1.2). Now define $h(\mathbf{r}) = \partial f(\mathbf{r})/\partial x$ and put $g(\mathbf{r}) = h(\mathbf{r}')$, where $\mathbf{r}' = \mathbf{R}(T)\mathbf{r}$. Then, by Equation (1.17),

$$P(T)\{\partial f(\mathbf{r}')/\partial x'\} = P(T)h(\mathbf{r}') = P(T)g(\mathbf{r}) = g(\mathbf{R}(T)^{-1}\mathbf{r}) = h(\mathbf{r}) = \partial f(\mathbf{r})/\partial x.$$

A similar argument applied to the second and third terms of Equation (5.36) then gives the first of Equations (5.35) immediately.

Inspection of the matrices $\mathbf{R}(T)$ for D_4 (see Example III of Chapter 1, Section 2) shows that $R(T)_{12} = R(T)_{21} = R(T)_{23} = R(T)_{32} = 0$ for all T of D_4 and $R(T)_{22} = \Gamma^3(T)_{11}$ $(=\chi^3(T))$, where $\mathbf{\Gamma}^3$ is the one-dimensional irreducible representation of D_4 given in the character table, Table 4.1. Thus, from Equations (5.35),

$$P(T)\,\partial/\partial y P(T)^{-1} = \Gamma^3(T)_{11}\,\partial/\partial y,$$

so that $\partial/\partial y$ is an irreducible tensor operator transforming as $\mathbf{\Gamma}^3$.

Inspection also shows that

$$\mathbf{\Gamma}^5(T) = \begin{bmatrix} R(T)_{11} & R(T)_{13} \\ R(T)_{31} & R(T)_{33} \end{bmatrix},$$

where $\mathbf{\Gamma}^5$ is the two-dimensional irreducible representation of D_4 (see Example I of Section 1). Thus Equations (5.35) show that Equation (5.34) is satisfied with $q = 5$ and $Q_1^5 = \partial/\partial x$, $Q_2^5 = \partial/\partial z$, so $\partial/\partial x$ and $\partial/\partial z$ are a set of irreducible tensor operators for $\mathbf{\Gamma}^5$.

Example IV *Multiplication by a basis function as an irreducible tensor operator*

Let $\psi_j^q(\mathbf{r})$, $j = 1, 2, \ldots, d_q$, be a set of basis functions for the irreducible representation $\mathbf{\Gamma}^q$, and define Q_j^q by

$$Q_j^q f(\mathbf{r}) = \psi_j^q(\mathbf{r}) f(\mathbf{r})$$

for $j = 1, 2, \ldots, d_q$, i.e. Q_j^q is the operation of multiplication by $\psi_j^q(\mathbf{r})$.

Then $Q_1^q, Q_2^q, \ldots$ are a set of irreducible tensor operators of Γ^q, for

$$\begin{aligned} P(T)Q_j^q P(T)^{-1} f(\mathbf{r}) &= P(T)[\psi_j^q(\mathbf{r})\{P(T)^{-1} f(\mathbf{r})\}] \\ &= \{P(T)\psi_j^q(\mathbf{r})\}\{P(T)P(T)^{-1} f(\mathbf{r})\} \\ &= \left\{\sum_{k=1}^{d_q} \Gamma^q(T)_{kj}\psi_k^q(\mathbf{r})\right\} f(\mathbf{r}) \\ &= \sum_{k=1}^{d_q} \Gamma^q(T)_{kj} Q_k^q f(\mathbf{r}). \end{aligned}$$

Theorem II *The Wigner–Eckart Theorem for a group of coordinate transformations in* $\mathbb{R}^3$

Let $\mathcal{G}$ be a group of coordinate transformations that is either a finite group or a compact Lie group. Let Γ^p, Γ^q and Γ^r be unitary irreducible representations of $\mathcal{G}$ of dimensions d_p, d_q and d_r respectively, and suppose that $\phi_j^p(\mathbf{r})$, $j = 1, 2, \ldots, d_p$, and $\psi_l^r(\mathbf{r})$, $l = 1, 2, \ldots, d_r$, are sets of basis functions for Γ^p and Γ^r respectively. Finally, let Q_k^q, $k = 1, 2, \ldots, d_q$, be a set of irreducible tensor operators of Γ^q. Then

$$(\psi_l^r, Q_k^q \phi_j^p) = \sum_{\alpha=1}^{n_{pq}^r} \begin{pmatrix} p & q \\ j & k \end{pmatrix}\!\!\left|\begin{matrix} r, & \alpha \\ l & \end{matrix}\right)^{\!*} (r\,|Q^q|\,p)_\alpha \tag{5.37}$$

for *all* $j = 1, 2, \ldots, d_p$, $k = 1, 2, \ldots, d_q$, and $l = 1, 2, \ldots, d_r$, where $(r\,|Q^q|\,p)_\alpha$ are a set of n_{pq}^r "reduced matrix elements" that are independent of j, k and l.

Proof The proof may be found in Appendix C, Section 6. It should be noted that it does *not* require the matrix of Clebsch–Gordan coefficients to be unitary.

The Wigner–Eckart Theorem provides both the most succinct and the most powerful expression in the whole field of application of group theory in physical problems. Indeed, most physical applications depend directly on it. It shows that the j, k, l dependence of the quantities $(\psi_l^r, Q_k^q \phi_j^p)$ is given completely by the Clebsch–Gordan coefficients. Moreover, the whole set of $d_p d_q d_r$ elements $(\psi_l^r, Q_k^q \phi_j^p)$ depend *only* on n_{pq}^r reduced matrix elements.

The theorem has been stated here for the case in which $\mathcal{G}$ is either a finite group or a compact Lie group. However, it may be generalized quite easily to any non-compact, semi-simple Lie group, both for the case in which the representations are finite-dimensional (Klimyk 1975) and the case in which they are unitary but infinite-dimensional (Klimyk 1971). Further generalization to unitary representations of non-semi-simple, non-compact Lie groups has also been achieved (Klimyk 1972). See also Agrawala (1980).

The actual definition of the reduced matrix elements is given in Equation (C.26), but in practice the simplest way of determining them is to find n^r_{pq} non-zero elements $(\psi^r_l, Q^q_k\phi^p_j)$ (either by direct evaluation or by fitting to experimental data) and then regard the n^r_{pq} equations (Equations (5.37)) in which these elements appear on the left-hand side as a set of simultaneous equations in $(r\,|\,Q^q\,|\,p)_\alpha$, $\alpha = 1, 2, \ldots, n^r_{pq}$.

The application of the Wigner–Eckart Theorem to a number of physical problems is described in detail in Chapter 6, particularly in Sections 2 and 3. Frequent use is also made of the following special case.

Theorem III If $\phi^p_1(\mathbf{r}), \phi^p_2(\mathbf{r}), \ldots$ and $\psi^q_1(\mathbf{r}), \psi^q_2(\mathbf{r}), \ldots$ are respectively basis functions for the unitary irreducible representations Γ^p and Γ^q of the group of the Schrödinger equation $\mathcal{G}$ that is either a finite group or a compact Lie group, and Γ^p and Γ^q are not equivalent if $p \neq q$ (but are identical if $p = q$), and if $H(\mathbf{r})$ is the Hamiltonian operator, then

$$(\phi^p_m, H\psi^q_n) = 0$$

unless $p = q$ and $m = n$. Moreover, if $p = q$ and $m = n$, then $(\phi^p_m, H\psi^p_m)$ is a constant independent of m.

Proof As noted in Example II above, $H(\mathbf{r})$ is an irreducible tensor operator of the one-dimensional identity representation Γ^1 of $\mathcal{G}$. The required result then follows immediately from the Wigner–Eckart Theorem on using Equations (5.32) and (5.33). Alternatively, this theorem may be proved by a simple generalization of the proof of Theorem V of Chapter 4, Section 5, of which it is an obvious extension.

4 The Wigner–Eckart Theorem generalized

It will now be shown how the developments of the previous section can be expressed in terms of the linear operators and carrier spaces first introduced in Chapter 4, Section 1, thereby enabling the theory to apply to any group $\mathcal{G}$ and not merely to groups of coordinate transformations in $\mathbb{R}^3$.

Suppose that the irreducible representations Γ^p and Γ^q of $\mathcal{G}$ have dimensions d_p and d_q and that ψ^p_j $(j = 1, 2, \ldots, d_p)$ and ψ^q_s $(s =$

$1, 2, \ldots, d_{\mathrm{q}}$) are ortho-normal bases for the two corresponding abstract inner product spaces V^{p} and V^{q}. A $d_{\mathrm{p}}d_{\mathrm{q}}$-dimensional "direct product space" $\mathrm{V}^{\mathrm{p}} \otimes \mathrm{V}^{\mathrm{q}}$ may be defined as the set of all quantities ϕ of the form

$$\phi = \sum_{j=1}^{d_{\mathrm{p}}} \sum_{s=1}^{d_{\mathrm{q}}} a_{js} \psi_j^{\mathrm{p}} \otimes \psi_s^{\mathrm{q}},$$

where a_{js} are a set of complex numbers. (This concept is developed in more detail in Appendix B, Section 7.) With an inner product in $\mathrm{V}^{\mathrm{p}} \otimes \mathrm{V}^{\mathrm{q}}$ defined by

$$(\phi, \psi) = \sum_{j=1}^{d_{\mathrm{p}}} \sum_{s=1}^{d_{\mathrm{q}}} a_{js}^* b_{js},$$

where

$$\psi = \sum_{j=1}^{d_{\mathrm{p}}} \sum_{s=1}^{d_{\mathrm{q}}} b_{js} \psi_j^{\mathrm{p}} \otimes \psi_s^{\mathrm{q}},$$

the products $\psi_j^{\mathrm{p}} \otimes \psi_s^{\mathrm{q}}$ for $j = 1, 2, \ldots, d_{\mathrm{p}}$, and $s = 1, 2, \ldots, d_{\mathrm{q}}$, form an ortho-normal basis for $\mathrm{V}^{\mathrm{p}} \otimes \mathrm{V}^{\mathrm{q}}$.

Now define the linear operators $\Phi^{\mathrm{p}}(\mathrm{T})$ and $\Phi^{\mathrm{q}}(\mathrm{T})$ for all $\mathrm{T} \in \mathcal{G}$ acting on the bases of V^{p} and V^{q} respectively by

$$\Phi^{\mathrm{p}}(\mathrm{T})\psi_j^{\mathrm{p}} = \sum_{k=1}^{d_{\mathrm{p}}} \Gamma^{\mathrm{p}}(\mathrm{T})_{kj} \psi_k^{\mathrm{p}}$$

for $j = 1, 2, \ldots, d_{\mathrm{p}}$, and

$$\Phi^{\mathrm{q}}(\mathrm{T})\psi_s^{\mathrm{q}} = \sum_{t=1}^{d_{\mathrm{q}}} \Gamma^{\mathrm{q}}(\mathrm{T})_{ts} \psi_t^{\mathrm{q}}$$

for $s = 1, 2, \ldots, d_{\mathrm{q}}$. These are essentially just Equations (4.1) embellished with extra indices, so $\Phi^{\mathrm{p}}(\mathrm{T})$ and $\Phi^{\mathrm{q}}(\mathrm{T})$ may be extended to the whole of V^{p} and V^{q} respectively. For each $\mathrm{T} \in \mathcal{G}$, a further linear operator $\Phi(\mathrm{T})$ acting on $\mathrm{V}^{\mathrm{p}} \otimes \mathrm{V}^{\mathrm{q}}$ may be defined by

$$\Phi(\mathrm{T})\{\psi_j^{\mathrm{p}} \otimes \psi_s^{\mathrm{q}}\} = \{\Phi^{\mathrm{p}}(\mathrm{T})\psi_j^{\mathrm{p}}\} \otimes \{\Phi^{\mathrm{q}}(\mathrm{T})\psi_s^{\mathrm{q}}\} \tag{5.38}$$

and again extended to the whole of $\mathrm{V}^{\mathrm{p}} \otimes \mathrm{V}^{\mathrm{q}}$, so that

$$\Phi(\mathrm{T})\{\psi_j^{\mathrm{p}} \otimes \psi_s^{\mathrm{q}}\} = \sum_{k=1}^{d_{\mathrm{p}}} \sum_{t=1}^{d_{\mathrm{q}}} (\boldsymbol{\Gamma}^{\mathrm{p}}(\mathrm{T}) \otimes \Gamma^{\mathrm{q}}(\mathrm{T}))_{kt,js} \{\psi_k^{\mathrm{p}} \otimes \psi_t^{\mathrm{q}}\} \tag{5.39}$$

for all $j = 1, 2, \ldots, d_{\mathrm{p}}$, and $s = 1, 2, \ldots, d_{\mathrm{q}}$. Thus the operators $\Phi(\mathrm{T})$ are the linear operators corresponding to the direct product representation $\boldsymbol{\Gamma}^{\mathrm{p}} \otimes \boldsymbol{\Gamma}^{\mathrm{q}}$ of $\mathcal{G}$.

As the Clebsch–Gordan coefficients $\left(\begin{smallmatrix} \mathrm{p} & \mathrm{q} \\ j & k \end{smallmatrix} \middle| \begin{smallmatrix} \mathrm{r} & \alpha \\ l & \end{smallmatrix}\right)$ are the matrix elements of a matrix **C** that completely reduces $\mathbf{\Gamma}^p \otimes \mathbf{\Gamma}^q$ (see Equation (5.16)), it follows that for $n_{pq}^r \neq 0$ the elements of $V^p \otimes V^q$ defined by

$$\theta_l^{r,\alpha} = \sum_{j=1}^{d_p} \sum_{k=1}^{d_q} \left(\begin{smallmatrix} p & q \\ j & k \end{smallmatrix} \middle| \begin{smallmatrix} r, & \alpha \\ l & \end{smallmatrix}\right) \psi_j^p \otimes \psi_k^q \tag{5.40}$$

satisfy

$$\Phi(T)\theta_l^{r,\alpha} = \sum_{u=1}^{d_r} \Gamma^r(T)_{ul}\theta_u^{r,\alpha} \tag{5.41}$$

for all $T \in \mathcal{G}$, $l = 1, 2, \ldots, d_r$, and $\alpha = 1, 2, \ldots, n_{pq}^r$. That is, again the Clebsch–Gordan coefficients give the appropriate linear combinations that form bases for the various irreducible representations of $\mathbf{\Gamma}^p \otimes \mathbf{\Gamma}^q$, the similarities between Equations (5.40) and (5.22) and between Equations (5.41) and (5.21) being particularly significant.

There is no difficulty in generalizing the derivation of the Clebsch–Gordan coefficients given in the previous section so as to make it apply to *any* finite group or compact Lie group $\mathcal{G}$. (All that is required is to redefine the projection operator $\mathcal{P}_{mn}^r$ to be $(d_r/g)\sum_{T\in\mathcal{G}} \Gamma^r(T)_{mn}^* \Phi(T)$ (where $\Phi(T)$ is defined in Equation (5.39)) for a finite group, with the obvious modification for a compact Lie Group.) In fact Equations (5.27) to (5.31) remain valid for any such groups.

(In comparing the developments of this section with those of Section 3, it must be observed that the products $\phi_j^p(\mathbf{r})\psi_s^q(\mathbf{r})$ of the basis functions $\phi_j^p(\mathbf{r})$ of $\mathbf{\Gamma}^p$ and $\psi_s^q(\mathbf{r})$ of $\mathbf{\Gamma}^q$ form a basis for a $d_p d_q$-dimensional subspace of L^2 *only* if they are linearly independent, and even then these products do *not* necessarily form an ortho-normal set with respect to the usual inner product of L^2. By contrast, the products $\psi_j^p \otimes \psi_s^q$ of basis vectors ψ_j^p and ψ_s^q of V^p and V^q *always* form an ortho-normal basis of $V^p \otimes V^q$ with the inner product defined as above. Thus for basis functions in general one *cannot* identify $\phi_j^p(\mathbf{r}) \otimes \psi_s^q(\mathbf{r})$ with $\phi_j^p(\mathbf{r})\psi_s^q(\mathbf{r})$ and at the same time take the inner product of $V^p \otimes V^q$ to be that of L^2.)

To proceed further it is necessary to redefine the concept of a set of irreducible tensor operators. To this end let Q be a linear mapping of V^p into V^r (V^p and V^r being carrier spaces for the irreducible representations $\mathbf{\Gamma}^p$ and $\mathbf{\Gamma}^r$ of $\mathcal{G}$) so that $Q\psi \in V^r$ for all $\psi \in V^p$. Defining the sum $(Q_1 + Q_2)$ of two such operators Q_1 and Q_2 by $(Q_1 + Q_2)\psi = Q_1\psi + Q_2\psi$ (for all $\psi \in V^p$), the product aQ for any complex number a by $(aQ)\psi = a(Q\psi)$ (for all $\psi \in V^p$), and the "zero" mapping O by $O\psi = 0$ (for all $\psi \in V^p$, where the 0 on the right-hand side here is the "zero" element of V^r), it follows that the set of all linear mappings Q from V^p to V^r form a vector space, which will be denoted by $L(V^p, V^r)$. If V^p and V^r are of dimensions d_p and d_r respectively, then $L(V^p, V^r)$ is of dimension $d_p d_r$ (Shepherd 1966).

Now define for each $T \in \mathscr{G}$ an operator $\Phi'(T)$ acting on $L(V^p, V^r)$ by

$$\Phi'(T)Q = \Phi^r(T)Q\Phi^p(T)^{-1}$$

for all $Q \in L(V^p, V^r)$, $\Phi^p(T)$ and $\Phi^r(T)$ the operators acting in V^p and V^r belonging (in the manner described above) to the irreducible representations Γ^p and Γ^r. Then $\Phi'(T)$ is a linear operator, and for any $T_1, T_2 \in \mathscr{G}$

$$\Phi'(T_1)\Phi'(T_2) = \Phi'(T_1T_2),$$

so that the set of operators $\Phi'(T)$ correspond to a representation of $\mathscr{G}$ for which the carrier space is $L(V^p, V^r)$. (The proof of this statement is as follows. For any $T_1, T_2 \in \mathscr{G}$ and any $Q \in L(V^p, V^r)$,

$$\begin{aligned}\Phi'(T_1)\Phi'(T_2)Q &= \Phi^r(T_1)\{\Phi^r(T_2)Q\Phi^p(T_2)^{-1}\}\Phi^p(T_1)^{-1}\\ &= \Phi^r(T_1T_2)Q\Phi^p(T_1T_2)^{-1}\\ &= \Phi'(T_1T_2)Q.)\end{aligned}$$

Let Γ' be the representation of $\mathscr{G}$ for which the operators $\Phi'(T)$ and the carrier space $L(V^p, V^r)$ form a module. That is, if Q_1, $Q_2, \ldots$ are a basis of the vector space $L(V^p, V^r)$, the matrix elements $\Gamma'(T)_{mn}$ are defined by

$$\Phi'(T)Q_n = \sum_{m=1}^{d_pd_r} \Gamma'(T)_{mn} Q_m$$

for all $T \in \mathscr{G}$. In general Γ' is reducible. Suppose that Γ' is completely reducible and that Γ^q is an irreducible representation that appears in its reduction, and let $Q_1^q, Q_2^q, \ldots$ be a basis for the corresponding subspace of $L(V^p, V^r)$. Then

$$\Phi'(T)Q_n^q = \sum_{m=1}^{d_q} \Gamma^q(T)_{mn} Q_m^q$$

for $n = 1, 2, \ldots, d_q$ and all $T \in \mathscr{G}$. That is, by the definition of $\Phi'(T)$,

$$\Phi^r(T)Q_n^q\Phi^p(T)^{-1} = \sum_{m=1}^{d_q} \Gamma^q(T)_{mn} Q_m^q \tag{5.42}$$

for $n = 1, 2, \ldots, d_q$, and all $T \in \mathscr{G}$. This set of operators will be called "irreducible tensor operators of the irreducible representation Γ^q of $\mathscr{G}$".

Theorem I *The generalized Wigner–Eckart Theorem*
Let $\mathscr{G}$ be a finite group or a compact Lie group. Let Γ^p, Γ^q and Γ^r be

unitary irreducible representations of $\mathcal{G}$ of dimensions d_p, d_q and d_r respectively, and suppose that ϕ_j^p $(j=1,2,\ldots,d_p)$ and ψ_l^r $(l=1,2,\ldots,d_r)$ are basis vectors of ortho-normal bases of the carrier spaces V^p and V^r of Γ^p and Γ^r respectively. Finally, let Q_k^q $(k=1,2,\ldots,d_q)$ be a set of irreducible tensor operators of Γ^q, defined as above. Then

$$(\psi_l^r, Q_k^q\phi_j^p) = \sum_{\alpha=1}^{n_{pq}^r} \begin{pmatrix} p & q \\ j & k \end{pmatrix}\begin{vmatrix} r, & \alpha \\ l & \end{vmatrix}^* (r\,|Q^q|\,p)_\alpha \tag{5.43}$$

for *all* $j=1,2,\ldots,d_p$, $k=1,2,\ldots,d_q$ and $l=1,2,\ldots,d_r$, where $(r\,|Q^q|\,p)_\alpha$ are a set of n_{pq}^r "reduced matrix elements" that are independent of j, k and l.

Proof As the operators $\Phi^r(T)$ are unitary (see Chapter 4, Section 3)

$$\begin{aligned}(\psi_l^r, Q_k^q\phi_j^p) &= (\Phi^r(T)\psi_l^r, \Phi^r(T)Q_k^q\phi_j^p)\\ &= (\Phi^r(T)\psi_l^r, \{\Phi^r(T)Q_k^q\Phi^p(T)^{-1}\}\Phi^p(T)\phi_j^p)\\ &= \sum_{s=1}^{d_p}\sum_{t=1}^{d_q}\sum_{u=1}^{d_r} \Gamma^p(T)_{sj}\Gamma^q(T)_{tk}\Gamma^r(T)_{ul}^*(\psi_u^r, Q_t^q\phi_s^p).\end{aligned}$$

The remainder of the proof is then exactly as for the case in which $\mathcal{G}$ is a group of coordinate transformations in $\mathbb{R}^3$ (see Appendix C, Section 6).

It should be noted that the appropriate inner product on the left-hand side of Equation (5.43) is that of V^r, as ψ_l^r and $Q_k^q\phi_j^p$ are both members of V^r. The remarks made in Section 3 about the Wigner–Eckart Theorem for a group of transformations in $\mathbb{R}^3$ apply equally to the theorem as generalized above. In particular, although the theorem is stated and proved here for the case in which $\mathcal{G}$ is a finite group or compact Lie group, the conclusion is valid much more generally. A detailed discussion of the range of validity has been given by Agrawala (1980).

5 Representations of direct product groups

The concept of direct product groups was discussed in some detail in Chapter 2, Section 7. In studies of their representations it is most convenient (and quite sufficient) to revert to the original formulation in terms of pairs.

Theorem I Let $\boldsymbol{\Gamma}_1$ and $\boldsymbol{\Gamma}_2$ be representations of $\mathcal{G}_1$ and $\mathcal{G}_2$ respectively. Then the set of matrices $\boldsymbol{\Gamma}((T_1, T_2))$ defined for all $T_1 \in \mathcal{G}_1$ and $T_2 \in \mathcal{G}_2$ by

$$\boldsymbol{\Gamma}((T_1, T_2)) = \boldsymbol{\Gamma}_1(T_1) \otimes \boldsymbol{\Gamma}_2(T_2) \tag{5.44}$$

provides a representation of $\mathcal{G}_1 \otimes \mathcal{G}_2$. This representation of $\mathcal{G}_1 \otimes \mathcal{G}_2$ is unitary if $\boldsymbol{\Gamma}_1$ and $\boldsymbol{\Gamma}_2$ are unitary representations and is faithful if $\boldsymbol{\Gamma}_1$ and $\boldsymbol{\Gamma}_2$ are faithful representations.

Proof For any $T_1, T_1' \in \mathcal{G}_1$ and any $T_2, T_2' \in \mathcal{G}_2$, from Equation (5.44),

$$\begin{aligned}\boldsymbol{\Gamma}((T_1, T_2))\boldsymbol{\Gamma}((T_1', T_2')) &= \{\boldsymbol{\Gamma}_1(T_1) \otimes \boldsymbol{\Gamma}_2(T_2)\}\{\boldsymbol{\Gamma}_1(T_1') \otimes \boldsymbol{\Gamma}_2(T_2')\} \\ &= \{\boldsymbol{\Gamma}_1(T_1)\boldsymbol{\Gamma}_1(T_1')\} \otimes \{\boldsymbol{\Gamma}_2(T_2)\boldsymbol{\Gamma}_2(T_2')\}\end{aligned}$$

(on using Equation (A.9))

$$= \boldsymbol{\Gamma}_1(T_1T_1') \otimes \boldsymbol{\Gamma}_2(T_2T_2')$$

(as $\boldsymbol{\Gamma}_1$ and $\boldsymbol{\Gamma}_2$ are assumed representations of $\mathcal{G}_1$ and $\mathcal{G}_2$ respectively)

$$\begin{aligned}&= \boldsymbol{\Gamma}((T_1T_1', T_2T_2')) \\ &= \boldsymbol{\Gamma}((T_1, T_2)(T_1', T_2'))\end{aligned}$$

(on using Equation (2.9)). Consequently the matrices $\boldsymbol{\Gamma}$ of Equation (5.44) form a representation of $\mathcal{G}_1 \otimes \mathcal{G}_2$. The unitary property follows from the fact that the direct product of two unitary matrices is itself unitary (see Appendix A, Section 1), while the faithful property is obvious.

This theorem allows the nature of $\mathcal{G}_1 \otimes \mathcal{G}_2$ to be investigated when $\mathcal{G}_1$ and $\mathcal{G}_2$ are finite groups or linear Lie groups. There are essentially three distinct cases:

(i) $\mathcal{G}_1$ and $\mathcal{G}_2$ are both finite groups.
In this case clearly $\mathcal{G}_1 \otimes \mathcal{G}_2$ is a finite group whose order is the product of those of $\mathcal{G}_1$ and $\mathcal{G}_2$ separately.

(ii) $\mathcal{G}_1$ is a finite group and $\mathcal{G}_2$ is a linear Lie group.
Suppose that $\mathcal{G}_1$ has order g_1 and has a faithful finite-dimensional representation $\boldsymbol{\Gamma}_1$. Suppose that $\mathcal{G}_2$ has a faithful finite-dimensional representation $\boldsymbol{\Gamma}_2$, that the elements of $\mathcal{G}_2$ near the identity are specified by n real parameters $x_1, x_2, \ldots, x_n$, and that $\mathcal{G}_2$ has N connected components. Then the faithful finite-dimensional representation of Equation (5.44) can be used to show that $\mathcal{G}_1 \otimes \mathcal{G}_2$ is a linear Lie group with Ng_1 connected components, whose connected subgroup is isomorphic

to the set of matrices $\boldsymbol{\Gamma}_1(\mathrm{E}_1)\otimes\boldsymbol{\Gamma}_2(\mathrm{T}_2)$ for all T_2 of the connected subgroup of $\mathcal{G}_2$. Moreover, the elements of $\mathcal{G}_1\otimes\mathcal{G}_2$ near the identity of $\mathcal{G}_1\otimes\mathcal{G}_2$ may be specified by the same n real parameters as for $\mathcal{G}_2$. As the "invariant integral" of $\mathcal{G}_1\otimes\mathcal{G}_2$ involves an integral over n variables with the same weight function as for $\mathcal{G}_2$ and a sum over the Ng_1 connected components, it is obvious that $\mathcal{G}_1\otimes\mathcal{G}_2$ is compact if and only if $\mathcal{G}_2$ is compact.

(iii) $\mathcal{G}_1$ and $\mathcal{G}_2$ are both linear Lie groups.
Suppose that $\boldsymbol{\Gamma}_j$ is a faithful finite-dimensional representation of $\mathcal{G}_j$, that the elements of $\mathcal{G}_j$ near the identity of $\mathcal{G}_j$ are specified by n_j real parameters, and that $\mathcal{G}_j$ has N_j components $(j=1,2)$. Then the faithful finite-dimensional representation (Equation (5.44)) of $\mathcal{G}_1\otimes\mathcal{G}_2$ can be employed to prove that $\mathcal{G}_1\otimes\mathcal{G}_2$ is a linear Lie group with N_1N_2 connected components and that the elements of $\mathcal{G}_1\otimes\mathcal{G}_2$ near the identity of $\mathcal{G}_1\otimes\mathcal{G}_2$ are specified by (n_1+n_2) real parameters. The "invariant integral" of $\mathcal{G}_1\otimes\mathcal{G}_2$ therefore involves an integral over (n_1+n_2) variables (whose weight function is the product of those of $\mathcal{G}_1$ and $\mathcal{G}_2$ separately) and a sum over the N_1N_2 components, so that $\mathcal{G}_1\otimes\mathcal{G}_2$ is compact if and only if $\mathcal{G}_1$ and $\mathcal{G}_2$ are both compact.

Theorem II If $\mathcal{G}_1\otimes\mathcal{G}_2$ is a finite group or a compact linear Lie group and $\boldsymbol{\Gamma}_1$ and $\boldsymbol{\Gamma}_2$ are irreducible representations of $\mathcal{G}_1$ and $\mathcal{G}_2$ respectively, then the representation $\boldsymbol{\Gamma}'$ defined by Equation (5.44) is an *irreducible* representation of $\mathcal{G}_1\otimes\mathcal{G}_2$. Moreover, *every* irreducible representation of $\mathcal{G}_1\otimes\mathcal{G}_2$ is equivalent to a representation constructed in this way.

Proof This may be found in Appendix C, Section 7.

It is interesting to view these developments in terms of modules (along the lines of the previous section). For $\beta=1,2$, let $\psi_{j,\beta}$ $(j=1,2,\ldots,d_\beta)$ be an ortho-normal basis for the carrier space V_β of the d_β-dimensional representation $\boldsymbol{\Gamma}_\beta$ of $\mathcal{G}_\beta$ and let $\Phi_\beta(\mathrm{T}_\beta)$ be the linear operator corresponding to the element $\mathrm{T}_\beta\in\mathcal{G}_\beta$ that acts on V_β, so that

$$\Phi_\beta(\mathrm{T}_\beta)\psi_{j,\beta}=\sum_{k=1}^{d_\beta}\Gamma_\beta(\mathrm{T}_\beta)_{kj}\psi_{k,\beta}$$

for $j=1,2,\ldots,d_\beta$. Then, with the operator $\Phi((\mathrm{T}_1,\mathrm{T}_2))$ defined by

$$\Phi((\mathrm{T}_1,\mathrm{T}_2))\{\psi_{j,1}\otimes\psi_{s,2}\}=\{\Phi_1(\mathrm{T}_1)\psi_{j,1}\}\otimes\{\Phi_2(\mathrm{T}_2)\psi_{s,2}\}$$

and its extension to the whole of $V_1 \otimes V_2$,

$$\Phi((T_1, T_2))\{\psi_{j,1} \otimes \psi_{s,2}\} = \sum_{k=1}^{d_1} \sum_{t=1}^{d_2} (\mathbf{\Gamma}_1(T_1) \otimes \mathbf{\Gamma}_2(T_2))_{kt,js}\{\psi_{k,1} \otimes \psi_{t,2}\} \tag{5.45}$$

for all $j = 1, 2, \ldots, d_1$, $s = 1, 2, \ldots, d_2$, and $(T_1, T_2) \in \mathcal{G}_1 \otimes \mathcal{G}_2$. Thus the operators $\Phi((T_1, T_2))$ are the linear operators corresponding to the direct product representation $\mathbf{\Gamma}_1 \otimes \mathbf{\Gamma}_2$ of $\mathcal{G}_1 \otimes \mathcal{G}_2$ and they act on the carrier space $V_1 \otimes V_2$.

These results are particularly important in the special case in which $\mathcal{G}_1$ and $\mathcal{G}_2$ are both isomorphic to the *same* group $\mathcal{G}$. The essential observation (made in Chapter 2, Section 7) is that $\mathcal{G} \otimes \mathcal{G}$ possess a "diagonal subgroup" isomorphic to $\mathcal{G}$ consisting of the pairs (T, T), where $T \in \mathcal{G}$. Then, if $\mathbf{\Gamma}^p$ and $\mathbf{\Gamma}^q$ are any two irreducible representations of $\mathcal{G}$, the above theorems show that Equation (5.44), which can be written in this case as

$$\mathbf{\Gamma}((T_1, T_2)) = \mathbf{\Gamma}^p(T_1) \otimes \mathbf{\Gamma}^q(T_2)$$

for all $T_1, T_2 \in \mathcal{G}$, is an irreducible representation of $\mathcal{G} \otimes \mathcal{G}$. On the diagonal subgroup this provides the representation

$$\mathbf{\Gamma}((T, T)) = \mathbf{\Gamma}^p(T) \otimes \mathbf{\Gamma}^q(T) \tag{5.46}$$

of $\mathcal{G}$, which is clearly identical to the direct product representation of Equation (5.13). In general this representation (Equation (5.46)) of $\mathcal{G}$ is reducible. Thus, with $\mathbf{\Gamma}_1 = \Gamma^p$ and $\mathbf{\Gamma}_2 = \mathbf{\Gamma}^q$, and with ψ_j^p and ψ_s^q denoting the elements of the carrier spaces V^p and V^q corresponding to $\mathbf{\Gamma}^p$ and $\mathbf{\Gamma}^q$ respectively, Equation (5.45) becomes

$$\Phi((T, T))\{\psi_j^p \otimes \psi_s^q\} = \sum_{k=1}^{d_p} \sum_{t=1}^{d_q} (\mathbf{\Gamma}^p(T) \otimes \mathbf{\Gamma}^q(T))_{kt,js}\{\psi_k^p \otimes \psi_t^q\}.$$

As this is exactly the same as Equation (5.39), but with $\Phi(T)$ replaced by $\Phi((T, T))$, it follows that Equation (5.41) becomes

$$\Phi((T, T))\theta_l^{r,\alpha} = \sum_{u=1}^{d_r} \Gamma^r(T)_{ul}\theta_u^{r,\alpha} \tag{5.47}$$

for all $T \in \mathcal{G}$, $l = 1, 2, \ldots, d_r$, and $\alpha = 1, 2, \ldots, n_{pq}^r$, $\theta_l^{r,\alpha}$ being given by Equation (5.40) as linear combinations of products determined by the Clebsch–Gordan coefficients.

It is essentially through the Lie algebraic extension of this formulation that the Clebsch–Gordan coefficients appear in the quantum theory of angular momentum (see Chapter 12, Section 5).

6 Irreducible representations of finite Abelian groups

The irreducible representations of every *finite Abelian* group $\mathcal{G}$ may now be found very easily. It should be recalled that Theorem III of Chapter 4, Section 5 shows that these representations must be one-dimensional, and as every representation of a finite group is equivalent to a unitary representation (equivalence implying identity for one-dimensional representations), all these irreducible representations are automatically unitary. Moreover, Theorem VIII of Chapter 4, Section 6 and Theorem II of Chapter 2, Section 2 together imply that the number of inequivalent irreducible representations of $\mathcal{G}$ is equal to the order of $\mathcal{G}$.

The first stage is to consider a special type of Abelian group.

Definition *Cyclic group*
A group is said to be "cyclic" if every element can be expressed as a power of a single element.

The most general form of a finite cyclic group of order g is therefore $\{E, T, T^2, \ldots, T^{g-1}\}$, with $T^g = E$, the element T being called the "generator" of $\mathcal{G}$. Obviously every cyclic group is Abelian. It is easily shown that all cyclic groups of the same order are isomorphic.

Theorem I The set of all unitary irreducible representations of a cyclic group of order g is given by

$$\Gamma^p(T^m) = [\exp\{2\pi i m(p-1)/g\}] \tag{5.48}$$

for $m = 1, 2, \ldots, g$. Here p takes values $p = 1, 2, \ldots, g$, and T is the generator of $\mathcal{G}$.

Proof Suppose that Γ is an irreducible representation and $\Gamma(T) = [\gamma]$, where γ is some complex number. Then $\gamma^g = 1$ as $T^g = E$, so γ can take any of the g possible values $\gamma = \exp\{2\pi i(p-1)/g\}$, where $p = 1, 2, \ldots, g$. These g values of γ then give the g inequivalent irreducible representations, which may therefore be labelled by p. As $\Gamma^p(T^m) = \{\Gamma^p(T)\}^m = [\gamma^m]$, Equation (5.48) follows immediately.

The factor $2\pi i m(p-1)/g$ has been introduced in Equation (5.48) instead of the factor $2\pi i m p/g$ simply to ensure conformity with the usual convention that Γ^1 is the identity representation.

The following theorem shows that any finite Abelian group is made up of cyclic groups and so enables all its irreducible representations to be calculated immediately.

Theorem II Every finite Abelian group is either a finite cyclic group or is isomorphic to a direct product of a set of finite cyclic groups.

Proof See, for example, Rotman (1965) (pages 58 to 62).

Example III *Irreducible representations of groups isomorphic to* $C_r \otimes C_s$, C_r *and* C_s *being cyclic groups of order r and s respectively*

Let Γ_1^p be an irreducible representation of C_r, so that, by Equation (5.48), $\Gamma_1^p(T_1^m) = [\exp\{2\pi i m(p-1)/r\}]$, where T_1 is the generator of C_r. Similarly, let Γ_2^q be an irreducible representation of C_s, with generator T_2, so that $\Gamma_2^q(T_2^n) = [\exp\{2\pi i n(q-1)/s\}]$. Then, by the theorems of the previous section, the irreducible representations of every group isomorphic to $C_r \otimes C_s$ may be labelled by a pair (p, q), where $p = 1, 2, \dots, r$, and $q = 1, 2, \dots, s$, and, from Equation (5.44),

$$\Gamma^{p,q}((T_1^m, T_2^n)) = [\exp 2\pi i(\{m(p-1)/r\} + \{n(q-1)/s\})]$$

for all $m = 1, 2, \dots, r$ and $n = 1, 2, \dots, s$.

The crystallographic point group D_2 (see Appendix D) is an example having this structure, as it is isomorphic to $C_2 \otimes C_2$.

7 Induced representations

The method of "induction" provides a very powerful technique for constructing representations of a group from representations of its subgroups. It will be described here for the case in which the group is finite and the results obtained will be applied in Chapter 9 to the crystallographic space groups. However, the technique is not restricted to finite groups. Indeed, one of the most significant developments of the last few years has been the generalization to arbitrary, locally compact topological groups, including particularly Lie groups. This development has been largely pioneered by the work of Mackey (1963b, 1968, 1976). It has proved extremely valuable in the construction of the infinite-dimensional unitary representations of non-compact, semi-simple Lie groups (Stein 1965, Lipsman 1974, Barut and Raczka 1977), thereby putting into a general context the original work on the homogeneous Lorentz group (Gel'fand *et al.* 1963, Naimark 1957, 1964). Other physically important non-compact Lie groups that are particularly well suited to treatment by the induced representation method include the Poincaré group (Wigner 1939, Bertrand 1966, Halpern 1968, Niederer and O'Raifeartaigh 1974a,b), the Galilei group (Inönü and Wigner 1952, Voisin 1965a,b, 1966, Brennich 1970, Neiderer and O'Raifeartaigh 1974a) and the Euclidean group of $\mathbb{R}^3$ (Miller 1964, Niederer and O'Raifeartaigh 1974a).

Most of the results to be derived in this section for finite groups carry over to the general case of locally compact topological groups with their group theoretical content essentially unchanged. The complications of the general case lie in the measure theoretic questions involved, together with the fact that nearly all the representations that appear are infinite-dimensional. Coleman (1968) has given a very readable introduction to these matters.

The basic theorem on induced representation is easily stated and proved:

Theorem I Let $\mathcal{S}$ be a subgroup, of order s, of a group $\mathcal{G}$ of order g, and let $T_1, T_2, \ldots$ be a set of M (=g/s) coset representatives for the decomposition of $\mathcal{G}$ into right cosets with respect to $\mathcal{S}$. Let $\boldsymbol{\Delta}$ be a d-dimensional unitary representation of $\mathcal{S}$. Then the set of $Md \times Md$ matrices $\boldsymbol{\Gamma}(T)$, defined for all $T \in \mathcal{G}$ by

$$\Gamma(T)_{kt,jr} = \begin{cases} \Delta(T_k T T_j^{-1})_{tr}, & \text{if } T_k T T_j^{-1} \in \mathcal{S}, \\ 0, & \text{if } T_k T T_j^{-1} \notin \mathcal{S}, \end{cases} \tag{5.49}$$

provide an Md-dimensional unitary representation of $\mathcal{G}$. If $\psi(S)$ are the characters of the representation $\boldsymbol{\Delta}$ of $\mathcal{S}$, then the characters $\chi(T)$ of the representation $\boldsymbol{\Gamma}$ of $\mathcal{G}$ are given by

$$\chi(T) = \sum_j \psi(T_j T T_j^{-1}), \tag{5.50}$$

where the sum is over all coset representatives T_j such that $T_j T T_j^{-1} \in \mathcal{S}$.

Proof As the proof is by direct verification it is relegated to Appendix C, Section 8.

This representation $\boldsymbol{\Gamma}$ of $\mathcal{G}$ is said to be "induced" from the representation $\boldsymbol{\Delta}$ of the subgroup $\mathcal{S}$, this being indicated by writing

$$\boldsymbol{\Gamma} = \boldsymbol{\Delta}(\mathcal{S}) \uparrow \mathcal{G}.$$

In Equations (5.49) the rows and the columns of $\boldsymbol{\Gamma}(T)$ are each separately labelled by a *pair* of indices, exactly as in the theory of direct product representations (see Section 2 and Appendix A, Section 1). The theorem (and proof) is also valid when $\mathcal{G}$ and $\mathcal{S}$ are compact Lie groups such that the decomposition of $\mathcal{G}$ into right cosets with respect to $\mathcal{S}$ contains only a *finite* number M of distinct cosets.

The following rather trivial example is included to demonstrate that in general $\mathbf{\Gamma}$ is a *reducible* representation of $\mathcal{G}$ even if $\mathbf{\Delta}$ is an *irreducible* representation of $\mathcal{S}$.

Example I *The regular representation as an induced representation*

Let $\mathcal{G}$ be a finite group of order g and choose $\mathcal{S}$ to consist only of the identity E. Let $\mathbf{\Delta}$ be the one-dimensional irreducible representation of $\mathcal{S}$ specified by $\mathbf{\Delta}(E)=[1]$. The set of coset representatives $T_1, T_2, \ldots$ comprises the whole of $\mathcal{G}$ and $T_j T T_j^{-1} \in \mathcal{S}(=\{E\})$ if and only if $T=E$. Thus Equation (5.50) gives

$$\chi(T)=\begin{cases} g, & \text{if} \quad T=E, \\ 0, & \text{if} \quad T\neq E. \end{cases}$$

Comparison with Equation (C.17) shows that $\mathbf{\Gamma}$ must be equivalent to the regular representation $\mathbf{\Gamma}^{\text{reg}}$ defined in Equation (C.16), which was shown to be reducible (for $g>1$).

However, for one physically important type of group the induced representation method not only produces irreducible representations of the group, but it generates the *whole* set of such representations. This satisfactory situation occurs when $\mathcal{G}$ has the *semi-direct product* structure $\mathcal{A} ⓢ \mathcal{B}$ and the invariant subgroup $\mathcal{A}$ is *Abelian* (see Chapter 2, Section 7). Physically important groups with this structure include the Euclidean group of $\mathbb{R}^3$ and the Poincaré group (see Examples IV and V of Chapter 2, Section 7), the linear infinite point groups $D_{\infty h}$ and $C_{\infty h}$ (see Example VI of Chapter 2, Section 7), together with the Galilei groups and the symmorphic crystallographic space groups (see Chapter 9). Of these only the latter are finite but all the results to be described can be generalized easily to the other groups. The construction of the unitary irreducible representations of $\mathcal{G}$ involves a number of stages which will now be described in detail. It will be assumed that the orders of $\mathcal{G}$, $\mathcal{A}$ and $\mathcal{B}$ are g, a and b respectively, so that $g=ab$.

(*a*) As $\mathcal{A}$ is Abelian it possesses a inequivalent irreducible representations, all of which are one-dimensional and therefore completely specified by their characters. Let these characters be denoted by $\chi^q_{\mathcal{A}}(A)$, $q=1,2,\ldots,a$, for all $A\in\mathcal{A}$.

(*b*) Let $\mathcal{B}(q)$ be the subset of elements B of $\mathcal{B}$ such that

$$\chi^q_{\mathcal{A}}(BAB^{-1})=\chi^q_{\mathcal{A}}(A) \tag{5.51}$$

for all $A\in\mathcal{A}$. Then $\mathcal{B}(q)$ is a subgroup of B, $\mathcal{B}(q)$ is called a "little group". (As $\mathcal{A}$ is an invariant subgroup of $\mathcal{G}$, $BAB^{-1}\in\mathcal{A}$ for all $A\in\mathcal{A}$ and all $B\in\mathcal{B}$, so $\mathcal{B}(q)$ is well defined. The subgroup property follows because, if B and B′ are members of $\mathscr{B}(q)$, then for any $A \in \mathscr{A}$, from Equation (5.51),

$$\chi^q_{\mathscr{A}}((B'B^{-1})A(B'B^{-1})^{-1}) = \chi^q_{\mathscr{A}}(B^{-1}AB) = \chi^q_{\mathscr{A}}(A),$$

so that $B'B^{-1}$ is also a member of $\mathscr{B}(q)$.) Let $b(q)$ be the order of $\mathscr{B}(q)$.

(*c*) Let $B_1, B_2, \ldots$ be the set of $M(q)$ $(=b/b(q))$ coset representatives for the decomposition of $\mathscr{B}$ into right cosets with respect to $\mathscr{B}(q)$.

(*d*) For each $B \in \mathscr{B}$ define the quantities $\chi^{B(q)}_{\mathscr{A}}(A)$ for all $A \in \mathscr{A}$ by

$$\chi^{B(q)}_{\mathscr{A}}(A) = \chi^q_{\mathscr{A}}(BAB^{-1}). \tag{5.52}$$

Then, for each fixed B, the set $\chi^{B(q)}_{\mathscr{A}}(A)$ are a set of characters of a one-dimensional irreducible representation of $\mathscr{A}$, so that $B(q)$ is an integer in the set $1, 2, \ldots, a$. (That $\chi^{B(q)}_{\mathscr{A}}(A)$ are such characters can be demonstrated as follows. Let A and A′ be any two elements of $\mathscr{A}$. Then

$$\begin{aligned}\chi^{B(q)}_{\mathscr{A}}(A)\chi^{B(q)}_{\mathscr{A}}(A') &= \chi^q_{\mathscr{A}}(BAB^{-1})\chi^q_{\mathscr{A}}(BA'B^{-1})\\ &= \chi^q_{\mathscr{A}}((BAB^{-1})(BA'B^{-1}))\\ &= \chi^q_{\mathscr{A}}(B(AA')B^{-1})\\ &= \chi^{B(q)}_{\mathscr{A}}(AA').)\end{aligned}$$

(*e*) Obviously Equations (5.51) and (5.52) imply that $B(q) = q$ for all $B \in \mathscr{B}(q)$. More generally, $B(q) = B_j(q)$ for every $B \in \mathscr{B}$ belonging to the right coset $\mathscr{B}(q)B_j$. (If $B \in \mathscr{B}(q)B_j$ then there exists an element B′ of $\mathscr{B}(q)$ such that $B = B'B_j$. Then for any $A \in \mathscr{A}$

$$\begin{aligned}\chi^{B(q)}_{\mathscr{A}}(A) &= \chi^q_{\mathscr{A}}((B'B_j)A(B'B_j)^{-1}) && \text{(by Equation (5.52))}\\ &= \chi^q_{\mathscr{A}}(B_jAB_j^{-1}) && \text{(by Equation (5.51))}\\ &= \chi^{B_j(q)}_{\mathscr{A}}(A) && \text{(by Equation (5.52)).}\end{aligned}$$

(*f*) The set of $M(q)(=b/b(q))$ integers $\{q(=B_1(q)), B_2(q), B_3(q), \ldots\}$ is known as the "orbit" of q.

(*g*) The groups $\mathscr{B}(B_j(q))$ are all isomorphic to $\mathscr{B}(q)$ for $j = 1, 2, \ldots, M(q)$. That is, all members of the orbit of q are associated with essentially the same group $\mathscr{B}(q)$. (Equation (5.51) implies that $\mathscr{B}(B_j(q))$ is the subgroup of $\mathscr{B}$ consisting of all $B \in \mathscr{B}$ such that

$$\chi^{B_j(q)}_{\mathscr{A}}(BAB^{-1}) = \chi^{B_j(q)}_{\mathscr{A}}(A)$$

for all $A \in \mathscr{A}$. By Equation (5.52) this can be rewritten as

$$\chi^q_{\mathscr{A}}(B_jBAB^{-1}B_j^{-1}) = \chi^q_{\mathscr{A}}(B_jAB_j^{-1}) \tag{5.53}$$

for all $A \in \mathscr{A}$. Now consider the automorphic mapping ϕ_j of $\mathscr{G}$ onto itself defined by $\phi_j(T) = B_jTB_j^{-1}$. As $\mathscr{A}$ is an invariant subgroup of

$\mathscr{G}$, this provides a one-to-one mapping of $\mathscr{A}$ onto itself. Consequently, let A′ be any element of $\mathscr{A}$ and let $A = B_j^{-1}A'B_j$. Then Equation (5.53) can be further rewritten as

$$\chi^q_{\mathscr{A}}((B_jBB_j^{-1})A'(B_jBB_j^{-1})^{-1}) = \chi^q_{\mathscr{A}}(A')$$

for all $A' \in \mathscr{A}$. Thus $\phi_j(B) = B_jBB_j^{-1}$ maps $\mathscr{B}(B_j(q))$ onto $\mathscr{B}(q)$ and, as ϕ_j is an isomorphic mapping, $\mathscr{B}(q)$ is isomorphic to $\mathscr{B}(B_j(q))$.)

(*h*) Let $\mathscr{S}(q)$ be the set of all products AB, where $A \in \mathscr{A}$ and $B \in \mathscr{B}(q)$. Then $\mathscr{S}(q)$ is a subgroup of $\mathscr{G}$ with the semi-direct product structure $\mathscr{A} \circledS \mathscr{B}(q)$. (If $A, A' \in \mathscr{A}$ and $B, B' \in \mathscr{B}(q)$, then, as $\mathscr{A}$ is an invariant subgroup of $\mathscr{G}$, there exists an $A'' \in \mathscr{A}$ such that $(B'B^{-1})A^{-1} = A''(B'B^{-1})$. Consequently $(A'B')(AB)^{-1} = A'B'B^{-1}A^{-1} = AA''B'B^{-1}$, which is a member of $\mathscr{S}(q)$, as $AA'' \in \mathscr{A}$ and $B'B^{-1} \in \mathscr{B}(q)$. The semi-direct product structure of $\mathscr{S}(q)$ follows directly from that of $\mathscr{G}$.)

(*i*) Let $\boldsymbol{\Gamma}^p_{\mathscr{B}(q)}$ be a unitary irreducible representation of $\mathscr{B}(q)$ of dimension d_p. Then the set of $d_p \times d_p$ matrices $\boldsymbol{\Delta}^{q,p}(AB)$ defined by

$$\boldsymbol{\Delta}^{q,p}(AB) = \chi^q_{\mathscr{A}}(A)\boldsymbol{\Gamma}^p_{\mathscr{B}(q)}(B) \tag{5.54}$$

for all $A \in \mathscr{A}$ and $B \in \mathscr{B}(q)$ form a d_p-dimensional unitary representation of $\mathscr{S}(q)$. (That $\boldsymbol{\Delta}^{q,p}$ is a representation $\mathscr{S}(q)$ can be proved as follows. Let A, A′ be any two elements of $\mathscr{A}$ and B, B′ any two elements of $\mathscr{B}(q)$. Then there exists an $A'' \in \mathscr{A}$ such that $BA'B^{-1} = A''$, so, from Equation (5.51), $\chi^q(A') = \chi^q(A'')$. Thus, from Equation (5.54),

$$\begin{aligned}\boldsymbol{\Delta}^{q,p}((AB)(A'B')) &= \boldsymbol{\Delta}^{q,p}(AA''BB')\\ &= \chi^q_{\mathscr{A}}(AA'')\boldsymbol{\Gamma}^p_{\mathscr{B}(q)}(BB')\\ &= \chi^q_{\mathscr{A}}(A)\chi^q_{\mathscr{A}}(A')\boldsymbol{\Gamma}^p_{\mathscr{B}(q)}(B)\boldsymbol{\Gamma}^p_{\mathscr{B}(q)}(B')\\ &= \boldsymbol{\Delta}^{q,p}(AB)\boldsymbol{\Delta}^{q,p}(A'B').\end{aligned}$$

The unitary property is obvious.)

(*j*) The set of $M(q)(=b/b(q))$ coset representatives $B_1, B_2, \ldots$ for the decomposition of $\mathscr{B}$ into right cosets with respect to $\mathscr{B}(q)$ also serve as coset representatives for the decomposition of $\mathscr{G}$ into right cosets with respect to $\mathscr{S}(q)$. (The numbers of distinct right cosets in the two decompositions are equal, as the number in the latter decomposition is $g/s(q) = (ab)/(ab(q)) = M(q)$ ($s(q)$ being the order of $\mathscr{S}(q)$). Moreover, $\mathscr{S}(q)B_j$ and $\mathscr{S}(q)B_k$ are distinct if and only if $\mathscr{B}(q)B_j$ and $\mathscr{B}(q)B_k$ are distinct. (To verify this, first suppose that $\mathscr{S}(q)B_j$ and $\mathscr{S}(q)B_k$ possess a common element. Then there exist $A, A' \in \mathscr{A}$ and $B, B' \in \mathscr{B}(q)$ such that $ABB_j = A'B'B_k$. However, as $\mathscr{S}(q)$ is a semi-direct product of $\mathscr{A}$ and $\mathscr{B}(q)$, $A = A'$

and $BB_j = B'B_k$, so that $\mathscr{B}(q)B_j$ and $\mathscr{B}(q)B_k$ possess a common element. The demonstration of the converse proposition is then obvious.)

(*k*) Unitary representations $\Gamma^{q,p}$ of $\mathscr{G}$ of dimensions $M(q)d_p$ may be induced from the unitary representations $\Delta^{q,p}$ of $\mathscr{S}(q)$ by applying the previous theorem with $\mathscr{S} = \mathscr{S}(q)$ and $\Delta = \Delta^{q,p}$. That is, symbolically,

$$\Gamma^{q,p} = \Delta^{q,p}(\mathscr{S}(q)) \uparrow \mathscr{G}.$$

Let T be any element of $\mathscr{G}$ and suppose that $T = AB$, where $A \in \mathscr{A}$ and $B \in \mathscr{B}$. By (*j*) the coset representatives $T_1, T_2, \ldots$ of the theorem may be identified with $B_1, B_2, \ldots$. Then $B_k(AB)B_j^{-1} = (B_k AB_k^{-1})(B_k BB_j^{-1})$, where $B_k AB_k^{-1} \in \mathscr{A}$, so $B_k(AB)B_j^{-1} \in \mathscr{S}(q)$ if and only if $B_k BB_j^{-1} \in \mathscr{B}(q)$. When $B_k BB_j^{-1} \in \mathscr{B}(q)$, Equations (5.52) and (5.54) give

$$\Delta^{q,p}(B_k ABB_j^{-1}) = \chi_{\mathscr{A}}^{B_k(q)}(A)\Gamma^p_{\mathscr{B}(q)}(B_k BB_j^{-1}).$$

Thus, from Equations (5.49),

$$\Gamma^{q,p}(AB)_{kt,jr} = \begin{cases} \chi_{\mathscr{A}}^{B_k(q)}(A)\Gamma^p_{\mathscr{B}(q)}(B_k BB_j^{-1})_{tr}, & \text{if} \quad B_k BB_j^{-1} \in \mathscr{B}(q), \\ 0, & \text{if} \quad B_k BB_j^{-1} \notin \mathscr{B}(q). \end{cases} \tag{5.55}$$

Similarly, Equation (5.50) implies that the characters $\chi^{q,p}(AB)$ of $\Gamma^{q,p}$ are given by

$$\chi^{q,p}(AB) = \sum_j \chi_{\mathscr{A}}^{B_j(q)}(A)\chi^p_{\mathscr{B}(q)}(B_j BB_j^{-1}), \tag{5.56}$$

where the sum is over all coset representatives B_j such that $B_j BB_j^{-1} \in \mathscr{B}(q)$, and where $\chi^p_{\mathscr{B}(q)}(B)$ are the characters of $\Gamma^p_{\mathscr{B}(q)}$.

The remarkable properties of these representations $\Gamma^{q,p}$ of $\mathscr{G}$ are summarized in the following theorem.

Theorem II Let $\Gamma^{q,p}$ be the unitary representation of the semi-direct product group $\mathscr{G}(= \mathscr{A} \circledS \mathscr{B})$ defined by Equations (5.55). Then

(*a*) $\Gamma^{q,p}$ is an *irreducible* representation of $\mathscr{G}$; and

(*b*) the *complete* set of unitary irreducible representations of $\mathscr{G}$ may be determined (up to equivalence) by choosing one q in each orbit and then constructing $\Gamma^{q,p}$ for each inequivalent $\Gamma^p_{\mathscr{B}(q)}$ of $\mathscr{B}(q)$.

Proof This may be found in Appendix C, Section 8.

This construction will be used in Chapter 9 in the discussion of irreducible representations of symmorphic crystallographic space groups. (It will also be shown in Chapter 9 that, although the non-symmorphic crystallographic space groups do *not* have the necessary semi-direct product structure, nevertheless all their unitary irreducible representations may also be found by the induced representation method.)

The procedure just described is particularly valuable when $\mathcal{G}$ has large order but the little groups $\mathcal{B}(q)$ have small order. This is the case when $\mathcal{G}$ is a symmorphic crystallographic space group, for then $\mathcal{G}$ has arbitrarily large order but all the little groups are crystallographic point groups and so have order 48 or less. However, because the construction process is quite complicated, to enable the various steps to be fully appreciated the method will now be worked out in detail for a group $\mathcal{G}$ of small order, the crystallographic point group D_4, whose characters were found more simply and directly in Example II of Chapter 4, Section 6.

Example II *Irreducible representations of the crystallographic point group* D_4 *by the induced representation method*

As noted in Example III of Chapter 2, Section 7, $\mathcal{G}=D_4$ has the semi-direct product structure $\mathcal{A} \circledS \mathcal{B}$ with $\mathcal{A}=\{E, C_{4y}, C_{2y}, C_{4y}^{-1}\}$ and $\mathcal{B}=\{E, C_{2x}\}$. As $C_{2y}=(C_{4y})^2$ and $C_{4y}^{-1}=(C_{4y})^3$, $\mathcal{A}$ is a cyclic group of order $a=4$. Its irreducible characters, as given by Equation (5.48), are displayed in Table 5.1. Table 1.2 shows that

$$\left.\begin{aligned} C_{2x}EC_{2x}^{-1}&=E, & C_{2x}C_{4y}C_{2x}^{-1}&=C_{4y}^{-1},\\ C_{2x}C_{2y}C_{2x}^{-1}&=C_{2y}, & C_{2x}C_{4y}^{-1}C_{2x}^{-1}&=C_{4y}. \end{aligned}\right\} \tag{5.57}$$

The representations of $\mathcal{A}$ will now be considered in turn:

(*a*) With $q=1$, Equations (5.51) and (5.57) and Table 5.1 imply that $\mathcal{B}(1)=\{E, C_{2x}\}$, so that $b(1)=2$, $M(1)=1$ and $B_1=E$. The little group $\mathcal{B}(1)$ is identical to $\mathcal{B}$, whose character table is displayed in Table 5.2. As $M(1)d_p=1$ for $p=1,2$, both of the irreducible representations $\Gamma^{1,p}$ of D_4 are one-dimensional. As $B_1BB_1^{-1}=B$ for all $B\in\mathcal{B}(1)$, Equation (5.55) reduces to

$$\Gamma^{1,p}(AB)_{11,11}=\chi^1_{\mathcal{A}}(A)\Gamma^p_{\mathcal{B}(1)}(B)_{11}$$

	E	C_{4y}	C_{2y}	C_{4y}^{-1}
$\Gamma^1_{\mathcal{A}}$	1	1	1	1
$\Gamma^2_{\mathcal{A}}$	1	i	-1	$-i$
$\Gamma^3_{\mathcal{A}}$	1	-1	1	-1
$\Gamma^4_{\mathcal{A}}$	1	$-i$	-1	i

Table 5.1 Character table for $\mathcal{A}=\{E, C_{4y}, C_{2y}, C_{4y}^{-1}\}$.

	E	C_{2x}
$\Gamma^1_{\mathscr{B}}$	1	1
$\Gamma^2_{\mathscr{B}}$	1	−1

Table 5.2 Character table for $\mathscr{B} = \{E, C_{2x}\}$.

for all $A \in \mathscr{A}$ and $B \in \mathscr{B}$ and $p = 1, 2$. Direct substitution then shows that the induced representations $\Gamma^{1,1}$ and $\Gamma^{1,2}$ of D_4 are identical to the representations called Γ^1 and Γ^3 respectively in Table 4.1.

(*b*) With $q = 2$, Equations (5.51) and (5.57) and Table 5.1 show that $\mathscr{B}(2) = \{E\}$. Thus $b(2) = 1$ and so $M(2) = 2$, a convenient choice of coset representatives being $B_1 = E$ and $B_2 = C_{2x}$. Then $B_1(2) = 2$ and $B_2(2) = 4$ (on using Equations (5.52) and (5.57) and Table 5.1). The only irreducible representation of $\mathscr{B}(2)$ is specified by $\Gamma^1_{\mathscr{B}(2)}(E) = [1]$. As $M(2)d_1 = 2$, the corresponding representation $\Gamma^{2,1}$ of D_4 is two-dimensional.

Consider for example $T = C_{2c} = C_{4y}C_{2x}$. Then, from Table 1.2, $B_1C_{2x}B_1^{-1} = C_{2x}$, $B_1C_{2x}B_2^{-1} = E$, $B_2C_{2x}B_1^{-1} = E$, $B_2C_{2x}B_2^{-1} = C_{2x}$. Thus, from Equation (5.55),

$$\left.\begin{aligned}
&\Gamma^{2,1}(C_{2c})_{11,11} = 0 \quad (\text{as } B_1C_{2x}B_1^{-1} \notin \mathscr{B}(1)),\\
&\Gamma^{2,1}(C_{2c})_{11,21} = \chi_{\mathscr{A}}^{B_1(2)}(C_{4y})\Gamma^1_{\mathscr{B}(1)}(E)_{11} = \chi^2_{\mathscr{A}}(C_{4y}).\,1 = i,\\
&\Gamma^{2,1}(C_{2c})_{21,11} = \chi_{\mathscr{A}}^{B_2(2)}(C_{4y})\Gamma^1_{\mathscr{B}(1)}(E)_{11} = \chi^4_{\mathscr{A}}(C_{4y}).\,1 = -i,\\
&\Gamma^{2,1}(C_{2c})_{21,21} = 0 \quad (\text{as } B_2C_{2x}B_2^{-1} \notin \mathscr{B}(1)).
\end{aligned}\right\}$$

The indices t and r are clearly redundant, so one can write $\Gamma^{2,1}(T)_{kt,jr} = \Gamma^{2,1}(T)_{kj}$. Thus

$$\Gamma^{2,1}(C_{2c}) = \begin{bmatrix} 0 & i \\ -i & 0 \end{bmatrix}.$$

A similar argument applied to the other elements of D_4 gives

$$\Gamma^{2,1}(E) = \begin{bmatrix} 1 & 0 \\ 0 & 1 \end{bmatrix}, \quad \Gamma^{2,1}(C_{2x}) = \begin{bmatrix} 0 & 1 \\ 1 & 0 \end{bmatrix}, \quad \Gamma^{2,1}(C_{2y}) = \begin{bmatrix} -1 & 0 \\ 0 & -1 \end{bmatrix},$$

$$\Gamma^{2,1}(C_{2z}) = \begin{bmatrix} 0 & -1 \\ -1 & 0 \end{bmatrix}, \quad \Gamma^{2,1}(C_{4y}) = \begin{bmatrix} i & 0 \\ 0 & -i \end{bmatrix}, \quad \Gamma^{2,1}(C_{4y}^{-1}) = \begin{bmatrix} -i & 0 \\ 0 & i \end{bmatrix},$$

$$\Gamma^{2,1}(C_{2d}) = \begin{bmatrix} 0 & -i \\ i & 0 \end{bmatrix}.$$

Thus the irreducible representation $\Gamma^{2,1}$ of D_4 is equivalent to the representation previously referred to as Γ^5. ($\Gamma^{2,1}$ differs by a similarity transformation from the explicit form of Γ^5 quoted in Example I of Section 1.)

(c) With $q=3$, $\mathscr{B}(q)$ again coincides with $\mathscr{B}$. As in (a) the induced representation method produces two one-dimensional irreducible representations $\Gamma^{3,1}$ and $\Gamma^{3,2}$ for D_4. These are identical to the representations Γ^2 and Γ^4 of Table 4.1 respectively.

(d) As $q=4$ is in the orbit of $q=2$, the induced representation of D_4 corresponding to $q=4$ is equivalent to that for $q=2$ found in (b) above.

There exist a number of general theorems dealing with such matters as direct products of induced representations and with the reduction of induced representations of $\mathscr{G}$ on subgroups of $\mathscr{G}$. Details may be found in the reviews of Coleman (1968) and Mackey (1963b, 1976).

This section will be concluded with a theorem that expresses the general induced representation construction of Theorem I of this section in the carrier space formulation. It is only in this form that it can be generalized to arbitrary locally compact topological groups.

Theorem III With the notation of Theorem I of this section, let V_Δ be a carrier space for the representation $\boldsymbol{\Delta}$ of $\mathscr{S}$ and let $\Phi_\Delta(S)$ be a set of operators defined for all $S \in \mathscr{S}$ to act on V_Δ in such a way that

$$\Phi_\Delta(S)\psi_n = \sum_{m=1}^{d} \Delta(S)_{mn}\psi_m, \tag{5.58}$$

where $\psi_1, \psi_2, \ldots, \psi_d$ are a basis for V_Δ. Let V be the vector space of all mappings $\phi(\mathscr{S}T_j)$ into V_Δ of right cosets $\mathscr{S}T_1, \mathscr{S}T_2, \ldots$ of $\mathscr{G}$ with respect to $\mathscr{S}$. For each $T \in \mathscr{G}$ define the operator $\Phi(T)$ by

$$\Phi(T)\phi(\mathscr{S}T_k) = \Phi_\Delta(T_k T T_j^{-1})\phi(\mathscr{S}T_j), \tag{5.59}$$

where ϕ is any member of V and T_j is the coset representative such that $T_k T \in \mathscr{S}T_j$. Then

(a) for any $T, T' \in \mathscr{G}$

$$\Phi(TT') = \Phi(T)\Phi(T'),$$

and

(b) there exists a basis ϕ_{kt} of V ($k=1,2,\ldots,M$; $t=1,2,\ldots,d$) such that for all $T \in \mathscr{G}$ and $j=1,2,\ldots,M$; $r=1,2,\ldots,d$,

$$\Phi(T)\phi_{jr} = \sum_{k=1}^{M}\sum_{t=1}^{d} \Gamma(T)_{kt,jr}\phi_{kt},$$

where $\mathbf{\Gamma}$ is the induced representation of $\mathcal{G}$ defined in Equations (5.49). Thus V is a carrier space for the induced representation $\mathbf{\Gamma}$ of $\mathcal{G}$ and the $\Phi(T)$ are the corresponding operators acting in V.

Proof The proof, together with some comments on aspects of the definitions in the theorem, may be found in Appendix C, Section 8.

8 The reality of representations

Let $\mathbf{\Gamma}$ be a d-dimensional unitary irreducible representation of a group $\mathcal{G}$ and define for each $T \in \mathcal{G}$ the $d \times d$ matrix $\mathbf{\Gamma}^*(T)$ by $\mathbf{\Gamma}^*(T) = \{\mathbf{\Gamma}(T)\}^*$. Then these matrices also form a d-dimensional unitary irreducible representation of $\mathcal{G}$.

If all the matrices of a representation are real, then the representation itself is said to be real. If $\mathbf{\Gamma}$ is equivalent to a real representation $\mathbf{\Gamma}'$, then $\mathbf{\Gamma}$ is equivalent to $\mathbf{\Gamma}^*$. (This follows because, if $\mathbf{\Gamma}$ is equivalent to $\mathbf{\Gamma}'$, there must exist a $d \times d$ non-singular matrix $\mathbf{S}$ such that $\mathbf{\Gamma}'(T) = \mathbf{S}^{-1}\mathbf{\Gamma}(T)\mathbf{S}$ for all $T \in \mathcal{G}$. Taking the complex conjugate gives $\mathbf{\Gamma}'(T) = (\mathbf{S}^*)^{-1}\mathbf{\Gamma}^*(T)\mathbf{S}^*$, so that $\mathbf{\Gamma}^*(T) = (\mathbf{S}\mathbf{S}^{*-1})^{-1}\mathbf{\Gamma}(T)(\mathbf{S}\mathbf{S}^{*-1})$ for all $T \in \mathcal{G}$.) However, examples can be constructed which show that it is possible to have $\mathbf{\Gamma}$ equivalent to $\mathbf{\Gamma}^*$ without $\mathbf{\Gamma}$ being equivalent to a real representation.

Thus there are three possibilities regarding the relationship of $\mathbf{\Gamma}$ to $\mathbf{\Gamma}^*$:

(*a*) $\mathbf{\Gamma}$ equivalent to a real representation; in this case $\mathbf{\Gamma}$ may be called "potentially real";
(*b*) $\mathbf{\Gamma}$ equivalent to $\mathbf{\Gamma}^*$, but $\mathbf{\Gamma}$ not equivalent to a real representation; in this case $\mathbf{\Gamma}$ is called "pseudo-real";
(*c*) $\mathbf{\Gamma}$ not equivalent to $\mathbf{\Gamma}^*$; in such a case $\mathbf{\Gamma}$ will be called "essentially complex".

This classification plays an important role in the theory of time-reversal symmetry of the Schrödinger equation (see Chapter 6, Section 5) and in the investigation of extra degeneracies of frequencies of vibrations (see Chapter 7, Section 3(*f*)). (Obviously a potentially real representation need not be real itself, but in the literature it is common to find that such a representation is called a "real" representation, thereby creating the possibility of some confusion. Likewise representations of type (*c*) are often simply called "complex" representations, although those of type (*a*) and (*b*) may also be complex in the sense of having complex matrix elements. Indeed this must be so for representations of type (*b*).)

Henceforth it will be assumed that $\mathcal{G}$ is either a *finite* group or a *compact* Lie group. There then exists a very simple criterion for distinguishing essentially complex irreducible representations from the other two types:

Theorem I An irreducible representation $\boldsymbol{\Gamma}$ is potentially real or pseudo-real if and only if its characters are all real.

Proof Let $\chi(T)$ and $\chi^*(T)$ be the characters of $\boldsymbol{\Gamma}$ and $\boldsymbol{\Gamma}^*$ respectively. Then, by Theorems I and IV of Chapter 4, Section 6, a necessary and sufficient condition for $\boldsymbol{\Gamma}$ to be equivalent to $\boldsymbol{\Gamma}^*$ is that $\chi(T)=\chi^*(T)$ for all $T\in\mathcal{G}$. But $\chi^*(T)=\{\chi(T)\}^*$ for all $T\in\mathcal{G}$, so $\boldsymbol{\Gamma}$ is equivalent to $\boldsymbol{\Gamma}^*$ if and only if $\chi(T)$ is real for every $T\in\mathcal{G}$.

Distinguishing a potentially real representation from a pseudo-real representation is slightly more difficult, but the following theorem provides a result that is useful in its own right as well as being a stage in the development of a more powerful criterion.

Theorem II Suppose that the unitary irreducible representations $\boldsymbol{\Gamma}$ and $\boldsymbol{\Gamma}^*$ are equivalent, the similarity transformation relating them being

$$\boldsymbol{\Gamma}^*(T)=\mathbf{Z}^{-1}\boldsymbol{\Gamma}(T)\mathbf{Z} \tag{5.60}$$

for all $T\in\mathcal{G}$, $\mathbf{Z}$ being a $d\times d$ non-singular matrix. Then

$$\mathbf{Z}^*\mathbf{Z}=c_Z\mathbf{1}, \tag{5.61}$$

where c_Z is a real non-zero number. Moreover, $\boldsymbol{\Gamma}$ is potentially real if $c_Z>0$ and is pseudo-real if $c_Z<0$. In particular, if $\mathbf{Z}$ is chosen to be unitary then $(c_Z)^2=1$, $\boldsymbol{\Gamma}$ being potentially real if $c_z=1$ and pseudo-real if $c_Z=-1$.

Proof This may be found in Appendix C, Section 9.

Corollary Every pseudo-real irreducible representation has *even* dimension.

Proof Suppose that $\boldsymbol{\Gamma}$ is a pseudo-real irreducible representation of dimension d. From Equation (5.61), $|\det\mathbf{Z}|^2=(c_Z)^d$. As $|\det\mathbf{Z}|^2>0$ and $c_Z<0$, d must be even.

Frobenius and Schur (1906) established the following simple test.

Theorem III If $\mathscr{G}$ is a finite group of order g and $\chi(T)$ is the character of $T \in \mathscr{G}$ in the irreducible representation $\boldsymbol{\Gamma}$, then

$$(1/g) \sum_{T \in \mathscr{G}} \chi(T^2) = \begin{cases} 1, & \text{if } \boldsymbol{\Gamma} \text{ is potentially real,} \\ 0, & \text{if } \boldsymbol{\Gamma} \text{ is essentially complex,} \\ -1, & \text{if } \boldsymbol{\Gamma} \text{ is pseudo-real.} \end{cases}$$

Similarly, if $\mathscr{G}$ is a compact Lie group,

$$\int_{\mathscr{G}} \chi(T^2)\, dT = \begin{cases} 1, & \text{if } \boldsymbol{\Gamma} \text{ is potentially real,} \\ 0, & \text{if } \boldsymbol{\Gamma} \text{ is essentially complex,} \\ -1, & \text{if } \boldsymbol{\Gamma} \text{ is pseudo-real.} \end{cases}$$

Proof Here T^2 indicates the product TT. The proof is given in Appendix C, Section 9.

Mehta (1966) and Mehta and Strivastava (1966) have given a complete classification of all the irreducible representations of the simple compact Lie groups in terms of these three types using a Lie algebraic technique.

One final result will be used in Chapter 7, Section 3(*f*).

Theorem IV If $\boldsymbol{\Gamma}$ is a representation that is real and reducible, then any pseudo-real irreducible representation that appears in its reduction can only occur an even number of times. Moreover, if an essentially complex irreducible representation $\boldsymbol{\Gamma}^p$ is contained in the reduction of $\boldsymbol{\Gamma}$, then its complex conjugate representation $\boldsymbol{\Gamma}^{p*}$ occurs the same number of times as $\boldsymbol{\Gamma}^p$.

Proof This too may be found in Appendix C, Section 9.

6

Group Theory in Quantum Mechanical Calculations

1 The solution of the Schrödinger equation

One of the most valuable applications of group theory is to the solution of the Schrödinger equation. Only for a small number of very simple systems, such as the hydrogen atom, is it possible to obtain an exact analytic solution. For all other systems it is necessary to resort to numerical calculations, but the work involved can be shortened considerably by the application of group representation theory. This is particularly true in electronic energy band calculations in solid state physics, where accurate calculations are only feasible when group theoretical arguments are used to exploit the symmetry of the system to the full.

For simplicity consider the "single-particle" time-independent Schrödinger equation described in Chapter 1, Section 3(a), namely

$$H(\mathbf{r})\psi(\mathbf{r}) = \varepsilon\psi(\mathbf{r}), \tag{6.1}$$

$H(\mathbf{r})$ being the Hamiltonian operator (see Equation (1.10)). It is required to find the low-lying energy eigenvalues ε and their corresponding eigenfunctions $\psi(\mathbf{r})$. The unknown function $\psi(\mathbf{r})$ can be expanded in terms of a complete set of known functions $\psi_1(\mathbf{r}), \psi_2(\mathbf{r}), \ldots$ that form a basis for L^2 (see Appendix B, Section 3), that is,

$$\psi(\mathbf{r}) = \sum_{j=1}^{\infty} a_j\psi_j(\mathbf{r}), \tag{6.2}$$

where $a_1, a_2, \ldots$ are a set of complex numbers whose values are unknown at this stage. The assumption is now made that the series (Equation (6.2)) converges sufficiently rapidly that only the first N terms need be retained. Then it can be replaced by the approximation

$$\psi(\mathbf{r}) = \sum_{j=1}^{N} a_j \psi_j(\mathbf{r}). \tag{6.3}$$

Some judgement is required as to the best choice of the set $\psi_1(\mathbf{r}), \psi_2(\mathbf{r}), \ldots$ that will ensure the validity of this approximation. Indeed, the different types of energy band calculation, described for instance in the article of Reitz (1955), essentially differ merely in this choice. For example, in solid state problems where the valence electrons are expected to be tightly bound to the ions, it is natural to take the $\psi_j(\mathbf{r})$ to be atomic orbitals, thereby giving the so-called "method of linear combinations of atomic orbitals", often described more briefly as the "L.C.A.O. method". At the other extreme, when the valence electrons are nearly free, it is natural to form the $\psi_j(\mathbf{r})$ from plane waves (orthogonalized to the ionic electronic energy eigenfunctions to prevent the expansion giving ionic electron energy eigenfunctions), thereby giving the so-called "orthogonalized plane wave method", or "O.P.W. method" for short.

Substituting Equation (6.3) into Equation (6.1) and forming the inner product of Equation (B.18) with $\psi_k(\mathbf{r})$ gives

$$\sum_{j=1}^{N} a_j \{(\psi_k, \mathrm{H}\psi_j) - \varepsilon(\psi_k, \psi_j)\} = 0. \tag{6.4}$$

This is a matrix eigenvalue equation of the form of Equation (A.10), in which the matrix elements $(\psi_h, \mathrm{H}\psi_j)$ are known but the eigenvalues ε and the elements a_j of the eigenvector are to be determined. Equation (6.4) has a non-trivial solution if and only if

$$\det\{(\psi_k, \mathrm{H}\psi_j) - \varepsilon(\psi_k, \psi_j)\} = 0, \tag{6.5}$$

in which the matrix involved is of dimension $N \times N$ (cf. Equation (A.11)). The left-hand side of the scalar equation derived from Equation (6.5) is a polynomial of degree N, the roots of which are the eigenvalues ε. Both the explicit determination of the polynomial and the calculation of its roots are very lengthy processes if N is large. With the roots obtained it is possible to go back to Equation (6.4), regarded now as a system of N linear algebraic equations, and for each root ε obtain the corresponding set of complex numbers a_j, thereby giving by Equation (6.3) an approximation to the corresponding eigenfunction $\psi(\mathbf{r})$.

The number N is here quite arbitrary, but clearly, as N is increased, two effects follow. Firstly, the accuracy of the approximations to the lower energy eigenvalues is improved, which is very desirable. Secondly, more eigenvalues at higher energies appear, although these are usually less important. However, as noted above, the numerical work involved increases rapidly as N increases, this work being roughly proportional to $N!$.

By invoking group representation theory this numerical work can be cut tremendously without any loss of accuracy. All that has to be done is to arrange that the members of the complete set of known functions $\psi_j(\mathbf{r})$ of Equation (6.3) are each basis functions of the various irreducible representations of the group of the Schrödinger equation, $\mathscr{G}$. In practice this is achieved by applying the projection operators of Chapter 5, Section 1 to the atomic orbitals, orthogonalized plane waves, or other given functions that are judged appropriate to the particular system under consideration. An extra pair of indices m and p has now to be included in the designation of the functions, so that $\psi^p_{jm}(\mathbf{r})$ transforms as the mth row of the irreducible representation Γ^p of $\mathscr{G}$, the index j distinguishing linearly independent basis functions having this particular symmetry. Equation (6.3) is then rewritten as

$$\psi(\mathbf{r}) = \sum_j \sum_p \sum_{m=1}^{d_p} a^p_{jm} \psi^p_{jm}(\mathbf{r}), \tag{6.6}$$

and Equation (6.5) becomes

$$\det\{(\psi^q_{kn}, \mathrm{H}\psi^p_{jm}) - \varepsilon(\psi^q_{kn}, \psi^p_{jm})\} = 0. \tag{6.7}$$

If $\mathscr{G}$ is a finite group or a compact Lie group, each irreducible representation Γ^p may be taken to be unitary. Theorem V of Chapter 4, Section 5 then shows that

$$(\psi^q_{kn}, \psi^p_{jm}) = \delta_{qp}\delta_{nm}(\psi^p_{km}, \psi^p_{jm}). \tag{6.8}$$

Similarly, Theorem III of Chapter 5, Section 3, implies that

$$(\psi^q_{kn}, \mathrm{H}\psi^p_{jm}) = \delta_{qp}\delta_{nm}(\psi^p_{km}, \mathrm{H}\psi^p_{jm}). \tag{6.9}$$

The rows and columns of the determinant of Equation (6.7) may be rearranged so that all the terms corresponding to a particular row of a particular irreducible representation of $\mathscr{G}$ are grouped together. (This can be achieved by successively interchanging pairs of rows of the determinant and then pairs of columns. Such interchanges in general change the sign of a determinant. However, here the value of the determinant is zero, so its value is unchanged by such a rearrangement.) Equations (6.8) and (6.9) then imply that the determinant of Equation (6.7) takes the "block form"

$$\det\begin{bmatrix} \mathbf{D}(1,1) & \mathbf{0} & \mathbf{0} & \cdots & \mathbf{0} & \mathbf{0} & \mathbf{0} & \cdots \\ \mathbf{0} & \mathbf{D}(1,2) & \mathbf{0} & \cdots & \mathbf{0} & \mathbf{0} & \mathbf{0} & \cdots \\ \mathbf{0} & \mathbf{0} & \mathbf{D}(1,3) & \cdots & \mathbf{0} & \mathbf{0} & \mathbf{0} & \cdots \\ \vdots & \vdots & \vdots & & & & & \\ \mathbf{0} & \mathbf{0} & \mathbf{0} & \cdots & \mathbf{D}(1,d_1) & \mathbf{0} & \mathbf{0} & \cdots \\ \mathbf{0} & \mathbf{0} & \mathbf{0} & \cdots & \mathbf{0} & \mathbf{D}(2,1) & \mathbf{0} & \cdots \\ \mathbf{0} & \mathbf{0} & \mathbf{0} & \cdots & \mathbf{0} & \mathbf{0} & \mathbf{D}(2,2) & \cdots \\ \vdots & \vdots & \vdots & & \vdots & \vdots & \vdots & \end{bmatrix} = 0, \tag{6.10}$$

where $\mathbf{D}(p, m)$ is a submatrix defined by

$$\mathrm{D}(p, m)_{kj} = (\psi^p_{km}, \mathrm{H}\psi^p_{jm}) - \varepsilon(\psi^p_{km}, \psi^p_{jm}).$$

Thus $\mathbf{D}(p, m)$ involves *only* basis functions corresponding to the mth row of $\boldsymbol{\Gamma}^p$. The matrices $\mathbf{0}$ consist entirely of zero elements. Equation (6.10) can be factorized to give

$$\prod_p \prod_{m=1}^{d_p} \det \mathbf{D}(p, m) = 0.$$

The complete set of eigenvalues of Equation (6.10) are then obtained by taking

$$\det \mathbf{D}(p, m) = 0 \tag{6.11}$$

for every p and every $m = 1, 2, \ldots, d_p$. *The energy eigenvalues corresponding to the mth row of* $\boldsymbol{\Gamma}^p$ *are therefore given by the secular equation, Equation* (6.11), *which only involves basis functions corresponding to the mth row of* $\boldsymbol{\Gamma}^p$. As the dimensions of $\mathbf{D}(p, m)$ are usually very much smaller than those of the determinant of Equation (6.7), very much less numerical work is now needed to find the energy eigenvalues and eigenfunctions for the same degree of accuracy.

A further valuable saving of effort is provided by noting that Equations (6.8) and (6.9) also imply that

$$\mathbf{D}(p, 1) = \mathbf{D}(p, 2) = \ldots = \mathbf{D}(p, d_p) \tag{6.12}$$

for each irreducible representation $\boldsymbol{\Gamma}^p$. Thus *only one* secular equation (Equation (6.11)) has to be solved for each irreducible representation $\boldsymbol{\Gamma}^p$ and each of the resulting energy eigenvalues can be taken to be d_p-fold degenerate.

It is very interesting to relate these results to the general conclusion drawn in Chapter 1, Section 4, that every d-fold degenerate energy eigenvalue is associated with a d-dimensional representation Γ of the group of the Schrödinger equation, the corresponding d linearly independent eigenfunctions being basis functions of this representation.

Suppose first that this representation Γ is irreducible and is identical to Γ^p. Then the d-fold degeneracy of the energy eigenvalue follows automatically from the identities in Equations (6.12). It is not unexpected that the energy eigenfunction $\psi(\mathbf{r})$ transforming as the mth row of Γ^p involves *only* basis functions transforming the same way. That is, $a^p_{kn}=0$ for $q\neq p$ and $n\neq m$ in the expansion in Equation (6.6).

The situation when Γ is reducible is more complicated. Suppose that Γ is the direct sum of two inequivalent irreducible representations Γ^p and Γ^q of dimensions d_p and d_q, so that $d=d_p+d_q$. Then d_p of the energy eigenfunctions can be taken as basis functions of Γ^p, the remaining d_q eigenfunctions being basis functions of Γ^q. The identities in Equations (6.12) produce a d_p-fold degeneracy in the energy eigenvalue. Similarly, the corresponding identities with p replaced by q give rise to a d_q-fold degeneracy. The overall $d(=d_p+d_q)$-fold degeneracy must be a consequence of the secular equations $\det \mathbf{D}(p,m)=0$ and $\det \mathbf{D}(q,n)=0$ possessing a common eigenvalue, but no reason for this can be attributed to the symmetry of the system. Consequently the extra degeneracy associated with this common eigenvalue is called an "accidental" degeneracy, and the energy levels corresponding to Γ^p and Γ^q are said to "stick together". In general, an arbitrarily small change in the potential that preserves its symmetry will break the accidental degeneracy.

It is to be expected that accidental degeneracies occur only very rarely, the normal situation being that in which the representation Γ is irreducible. However, in some exceptional systems, such as the hydrogen atom, these accidental degeneracies occur so extensively and in such a regular fashion that they cannot be truly coincidental. Their origin lies in a "hidden" symmetry which gives rise to an invariance group that is larger than the obvious invariance group. For the hydrogen atom the situation will be examined in detail in Chapter 12, Section 8.

2 Transition probabilities and selection rules

Suppose that a small time-dependent perturbation $H'(t)$ is applied to a system whose time-independent "unperturbed" Hamiltonian is H_0, so that the total Hamiltonian becomes

$$H(t) = H_0 + H'(t).$$

Suppose that before the perturbation is applied (that is, at time $t = -\infty$) the system is in an eigenstate ϕ_i of H_0 with energy eigenvalue ε_i. Then, according to first-order perturbation theory (Schiff 1968), the probability of finding the system at time t in another eigenstate ϕ_f of H_0 (whose energy eigenvalue is ε_f) is given by

$$\left|(i\hbar)^{-1}\int_{-\infty}^{t} (\phi_f, H'(t')\phi_i)\exp\{i(\varepsilon_f - \varepsilon_i)t'/\hbar\}\,dt'\right|^2.$$

(Here it is assumed that ϕ_i and ϕ_f have been appropriately normalized.)

The significant part of this expression is the inner product $(\phi_f, H'(t')\phi_i)$. The analysis of Chapter 1, Section 4 shows that ϕ_i and ϕ_f must be basis functions of some representations of the invariance group $\mathcal{G}_0$ of the unperturbed Hamiltonian H_0. With the perturbation $H'(t)$ expressed in terms of irreducible tensor operators of $\mathcal{G}_0$, such inner products can be studied using the Wigner–Eckart Theorem of Chapter 5, Section 3. In particular, it is possible to deduce when symmetry requires that $(\phi_f, H'(t')\phi_i) = 0$. When this is so, transitions from ϕ_i to ϕ_f are *forbidden* (at least in first-order perturbation theory).

This basic idea will now be developed for the very important case in which the system interacts with an external electromagnetic field, resulting in absorption or induced emission of radiation. It will be assumed that the unperturbed system is described by a "single-particle" Schrödinger equation (in the sense of Chapter 1, Section 3(*a*)), so that (cf. Equation (1.10))

$$H_0(\mathbf{r}) = -\frac{\hbar^2}{2\mu}\left(\frac{\partial^2}{\partial x^2} + \frac{\partial^2}{\partial y^2} + \frac{\partial^2}{\partial z^2}\right) + V(\mathbf{r}),$$

where μ is the mass of the particle and $V(\mathbf{r})$ the potential that it experiences. The perturbing operator $H'(\mathbf{r}, t)$ may be taken to be

$$H'(\mathbf{r}, t) = (ie\hbar/\mu c)\mathbf{A}(\mathbf{r}, t)\,.\,\mathbf{grad}, \tag{6.13}$$

where c is the speed of light and the vector potential $\mathbf{A}(\mathbf{r}, t)$ for a plane electromagnetic wave with wave vector $\mathbf{k}$ and frequency $\omega/2\pi$ is of the form

$$\mathbf{A}(\mathbf{r}, t) = 2\mathbf{A}_0 \cos(\mathbf{k}\,.\,\mathbf{r} - \omega t + \alpha).$$

Here $\mathbf{A}_0$ is a constant vector, with real components, that specifies the polarization of the radiation, α gives its phase, $\mathbf{A}_0 . \mathbf{k}=0$, and $\omega = |\mathbf{k}|\, c$. Then the transition probability for *absorption* of a photon from the radiation field causing a transition from $\phi_i(\mathbf{r})$ to $\phi_f(\mathbf{r})$ is given by

$$\frac{4\pi^2 e^2 \hbar^2 I((\varepsilon_f - \varepsilon_i)/\hbar)}{\mu^2 c^2 (\varepsilon_f - \varepsilon_i)^2} |(\phi_f, \mathbf{A}_0 . \mathbf{grad}\, \phi_i)|^2, \tag{6.14}$$

the angular frequency of the absorbed radiation being $(\varepsilon_f - \varepsilon_i)/\hbar$. Similarly, the transition probability for *induced emission* of a photon associated with a transition from $\phi_i(\mathbf{r})$ to $\phi_f(\mathbf{r})$ is given by

$$\frac{4\pi^2 e^2 \hbar^2 I((\varepsilon_i - \varepsilon_f)/\hbar)}{\mu^2 c^2 (\varepsilon_i - \varepsilon_f)^2} |(\phi_f, \mathbf{A}_0 . \mathbf{grad}\, \phi_i)|^2, \tag{6.15}$$

the angular frequency of the emitted radiation being $(\varepsilon_i - \varepsilon_f)/\hbar$. In Expressions (6.14) and (6.15) $I(\omega)$ is defined to be such that the intensity of the radiation field in the angular frequency range ω to $\omega + d\omega$ is $I(\omega)\, d\omega$. In deriving Expressions (6.14) and (6.15) it has been assumed that the wavelength of the radiation is much greater than the dimensions of the regions in which the eigenfunctions $\phi_i(\mathbf{r})$ and $\phi_f(\mathbf{r})$ are significantly different from zero. As the inner products of Expressions (6.14) and (6.15) can be rewritten in the form

$$(\phi_f, \mathbf{A}_0 . \mathbf{grad}\, \phi_i) = -\{\mu(\varepsilon_f - \varepsilon_i)/\hbar^2\}(\phi_f, \mathbf{A}_0 . \mathbf{r}\phi_i), \tag{6.16}$$

the transitions are often called "electric dipole transitions". (The detailed derivations of Expressions (6.13), (6.14), (6.15) and (6.16) may be found in the book by Schiff (1968).)

It is interesting to note (see Schiff (1968)) that the transition probability for *spontaneous* emission of radiation of frequency $(\varepsilon_i - \varepsilon_f)/\hbar$ polarized in the direction of the unit vector $\mathbf{n}$ in a transition from $\phi_i(\mathbf{r})$ to $\phi_f(\mathbf{r})$ is given by

$$\frac{4e^2(\varepsilon_i - \varepsilon_f)}{3c^2\mu^2} |(\phi_f, \mathbf{n} . \mathbf{grad}\, \phi_i)|^2$$

Equation (6.16) allows this too to be expressed in terms of "dipole" inner products.

Let $Q = \mathbf{A}_0 . \mathbf{grad}$ for absorption or induced emission and $Q = \mathbf{n} . \mathbf{grad}$ for spontaneous emission. Then Example III of Chapter 5, Section 3 shows that Q is easily expressed as a linear combination of irreducible tensor operators. There is no intrinsic difficulty in carrying out the analysis for the most general case, but for simplicity it will be assumed that Q is an irreducible tensor operator transforming as the kth row of the unitary irreducible representation Γ_0^q of $\mathcal{G}_0$. Suppose also that $\phi_i(\mathbf{r}) = \phi_j^p(\mathbf{r})$ and $\phi_f(\mathbf{r}) = \psi_l^r(\mathbf{r})$, where $\phi_j^p(\mathbf{r})$ and

$\psi_l^r(\mathbf{r})$ are respectively basis functions transforming as the jth and lth rows of the unitary irreducible representations Γ_0^p and Γ_0^r of $\mathcal{G}_0$. If $\mathcal{G}_0$ is a finite group or a compact Lie group, the Wigner–Eckart Theorem shows that

$$(\phi_f, Q\phi_i) = 0$$

if $n_{pq}^r = 0$, that is, if Γ_0^r does *not* appear in the reduction of $\Gamma_0^p \otimes \Gamma_0^q$. Thus a list of forbidden transitions can be found from the Clebsch–Gordan *series* alone. However, a *complete* list requires a knowledge of the Clebsch–Gordan coefficients, for if $n_{pq}^r \neq 0$ the Wigner–Eckart Theorem implies that

$$(\phi_f, Q\phi_i) = \sum_{\alpha=1}^{n_{pq}^r} \begin{pmatrix} p & q \\ j & k \end{pmatrix} \begin{vmatrix} r, & \alpha \\ l & \end{pmatrix}^* (r\,|Q|\,p)_\alpha, \tag{6.17}$$

and it can happen that $\begin{pmatrix} p & q \\ j & k \end{pmatrix} \begin{vmatrix} r, & \alpha \\ l & \end{pmatrix} = 0$ for all $\alpha = 1, 2, \ldots, n_{pq}^r$, thereby giving again a zero transition probability. (This situation occurs in the example that will be given shortly.)

The Wigner–Eckart Theorem can also be employed in the analysis of the magnitudes of the transition probabilities of the allowed transitions. If $n_{pq}^r \neq 0$, Equation (6.17) shows that $(\phi_f, Q\phi_i)$ depends on n_{pq}^r reduced matrix elements. The transition probabilities for *all* other transitions involving initial states that are partners of $\phi_i(\mathbf{r})$ in the same basis, final states that are partners of $\phi_f(\mathbf{r})$ in the same basis, and other directions of polarization whose operators form part of the same set of irreducible tensor operators as Q also depend on the Clebsch–Gordan coefficients and these n_{pq}^r reduced matrix elements. Thus a complete description of $d_p d_q d_r$ possible transitions depends only on n_{pq}^r reduced matrix elements, where n_{pq}^r can be considerably smaller than $d_p d_q d_r$.

The most important case is that in which $\mathcal{G}_0$ is the group of all rotations in $\mathbb{R}^3$. This will be considered in detail in Chapter 12, Section 6, after the irreducible representations of this group have been derived. The following simple example will serve until then to illustrate the power of the technique.

Example I *Optical selection rules associated with the crystallographic point group* D_4

Let $\mathcal{G}_0 = D_4$ and let Γ_0^p denote the irreducible representations of D_4 previously referred to as Γ^p, $p = 1, 2, 3, 4, 5$. Consider absorption from an initial state $\phi_i(\mathbf{r})$ that transforms as the first row of the two-dimensional irreducible representation Γ_0^5 (the explicit matrices being as in Example I of Chapter 5, Section 1).

For polarization vector $\mathbf{A}_0$ in the x-direction, $\mathbf{A}_0 \cdot \mathbf{grad} = A_{01}\, \partial/\partial x$, which (by Example III of Chapter 5, Section 3) is an irreducible tensor operator transforming as the first row of $\boldsymbol{\Gamma}_0^5$. As $\boldsymbol{\Gamma}_0^5 \otimes \boldsymbol{\Gamma}_0^5 \approx \boldsymbol{\Gamma}_0^1 \oplus \boldsymbol{\Gamma}_0^2 \oplus \boldsymbol{\Gamma}_0^3 \oplus \boldsymbol{\Gamma}_0^4$ (see Example I of Chapter 5, Section 2), the Clebsch–Gordan series implies that the final state $\phi_f(\mathbf{r})$ cannot transform as either row of $\boldsymbol{\Gamma}_0^5$. However, the Clebsch–Gordan coefficients (see Example I of Chapter 5, Section 3) show that $\phi_f(\mathbf{r})$ can only transform as $\boldsymbol{\Gamma}_0^1$ or $\boldsymbol{\Gamma}_0^2$.

Similarly, for $\mathbf{A}_0$ in the y-direction, $\mathbf{A}_0 \cdot \mathbf{grad} = A_{02}\, \partial/\partial y$, which is an irreducible tensor operator transforming as $\boldsymbol{\Gamma}_0^3$. As $\boldsymbol{\Gamma}_0^5 \otimes \boldsymbol{\Gamma}_0^3 \approx \boldsymbol{\Gamma}_0^5$, transitions can only occur to final states $\phi_f(\mathbf{r})$ transforming as $\boldsymbol{\Gamma}_0^5$. Examination of the Clebsch–Gordan coefficients shows that $\phi_f(\mathbf{r})$ must transform as the second row of $\boldsymbol{\Gamma}_0^5$.

Finally, for $\mathbf{A}_0$ in the z-direction, $\mathbf{A}_0 \cdot \mathbf{grad} = A_{03}\, \partial/\partial z$, an irreducible tensor operator transforming as the second row of $\boldsymbol{\Gamma}_0^5$. Again the Clebsch–Gordan series indicates that $\phi_f(\mathbf{r})$ cannot transform as either row of $\boldsymbol{\Gamma}_0^5$, while the Clebsch–Gordan coefficients show that $\phi_f(\mathbf{r})$ can only transform as $\boldsymbol{\Gamma}_0^3$ or $\boldsymbol{\Gamma}_0^4$.

3 Time-independent perturbation theory

Suppose that the Hamiltonian H of a system is time-independent and is made up of two time-independent parts H_0 and H', where H_0 is sufficiently simple that its eigenfunctions and eigenvalues are known, and H' is sufficiently small that its effect can be considered to be a perturbation on H_0. The problem is then to find the eigenfunctions and eigenvalues of $H = H_0 + H'$ in terms of those of H_0. It is assumed that the eigenfunctions of H and H_0 can be put into a one-to-one correspondence, in the sense that they are the limits as $\lambda \to 0$ and $\lambda \to 1$ respectively of the eigenfunctions of the operator $H_0 + \lambda H_1$, $0 \leq \lambda \leq 1$. If $\phi_1, \phi_2, \ldots, \phi_d$ are a set of eigenfunctions of H_0 corresponding to the d-fold degenerate energy eigenvalue ε_0 and the associated eigenfunctions $\psi_1, \psi_2, \ldots, \psi_d$ of H correspond to eigenvalues that are *not* equal to each other, then the perturbation H' may be said to "split" ε_0.

It will be shown that a considerable amount of information about such splittings can be found very simply using group representation theory. For simplicity it will be assumed that H and H_0 give "single-particle" Schrödinger equations (in the sense of Chapter 1, Section 3(a)). Let $\mathscr{G}_0$ and $\mathscr{G}$ be the invariance groups of H_0 and H respectively. Usually there exists a coordinate transformation T of $\mathscr{G}_0$ such that

$$H'(\{\mathbf{R}(T) \mid \mathbf{t}(T)\}\mathbf{r}) \neq H'(\mathbf{r}),$$

so that Equation (1.13) implies that $\mathcal{G}$ is a proper subgroup of $\mathcal{G}_0$. Suppose that the unperturbed energy eigenvalue ε_0 corresponds to a representation Γ_0 of $\mathcal{G}_0$. Then Γ_0 provides a representation of the subgroup $\mathcal{G}$. However, even if Γ_0 is an irreducible representation of $\mathcal{G}_0$, as a representation of $\mathcal{G}$ Γ_0 may be *reducible*. For convenience it will be assumed that $\mathcal{G}_0$ is a finite group or a compact Lie group, so that Γ_0 is then completely reducible (see Chapter 4, Section 4). Thus suppose that as a representation of $\mathcal{G}_0$

$$\Gamma_0 \approx \sum_r \oplus n_r \Gamma^r, \tag{6.18}$$

where the sum is over all inequivalent unitary irreducible representations Γ^r of $\mathcal{G}$, n_r being the number of times that Γ^r appears in the reduction of Γ_0. Here n_r may be zero for some Γ^r. These integers are easily evaluated from a knowledge of the characters of $\mathcal{G}_0$ and $\mathcal{G}$ alone, for Theorem V of Chapter 4, Section 6 gives

$$n_r = (1/g) \sum_{T \in \mathcal{G}} \chi_0(T) \chi^r(T)^*$$

for the case in which $\mathcal{G}$ is finite and of order g. Here $\chi_0(T)$ and $\chi^r(T)$ denote the characters of Γ_0 and Γ^r respectively. Similarly, if $\mathcal{G}$ is a compact Lie group,

$$n_r = \int_{\mathcal{G}} \chi_0(T) \chi^r(T)^* \, dT.$$

Suppose that $\sum_r n_r = n$, so that Equation (6.18) contains n irreducible representations of $\mathcal{G}$. Then the d perturbed energy eigenfunctions $\psi_1, \psi_2, \ldots, \psi_d$ in general belong to n different eigenvalues of H. Thus ε_0 is split by the perturbation H′ into n different values. If d_r is the dimension of Γ^r, then the perturbed energy eigenvalue corresponding to Γ^r is d_r-fold degenerate. Naturally $\sum_r n_r d_r = d$. As $d_r = \chi^r(E)$, both the number of perturbed energy eigenvalues and their degeneracies can be predicted using the *characters* of $\mathcal{G}_0$ and $\mathcal{G}$ alone.

In the special case in which Γ_0 provides an *irreducible* representation of $\mathcal{G}$, the unperturbed eigenvalue ε_0 is not split. (In particular, this happens when $\mathcal{G}$ coincides with $\mathcal{G}_0$ and Γ_0 is an irreducible representation of $\mathcal{G}_0$.)

As all expressions for the perturbed eigenvalues and eigenfunctions involve only the unperturbed energy eigenvalues and matrix elements of H′ between unperturbed energy eigenvalues (Schiff 1968), the Wigner–Eckart Theorem of Chapter 5, Section 3, can be brought into use again. For example, suppose that Γ_0 is equal to Γ_0^p, a unitary irreducible representation of $\mathcal{G}_0$, and $\phi_1^p(\mathbf{r}), \phi_2^p(\mathbf{r}), \ldots, \phi_d^p(\mathbf{r})$

are a set of linearly independent eigenfunctions of H_0, with eigenvalue ε_0, that are basis functions for Γ_0^p. Suppose also that $H'(\mathbf{r})$ is an irreducible tensor operator transforming as the kth row of a unitary irreducible representation Γ_0^q of $\mathcal{G}_0$. Then, if $d=1$ (that is, if ε_0 is "non-degenerate"), the corresponding perturbed eigenvalue ε is given to first order by

$$\varepsilon = \varepsilon_0 + (\phi_1^p, H'\phi_1^p),$$

while if $d>1$ the corresponding first-order perturbed eigenvalues are the eigenvalues of the $d\times d$ matrix $\mathbf{A}$ whose elements are given by

$$A_{lj} = \varepsilon_0\delta_{lj} + (\phi_l^p, H'\phi_j^p), \tag{6.19}$$

$j, l = 1, 2, \ldots, d$. In both cases there is a first-order effect only if $\Gamma_0^p \otimes \Gamma_0^q$ contains Γ_0^p (that is, when $n_{pq}^p \neq 0$). When this is so the Wigner–Eckart Theorem shows that

$$(\phi_l^p, H'\phi_j^p) = \sum_{\alpha=1}^{n_{pq}^p} \begin{pmatrix} p & q \\ j & k \end{pmatrix} \begin{vmatrix} p, & \alpha \\ l & \end{vmatrix}(p\,|H'|\,p)_\alpha, \tag{6.20}$$

so that the matrix elements depend on the Clebsch–Gordan coefficients and n_{pq}^p reduced matrix elements.

The case $d>1$ can be investigated further. (There will be no need to assume in what follows that $H'(\mathbf{r})$ is a single irreducible tensor operator.) Let $\mathbf{S}$ be a $d\times d$ matrix such that on the subgroup $\mathcal{G}$

$$\mathbf{S}^{-1}\Gamma_0^p\mathbf{S} = \sum_r \oplus n_r\Gamma^r, \tag{6.21}$$

that is, $\mathbf{S}$ induces a similarity transformation that puts Γ_0^p into direct sum form on $\mathcal{G}$. Then, as will be shown below,

$$\mathbf{S}^{-1}\mathbf{A}\mathbf{S} = \sum_{r \text{ such that } n_r\neq 0} \oplus\, d_r\mathbf{A}'_r, \tag{6.22}$$

where $\mathbf{A}'_r$ is a $n_r\times n_r$ matrix. (For example, for $d=7$, if

$$\mathbf{S}^{-1}\Gamma_0^p\mathbf{S} = 2\Gamma^2 \oplus \Gamma^3 \tag{6.23}$$

and if $d_2=3$ and $d_3=1$, then Equation (6.22) becomes

$$\mathbf{S}^{-1}\mathbf{A}\mathbf{S} = \left[\begin{array}{c:c:c:c} \mathbf{A}'_2 & \mathbf{0} & \mathbf{0} & \mathbf{0} \\ \hdashline \mathbf{0} & \mathbf{A}'_2 & \mathbf{0} & \mathbf{0} \\ \hdashline \mathbf{0} & \mathbf{0} & \mathbf{A}'_2 & \mathbf{0} \\ \hdashline \mathbf{0} & \mathbf{0} & \mathbf{0} & \mathbf{A}'_3 \end{array}\right],$$

where $\mathbf{A}'_2$ and $\mathbf{A}'_3$ are 2×2 and 1×1 matrices respectively.) As the sets of eigenvalues of $\mathbf{A}$ and $\mathbf{S}^{-1}\mathbf{AS}$ coincide, *the first-order perturbed energy eigenvalues are given by the eigenvalues of the submatrices* $\mathbf{A}'_r$. In fact, when $n_r = 1$, $\mathbf{A}'_r$ only has one eigenvalue which is equal to its diagonal matrix element. Indeed, when $n_r = 0$ or 1 for all Γ^r of $\mathscr{G}$, $\mathbf{S}^{-1}\mathbf{AS}$ is *completely* diagonal. (The d_r-fold degeneracy of the perturbed energy eigenvalue associated with Γ^r is a consequence of the appearance of the *same* submatrix $\mathbf{A}'_r$ d_r times in Equation (6.22).)

The powerful matrix $\mathbf{S}$ is easily found by the requirement that if $\psi_1(\mathbf{r}), \psi_2(\mathbf{r}), \ldots, \psi_d(\mathbf{r})$ are *any* set of basis functions transforming as Γ_0^p and $\psi'_1(\mathbf{r}), \psi'_2(\mathbf{r}), \ldots, \psi'_d(\mathbf{r})$ are defined by

$$\psi'_n(\mathbf{r}) = \sum_{m=1}^{d} S_{mn}\psi_m(\mathbf{r}),$$

then $\psi'_1(\mathbf{r}), \psi'_2(\mathbf{r}), \ldots, \psi'_d(\mathbf{r})$ must be basis functions for the irreducible representations of $\mathscr{G}$ that appear on the right-hand side of Equation (6.18) (see Theorem II of Chapter 4, Section 2). When $\mathscr{G}$ is finite, the appropriate linear combinations are most easily found using the projection operators of Chapter 5, Section 1. (In such an analysis the functions $\psi_m(\mathbf{r})$ may be chosen to be as simple as possible. They need not be the actual unperturbed energy eigenfunctions $\phi_m^p(\mathbf{r})$.)

It remains only to prove Equation (6.22). Let $\phi'_n(\mathbf{r}) = \sum_{m=1}^{d} S_{mn}\phi_m^p(\mathbf{r})$, so that $\phi'_1(\mathbf{r}), \phi'_2(\mathbf{r}), \ldots, \phi'_d(\mathbf{r})$ transform as basis functions for the irreducible representations of $\mathscr{G}$ that appear on the right-hand side of Equation (6.18). Then

$$(\mathbf{S}^{-1}\mathbf{AS})_{mn} = \varepsilon_0(\phi'_m, \phi'_n) + (\phi'_m, H'\phi'_n).$$

Then Theorem V of Chapter 4, Section 5 and Theorem III of Chapter 5, Section 3 imply that $(\mathbf{S}^{-1}\mathbf{AS})_{mn}$ is non-zero only when $\phi'_m(\mathbf{r})$ and $\phi'_n(\mathbf{r})$ transform as the same row of the same representation Γ^r of $\mathscr{G}$. These theorems also show that, when $d_r > 1$, $(\mathbf{S}^{-1}\mathbf{AS})_{mn} = (\mathbf{S}^{-1}\mathbf{AS})_{st}$, where $\phi'_m(\mathbf{r}), \phi'_n(\mathbf{r}), \phi'_s(\mathbf{r})$ and $\phi'_t(\mathbf{r})$ all transform as the same row of Γ^r. For example, with the decomposition of Equation (6.23), let $\phi'_1 = \psi^2_{1,1}$, $\phi'_2 = \psi^2_{1,2}$, $\phi'_3 = \psi^2_{2,1}$, $\phi'_4 = \psi^2_{2,2}$, $\phi'_5 = \psi^2_{3,1}$, $\phi'_6 = \psi^2_{3,2}$, $\phi'_7 = \psi^3_1$, where $(\psi^2_{1,1}, \psi^2_{2,1}, \psi^2_{3,1})$ and $(\psi^2_{1,2}, \psi^2_{2,2}, \psi^2_{3,2})$ are two linearly independent sets of basis transforming as Γ^2, and ψ^3_1 transforms as Γ^3. Then, for instance,

$$\varepsilon_0(\psi^2_{m,j}, \psi^2_{n,k}) + (\psi^2_{m,j}, H'\psi^2_{n,k}) \tag{6.24}$$

is zero if $m \neq n$ and, as the value of Expression (6.24) is independent of m when $m = n$, it may be denoted by $(\mathbf{A}'_2)_{jk}$, thereby defining the matrix $\mathbf{A}'_2$.

Example I *The crystallographic point group* D_4
Suppose that $\mathscr{G}_0 = D_4$ and that Γ_0 is the two-dimensional irreducible representation Γ_0^5 of D_4 (known in previous examples as Γ^5). Suppose also that $H'(\mathbf{r}) = axz$, where a is a real number. Then $\mathscr{G} = \{E, C_{2y}, C_{2c}, C_{2d}\}$, the crystallographic point group D_2, whose characters are displayed in Table D.22 of Appendix D (with $\mathscr{C}_1 = E$, $\mathscr{C}_2 = C_{2y}$, $\mathscr{C}_3 = C_{2c}$, $\mathscr{C}_4 = C_{2d}$). Then

$$\Gamma_0^5 \approx \Gamma^3 \oplus \Gamma^4,$$

so the two-fold degenerate unperturbed energy eigenvalue ε_0 associated with Γ_0^5 will be split by $H'(\mathbf{r})$ into two non-degenerate eigenvalues. These will now be determined.

As xz is a basis function of the one-dimensional irreducible representation Γ_0^4 of D_4 (referred to in previous examples as Γ^4), Example IV of Chapter 5, Section 3, shows that $H'(\mathbf{r})$ is an irreducible tensor operator transforming as this representation. Thus, from Equations (6.19) and (6.20),

$$\mathbf{A} = \begin{bmatrix} \varepsilon_0 + \left(\begin{smallmatrix}5 & 4\\ 1 & 1\end{smallmatrix}\middle|\begin{smallmatrix}5, & 1\\ 1 & \end{smallmatrix}\right)h' & \left(\begin{smallmatrix}5 & 4\\ 2 & 1\end{smallmatrix}\middle|\begin{smallmatrix}5, & 1\\ 1 & \end{smallmatrix}\right)h' \\ \left(\begin{smallmatrix}5 & 4\\ 1 & 1\end{smallmatrix}\middle|\begin{smallmatrix}5, & 1\\ 2 & \end{smallmatrix}\right)h' & \varepsilon_0 + \left(\begin{smallmatrix}5 & 4\\ 2 & 2\end{smallmatrix}\middle|\begin{smallmatrix}5, & 1\\ 2 & \end{smallmatrix}\right)h' \end{bmatrix},$$

where $h' = (5\,|H'|\,5)_1$. Invoking the Clebsch–Gordan coefficients of Example I of Chapter 5, Section 3, gives

$$\mathbf{A} = \begin{bmatrix} \varepsilon_0 & h' \\ h' & \varepsilon_0 \end{bmatrix},$$

and $h' = (\phi_1^5, H'\phi_2^5)$. As noted in Example I of Chapter 5, Section 1, x and z are basis functions for the rows of Γ_0^5. But, from Table D.22, $\mathscr{P}_{11}^3 x = \frac{1}{2}(x - z)$ and $\mathscr{P}_{11}^4 x = \frac{1}{2}(x + z)$, so that one choice of $\mathbf{S}$ is

$$\mathbf{S} = \tfrac{1}{2}\begin{bmatrix} 1 & 1 \\ -1 & 1 \end{bmatrix}.$$

Then Equation (6.22) becomes

$$\mathbf{S}^{-1}\mathbf{A}\mathbf{S} = \begin{bmatrix} \varepsilon_0 - h' & 0 \\ 0 & \varepsilon_0 + h' \end{bmatrix},$$

so that the perturbed energy eigenvalues are $\varepsilon_0 - h'$ and $\varepsilon_0 + h'$ respectively.

4 Generalization of the theory to incorporate spin-$\frac{1}{2}$ particles

(a) *The spinor transformation operators*

In all the developments so far it has been assumed that the particles involved are spinless or, if they do possess a non-zero spin, that this spin does not appear in the Hamiltonian operator. (This latter assumption was discussed in some detail in Chapter 1, Section 3(*a*).) It is now time to remove this restriction, so that spin-dependent interactions, such as spin-orbit coupling, can be taken into account explicitly. The foregoing theory has to be modified, but the same basic features remain. The only major difference is the use of "double" groups instead of the "single" groups employed hitherto.

The first stage is to develop the transformation theory for two-component spinors along the same lines as those given in Chapter 1, Section 3 for scalars. This will be carried out in this subsection. In the next subsection it will be shown how this leads naturally to the concept of a double group. The representation theory of double groups will be outlined in subsection (*c*). Finally, in subsection (*d*) these ideas will be used to show how the effect of spin-orbit coupling on the degeneracies of energy eigenvalues can be predicted.

As discussed in Chapter 1, Section 3(*c*), a *scalar* field transforms under a coordinate transformation $\mathbf{r}' = \{\mathbf{R}(\mathrm{T}) \mid \mathbf{t}(\mathrm{T})\}\mathbf{r}$ according to the simple rule in Equation (1.14), namely $\psi'(\mathbf{r}') = \psi(\mathbf{r})$, where $\psi(\mathbf{r})$ and $\psi'(\mathbf{r}')$ specify the field relative to Ox, Oy, Oz and $O'x', O'y', O'z'$ respectively. For a *vector* field the transformation law is only slightly more complicated. Such a field may be defined as a three-component quantity whose components $A_1(\mathbf{r}), A_2(\mathbf{r}), A_3(\mathbf{r})$ relative to Ox, Oy, Oz and components $A_1'(\mathbf{r}'), A_2'(\mathbf{r}'), A_3'(\mathbf{r}')$ relative to $O'x', O'y', O'z'$ are related by the transformation law

$$A_\alpha'(\mathbf{r}') = \sum_{\beta=1}^{3} \mathrm{R(T)}_{\alpha\beta} A_\beta(\mathbf{r})$$

for $\alpha = 1, 2, 3$. Similarly, a ***second-order tensor*** field is a nine-component quantity for which the components $A_{\alpha\beta}(\mathbf{r})$ and $A'_{\alpha\beta}(\mathbf{r}')$ $(\alpha, \beta = 1, 2, 3)$ relative to Ox, Oy, Oz and $O'x', O'y', O'z'$ are related by the tensor transformation law

$$A'_{\alpha\beta}(\mathbf{r}') = \sum_{\rho=1}^{3} \sum_{\sigma=1}^{3} \mathrm{R(T)}_{\alpha\rho} \mathrm{R(T)}_{\beta\sigma} A_{\rho\sigma}(\mathbf{r})$$

for $\alpha, \beta = 1, 2, 3$. Higher-order tensor fields are defined in a similar way.

When it is recognized from Equation (A.8) that $\mathrm{R(T)}_{\alpha\rho}\mathrm{R(T)}_{\beta\sigma} = (\mathbf{R}(\mathrm{T}) \otimes \mathbf{R}(\mathrm{T}))_{\alpha\beta,\rho\sigma}$, the general rule becomes apparent. It is that all these quantities transform under the coordinate transformation $\mathbf{r}' = \{\mathbf{R}(\mathrm{T}) \mid \mathbf{t}(\mathrm{T})\}\mathbf{r}$ as

$$B_\alpha'(\mathbf{r}') = \sum_{\alpha=1}^{d} \Gamma(\mathbf{R}(\mathrm{T}))_{\alpha\beta} B_\beta(\mathbf{r}),$$

where $\mathbf{\Gamma}$ is some faithful d-dimensional *representation* of the group of all rotations in $\mathbb{R}^3$. (For *scalars* $d=1$, $B_1'=\psi'$, $B_1=\psi$ and $\mathbf{\Gamma}(\mathbf{R}(\mathrm{T}))=[1]$ for all rotations, while for *vectors* $d=3$, $B_\alpha'=A_\alpha'$, $B_\alpha=A_\alpha$ and $\mathbf{\Gamma}(\mathbf{R}(\mathrm{T}))=\mathbf{R}(\mathrm{T})$ for all rotations and so on.)

The scalar field describes spinless particles, the vector field corresponds to particles with spin 1, and higher-order tensor fields are associated with particles of higher integral spin. Unfortunately, these considerations leave out the important case of the electron, which, together with protons, neutrons and certain other elementary particles, has spin $\frac{1}{2}$. It is well known (Schiff 1968) that such particles may be described by two-component quantities, but the group of all rotations in $\mathbb{R}^3$ possesses *no* faithful two-dimensional representation (see Chapter 12, Section 4). However, as was shown in Chapter 3, Section 5, the group of all proper rotations in $\mathbb{R}^3$ does possess a "two-valued" two-dimensional representation, made up of the matrices $\pm\mathbf{u}(\mathbf{R})$ of SU(2). These have the property that

$$\mathbf{u}(\mathbf{R}_1)\mathbf{u}(\mathbf{R}_2)=\pm\mathbf{u}(\mathbf{R}_1\mathbf{R}_2)$$

for any $\mathbf{R}_1, \mathbf{R}_2 \in \mathrm{SO}(3)$. Accordingly, the transformation law under proper rotations for a spin-$\frac{1}{2}$ particle may be taken to be

$$\psi_\alpha'(\mathbf{r}')=\pm\sum_{\beta=1}^{2}\mathrm{u}(\mathbf{R}(\mathrm{T}))_{\alpha\beta}\psi_\beta(\mathbf{r}) \tag{6.25}$$

for $\alpha=1,2$. Here $\psi_\alpha(\mathbf{r})$ and $\psi_\alpha'(\mathbf{r}')$ $(\alpha=1,2)$ are the components of the "two-component spinor" defined relative to Ox, Oy, Oz and $O'x', O'y', O'z'$ respectively. This formulation is entirely equivalent to the usual Pauli theory, the precise connection being established in Chapter 12, Section 7. (Henceforth, for clarity, the different *components* of a spinor will be distinguished by Greek subscripts, with Latin subscripts and superscripts indicating different spinors.)

The sign of $\mathbf{u}(\mathbf{R}(\mathrm{T}))$ is entirely a matter of convention, there being no reason of principle for choosing one sign rather than the other. This point will be taken up later (in the next subsection), but for the moment it will be simply assumed that some specific choice has been made for each proper rotation.

The transformation law, Equation (6.25), can be generalized to arbitrary coordinate transformations $\mathbf{r}'=\{\mathbf{R}(\mathrm{T})\,|\,\mathbf{t}(\mathrm{T})\}\mathbf{r}$ in $\mathbb{R}^3$ by defining for every rotation $\mathbf{R}(\mathrm{T})$ a "proper part" $\mathbf{R}_p(\mathrm{T})$ by

$$\mathbf{R}_p(\mathrm{T})=\begin{cases}\mathbf{R}(\mathrm{T}), & \text{if } \mathbf{R}(\mathrm{T}) \text{ is a proper rotation,}\\ -\mathbf{R}(\mathrm{T}), & \text{if } \mathbf{R}(\mathrm{T}) \text{ is improper.}\end{cases}$$

Then Equation (6.25) generalizes to

$$\psi'_\alpha(\mathbf{r}') = \pm \sum_{\beta=1}^{2} u(\mathbf{R}_p(T))_{\alpha\beta}\psi_\beta(\mathbf{r}) \tag{6.26}$$

for $\alpha = 1, 2$. In essentially the same way that Equation (1.14) can be re-expressed as Equation (1.16), Equation (6.26) can be rewritten as

$$\psi'_\alpha(\mathbf{r}) = \pm \sum_{\beta=1}^{2} u(\mathbf{R}_p(T))_{\alpha\beta}\psi_\beta(\{\mathbf{R}(T) \mid \mathbf{t}(T)\}^{-1}\mathbf{r})$$

for $\alpha = 1, 2$.

By analogy with Equation (1.17), for every coordinate transformation T a *pair* of "spinor transformation operators" O(T) and O($\bar{T}$) may be defined by

$$\left.\begin{aligned} O(T)\psi_\alpha(\mathbf{r}) &= +\sum_{\beta=1}^{2} u(\mathbf{R}_p(T))_{\alpha\beta}\psi_\beta(\{\mathbf{R}(T) \mid \mathbf{t}(T)\}^{-1}\mathbf{r}),\\ O(\bar{T})\psi_\alpha(\mathbf{r}) &= -\sum_{\beta=1}^{2} u(\mathbf{R}_p(T))_{\alpha\beta}\psi_\beta(\{\mathbf{R}(T) \mid \mathbf{t}(T)\}^{-1}\mathbf{r}),\end{aligned}\right\} \tag{6.27}$$

for $\alpha = 1, 2$. It is useful to regard two-component spinors as 2×1 column matrices and to use bold type to distinguish them from scalars. Then

$$\boldsymbol{\psi}(\mathbf{r}) = \begin{bmatrix} \psi_1(\mathbf{r}) \\ \psi_2(\mathbf{r}) \end{bmatrix},$$

where $\psi_1(\mathbf{r})$, $\psi_2(\mathbf{r})$ are the components of $\boldsymbol{\psi}(\mathbf{r})$. No confusion ever arises if the operators O(T) and O($\bar{T}$) are considered (when convenient) to act on the whole spinor $\boldsymbol{\psi}(\mathbf{r})$ rather than on its components, in which case Equations (6.27) become

$$\left.\begin{aligned} O(T)\boldsymbol{\psi}(\mathbf{r}) &= +\mathbf{u}(\mathbf{R}_p(T))\boldsymbol{\psi}(\{\mathbf{R}(T) \mid \mathbf{t}(T)\}^{-1}\mathbf{r}),\\ O(\bar{T})\boldsymbol{\psi}(\mathbf{r}) &= -\mathbf{u}(\mathbf{R}_p(T))\boldsymbol{\psi}(\{\mathbf{R}(T) \mid \mathbf{t}(T)\}^{-1}\mathbf{r}).\end{aligned}\right\} \tag{6.28}$$

It is not difficult to extend the arguments for L^2 (see Appendix B) to show that, if the inner product of two two-component spinors $\boldsymbol{\phi}(\mathbf{r})$ and $\boldsymbol{\psi}(\mathbf{r})$ is defined by

$$(\boldsymbol{\phi}, \boldsymbol{\psi}) = \sum_{\alpha=1}^{2} \int_{-\infty}^{+\infty}\int_{-\infty}^{+\infty}\int_{-\infty}^{+\infty} \phi_\alpha^*(\mathbf{r})\psi_\alpha(\mathbf{r})\,dx\,dy\,dz \tag{6.29}$$

and the zero spinor is defined to have components that are identically zero almost everywhere, then the set of two-component spinors $\boldsymbol{\psi}(\mathbf{r})$ for which $\|\boldsymbol{\psi}\|$ is finite forms an infinite-dimensional complex Hilbert space.

The operators O(T) and O($\bar{T}$) then play for spinors the same role that the operators P(T) play for scalars, their properties being largely the same. In particular, O(T) and O($\bar{T}$) are *linear* operators acting in this Hilbert space, and $O(T_1) = O(T_2)$ and $O(\bar{T}_1) = O(\bar{T}_2)$ only if $T_1 = T_2$. Their other properties are summarized in the following theorems.

Theorem I Each of the operators O(T) and O($\bar{T}$) is a unitary operator in the Hilbert space of two-component spinors. That is, with $(\boldsymbol{\phi}, \boldsymbol{\psi})$ defined by Equation (6.29), for any coordinate transformation T,

$$(O(T)\boldsymbol{\phi}, O(T)\boldsymbol{\psi}) = (O(\bar{T})\boldsymbol{\phi}, O(\bar{T})\boldsymbol{\psi}) = (\boldsymbol{\phi}, \boldsymbol{\psi}),$$

for any two two-component spinors $\boldsymbol{\phi}(\mathbf{r})$ and $\boldsymbol{\psi}(\mathbf{r})$ of the space.

Proof Equation (6.29) can be rewritten as

$$(\boldsymbol{\phi}, \boldsymbol{\psi}) = \int_{-\infty}^{+\infty}\int_{-\infty}^{+\infty}\int_{-\infty}^{+\infty} \{\boldsymbol{\phi}(\mathbf{r})\}^{\dagger}\boldsymbol{\psi}(\mathbf{r})\, dx\, dy\, dz$$

Then, by Equations (6.28), with $\mathbf{r}''$ defined by $\mathbf{r}'' = \{\mathbf{R}(T) \mid \mathbf{t}(T)\}^{-1}\mathbf{r}$,

$$\begin{aligned}(O(T)\boldsymbol{\phi}, O(T)\boldsymbol{\psi}) &= (O(\bar{T})\boldsymbol{\phi}, O(\bar{T})\boldsymbol{\psi})\\ &= \int_{-\infty}^{+\infty}\int_{-\infty}^{+\infty}\int_{-\infty}^{+\infty} \{\boldsymbol{\phi}(\mathbf{r}'')\}^{\dagger}\mathbf{u}(\mathbf{R}_p(T))^{\dagger}\mathbf{u}(\mathbf{R}_p(T))\boldsymbol{\psi}(\mathbf{r}'')\, dx\, dy\, dz\\ &= \int_{-\infty}^{+\infty}\int_{-\infty}^{+\infty}\int_{-\infty}^{+\infty} \{\boldsymbol{\phi}(\mathbf{r}'')\}^{\dagger}\boldsymbol{\psi}(\mathbf{r}'')\, dx\, dy\, dz,\end{aligned}$$

as $\mathbf{u}(\mathbf{R}_p(T))$ is unitary. The rest of the argument is then essentially the same as in the proof of Theorem I of Chapter 1, Section 3.

Theorem II For any two coordinate transformations T_1 and T_2,

$$O(T_1)O(T_2) = O(\bar{T}_1)O(\bar{T}_2) = \begin{cases} O(T_1T_2), & \text{if } \mathbf{u}(\mathbf{R}_p(T_1))\mathbf{u}(\mathbf{R}_p(T_2)) = +\mathbf{u}(\mathbf{R}_p(T_1T_2)),\\ O(\overline{T_1T_2}), & \text{if } \mathbf{u}(\mathbf{R}_p(T_1))\mathbf{u}(\mathbf{R}_p(T_2)) = -\mathbf{u}(\mathbf{R}_p(T_1T_2)),\end{cases} \tag{6.30}$$

and

$$O(\bar{T}_1)O(T_2) = O(T_1)O(\bar{T}_2) = \begin{cases} O(\overline{T_1T_2}), & \text{if } \mathbf{u}(\mathbf{R}_p(T_1))\mathbf{u}(\mathbf{R}_p(T_2)) = -\mathbf{u}(\mathbf{R}_p(T_1T_2),\\ O(T_1T_2), & \text{if } \mathbf{u}(\mathbf{R}_p(T_1))\mathbf{u}(\mathbf{R}_p(T_2)) = +\mathbf{u}(\mathbf{R}_p(T_1T_2)).\end{cases} \tag{6.31}$$

Proof Consider first $O(T_1)O(T_2)$. By an immediate generalization of the proof given for Theorem II of Chapter 1, Section 3, for any $\boldsymbol{\psi}(\mathbf{r})$,

$$O(T_1)O(T_2)\boldsymbol{\psi}(\mathbf{r}) = \mathbf{u}(\mathbf{R}_p(T_1))\mathbf{u}(\mathbf{R}_p(T_2))\boldsymbol{\psi}(\{\mathbf{R}(T_1T_2) \mid \mathbf{t}(T_1T_2)\}^{-1}\mathbf{r})$$

As

$$\mathbf{u}(\mathbf{R}_p(T_1))\mathbf{u}(\mathbf{R}_p(T_2)) = \pm\mathbf{u}(\mathbf{R}_p(T_1)\mathbf{R}_p(T_2)) = \pm\mathbf{u}(\mathbf{R}_p(T_1T_2)),$$

the quoted result follows at once. The arguments for the other products are similar.

(*b*) *Double groups*

The group theoretical properties of the spinor transformation operators will now be investigated.

Theorem III If the set of coordinate transformations T forms a group $\mathcal{G}$, then the corresponding set of operators $O(T)$ and $O(\bar{T})$ also forms a group.

Proof All that has to be verified is that the four group axioms are satisfied. If $T_1, T_2 \in \mathcal{G}$, then, by Theorem II, $O(T_1)O(T_2)$, $O(\bar{T}_1)O(T_2)$, $O(T_1)O(\bar{T}_2)$ and $O(\bar{T}_1)O(\bar{T}_2)$ are each equal either to $O(T_1T_2)$ or to $O(\overline{T_1T_2})$ and, as $T_1T_2 \in \mathcal{G}$, the first axiom is satisfied. The associative law is automatic for operators, while the identity of the set of spinor transformation operators is $O(E)$, where E is the identity of $\mathcal{G}$ and $\mathbf{u}(\mathbf{R}_p(E)) = \mathbf{1}_2$. Finally, $O(T)^{-1}$ is either $O(T')$ or $O(\bar{T}')$, where $T' = T^{-1}$, the assignment depending on the sign conventions for $\mathbf{u}(\mathbf{R}_p(T))$ and $\mathbf{u}(\mathbf{R}_p(T^{-1}))$. The same is also true of $O(\bar{T})^{-1}$.

It is very convenient to develop the notation further. As $O(T)$ and $O(\bar{T})$ differ only in the treatment of the *rotational* part of T, it is possible to write $O(T) = O([\mathbf{R}(T) \mid \mathbf{t}(T)])$ and $O(\bar{T}) = O([\bar{\mathbf{R}}(T) \mid \mathbf{t}(T)])$. It is then fruitful to consider that for every "ordinary" transformation $\{\mathbf{R}(T) \mid \mathbf{t}(T)\}$ there correspond two "generalized" transformations $[\mathbf{R}(T) \mid \mathbf{t}(T)]$ and $[\bar{\mathbf{R}}(T) \mid \mathbf{t}(T)]$ that are defined to be isomorphic to the operators $O([\mathbf{R}(T) \mid \mathbf{t}(T)]$ and $O([\bar{\mathbf{R}}(T) \mid \mathbf{t}(T)])$ respectively. (These generalized transformations are distinguished from the ordinary transformations by writing them in square brackets.) Thus, for any T_1 and T_2, by Equations (6.30)

$$[\mathbf{R}(T_1) \mid \mathbf{t}(T_1)][\mathbf{R}(T_2) \mid \mathbf{t}(T_2)] = [\bar{\mathbf{R}}(T_1) \mid \mathbf{t}(T_1)][\bar{\mathbf{R}}(T_2) \mid \mathbf{t}(T_2)]$$
$$= \begin{cases} [\mathbf{R}(T_1T_2) \mid \mathbf{t}(T_1T_2)], & \text{if} \quad \mathbf{u}(\mathbf{R}_p(T_1))\mathbf{u}(\mathbf{R}_p(T_2)) = +\mathbf{u}(\mathbf{R}_p(T_1T_2)), \\ [\bar{\mathbf{R}}(T_1T_2) \mid \mathbf{t}(T_1T_2)], & \text{if} \quad \mathbf{u}(\mathbf{R}_p(T_1))\mathbf{u}(\mathbf{R}_p(T_2)) = -\mathbf{u}(\mathbf{R}_p(T_1T_2)), \end{cases} \quad (6.32)$$

and by Equations (6.31)

$$[\bar{\mathbf{R}}(T_1) \mid \mathbf{t}(T_1)][\mathbf{R}(T_2) \mid \mathbf{t}(T_2)] = [\mathbf{R}(T_1 \mid \mathbf{t}(T_1)][\bar{\mathbf{R}}(T_2) \mid \mathbf{t}(T_2)]$$
$$= \begin{cases} [\mathbf{R}(T_1T_2) \mid \mathbf{t}(T_1T_2)], & \text{if} \quad \mathbf{u}(\mathbf{R}_p(T_1))\mathbf{u}(\mathbf{R}_p(T_2)) = -\mathbf{u}(\mathbf{R}_p(T_1T_2)), \\ [\bar{\mathbf{R}}(T_1T_2) \mid \mathbf{t}(T_1T_2)], & \text{if} \quad \mathbf{u}(\mathbf{R}_p(T_1))\mathbf{u}(\mathbf{R}_p(T_2)) = +\mathbf{u}(\mathbf{R}_p(T_1T_2)), \end{cases} \quad (6.33)$$

where, as usual (cf. Equation (1.7)), $\mathbf{R}(T_1T_2)=\mathbf{R}(T_1)\mathbf{R}(T_2)$ and $\mathbf{t}(T_1T_2)=\mathbf{R}(T_1)\mathbf{t}(T_2)+\mathbf{t}(T_1)$.

If the set of coordinate transformations $\{\mathbf{R}(T)\,|\,\mathbf{t}(T)\}$ forms a group $\mathcal{G}$, then by construction the set of generalized coordinate transformations $[\mathbf{R}(T)\,|\,\mathbf{t}(T)]$ and $[\bar{\mathbf{R}}(T)\,|\,\mathbf{t}(T)]$ also forms a group that is isomorphic to the group of spinor transformation operators. This group is called the "double group" corresponding to $\mathcal{G}$ and will be denoted by $\mathcal{G}^D$.

If $\mathcal{G}$ is a group of finite order g, then $\mathcal{G}^D$ is a finite group of order 2g. As the group of all proper rotations in $\mathbb{R}^3$ is isomorphic to SO(3), its corresponding double group is isomorphic to SU(2).

Because there may exist $T_1, T_2 \in \mathcal{G}$ such that

$$[\mathbf{R}(T_1)\,|\,\mathbf{t}(T_1)][\mathbf{R}(T_2)\,|\,\mathbf{t}(T_2)]=[\bar{\mathbf{R}}(T_1T_2)\,|\,\mathbf{t}(T_1T_2)],$$

in general the set of "unbarred" generalized transformations does *not* form a group on its own. The precise relationship between $\mathcal{G}$ and $\mathcal{G}^D$ is made clear by the following theorem:

Theorem IV The two-to-one mapping ϕ of $\mathcal{G}^D$ onto $\mathcal{G}$ defined by

$$\phi([\mathbf{R}(T)\,|\,\mathbf{t}(T)])=\phi([\bar{\mathbf{R}}(T)\,|\,\mathbf{t}(T)])=\{\mathbf{R}(T)\,|\,\mathbf{t}(T)\} \tag{6.34}$$

is a *homomorphic* mapping. The kernel $\mathcal{K}$ of the mapping consists of $[\mathbf{R}(E)\,|\,\mathbf{0}]$ and $[\bar{\mathbf{R}}(E)\,|\,\mathbf{0}]$, E being the identity of $\mathcal{G}$. Thus, by the First Homomorphism Theorem (Theorem I of Chapter 2, Section 6), $\mathcal{G}$ is isomorphic to $\mathcal{G}^D/\mathcal{K}$, the isomorphic mapping being that which maps $\{\mathbf{R}(T)\,|\,\mathbf{t}(T)\}$ into the coset $[\mathbf{R}(T)\,|\,\mathbf{t}(T)]\mathcal{K}$.

Proof The only part that requires demonstration is the proposition that ϕ is a homomorphism. For any $T_1, T_2 \in \mathcal{G}$,

$$\begin{aligned}
&\phi([\mathbf{R}(T_1)\,|\,\mathbf{t}(T_1)][\mathbf{R}(T_2)\,|\,\mathbf{t}(T_2)])\\
&\quad=\phi([\mathbf{R}(T_1T_2)\,|\,\mathbf{t}(T_1T_2)]) \quad \text{or} \quad \phi([\bar{\mathbf{R}}(T_1T_2)\,|\,\mathbf{t}(T_1T_2)])\\
&\qquad\qquad\qquad\qquad\qquad\qquad\text{(by Equations (6.32))}\\
&\quad=\{\mathbf{R}(T_1T_2)\,|\,\mathbf{t}(T_1T_2)\} \quad \text{(in either case (by Equation (6.34)))}\\
&\quad=\{\mathbf{R}(T_1)\,|\,\mathbf{t}(T_1)\}\{\mathbf{R}(T_2)\,|\,\mathbf{t}(T_2)\}\\
&\quad=\phi([\mathbf{R}(T_1)\,|\,\mathbf{t}(T_1)])\phi([\mathbf{R}(T_2)\,|\,\mathbf{t}(T_2)]).
\end{aligned}$$

The proof is essentially the same for all the other types of product in Equations (6.32) and (6.33).

It is often convenient to abbreviate the notation even further and write the ordinary and generalized transformations corresponding to the composite operation consisting of a rotation C_{nj} followed by a translation $\mathbf{t}$ as $\{C_{nj} \mid \mathbf{t}\}$, $[C_{nj} \mid \mathbf{t}]$ and $[\bar{C}_{nj} \mid \mathbf{t}]$, improper rotations being expressed in a similar way. (Readers should be warned that in the solid state literature little attempt has been made to distinguish the generalized coordinate transformations from the ordinary coordinate transformations. Often the *same* bracket notation is used in *all* cases and in some papers dealing solely with pure rotations and pure generalized rotations the bracket notation is dispensed with completely. This can easily lead to errors.)

For each T the sign of $\mathbf{u}(\mathbf{R}_p(T))$ remains purely a matter of convention. Changing this sign corresponds to interchanging $[\mathbf{R}(T) \mid \mathbf{t}(T)]$ and $[\bar{\mathbf{R}}(T) \mid \mathbf{t}(T)]$ in the group multiplication table. Thus $[\mathbf{R}(T) \mid \mathbf{t}(T)]$ and $[\bar{\mathbf{R}}(T) \mid \mathbf{t}(T)]$ are only completely specified when an explicit choice has been made of $\mathbf{u}(\mathbf{R}_p(T))$. For the identity $[E \mid \mathbf{0}]$ of $\mathcal{G}$ the universally agreed choice is $\mathbf{u}(\mathbf{R}_p(E)) = \mathbf{1}_2$, which makes $[E \mid \mathbf{0}]$ the identity of $\mathcal{G}^D$. In Table D.26 a specific choice is made for all the pure rotations listed in Table D.1 that appear in the character tables for the crystallographic point groups of Appendix D. This choice is completely consistent with that made implicitly in the original papers on the application of double groups in solid state physics by Elliott (1954), Parmenter (1955) and Dresselhaus (1955) (see also Cornwell (1969), Chapter 8, Section 8).

(*c*) *The spinor Hamiltonian operator*

The "single particle" Hamiltonian operator for a particle with spin $\frac{1}{2}$ has the form of a 2×2 matrix whose components are operators. For example, the Hamiltonian operator $\mathbf{H}(\mathbf{r})$ for a particle of mass μ moving in a potential $V(\mathbf{r})$ is

$$\mathbf{H}(\mathbf{r}) = \mathbf{H}_0(\mathbf{r}) + \mathbf{H}_1(\mathbf{r}), \tag{6.35}$$

where

$$\mathbf{H}_0(\mathbf{r}) = \left\{ -\frac{\hbar^2}{2\mu}\left(\frac{\partial^2}{\partial x^2} + \frac{\partial^2}{\partial y^2} + \frac{\partial^2}{\partial z^2}\right) + V(\mathbf{r}) \right\} \mathbf{1}_2$$

and

$$\mathbf{H}_1(\mathbf{r}) = (\hbar^2/4i\mu^2c^2)\{\mathbf{grad}\, V(\mathbf{r})_\wedge \mathbf{grad}\} \,.\, \boldsymbol{\sigma}. \tag{6.36}$$

Here the non-zero elements of $\mathbf{H}_0(\mathbf{r})$ are given by the operator $H(\mathbf{r})$ of Equation (1.10), while $\mathbf{H}_1(\mathbf{r})$ is the "spin-orbit coupling" term which may be derived from the relativistic electron theory of Dirac (1928) (see Schiff (1968)). In Equation (6.36) c is the speed of light and $\boldsymbol{\sigma} = (\boldsymbol{\sigma}_1, \boldsymbol{\sigma}_2, \boldsymbol{\sigma}_3)$, $\boldsymbol{\sigma}_1$, $\boldsymbol{\sigma}_2$ and $\boldsymbol{\sigma}_3$ being the Pauli spin matrices (Equations (3.24)). The corresponding time-independent Schrödinger equation then has the form

$$\mathbf{H}(\mathbf{r})\boldsymbol{\psi}(\mathbf{r}) = \varepsilon\boldsymbol{\psi}(\mathbf{r}), \tag{6.37}$$

the eigenfunction $\boldsymbol{\psi}(\mathbf{r})$ being a two-component spinor.

The group of the Schrödinger equation, sometimes called "the invariance group of $\mathbf{H}(\mathbf{r})$", may be defined as the group of generalized transformations $[\mathbf{R}(T) \mid \mathbf{t}(T)]$ and $[\bar{\mathbf{R}}(T) \mid \mathbf{t}(T)]$ with the property that the associated operators $O([\mathbf{R}(T) \mid \mathbf{t}(T)])$ and $O([\bar{\mathbf{R}}(T) \mid \mathbf{t}(T)])$ *commute* with $\mathbf{H}(\mathbf{r})$.

Theorem V The operators $O([\mathbf{R}(T) \mid \mathbf{t}(T)])$ and $O([\bar{\mathbf{R}}(T) \mid \mathbf{t}(T)])$ commute with $\mathbf{H}(\mathbf{r})$ if and only if

$$\mathbf{H}(\mathbf{r}) = \mathbf{u}(\mathbf{R}_p(T))^{-1}\mathbf{H}(\{\mathbf{R}(T) \mid \mathbf{t}(T)\}\mathbf{r})\mathbf{u}(\mathbf{R}_p(T)). \tag{6.38}$$

Proof This is exactly the same as for Theorem V of Chapter 1, Section 3, except that the scalar $\psi(\mathbf{r})$ must be replaced by a two-component spinor $\boldsymbol{\psi}(\mathbf{r})$.

Theorem VI Let $\mathscr{G}$ be the group of ordinary coordinate transformations $\{\mathbf{R}(T) \mid \mathbf{t}(T)\}$ such that

$$\mathbf{H}_0(\mathbf{r}) = \mathbf{H}_0(\{\mathbf{R}(T) \mid \mathbf{t}(T)\}\mathbf{r}). \tag{6.39}$$

Then the invariance group of the Hamiltonian $\mathbf{H}(\mathbf{r})$ of Equation (6.35) with the spin-orbit coupling term included is the double group $\mathscr{G}^D$ of $\mathscr{G}$.

Proof It has to be shown that Equation (6.38) is true for the Hamiltonian $\mathbf{H}(\mathbf{r})$ of Equation (6.35) if and only if Equation (6.39) is valid. It is obvious that Equation (6.38) holds with $\mathbf{H}(\mathbf{r})$ replaced by $\mathbf{H}_0(\mathbf{r})$ if and only if Equation (6.39) is true, so it remains only to establish that this is also the case if $\mathbf{H}(\mathbf{r})$ is replaced by $\mathbf{H}_1(\mathbf{r})$ in Equation (6.38).

The scalar triple product of Equation (6.36) can be written as

$$\{\mathbf{grad}\, V(\mathbf{r}) \,.\, \mathbf{grad}\} \,.\, \boldsymbol{\sigma} = \det \begin{bmatrix} \partial V(\mathbf{r})/\partial x & \partial V(\mathbf{r})/\partial y & \partial V(\mathbf{r})/\partial z \\ \partial/\partial x & \partial/\partial y & \partial/\partial z \\ \boldsymbol{\sigma}_1 & \boldsymbol{\sigma}_2 & \boldsymbol{\sigma}_3 \end{bmatrix}$$

Let $\mathbf{r}' = \{\mathbf{R}(T) \mid \mathbf{t}(T)\}\mathbf{r}$. Then, if $\mathbf{R}(T)$ is a proper rotation,

$$\partial/\partial x' = R_p(T)_{11}\, \partial/\partial x + R_p(T)_{12}\, \partial/\partial y + R_p(T)_{13}\, \partial/\partial z. \tag{6.40}$$

Consequently, as Equation (6.39) implies $V(\mathbf{r}')=V(\mathbf{r})$,

$$\partial V(\mathbf{r}')/\partial x' = R_p(T)_{11}\,\partial V(\mathbf{r})/\partial x + R_p(T)_{12}\,\partial V(\mathbf{r})/\partial y + R_p(T)_{13}\,\partial V(\mathbf{r})/\partial z. \tag{6.41}$$

Moreover, Equations (3.27) and (3.31) with $y=z=0$ give

$$\mathbf{u}(\mathbf{R}_p(T))\boldsymbol{\sigma}_1\mathbf{u}(\mathbf{R}_p(T))^{-1} = R_p(T)_{11}\boldsymbol{\sigma}_1 + R_p(T)_{21}\boldsymbol{\sigma}_2 + R_p(T)_{31}\boldsymbol{\sigma}_3,$$

so that, replacing T by T^{-1} and using the fact that $\mathbf{R}_p(T)^{-1} = \tilde{\mathbf{R}}_p(T)$,

$$\mathbf{u}(\mathbf{R}_p(T))^{-1}\boldsymbol{\sigma}_1\mathbf{u}(\mathbf{R}_p(T)) = R_p(T)_{11}\boldsymbol{\sigma}_1 + R_p(T)_{12}\boldsymbol{\sigma}_2 + R_p(T)_{13}\boldsymbol{\sigma}_3.$$

Thus

$$\mathbf{u}(\mathbf{R}_p(T))^{-1}\mathbf{H}_1(\mathbf{r})\mathbf{u}(\mathbf{R}_p(T))$$

$$= \frac{\hbar^2}{4i\mu^2c^2}\det\left\{\begin{bmatrix} \partial V(\mathbf{r})/\partial x & \partial V(\mathbf{r})/\partial y & \partial V(\mathbf{r})/\partial z \\ \partial/\partial x & \partial/\partial y & \partial/\partial z \\ \boldsymbol{\sigma}_1 & \boldsymbol{\sigma}_2 & \boldsymbol{\sigma}_3 \end{bmatrix}\tilde{\mathbf{R}}_p(T)\right\} \tag{6.42}$$

$$= \mathbf{H}_1(\mathbf{r}), \quad \text{as} \quad \det\tilde{\mathbf{R}}_p(T) = 1.$$

In the case in which $\mathbf{R}(T)$ is an improper rotation there is a change of sign in the right-hand sides of Equations (6.40) and (6.41), leaving the value of the determinant in Equation (6.42) unchanged.

In the absence of spin-orbit coupling $\mathbf{H}(\mathbf{r}) = \mathbf{H}_0(\mathbf{r}) = H(\mathbf{r})\mathbf{1}_2$, where $H(\mathbf{r})$ is the "scalar" Hamiltonian of Equation (1.10). In this approximation $\boldsymbol{\psi}(\mathbf{r})$ is a spinor eigenfunction of $\mathbf{H}(\mathbf{r})$ with eigenvalue ε if and only if the two components of $\boldsymbol{\psi}(\mathbf{r})$ are both eigenfunctions of $H(\mathbf{r})$ with the same eigenvalue ε and $\mathcal{G}$ is the invariance group of the scalar Hamiltonian $H(\mathbf{r})$. Thus the theorem shows that the invariance group with spin-orbit coupling included is the double group of the invariance group belonging to the same system but with spin-orbit coupling neglected.

(*d*) *Spinor basis functions and energy eigenfunctions*

The definition (Equation (1.26)) of scalar basis functions can be generalized to two-component spinors as follows.

Definition *Spinor basis functions of a double group* $\mathcal{G}^D$
A set of d linearly independent two-component spinor functions $\boldsymbol{\psi}_1(\mathbf{r}), \boldsymbol{\psi}_2(\mathbf{r}), \ldots, \boldsymbol{\psi}_d(\mathbf{r})$ forms a basis for a d-dimensional representation $\boldsymbol{\Gamma}$ of $\mathcal{G}^D$ if, for every coordinate transformation T of the corresponding single group $\mathcal{G}$,

$$O(T)\boldsymbol{\psi}_n(\mathbf{r}) = \sum_{m=1}^{d} \Gamma(T)_{mn}\boldsymbol{\psi}_m(\mathbf{r}), \qquad n = 1, 2, \ldots, d, \tag{6.43}$$

and

$$O(\bar{T})\boldsymbol{\psi}_n(\mathbf{r}) = \sum_{m=1}^{d} \Gamma(\bar{T})_{mn}\boldsymbol{\psi}_m(\mathbf{r}), \qquad n = 1, 2, \ldots, d. \tag{6.44}$$

The spinor $\boldsymbol{\psi}_n(\mathbf{r})$ is then said to transform as the nth row of the representation $\boldsymbol{\Gamma}$.

In terms of components Equation (6.43) is

$$O(T)\psi_{n\alpha}(\mathbf{r}) = \sum_{m=1}^{d} \Gamma(T)_{mn}\psi_{m\alpha}(\mathbf{r})$$

for $n = 1, 2, \ldots, d$ and $\alpha = 1, 2$, a similar expression holding with T replaced by $\bar{T}$.

Theorem I of Chapter 1, Section 4, immediately generalizes to:

Theorem VII The two-component eigenfunctions $\boldsymbol{\psi}(\mathbf{r})$ of a d-fold degenerate eigenvalue ε of the time-independent Schrödinger equation

$$\mathbf{H}(\mathbf{r})\boldsymbol{\psi}(\mathbf{r}) = \varepsilon\boldsymbol{\psi}(\mathbf{r})$$

form a basis for a d-dimensional representation of the double group $\mathscr{G}^D$ that is the invariance group of $\mathbf{H}(\mathbf{r})$.

It follows that the *whole* theory developed previously in Chapters 4, 5 and 6 for scalar functions can be applied immediately to spinors with only two simple modifications:

(*a*) the group of operators P(T) is replaced by the group of operators O(T) and O($\bar{T}$); and
(*b*) every scalar function is replaced by a two-component spinor function.

For example, the projection operator $\mathscr{P}^p_{mn}$ originally defined in Equation (5.2) for a finite single group becomes

$$\mathscr{P}^p_{mn} = (d_p/2g) \sum_{T\in\mathscr{G}} \{\Gamma^p(T)^*_{mn}O(T) + \Gamma^p(\bar{T})^*_{mn}O(\bar{T})\}, \tag{6.45}$$

where d_p is the dimension of the unitary irreducible representation Γ^p of $\mathscr{G}^D$ and the sum is over the g coordinate transformations of the corresponding single group $\mathscr{G}$. Naturally this projection operator now acts on two-component spinors.

Of particular importance are the generalizations of Theorem V of Chapter 4, Section 5, which deals with inner products $(\boldsymbol{\phi}^p_m, \boldsymbol{\psi}^q_n)$, the Wigner–Eckart Theorem of Chapter 5, Section 3, dealing with inner products $(\boldsymbol{\psi}_l^r, \mathbf{Q}_k^q \boldsymbol{\phi}_j^p)$, and its corollary that gives properties of $(\boldsymbol{\phi}_m^p, \mathbf{H}\boldsymbol{\psi}_n^q)$. In all these generalizations the appropriate inner product is that in Equation (6.29).

(e) *Irreducible representations of double groups*

Equations (6.32) and (6.33) show that $[\bar{\mathrm{E}} \,|\, \mathbf{0}]$ commutes with every element of $\mathscr{G}^D$. Schur's Lemma (see Chapter 4, Section 5) therefore implies that, in every irreducible representation $\boldsymbol{\Gamma}$ of $\mathscr{G}^D$, $\boldsymbol{\Gamma}([\bar{\mathrm{E}} \,|\, \mathbf{0}]) = \kappa\mathbf{1}$, where κ is some complex number. Equation (6.33) also shows that, for any $[\mathbf{R}(\mathrm{T}) \,|\, \mathbf{t}(\mathrm{T})]$ of $\mathscr{G}^D$,

$$[\bar{\mathbf{R}}(\mathrm{T}) \,|\, \mathbf{t}(\mathrm{T})] = [\bar{\mathrm{E}} \,|\, \mathbf{0}][\mathbf{R}(\mathrm{T}) \,|\, \mathbf{t}(\mathrm{T})],$$

so that

$$\boldsymbol{\Gamma}([\bar{\mathbf{R}}(\mathrm{T}) \,|\, \mathbf{t}(\mathrm{T})]) = \kappa\boldsymbol{\Gamma}([\mathbf{R}(\mathrm{T}) \,|\, \mathbf{t}(\mathrm{T})]).$$

However, Equation (6.32) implies that $[\bar{\mathrm{E}} \,|\, \mathbf{0}][\bar{\mathrm{E}} \,|\, \mathbf{0}] = [\mathrm{E} \,|\, \mathbf{0}]$, so $\kappa^2 = 1$ and hence the only possible values of κ are $+1$ and -1, each of which gives rise to a family of irreducible representations of $\mathscr{G}^D$.

To every irreducible representation $\boldsymbol{\Gamma}$ of $\mathscr{G}^D$ with $\kappa = +1$ there corresponds an irreducible representation $\boldsymbol{\Gamma}'$ of $\mathscr{G}$ of the same dimension, given by

$$\boldsymbol{\Gamma}'(\{\mathbf{R}(\mathrm{T}) \,|\, \mathbf{t}(\mathrm{T})\}) = \boldsymbol{\Gamma}([\mathbf{R}(\mathrm{T}) \,|\, \mathbf{t}(\mathrm{T})]) = \boldsymbol{\Gamma}([\bar{\mathbf{R}}(\mathrm{T}) \,|\, \mathbf{t}(\mathrm{T})]). \tag{6.46}$$

Conversely, if $\boldsymbol{\Gamma}'$ is any irreducible representation of $\mathscr{G}$ and $\boldsymbol{\Gamma}$ is defined for all elements of $\mathscr{G}^D$ by Equation (6.46), then $\boldsymbol{\Gamma}$ provides an irreducible representation of $\mathscr{G}^D$ with $\kappa = +1$.

By contrast, the irreducible representations of $\mathscr{G}^D$ with $\kappa = -1$ are *not* directly related to any irreducible representations of $\mathscr{G}$ and so they are called the "extra" irreducible representations of $\mathscr{G}^D$. *In fact they are the representations of most interest,* for two-component spinors transform as basis functions *only* for the *extra* representations. This is most easily seen by noting that Equations (6.28) imply $\mathrm{O}(\mathrm{T})\boldsymbol{\psi}_n(\mathbf{r}) = -\mathrm{O}(\bar{\mathrm{T}})\boldsymbol{\psi}_n(\mathbf{r})$ for any two-component spinor $\boldsymbol{\psi}_n(\mathbf{r})$, whence Equations (6.43) and (6.44) show that, if $\boldsymbol{\psi}_n(\mathbf{r})$ is a basis spinor for $\boldsymbol{\Gamma}$, then $\boldsymbol{\Gamma}(\mathrm{T}) = -\boldsymbol{\Gamma}(\bar{\mathrm{T}})$.

The double crystallographic point groups were investigated first by Bethe (1929) and subsequently examined more thoroughly by Opechowski (1940). The construction of their extra representations is facilitated by the following simple rules derived by Opechowski:

(i) If a set of proper or improper rotations $\{\mathbf{R}(\mathrm{T}) \,|\, \mathbf{0}\}$ through $2\pi/n$ form a class in the single group, then the sets of generalized rotations $[\mathbf{R}(T)\,|\,\mathbf{0}]$ and $[\bar{\mathbf{R}}(T)\,|\,\mathbf{0}]$ form two separate classes in the double group.

(ii) There is one exception to (i), namely that if $n=2$, then $[\mathbf{R}(T)\,|\,\mathbf{0}]$ and $[\bar{\mathbf{R}}(T)\,|\,\mathbf{0}]$ lie in the same class of the double group if and only if there exists in the single group a proper or improper rotation through π about an axis perpendicular to the axis of $\{\mathbf{R}(T)\,|\,\mathbf{0}\}$.

(The characters of the extra representations must satisfy the relations $\chi([\bar{\mathbf{R}}(T)\,|\,\mathbf{0}]) = -\chi([\mathbf{R}(T)\,|\,\mathbf{0}])$ for every $\mathbf{R}(T)$ of $\mathcal{G}$. As the characters of all the elements in a class are equal, it follows that $\chi([C_{2i}\,|\,\mathbf{0}]) = 0$ in this exceptional case.)

Example I *Irreducible representations of the double crystallographic point group* D_4^D

D_4^D is the double group of D_4. Its classes are:

$\mathscr{C}_1 = \{[E\,|\,\mathbf{0}]\}$; $\mathscr{C}_2 = \{[\bar{E}\,|\,\mathbf{0}]\}$;

$\mathscr{C}_3 = \{[C_{2x}\,|\,\mathbf{0}], [C_{2z}\,|\,\mathbf{0}], [\bar{C}_{2x}\,|\,\mathbf{0}], [\bar{C}_{2z}\,|\,\mathbf{0}]\}$;

$\mathscr{C}_4 = \{[C_{2y}\,|\,\mathbf{0}], [\bar{C}_{2y}\,|\,\mathbf{0}]\}$; $\mathscr{C}_5 = \{[C_{4y}\,|\,\mathbf{0}], [C_{4y}^{-1}\,|\,\mathbf{0}]\}$;

$\mathscr{C}_6 = \{[\bar{C}_{4y}\,|\,\mathbf{0}], [\bar{C}_{4y}^{-1}\,|\,\mathbf{0}]\}$; $\mathscr{C}_7 = \{[C_{2c}\,|\,\mathbf{0}], [C_{2d}\,|\,\mathbf{0}], [\bar{C}_{2c}\,|\,\mathbf{0}], [\bar{C}_{2d}\,|\,\mathbf{0}]\}$.

The character table is displayed in Table 6.1. Clearly $\mathbf{\Gamma}^6$ and $\mathbf{\Gamma}^7$ are the "extra" representations.

The character tables for the extra representations of all the double crystallographic point groups are listed in Appendix D. Spinor basis functions formed from spherical harmonics that transform as the extra representations of the double crystallographic point groups O_h^D, T_d^D, O^D, D_{4h}^D, D_{3d}^D, C_{4v}^D, C_{3v}^D, and C_{2v}^D have been tabulated by Onodera and Okazaki (1966), and those for O_h^D and D_{3h}^D have been tabulated by Teleman and Glodeanu (1967).

As noted in subsection (*b*), the double group corresponding to the

	$\mathscr{C}_1$	$\mathscr{C}_2$	$\mathscr{C}_3$	$\mathscr{C}_4$	$\mathscr{C}_5$	$\mathscr{C}_6$	$\mathscr{C}_7$
$\mathbf{\Gamma}^1$	1	1	1	1	1	1	1
$\mathbf{\Gamma}^2$	1	1	1	1	−1	−1	−1
$\mathbf{\Gamma}^3$	1	1	−1	1	1	1	−1
$\mathbf{\Gamma}^4$	1	1	−1	1	−1	−1	1
$\mathbf{\Gamma}^5$	2	2	0	−2	0	0	0
$\mathbf{\Gamma}^6$	2	−2	0	0	$\sqrt{2}$	$-\sqrt{2}$	0
$\mathbf{\Gamma}^7$	2	−2	0	0	$-\sqrt{2}$	$\sqrt{2}$	0

Table 6.1 Character table for the double crystallographic point group D_4^D.

group of all proper rotations in $\mathbb{R}^3$ is isomorphic to SU(2). The irreducible representations of SU(2) will be studied in detail in Chapter 12, Section 4.

If $\mathbf{\Gamma}$ is an *extra* representation of $\mathscr{G}^D$, then the set of matrices $\mathbf{\Gamma}'(\mathrm{T})$ defined for all T of $\mathscr{G}$ by

$$\mathbf{\Gamma}'(\mathrm{T}) = \mathbf{\Gamma}([\mathbf{R}(\mathrm{T}) \mid \mathbf{t}(\mathrm{T})]) \tag{6.47}$$

have the property that

$$\mathbf{\Gamma}'(\mathrm{T}_1)\mathbf{\Gamma}'(\mathrm{T}_2) = \pm\mathbf{\Gamma}'(\mathrm{T}_1\mathrm{T}_2). \tag{6.48}$$

The occurrence of the minus sign here indicates that the matrices $\mathbf{\Gamma}'$ may *not* form a representation of $\mathscr{G}$. However, they do form a "projective representation" or "ray representation", the general definition of which is as follows:

Definition *Projective representation or ray representation of a group*

If every element T of a group $\mathscr{G}$ can be assigned a non-singular $d \times d$ matrix $\mathbf{\Gamma}'(\mathrm{T})$ such that

$$\mathbf{\Gamma}'(\mathrm{T}_1)\mathbf{\Gamma}'(\mathrm{T}_2) = \omega(\mathrm{T}_1, \mathrm{T}_2)\mathbf{\Gamma}'(\mathrm{T}_1\mathrm{T}_2),$$

where $\omega(\mathrm{T}_1, \mathrm{T}_2)$ is complex number of modulus unity that depends on T_1 and T_2, then the matrices $\mathbf{\Gamma}'$ are said to form a "projective" or "ray" representation of $\mathscr{G}$.

In the case of Equation (6.48), $\omega(\mathrm{T}_1, \mathrm{T}_2) = \pm 1$.

Schur (1904, 1907) has shown that if $\mathscr{G}$ is any *finite* group, then *every* projective representation of $\mathscr{G}$ can be obtained from an ordinary representation of a larger group in essentially the same way that Equation (6.47) gives a projective representation $\mathbf{\Gamma}'$ of $\mathscr{G}$ from an ordinary representation of $\mathscr{G}^D$. For *Lie groups* the general situation is more complicated. A very thorough analysis has been given by Bargmann (1954), the subject being re-examined from a different angle by de Swart (1974).

In the physical situation considered in this section, although the single group $\mathscr{G}$ is fundamental in the sense that it corresponds to the geometric symmetry of the system, its relevant representations are projective representations. It is normally more convenient to work (as above) with the group whose ordinary representations are those that occur in the physical problem and to regard this group as being the more fundamental.

(f) *Splitting of degeneracies by spin-orbit coupling*

The introduction of the spin-orbit coupling term (Equation (6.36)) into a Hamiltonian operator reduces the symmetry of the Hamiltonian and, therefore, in general, causes degeneracies of energy levels to be split. In order to deduce exactly how the degeneracies are split, it is necessary to relate the eigenfunctions of the spin-independent Hamiltonian to those of the Hamiltonian of Equation (6.35) with the spin-orbit coupling term included. The following theorem gives the necessary relationship between the basis functions of the single groups and those of the corresponding double groups.

Theorem VIII Let $\psi_1^p(\mathbf{r}), \psi_2^p(\mathbf{r}), \ldots, \psi_d^p(\mathbf{r})$ be a set of scalar basis functions for a d-dimensional irreducible representation $\mathbf{\Gamma}^p$ of a single group $\mathcal{G}$. Define the $2d$ linearly independent two-component spinors $\boldsymbol{\phi}_{sm}^p(\mathbf{r})$ ($s = 1, 2, \ldots, d$; $m = 1, 2$) by

$$\boldsymbol{\phi}_{s1}^p(\mathbf{r}) = \begin{pmatrix} \psi_s^p(\mathbf{r}) \\ 0 \end{pmatrix}, \qquad \boldsymbol{\phi}_{s2}^p(\mathbf{r}) = \begin{pmatrix} 0 \\ \psi_s^p(\mathbf{r}) \end{pmatrix}. \tag{6.49}$$

Then this set of two-component spinors form a basis for the $2d$-dimensional direct product representation $\mathbf{\Gamma}^p \otimes \mathbf{u}$ of the corresponding double group $\mathcal{G}^D$. Here, by definition,

$$\left.\begin{aligned} \mathbf{u}([\mathbf{R}(\mathrm{T}) \mid \mathbf{t}(\mathrm{T})]) &= \mathbf{u}(\mathbf{R}_p(\mathrm{T})), \\ \mathbf{u}([\bar{\mathbf{R}}(\mathrm{T}) \mid \mathbf{t}(\mathrm{T})]) &= -\mathbf{u}(\mathbf{R}_p(\mathrm{T})), \end{aligned}\right\} \tag{6.50}$$

for all $\mathrm{T} \in \mathcal{G}$, and $\mathbf{\Gamma}^p$ provides a representation of $\mathcal{G}^D$ by Equation (6.46), that is,

$$\mathbf{\Gamma}^p([\mathbf{R}(\mathrm{T}) \mid \mathbf{t}(\mathrm{T})]) = \mathbf{\Gamma}^p([\bar{\mathbf{R}}(\mathrm{T}) \mid \mathbf{t}(\mathrm{T})]) = \mathbf{\Gamma}^p(\{\mathbf{R}(\mathrm{T}) \mid \mathbf{t}(\mathrm{T})\}). \tag{6.51}$$

Proof It is obvious that the matrices $\mathbf{u}$ defined by Equations (6.50) do form a representation of $\mathcal{G}^D$.

From Equations (6.49) $\phi_{sm\alpha}^p(\mathbf{r}) = \psi_s^p(\mathbf{r})\delta_{m\alpha}$ for $s = 1, 2, \ldots, d$; $m = 1, 2$; $\alpha = 1, 2$. Thus

$$\mathrm{O(T)}\phi_{sm\alpha}^p(\mathbf{r}) = \sum_{\beta=1}^{2} \mathrm{u}(\mathbf{R}_p(\mathrm{T}))_{\alpha\beta}\phi_{sm\beta}^p(\{\mathbf{R}(\mathrm{T}) \mid \mathbf{t}(\mathrm{T})\}^{-1}\mathbf{r})$$

(from Equations (6.27))

$$= \mathrm{u}([\mathbf{R}(\mathrm{T}) \mid \mathbf{t}(\mathrm{T})])_{\alpha m}\psi_s^p(\{\mathbf{R}(\mathrm{T}) \mid \mathbf{t}(\mathrm{T})\}^{-1}\mathbf{r})$$

$$= \sum_{t=1}^{d} \Gamma(\{\mathbf{R}(\mathrm{T}) \mid \mathbf{t}(\mathrm{T})\})_{ts}\,\mathrm{u}([\mathbf{R}(\mathrm{T}) \mid \mathbf{t}(\mathrm{T})])_{\alpha m}\psi_t^p(\mathbf{r})$$

$$= \sum_{t=1}^{d} \sum_{n=1}^{2} \{\mathbf{\Gamma}([\mathbf{R}(\mathrm{T}) \mid \mathbf{t}(\mathrm{T})]) \otimes \mathbf{u}([\mathbf{R}(\mathrm{T}) \mid \mathbf{t}(T)])\}_{tn,sm}\phi_{tn\alpha}^p(\mathbf{r})$$

as required. The proof with T replaced by $\bar{\mathrm{T}}$ is similar.

Now suppose that the functions $\psi_s^p(\mathbf{r})$ are eigenfunctions of the spin-independent Hamiltonian of Equation (1.10) belonging to a d-fold degenerate energy eigenvalue ε. If the representation $\mathbf{\Gamma}^p \otimes \mathbf{u}$ is a *reducible* representation of $\mathscr{G}^D$, as it often is, and $\mathbf{\Gamma}^q, \mathbf{\Gamma}^r, \ldots$ are the extra irreducible representations of $\mathscr{G}^D$ that appear in its reduction, then the spinors $\boldsymbol{\phi}_{sm}^p(\mathbf{r})$ of Equations (6.49) can be rearranged so as to form bases for $\mathbf{\Gamma}^q, \mathbf{\Gamma}^r, \ldots$. (Equations (6.50) and (6.51) imply that $\mathbf{\Gamma}^p \otimes \mathbf{u}$ contains *only extra* irreducible representations of $\mathscr{G}^D$.) Thus, in the absence of the spin-orbit coupling term (Equation (6.36)), the energy eigenvalues corresponding to $\mathbf{\Gamma}^q, \mathbf{\Gamma}^r, \ldots$ stick together, this example of an "accidental" degeneracy being the consequence of the special form of $\mathbf{H}_0(\mathbf{r})$ in Equation (6.35). When the spin-orbit coupling term is included in $\mathbf{H}(\mathbf{r})$, the energy eigenvalues corresponding to $\mathbf{\Gamma}^q, \mathbf{\Gamma}^r, \ldots$ will in general be different and the degeneracy will be split. Thus the Clebsch–Gordan series

$$\mathbf{\Gamma}^p \otimes \mathbf{u} \approx n_q \mathbf{\Gamma}^q \oplus n_r \mathbf{\Gamma}^r \oplus \ldots \qquad (6.52)$$

shows immediately how the splitting takes place. Here the number of times n_q that $\mathbf{\Gamma}^q$ appears in $\mathbf{\Gamma}^p \otimes \mathbf{u}$ is given by

$$n_q = (1/g) \sum_{T \in \mathscr{G}} \chi^p(\{\mathbf{R}(T) \mid \mathbf{t}(T)\}) \operatorname{tr} \mathbf{u}(\mathbf{R}_p(T)) \chi^q([\mathbf{R}(T) \mid \mathbf{t}(T)])^*$$

when $\mathscr{G}$ is a finite group of order g, with the obvious generalization for a compact Lie group. (This follows Theorem V of Chapter 4, Section 6 and Equations (6.50) and (6.51).) Thus the splitting caused by spin-orbit coupling is very easily obtained. Of course this argument does not tell the ordering of the energy levels after the splitting has taken place. This requires explicit calculation, for which time-independent perturbation theory would be the appropriate tool, provided the spin-orbit coupling is not too strong.

It will be shown in Section 5 that "time-reversal" symmetry may cause some of the eigenvalues associated with the irreducible representations of the right-hand side of Equation (6.52) to "stick together".

Example II *Splitting when $\mathscr{G}$ is the crystallographic point group* D_4

From Tables 6.1 and D.26 it follows that

$$\mathbf{\Gamma}^1 \otimes \mathbf{u} \approx \mathbf{\Gamma}^6 \qquad (6.53)$$

and

$$\mathbf{\Gamma}^5 \otimes \mathbf{u} \approx \mathbf{\Gamma}^6 \oplus \mathbf{\Gamma}^7. \qquad (6.54)$$

Consider first the spin-independent energy eigenvalue associated with Γ^1. As Γ^1 is one-dimensional, the "orbital" degeneracy of this level is one, so the overall degeneracy (taking into account the two possible spin functions that can be attached to an "orbital" function) is two. Then Equation (6.53) shows that this two-fold degeneracy is *not* split by spin-orbit coupling.

By contrast, the spin-independent energy eigenvalue associated with Γ^5 corresponds to an overall four-fold degeneracy, which Equation (6.54) shows is split into two two-fold degenerate levels.

5 Time-reversal symmetry

(a) *Spin-independent theory*

The spin-independent Hamiltonian of Equation (1.10) is real and so are all its eigenvalues. Consequently, if $\psi(\mathbf{r})$ is an eigenfunction of $H(\mathbf{r})$ with eigenvalue ε, $\psi(\mathbf{r})^*$ is also an eigenfunction with the same eigenvalue. The interesting question is whether $\psi(\mathbf{r})$ is linearly independent of $\psi(\mathbf{r})^*$. If it is then an extra degeneracy must exist.

The name "time-reversal symmetry" for this phenomenon is used because if the process of complex conjugation is applied to the time-dependent Schrödinger equation $H(\mathbf{r})\psi(\mathbf{r}, t) = i\hbar\, \partial\psi(\mathbf{r}, t)/\partial t$, the result is $H(\mathbf{r})\psi(\mathbf{r}, t)^* = i\hbar\, \partial\psi(\mathbf{r}, t)^*/\partial(-t)$. That is, the sign of the time t is reversed.

The analysis depends on the classification of irreducible representations into potentially real, pseudo-real, or essentially complex types, as discussed in detail in Chapter 5, Section 8.

Theorem I Suppose that $\psi_1^p(\mathbf{r}), \psi_2^p(\mathbf{r}), \ldots$ are a set of d_p linearly independent eigenfunctions of the spin-independent Hamiltonian of Equation (1.10), corresponding to the energy eigenvalue ε and transforming as basis functions for the d_p-dimensional unitary irreducible representation Γ^p of the group of the Schrödinger equation $\mathcal{G}$. Suppose that $\mathcal{G}$ is a finite group or a compact Lie group. Then

(*a*) if Γ^p is pseudo-real or essentially complex then the functions $\psi_1^p(\mathbf{r})^*, \psi_2^p(\mathbf{r})^*, \ldots$ are linearly independent of $\psi_1^p(\mathbf{r}), \psi_2^p(\mathbf{r}), \ldots$ and so the eigenvalue ε is $2d_p$-fold degenerate, but

(*b*) if Γ^p is potentially real there is *no* requirement that $\psi_1^p(\mathbf{r})^*, \psi_2^p(\mathbf{r})^*, \ldots$ be linearly independent of $\psi_1^p(\mathbf{r}), \psi_2^p(\mathbf{r}), \ldots$, and hence *no* requirement that ε be more than d_p-fold degenerate.

Proof The definition of the operator $P(T)$ (Equation (1.17)) implies that

$$\mathrm{P(T)}\psi(\mathbf{r})^* = \{\mathrm{P(T)}\psi(\mathbf{r})\}^*$$

for any T and any function $\psi(\mathbf{r})$. Thus, if $\psi_1^p(\mathbf{r}), \psi_2^p(\mathbf{r}), \ldots$ are basis functions for $\mathbf{\Gamma}^p$, then $\psi_1^p(\mathbf{r})^*, \psi_2^p(\mathbf{r})^*, \ldots$ are basis functions for the complex conjugate representation $\mathbf{\Gamma}^{p^*}$ defined by $\mathbf{\Gamma}^{p^*}(\mathrm{T}) = \{\mathbf{\Gamma}^p(\mathrm{T})\}^*$ for all $\mathrm{T} \in \mathscr{G}$.

Suppose first that $\psi_1^p(\mathbf{r})^*, \psi_2^p(\mathbf{r})^*, \ldots$ depend linearly on $\psi_1^p(\mathbf{r}), \psi_2^p(\mathbf{r}), \ldots$, so that there exists a $d_p \times d_p$ non-singular matrix $\mathbf{Z}$ such that

$$\psi_n^p(\mathbf{r})^* = \sum_{m=1}^{d_p} Z_{mn}\psi_m^p(\mathbf{r}), \qquad n = 1, 2, \ldots, d_p. \tag{6.55}$$

Then $\mathbf{\Gamma}^{p^*}(\mathrm{T}) = \mathbf{Z}^{-1}\mathbf{\Gamma}^p(\mathrm{T})\mathbf{Z}$ for all $\mathrm{T} \in \mathscr{G}$, so $\mathbf{\Gamma}^p$ cannot be essentially complex. Taking the complex conjugate of Equation (6.55) and substituting back into Equation (6.55) gives $\mathbf{Z}^*\mathbf{Z} = \mathbf{1}$. The second theorem of Chapter 5, Section 8, then shows that $\mathbf{\Gamma}^p$ must be potentially real.

Conversely, if $\mathbf{\Gamma}^p$ is potentially real, a similarity transformation will make $\mathbf{\Gamma}^p$ real. When this is so the basis functions of $\mathbf{\Gamma}^p$ that are eigenfunctions of $\mathrm{H}(\mathbf{r})$ can be taken to be purely real. There is thus no requirement that there be extra degeneracies.

In the special cases in which $\mathscr{G}$ is the group of all rotations in $\mathbb{R}^3$ or the group of all proper rotations in $\mathbb{R}^3$, every irreducible representation is potentially real, so that time-reversal causes *no* extra degeneracies. By contrast, extra degeneracies occur frequently in electronic energy bands in solids, where the original application of the above general theory to single crystallographic space groups was made by Herring (1937a). (A detailed account of the consequences made can be found in the book by Cornwell (1969).)

(*b*) *Spin-dependent theory*

The spin-orbit coupling Hamiltonian of Equation (6.35) is not real, so that the complex conjugate $\boldsymbol{\phi}(\mathbf{r})^*$ of a two-component spinor eigenfunction of $\mathbf{H}(\mathbf{r})$ is not, in general, another eigenfunction. However, it is easily verified that

$$\boldsymbol{\sigma}_2^{-1}\mathbf{H}(\mathbf{r})\boldsymbol{\sigma}_2 = \mathbf{H}(\mathbf{r})^*, \tag{6.56}$$

where $\boldsymbol{\sigma}_2$ is the Pauli spin matrix

$$\boldsymbol{\sigma}_2 = \begin{bmatrix} 0 & -i \\ i & 0 \end{bmatrix},$$

so that if

$$\mathbf{H}(\mathbf{r})\boldsymbol{\phi}(\mathbf{r}) = \varepsilon\boldsymbol{\phi}(\mathbf{r}),$$

then taking the complex conjugate and using Equation (6.56) gives

$$\mathbf{H}(\mathbf{r})\{\boldsymbol{\sigma}_2\boldsymbol{\phi}(\mathbf{r})^*\} = \varepsilon\{\boldsymbol{\sigma}_2\boldsymbol{\phi}(\mathbf{r})^*\}.$$

Thus $\boldsymbol{\sigma}_2\boldsymbol{\phi}(\mathbf{r})^*$ is an eigenspinor with the same energy eigenvalue ε.

The classification of representations in terms of their reality properties is again relevant *but* the degeneracies corresponding to the three types are quite *different* when the Hamiltonian $\mathbf{H}(\mathbf{r})$ is spin-dependent, as the following theorem shows.

Theorem II Suppose that $\boldsymbol{\psi}_1^p(\mathbf{r}), \boldsymbol{\psi}_2^p(\mathbf{r}), \ldots$ are a set of d_p linearly independent spinor eigenfunctions of the spin-dependent Hamiltonian of Equation (6.35) corresponding to the energy eigenvalue ε and transforming as spinor basis functions for the d_p-dimensional extra unitary irreducible representation $\boldsymbol{\Gamma}^p$ of the double group of the Schrödinger equation $\mathcal{G}^D$. Suppose that $\mathcal{G}^D$ is a finite group or a compact Lie group. Then

(*a*) if $\boldsymbol{\Gamma}^p$ is potentially real or essentially complex, the spinors $\boldsymbol{\sigma}_2\boldsymbol{\psi}_1^p(\mathbf{r})^*, \boldsymbol{\sigma}_2\boldsymbol{\psi}_2^p(\mathbf{r})^*, \ldots$ are linearly independent of $\boldsymbol{\psi}_1^p(\mathbf{r})$, $\boldsymbol{\psi}_2^p(\mathbf{r}), \ldots$ and so the eigenvalue ε is $2d_p$-fold degenerate, but

(*b*) if $\boldsymbol{\Gamma}^p$ is pseudo-real there is *no* requirement that $\boldsymbol{\sigma}_2\boldsymbol{\psi}_1^p(\mathbf{r})^*, \boldsymbol{\sigma}_2\boldsymbol{\psi}_2^p(\mathbf{r})^*, \ldots$ be linearly independent of $\boldsymbol{\psi}_1^p(\mathbf{r}), \boldsymbol{\psi}_2^p(\mathbf{r}), \ldots$ and hence *no* requirement that ε be more than d_p-fold degenerate.

Proof Let $\mathbf{u}$ be any matrix of the group SU(2). Then consideration of the parametrization of Equation (3.15) shows that $\boldsymbol{\sigma}_2^{-1}\mathbf{u}\boldsymbol{\sigma}_2 = \mathbf{u}^*$. As

$$\boldsymbol{\sigma}_2^*\boldsymbol{\sigma}_2 = -\mathbf{1}, \tag{6.57}$$

Theorem II of Chapter 5, Section 8 shows that $\mathbf{u}$ is a pseudo-real representation of SU(2). Also, from Equations (6.28), for any two-component spinor $\boldsymbol{\psi}(\mathbf{r})$ and any $\mathrm{T}\in\mathcal{G}$,

$$\left.\begin{aligned} O(\mathrm{T})\{\boldsymbol{\sigma}_2\boldsymbol{\psi}(\mathbf{r})^*\} &= \boldsymbol{\sigma}_2\{O(\mathrm{T})\boldsymbol{\psi}(\mathbf{r})\}^*, \\ O(\bar{\mathrm{T}})\{\boldsymbol{\sigma}_2\boldsymbol{\psi}(\mathbf{r})^*\} &= \boldsymbol{\sigma}_2\{O(\bar{\mathrm{T}})\boldsymbol{\psi}(\mathbf{r})\}^*. \end{aligned}\right\}$$

Thus, by Equations (6.43) and (6.44), if $\boldsymbol{\psi}_1^p(\mathbf{r}), \boldsymbol{\psi}_2^p(\mathbf{r}), \ldots$ for a basis for the d_p-dimensional irreducible representation $\boldsymbol{\Gamma}^p$ of $\mathcal{G}^D$, $\boldsymbol{\sigma}_2\boldsymbol{\psi}_1^p(\mathbf{r})^*, \boldsymbol{\sigma}_2\boldsymbol{\psi}_2^p(\mathbf{r})^*, \ldots$ form a basis for the complex conjugate representation $\boldsymbol{\Gamma}^{p^*}$.

Suppose that $\boldsymbol{\sigma}_2\psi_1^p(\mathbf{r})^*$, $\boldsymbol{\sigma}_2\psi_2^p(\mathbf{r})^*, \ldots$ depend linearly on $\psi_1^p(\mathbf{r})$, $\psi_2^p(\mathbf{r}), \ldots$ so that there exists a $d_p \times d_p$ non-singular matrix $\mathbf{Z}$ such that

$$\boldsymbol{\sigma}_2\psi_n^p(\mathbf{r})^* = \sum_{m=1}^{d_p} Z_{mn}\psi_m^p(\mathbf{r}), \qquad n = 1, 2, \ldots, d_p. \tag{6.58}$$

Then $\boldsymbol{\Gamma}^{p*}(T) = \mathbf{Z}^{-1}\boldsymbol{\Gamma}^p(T)\mathbf{Z}$ for all $T \in \mathscr{G}$, so that $\boldsymbol{\Gamma}^p$ cannot be essentially complex. Taking the complex conjugate of Equation (6.58), multiplying through by $\boldsymbol{\sigma}_2$, substituting back in Equation (6.58) and using Equation (6.57) gives $\mathbf{Z}^*\mathbf{Z} = -\mathbf{1}$. Theorem II of Chapter 5, Section 8, then shows that $\boldsymbol{\Gamma}^p$ must be pseudo-real.

As $[\mathbf{R}(T) \mid \mathbf{t}(T)]^2 = [\bar{\mathbf{R}}(T) \mid \mathbf{t}(T)]^2$, the Frobenius and Schur condition of Theorem III of Chapter 5, Section 8 for $\mathscr{G}^D$ can be written as

$$(1/g) \sum_{T \in \mathscr{G}} \chi([\mathbf{R}(T) \mid \mathbf{t}(T)]^2) = \begin{cases} 1, & \text{if } \Gamma \text{ is potentially real,} \\ 0, & \text{if } \Gamma \text{ is essentially complex,} \\ -1, & \text{if } \Gamma \text{ is pseudo-real,} \end{cases}$$

where the sum is over all g elements of the corresponding single group $\mathscr{G}$, the obvious modification applying for compact Lie groups.

When $\mathscr{G}$ is the group of all rotations in $\mathbb{R}^3$ or the group of all proper rotations in $\mathbb{R}^3$, the extra irreducible representations of $\mathscr{G}^D$ are all pseudo-real, so that again time-reversal symmetry produces *no* extra degeneracies. Elliott (1954) applied the above general theory to double crystallographic space groups and showed that time-reversal symmetry plays a very important role when spin-dependent terms are included in the calculation of electronic energy bands in crystals (see Cornwell (1969)).

Part B

Applications in Molecular and Solid State Physics

7

Group Theoretical Treatment of Vibrational Problems

1 Introduction

The nature of vibrations in molecules and in crystalline solids becomes much clearer when a group theoretical treatment is applied to the problem. In contrast with the physical applications considered so far, this application takes place at the classical level, that is when the vibrations are considered as a problem in classical mechanics. The quantities of primary interest are the frequencies of the normal modes of vibration. The group theoretical analysis provides both a classification of these frequencies in terms of irreducible representations of an invariance group and a method which greatly assists in their evaluation. After the transition to quantum mechanics these frequencies remain of fundamental importance, for they provide the quanta of energy associated with the vibrations.

In Section 2 a general treatment of vibrations is given, first in the Lagrangian and Hamiltonian formulations of classical mechanics and then in quantum mechanics. The simplifying effects of symmetry are then studied in detail in Section 3, where it will be shown that many of the features of the symmetry theory of the Schrödinger equation developed in Chapters 1 and 6 reappear in a form that is only superficially different. In Section 3 several examples of molecular vibrations will be examined. The corresponding application to lattice vibrations of crystalline solids is deferred until Chapter 8, Section 7, and Chapter 9.

2 General theory of vibrations

(a) *Classical theory*

It will be assumed that the reader is familiar with the Lagrangian and Hamiltonian formulations of classical mechanics. (An excellent account may be found in the monograph of Goldstein (1950).)

Suppose that a conservative mechanical system has D degrees of freedom and so can be described by D generalized coordinates $q_1, q_2, \ldots, q_D$. In all the examples to be considered later the natural choice of these generalized coordinates automatically makes the kinetic energy T have the form

$$T = \tfrac{1}{2} \sum_{m=1}^{D} \dot{q}_m^2, \tag{7.1}$$

where $\dot{q}_1, \dot{q}_2, \ldots, \dot{q}_D$ are the generalized velocities defined by $\dot{q}_m = dq_m/dt$, t denoting time. The potential energy V is a function of $q_1, q_2, \ldots, q_D$ alone.

In every vibrational problem there is a position of stable equilibrium about which the oscillations take place. It may be assumed that the generalized coordinates are chosen so that this equilibrium position corresponds to $q_1 = q_2 = \ldots = q_D = 0$. Then, at $q_1 = q_2 = \ldots = q_D = 0$, $\partial V/\partial q_m = 0$ for all $m = 1, 2, \ldots, D$, and $\partial^2 V/\partial q_m \partial q_n \geqslant 0$ for all $m, n = 1, 2, \ldots, D$. Moreover, the zero of energy may be chosen so that $V(0, 0, \ldots, 0) = 0$. Then, for small displacements from the equilibrium position,

$$V = \tfrac{1}{2} \sum_{m=1}^{D} \sum_{n=1}^{D} F_{mn} q_m q_n, \tag{7.2}$$

where $F_{mn} = \partial^2 V/\partial q_m \partial q_n$ evaluated at $q_1 = q_2 = \ldots = q_D = 0$. There is no loss in generality in assuming that $F_{mn} = F_{nm}$ for all $m, n = 1, 2, \ldots, D$. (For example, in the case $D = 2$, the form $V = Aq_1^2 + Bq_1q_2 + Cq_2^2$ can be rewritten as $V = \tfrac{1}{2}q_1^2 + \tfrac{1}{2}B(q_1q_2 + q_2q_1) + Cq_2^2$.)

It is very convenient to introduce a real $D \times 1$ column matrix $\mathbf{q}$, having q_m as its mth component, and to regard the F_{mn} as elements of a $D \times D$ matrix $\mathbf{F}$, called the "force matrix". Then $\mathbf{F}$ is a real symmetric matrix and Equations (7.1) and (7.2) can be rewritten as

$$T = \tfrac{1}{2}\tilde{\dot{\mathbf{q}}}\dot{\mathbf{q}} \tag{7.3}$$

and

$$V = \tfrac{1}{2}\tilde{\mathbf{q}}\mathbf{F}\mathbf{q}. \tag{7.4}$$

The Lagrangian L is defined by $L = T - V$, so

$$L = \frac{1}{2} \sum_{m=1}^{D} (\dot{q}_m)^2 - \frac{1}{2} \sum_{m=1}^{D} \sum_{n=1}^{D} F_{mn} q_m q_n.$$

Lagrange's equations, which have the general form

$$\frac{d}{dt}\left(\frac{\partial L}{\partial \dot{q}_m}\right) = \frac{\partial L}{\partial q_m}, \qquad m = 1, 2, \ldots, D,$$

then become

$$\ddot{q}_m + \sum_{n=1}^{D} F_{mn} q_n = 0, \qquad m = 1, 2, \ldots, D,$$

which may be expressed in matrix terms as

$$\ddot{\mathbf{q}} + \mathbf{F}\mathbf{q} = \mathbf{0}. \tag{7.5}$$

Now consider a particular solution of the form

$$\mathbf{q} = \mathbf{c} \cos (\omega t + \alpha),$$

where $\mathbf{c}$ is some constant real $D \times 1$ column matrix and α and ω are some real constants. This corresponds to a motion in which *all* the generalized coordinates oscillate with the *same* frequency $\omega/2\pi$. Substitution in Equation (7.5) gives

$$\mathbf{F}\mathbf{c} = \omega^2 \mathbf{c},$$

a matrix eigenvalue equation of the form of Equation (A.10) with eigenvalue $\lambda = \omega^2$.

As $\mathbf{F}$ is a real symmetric matrix, $\mathbf{F}$ is diagonalizable, so there exists a real orthogonal $D \times D$ matrix $\mathbf{S}$ such that

$$\mathbf{S}^{-1}\mathbf{F}\mathbf{S} = \begin{bmatrix} \lambda_1 & 0 & \ldots & 0 \\ 0 & \lambda_2 & & 0 \\ \vdots & \vdots & & \vdots \\ 0 & 0 & \ldots & \lambda_D \end{bmatrix}, \tag{7.6}$$

where $\lambda_1, \lambda_2, \ldots, \lambda_D$ are the eigenvalues of $\mathbf{F}$, some of which may be equal. Let $\mathbf{c}_m$ denote an eigenvector of $\mathbf{F}$ corresponding to the eigenvalue λ_m, so that $\mathbf{F}\mathbf{c}_m = \lambda_m \mathbf{c}_m$. Then a convenient choice of the normalization and phase of $\mathbf{c}_m$ gives

$$\mathbf{c}_m = \mathbf{S}\mathbf{v}_m,$$

where $\mathbf{v}_m$ is the $D \times 1$ column matrix whose mth component is unity and whose other elements are all zero. With this choice $\mathbf{c}_m$ is real and

$\tilde{\mathbf{c}}_m \mathbf{c}_n = \delta_{mn}$, $m, n = 1, 2, \ldots, D$. Also

$$\mathbf{S} = [\mathbf{c}_1 \mid \mathbf{c}_2 \mid \ldots \mid \mathbf{c}_D], \tag{7.7}$$

so that $\mathbf{S}$ is the matrix whose columns are provided by the eigenvectors of $\mathbf{F}$.

The solution

$$\mathbf{q} = \mathbf{c}_m \cos(\omega_m t + \alpha_m),$$

where $\omega_m^2 = \lambda_m$, is called a "normal mode" of oscillation. As Equations (7.5) are linear, the most general solution has the form

$$\mathbf{q} = \sum_{m=1}^{D} A_m \mathbf{c}_m \cos(\omega_m t + \alpha_m), \tag{7.8}$$

which represents a superposition of normal modes. Here A_m and α_m ($m = 1, 2, \ldots, D$) are a set of $2D$ real parameters whose values depend on the $2D$ initial conditions provided by the values of $\mathbf{q}$ and $\dot{\mathbf{q}}$ at time $t = 0$.

Now let $\mathbf{Q}$ be the real $D \times 1$ column matrix defined by

$$\mathbf{Q} = \tilde{\mathbf{S}}\mathbf{q}.$$

The components $Q_1, Q_2, \ldots, Q_D$ of $\mathbf{Q}$ are called the "normal coordinates" of the system. By virtue of Equation (7.7)

$$Q_m = \tilde{\mathbf{c}}_m \mathbf{q}, \tag{7.9}$$

so that there is a one-to-one correspondence between the normal coordinates and the real eigenvectors of $\mathbf{F}$. As $\mathbf{S}$ is orthogonal, Equations (7.3), (7.4) and (7.6) imply that

$$T = \tfrac{1}{2} \sum_{m=1}^{D} \dot{Q}_m^2$$

and

$$V = \tfrac{1}{2} \sum_{m=1}^{D} \lambda_m Q_m^2. \tag{7.10}$$

Thus

$$L = \tfrac{1}{2} \sum_{m=1}^{D} \{\dot{Q}_m^2 - \omega_m^2 Q_m^2\} \tag{7.11}$$

with $\omega_m^2 = \lambda_m$. This is the Lagrangian for a set of *uncoupled* simple harmonic oscillators of frequencies $\omega_m/2\pi$, $m = 1, 2, \ldots, D$. It is particularly convenient to use this form in the transition to the quantum mechanical treatment.

When degeneracies occur among the eigenvalues there is some degree of choice in the eigenvectors of $\mathbf{F}$ and thereby in the normal coordinates. Suppose, for example, that $\lambda_1 = \lambda_2 = \ldots = \lambda_d$ where $1 < d \leq D$. Then for *any* $d \times d$ real orthogonal matrix $\mathbf{Y}$, the set of d $D \times 1$ column matrices $\mathbf{c}'_1, \mathbf{c}'_2, \ldots, \mathbf{c}'_d$ defined by $\mathbf{c}'_n = \sum_{m=1}^{d} Y_{mn}\mathbf{c}_m$ is a linearly independent set of real ortho-normal eigenvectors of $\mathbf{F}$ with the same eigenvalue. Thus, if $Q'_1, Q'_2, \ldots, Q'_d$ are defined by $Q'_m = \tilde{\mathbf{c}}'_m\mathbf{q}$ $(m = 1, 2, \ldots, d)$ (cf. Equation (7.9)), then

$$L = \tfrac{1}{2}\sum_{m=1}^{d}\{(\dot{Q}'_m)^2 - \omega_m^2(Q'_m)^2\} + \tfrac{1}{2}\sum_{m=d+1}^{D}\{\dot{Q}_m^2 - \omega_m^2 Q_m^2\},$$

so that $Q'_1, Q'_2, \ldots, Q'_d, Q_{d+1}, \ldots, Q_D$ is an equally good set of normal coordinates.

The generalized momenta $P_1, P_2, \ldots, P_D$ canonical to the normal coordinates $Q_1, Q_2, \ldots, Q_D$ may be defined by $P_m = \partial L/\partial \dot{Q}_m$, $m = 1, 2, \ldots, D$, so that Equation (7.11) implies $P_m = \dot{Q}_m$, $m = 1, 2, \ldots, D$. The Hamiltonian function H is defined by

$$H = \sum_{m=1}^{D} P_m\dot{Q}_m - L.$$

Thus $H = \sum_{m=1}^{D} H_m$, where

$$H_m = \tfrac{1}{2}P_m^2 + \tfrac{1}{2}\omega_m^2 Q_m^2 \tag{7.12}$$

Again this corresponds to a set of D independent simple harmonic oscillators of frequencies $\omega_m/2\pi$, $m = 1, 2, \ldots, D$.

To summarize, the eigenvalues $\lambda_m (= \omega_m^2)$ of $\mathbf{F}$ are important because they appear in the general solution (Equation (7.8)) and in the uncoupled forms of the Lagrangian and Hamiltonian (Equations (7.11) and (7.12)). The corresponding eigenvectors $\mathbf{c}_m$ are likewise significant because of their appearance in Equation (7.8) and because, by Equation (7.9), they give the normal coordinates.

(*b*) *Quantum theory*

In quantum mechanics the generalized coordinates Q_m and P_m appearing in the Hamiltonian must be regarded as operators Q_m and P_m satisfying the commutation relations $[\mathrm{Q}_m, \mathrm{P}_n] = i\hbar\delta_{mn}$ $(m, n = 1, 2, \ldots, D)$. In the Schrödinger representation Q_m is merely the operation of multiplying by the variable Q_m, $\mathrm{P}_m = (\hbar/i)\,\partial/\partial Q_m$ $(m, n = 1, 2, \ldots, D)$ and the eigenfunctions of the Hamiltonian operator are functions of the normal coordinates $Q_1, Q_2, \ldots, Q_D$.

The most important deduction from the previous subsection is that the Hamiltonian operator H for a vibrating system can be written in the form $H = \sum_{m=1}^{D} H_m$, where

$$H_m = -\tfrac{1}{2}\hbar^2\, \partial^2/\partial Q_m^2 + \tfrac{1}{2}\omega_m^2 Q_m^2,$$

in which only one generalized coordinate Q_m appears. As H_m is the Hamiltonian operator for a simple harmonic oscillator of mass unity oscillating with frequency $\omega_m/2\pi$, all its eigenfunctions and eigenvalues are known and are described in detail in all the standard texts on quantum mechanics (see Schiff (1968)). The eigenvalues of H_m are

$$\varepsilon_{mn} = (n + \tfrac{1}{2})\hbar\omega_m,$$

where $n = 0, 1, 2, \ldots$. Each is non-degenerate with normalized eigenfunction

$$\phi_n(Q_m) = (\omega_m/\pi\hbar)^{1/4}(2^n n!)^{-1/2} H_n((\omega_m/\hbar)^{1/2} Q_m) \exp(-\omega_m Q_m^2/2\hbar),$$

where $H_n(x)$ is a Hermite polynomial. Thus the eigenfunctions of H are of the product form

$$\phi(Q_1, Q_2, \ldots, Q_D) = \phi_{n_1}(Q_1)\phi_{n_2}(Q_2) \ldots \phi_{n_D}(Q_D),$$

where $(n_1, n_2, \ldots, n_D)$ is any set of non-negative integers, the corresponding energy eigenvalue being

$$\varepsilon = \varepsilon_{1n_1} + \varepsilon_{2n_2} + \ldots + \varepsilon_{Dn_D}.$$

The set of integers $(n_1, n_2, \ldots, n_D)$ may be called the set of "occupation numbers".

In a transition between two energy eigenstates in which all the occupation numbers are unchanged except for the mth, which changes from n_m to n'_m, the initial and final states of the vibrating system differ in energy by $(n_m - n'_m)\hbar\omega_m$. Thus $\hbar\omega_m$ can be interpreted as the quantum of energy of the mth oscillator.

This analysis shows that the quantum mechanical problem is completely solved once the frequencies $\omega_m/2\pi$ and the normal coordinates are known, that is, immediately the corresponding classical problem is solved.

3 Symmetry considerations

(a) *Motivation*

In the approximation of Born and Oppenheimer (1927), the nuclei of a molecule or solid move in a potential field that is determined by the positions of the other nuclei and by the instantaneous electronic configuration. Each nucleus may be considered to be a point particle without internal structure, so that, if the molecule or solid is made up from A atoms, then the system consisting only of their nuclei has $3A$ $(=D)$ degrees of freedom.

As shown in Section 2, the key problem is the diagonalization of the $D \times D$ matrix $\mathbf{F}$, whose eigenvalues give the frequencies of the normal modes and whose eigenvectors yield the normal coordinates. Even for relatively simple molecules, D $(=3A)$ can be quite large and the diagonalization of $\mathbf{F}$ troublesome. For example, for the ammonium molecule NH_3, $A=4$ and so $D=12$. The symmetry considerations that follow enable $\mathbf{F}$ to be very largely diagonalized with very little effort.

The potential field V experienced by the nuclei is inherently extremely difficult to calculate accurately. However, the symmetry of V, being the same as that of the equilibrium positions of the molecule or solid, can be specified precisely. It will be shown that several important features of the vibrational spectrum can be predicted from this symmetry *alone*, without having a detailed numerical knowledge of V.

Not all the $D=3A$ degrees of freedom correspond to vibrations, that is, to "internal" motions of the system. In all cases three degrees of freedom are associated with the translational motion of the system as a whole, in which every nucleus is translated by the same amount. In the absence of external forces, the system as a whole behaves as if it were a free particle situated at the centre of mass, so that a translation is not accompanied by a change in potential energy. Consequently there are three "translational" normal modes with zero frequency.

For a non-linear molecule there are also three degrees of freedom associated with the rotational motion of the system as a whole. Thus, in the absence of external forces, there will be three "rotational" normal modes that also have zero frequency. Therefore, for a non-linear molecule there are only $3A-6$ vibrational normal modes, each of which corresponds to a non-zero frequency. The separation and analysis of the translational and rotational modes can be accomplished by a group theoretical analysis, full details being given in subsection (h).

For a linear molecule there are only two rotational normal modes, so such a system has $3A-5$ vibrational normal modes.

In principle, a crystalline solid must also have three rotational normal modes, but these are lost when the system is approximated by an idealized model having infinite extent on which Born cyclic boundary conditions are imposed (see Chapter 8, Section 7).

Further details of molecular vibration theory and a comprehensive account of its applications may be found in the monographs of Wilson *et al.* (1955) and Harris and Bertolucci (1978).

(*b*) *The invariance group*

In general a molecule or a crystalline solid contains more than one species of nuclei. For example, in the ammonium molecule NH_3 there are two species, one consisting of three hydrogen nuclei and the other of one nitrogen nucleus.

The symmetry transformations of the equilibrium positions may be envisaged as follows. Suppose first that the equilibrium positions of all the nuclei are noted. Then imagine that the system is concealed from view and is subjected to a coordinate transformation so that when the system is again revealed it is apparently unchanged. As all the nuclei of a particular species are identical, the transformation must have permuted nuclei of the same species. The formal definition is then as follows:

Definition *Symmetry transformation* T *of the equilibrium positions*

Let $\mathbf{r}_j^e$ $(j=1,2,\ldots,A)$ be the equilibrium position vectors of the nuclei of a molecule or solid. A coordinate transformation T is said to be a symmetry transformation if it causes a mapping of each equilibrium position vector into an equilibrium position vector of the same set belonging to a nucleus of the same species.

For a molecule this implies that, for every $k=1,2,\ldots,A$, there exists a position vector $\mathbf{r}_j^e$ in the same set belonging to the same species as that at $\mathbf{r}_k^e$ such that

$$\mathbf{r}_j^e=\mathbf{R}(\mathrm{T})\mathbf{r}_k^e. \tag{7.13}$$

For a crystalline solid, after Born cyclic boundary conditions have been imposed, the generalization of Equation (7.13) corresponding to the coordinate transformation $\mathbf{r}'=\{\mathbf{R}(\mathrm{T})\mid\mathbf{t}(\mathrm{T})\}\mathbf{r}$ is of the form

$$\mathbf{r}_j^e=\{\mathbf{R}(\mathrm{T})\mid\mathbf{t}_k(\mathrm{T})\}\mathbf{r}_k^e,$$

where $\mathbf{t}_k(\mathrm{T})$ depends both on $\mathbf{t}(\mathrm{T})$ *and* $\mathbf{r}_k^e$ (see Equations (8.10) and (9.1) below). (However, it should be noted the rotational part $\mathbf{R}(\mathrm{T})$ is exactly as in the coordinate transformation itself.)

It is obvious that each symmetry transformation of the equilibrium positions also transforms any non-equilibrium configuration into another configuration with the same potential energy and transforms any mode of vibration into another mode with the same frequency. (The precise mathematical formulation of these statements will be given later.) Consequently the symmetry transformations of the equilibrium positions are symmetry transformations of the system in any configuration.

It is easily shown that these transformations form a group, $\mathscr{G}$, which will be called the "invariance group of the system". In the case of a molecule, $\mathscr{G}$ contains only rotations and so is a point group. Although $\mathscr{G}$ need not be one of the 32 crystallographic point groups, $\mathscr{G}$ is always either finite or a compact Lie group. For crystalline solids translations must be included as well, so that $\mathscr{G}$ becomes a crystallographic space group. The Born cyclic boundary conditions (see Chapter 8, Section 2(*b*)) allow this to be treated as if it were a finite group.

Example I *The invariance group of the ammonium molecule* NH_3
The arrangement of equilibrium positions of the nuclei of the ammonium molecule is shown in Figure 7.1, the three H nuclei lying at the vertices of an equilateral triangle and the N nucleus being positioned "above" the mid-point of the triangle. With the coordinate axes set up in Figure 7.1, the position vectors of the three H nuclei

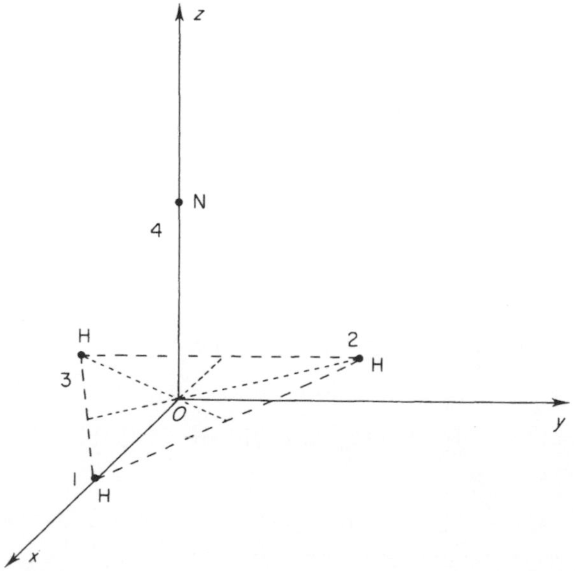

Figure 7.1 Equilibrium configuration of the ammonium molecule NH_3.

are

$$\mathbf{r}_1^e = (a/\sqrt{3}, 0, 0), \qquad \mathbf{r}_2^e = (-a/2\sqrt{3}, a/2, 0), \qquad \mathbf{r}_3^e = (-a/2\sqrt{3}, -a/2, 0),$$

and the position vector of the N nucleus is

$$\mathbf{r}_4^e = (0, 0, c).$$

$\mathscr{G}$ is the point group C_{3v}, consisting of the three classes:

$\mathscr{C}_1 = \{E\}$, E being the identity operation;
$\mathscr{C}_2 = \{C_{3z}, C_{3z}^{-1}\}$, C_{3z} and C_{3z}^{-1} being rotations through $2\pi/3$ about Oz in the right-hand and left-hand screw senses respectively;
$\mathscr{C}_3 = \{IC_{2y}, IC_{2C}, IC_{2D}\}$, IC_{2y}, IC_{2C}, IC_{2D} being reflections in the planes containing Oz whose normals are Oy, OC and OD. (Here OC and OD are axes in the plane Oxy making angles of $\pi/6$ with Ox, as shown in Figure 7.2.)

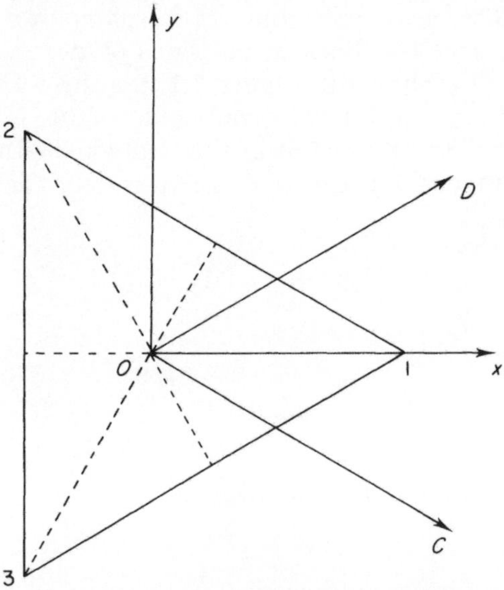

Figure 7.2 Axes of the crystallographic point group C_{3v}.

The matrices $\mathbf{R}(T)$ for $T \in C_{3v}$ are listed in Table D.1. Then, for example, for $T = C_{3z}$,

$$\mathbf{r}_3^e = \mathbf{R}(T)\mathbf{r}_1^e, \qquad \mathbf{r}_1^e = \mathbf{R}(T)\mathbf{r}_2^e, \qquad \mathbf{r}_2^e = \mathbf{R}(T)\mathbf{r}_3^e, \qquad \mathbf{r}_4^e = \mathbf{R}(T)\mathbf{r}_4^e,$$

and, for $T = IC_{2y}$,

$$\mathbf{r}_1^e = \mathbf{R}(T)\mathbf{r}_1^e, \qquad \mathbf{r}_3^e = \mathbf{R}(T)\mathbf{r}_2^e, \qquad \mathbf{r}_2^e = \mathbf{R}(T)\mathbf{r}_3^e, \qquad \mathbf{r}_4^e = \mathbf{R}(T)\mathbf{r}_4^e.$$

Example II *The invariance group of the acetylene molecule* C_2H_2 The acetylene molecule is linear, the equilibrium positions of the nuclei being as shown in Figure 7.3. The equilibrium position vectors

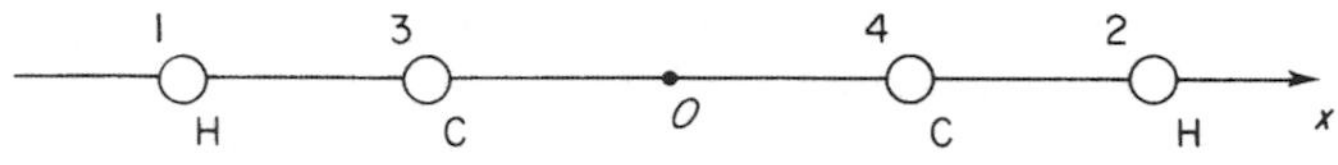

Figure 7.3 Equilibrium configuration of the acetylene molecule C_2H_2.

of the two H nuclei may be taken to be

$$\mathbf{r}_1^e = (-a, 0, 0), \qquad \mathbf{r}_2^e = (a, 0, 0),$$

those of the two C nuclei being

$$\mathbf{r}_3^e = (-b, 0, 0), \qquad \mathbf{r}_4^e = (b, 0, 0),$$

with $0 < b < a$.

The invariance group $\mathscr{G}$ is the linear infinite point group $D_{\infty h}$, with elements exactly as specified in Example VI of Chapter 2, Section 7. For $T = E$, C_θ and $IC_{2\Phi}$, $\mathbf{r}_j^e = \mathbf{R}(T)\mathbf{r}_j^e$ for $j = 1, 2, 3, 4$, whereas for $T = I$, IC_θ and $C_{2\Phi}$, $\mathbf{r}_2^e = \mathbf{R}(T)\mathbf{r}_1^e$, $\mathbf{r}_1^e = \mathbf{R}(T)\mathbf{r}_2^e$, $\mathbf{r}_4^e = \mathbf{R}(T)\mathbf{r}_3^e$ and $\mathbf{r}_3^e = \mathbf{R}(T)\mathbf{r}_4^e$.

(c) *The displacement representation*

For small oscillations, each nucleus can be distinguished from other nuclei of the same species by its equilibrium position. Therefore let $\mathbf{r}_k$ be the position vector at time t of the nucleus whose equilibrium position vector is $\mathbf{r}_k^e$. Let m_k be the mass of this nucleus. Then a very convenient set of generalized coordinates $q_1, q_2, \ldots, q_{3A}$ is defined by

$$\left.\begin{aligned} q_{3k-2} &= (m_k)^{1/2}(x_k - x_k^e), \\ q_{3k-1} &= (m_k)^{1/2}(y_k - y_k^e), \\ q_{3k} &= (m_k)^{1/2}(z_k - z_k^e), \end{aligned}\right\} \qquad k = 1, 2, \ldots, A \tag{7.14}$$

where $\mathbf{r}_k = (x_k, y_k, z_k)$ and $\mathbf{r}_k^e = (x_k^e, y_k^e, z_k^e)$. (That is

$$q_1 = (m_1)^{1/2}(x_1 - x_1^e), \qquad q_2 = (m_1)^{1/2}(y_1 - y_1^e), \qquad q_3 = (m_1)^{1/2}(z_1 - z_1^e),$$

$$q_4 = (m_2)^{1/2}(x_2 - x_2^e), \qquad q_5 = (m_2)^{1/2}(y_2 - y_2^e), \qquad q_6 = (m_2)^{1/2}(z_2 - z_2^e),$$

and so on.) Then $q_1 = q_2 = \ldots = q_{3A} = 0$ certainly corresponds to an equilibrium position. The "mass-weighting" in Equations (7.14) is introduced to give $T = \frac{1}{2}\sum_{k=1}^{A} m_k \dot{\mathbf{r}}_k^2$ the simple form of Equation (7.1). For small displacements V has the general form of Equation (7.2).

Suppose that $\mathbf{r}_k^e$ is transformed into $\mathbf{r}_j^e$ by the symmetry transformation $T \in \mathcal{G}$ according to Equation (7.13) or its generalization. Let $\mathbf{r}'_j$ be defined in terms of $\mathbf{r}_k$ be the same transformation. Then, from Equation (1.5),

$$\mathbf{r}'_j - \mathbf{r}_j^e = \mathbf{R}(T)(\mathbf{r}_k - \mathbf{r}_k^e), \tag{7.15}$$

which implies that $\mathbf{r}'_j = \mathbf{r}_j^e$ when $\mathbf{r}_k = \mathbf{r}_k^e$.

Let $q'_1, q'_2, \ldots, q'_{3A}$ be defined, by analogy with Equations (7.14), by

$$\left.\begin{aligned} q'_{3j-2} &= (m_j)^{1/2}(x'_j - x_j^e), \\ q'_{3j-1} &= (m_j)^{1/2}(y'_j - y_j^e), \\ q'_{3j} &= (m_j)^{1/2}(z'_j - z_j^e), \end{aligned}\right\} \quad j = 1, 2, \ldots, A$$

where $\mathbf{r}'_j = (x'_j, y'_j, z'_j)$. Then Equation (7.15) can be rewritten as

$$q'_{3j-3+s} = \sum_{t=1}^{3} R(T)_{st} q_{3k-3+t} \tag{7.16}$$

for $j = 1, 2, \ldots, A$ and $s = 1, 2, 3$. (The mass factors do not appear here because j and k refer to nuclei of the same species, which necessarily have the same mass.)

As noted in Section 2(a) it is particularly convenient to express the generalized coordinates in matrix form. Accordingly, let $\mathbf{q}$ and $\mathbf{q}'$ be the $3A \times 1$ real column vectors with components q_m and q'_m respectively ($m = 1, 2, \ldots, 3A$). Then Equation (7.16) has the form

$$\mathbf{q}' = \mathbf{\Gamma}^{disp}(T)\mathbf{q}, \tag{7.17}$$

where $\mathbf{\Gamma}^{disp}(T)$ is a $3A \times 3A$ real matrix whose elements are given for $s = 1, 2, 3$ by

$$\Gamma^{disp}(T)_{3j-3+s,m} = \begin{cases} R(T)_{st}, & \text{if} \quad m = 3k-3+t, t = 1, 2, \text{or } 3, \\ 0, & \text{if} \quad m \neq 3k-2, 3k-1, \text{or } 3k. \end{cases} \tag{7.18}$$

(Here, as above, j, k and T are related by Equation (7.13) or its generalization.)

The structure of $\mathbf{\Gamma}^{disp}(T)$ is clarified by introducing for each $T \in \mathcal{G}$ an $A \times A$ real matrix $\mathbf{p}(T)$, which may be called the "permutation matrix", defined for each $k = 1, 2, \ldots, A$ by

$$\mathbf{p}(T)_{jk} = \begin{cases} 1, & \text{if } \mathbf{r}_k^e \text{ is mapped into } \mathbf{r}_j^e \text{ by T}, \\ 0, & \text{otherwise.} \end{cases} \tag{7.19}$$

Then Equations (7.18) imply that

$$\mathbf{\Gamma}^{disp}(T) = \mathbf{p}(T) \otimes \mathbf{R}(T),$$

where the rows and columns of the direct product matrix are labelled by the prescription of Appendix A. It is easily established that the matrices $\mathbf{p}(\mathrm{T})$ form an A-dimensional representation of $\mathcal{G}$ that is both real and unitary. As the matrices $\mathbf{R}(\mathrm{T})$ form a three-dimensional representation of $\mathcal{G}$ that is also real and unitary, the argument given in Chapter 5, Section 2, shows that the matrices $\mathbf{\Gamma}^{disp}(\mathrm{T})$ form a $3A$-dimensional real unitary representation of $\mathcal{G}$. This is called the "displacement representation" of $\mathcal{G}$.

As an example, suppose that $A=3$ and T is such that

$$\mathbf{p}(\mathrm{T})=\begin{bmatrix}0 & 0 & 1\\ 1 & 0 & 0\\ 0 & 1 & 0\end{bmatrix}$$

Then Equations (7.18) imply that $\mathbf{\Gamma}^{disp}(\mathrm{T})$ may be partitioned so that

$$\mathbf{\Gamma}^{disp}(\mathrm{T})=\left[\begin{array}{c:c:c}\mathbf{0} & \mathbf{0} & \mathbf{R}(\mathrm{T})\\ \hdashline \mathbf{R}(\mathrm{T}) & \mathbf{0} & \mathbf{0}\\ \hdashline \mathbf{0} & \mathbf{R}(\mathrm{T}) & \mathbf{0}\end{array}\right]. \tag{7.20}$$

This shows the general rule that to every non-zero element in $\mathbf{p}(\mathrm{T})$ there corresponds a submatrix $\mathbf{R}(\mathrm{T})$ in $\mathbf{\Gamma}^{disp}(\mathrm{T})$, whereas to every zero element in $\mathbf{p}(\mathrm{T})$ there corresponds a zero submatrix in $\mathbf{\Gamma}^{disp}(\mathrm{T})$.

Even though $\mathbf{\Gamma}^{disp}(\mathrm{T})$ has a fairly complicated structure, its characters $\chi^{disp}(\mathrm{T})$ are very easily obtained. The argument leading to Equation (5.14) can be generalized to show that

$$\chi^{disp}(\mathrm{T})=\{\mathrm{tr}\,\mathbf{p}(\mathrm{T})\}\{\mathrm{tr}\,\mathbf{R}(\mathrm{T})\}, \tag{7.21}$$

but Equations (7.19) imply that $\mathrm{tr}\,\mathbf{p}(\mathrm{T})=u(\mathrm{T})$, the number of nuclei that are unmoved by the coordinate transformation T; thus, from Equations (7.21) and (12.6)

$$\chi^{disp}(\mathrm{T})=u(\mathrm{T})(1+2\cos\theta) \tag{7.22}$$

if $\mathbf{R}(\mathrm{T})$ is a proper rotation through an angle θ, while if $\mathbf{R}(\mathrm{T})$ is an improper rotation that is the product of the spatial inversion operation with a proper rotation through θ,

$$\chi^{disp}(\mathrm{T})=-u(\mathrm{T})(1+2\cos\theta). \tag{7.23}$$

Example III *The displacement representation for the ammonium molecule* NH_3

With the equilibrium positions as specified in Example I, the matrices of the permutation representation $\mathbf{p}(T)$ are:

$$\mathbf{p}(E)=\begin{bmatrix}1&0&0&0\\0&1&0&0\\0&0&1&0\\0&0&0&1\end{bmatrix},\qquad \mathbf{p}(C_{3z})=\begin{bmatrix}0&1&0&0\\0&0&1&0\\1&0&0&0\\0&0&0&1\end{bmatrix},$$

$$\mathbf{p}(C_{3z}^{-1})=\begin{bmatrix}0&0&1&0\\1&0&0&0\\0&1&0&0\\0&0&0&1\end{bmatrix},\qquad \mathbf{p}(IC_{2y})=\begin{bmatrix}1&0&0&0\\0&0&1&0\\0&1&0&0\\0&0&0&1\end{bmatrix},$$

$$\mathbf{p}(IC_{2C})=\begin{bmatrix}0&1&0&0\\1&0&0&0\\0&0&1&0\\0&0&0&1\end{bmatrix},\qquad \mathbf{p}(IC_{2D})=\begin{bmatrix}0&0&1&0\\0&1&0&0\\1&0&0&0\\0&0&0&1\end{bmatrix},$$

so that

$$\left.\begin{aligned}u(E)=u(C_{3z})=u(C_{3z}^{-1})=1,\\ u(IC_{2y})=u(IC_{2C})=u(IC_{2D})=2.\end{aligned}\right\}$$

Thus, from Equations (7.22) and (7.23), if $\mathscr{C}_1$, $\mathscr{C}_2$, $\mathscr{C}_3$ are the three classes of C_{3v}

$$\chi^{disp}(\mathscr{C}_1)=12,\qquad \chi^{disp}(\mathscr{C}_2)=0,\qquad \chi^{disp}(\mathscr{C}_3)=2.$$

(The matrices $\mathbf{R}(T)$ for $T\in C_{3v}$ are contained in Table D.1.)

Example IV *The displacement representation for the acetylene molecule* C_2H_2

With the equilibrium positions as specified in Example II, the matrices of the permutation representation $\mathbf{p}(T)$ are given by

$$\mathbf{p}(E)=\mathbf{p}(C_\theta)-\mathbf{p}(IC_{2\Phi})=\begin{bmatrix}1&0&0&0\\0&1&0&0\\0&0&1&0\\0&0&0&1\end{bmatrix},$$

$$\mathbf{p}(I)=\mathbf{p}(IC_{2\theta})=\mathbf{p}(C_{2\Phi})=\begin{bmatrix}0&1&0&0\\1&0&0&0\\0&0&0&1\\0&0&1&0\end{bmatrix},$$

so that

$$u(\mathrm{E}) = u(\mathrm{C}_\theta) = u(\mathrm{IC}_{2\Phi}) = 4,$$
$$u(\mathrm{I}) = u(\mathrm{IC}_\theta) = u(\mathrm{C}_{2\Phi}) = 0.$$

Thus

$$\chi^{disp}(\mathrm{E}) = 12, \quad \chi^{disp}(\mathrm{C}_\theta) = 4(1+2\cos\theta), \quad \chi^{disp}(\mathrm{C}_{2\Phi}) = 0,$$
$$\chi^{disp}(\mathrm{I}) = 0, \quad \chi^{disp}(\mathrm{IC}_\theta) = 0, \quad \chi^{disp}(\mathrm{IC}_{2\Phi}) = 4.$$

Formulae giving $u(\mathrm{T})$ for the case in which $\mathcal{G}$ is a crystallographic space group have been given by de Angelis *et al.* (1972).

The dynamical symmetry of the system is succinctly expressed by the following theorem, which shows the vital role played by the displacement representation.

Theorem I For every symmetry transformation T of the invariance group $\mathcal{G}$

$$\mathbf{\Gamma}^{disp}(\mathrm{T})\mathbf{F} = \mathbf{F}\mathbf{\Gamma}^{disp}(\mathrm{T}), \tag{7.24}$$

$\mathbf{F}$ being the force matrix of Equation (7.4).

Proof The invariance of the system under the coordinate transformation means that the potential energy V for the configuration $\mathbf{q}$ must be the same as for the configuration $\mathbf{q}'$, that is, $V(q_1, q_2, \ldots, q_{3A}) = V(q_1', q_2', \ldots, q_{3A}')$. Thus, from Equation (7.4), $V = \frac{1}{2}\tilde{\mathbf{q}}\mathbf{F}\mathbf{q} = \frac{1}{2}\tilde{\mathbf{q}}'\mathbf{F}\mathbf{q}'$ and hence, by Equation (7.17), $\frac{1}{2}\tilde{\mathbf{q}}\mathbf{F}\mathbf{q} = \frac{1}{2}\tilde{\mathbf{q}}\tilde{\mathbf{\Gamma}}^{disp}(\mathrm{T})\mathbf{F}\mathbf{\Gamma}^{disp}(\mathrm{T})\mathbf{q}$. As this must be valid for all small $\mathbf{q}$, $\mathbf{F} = \tilde{\mathbf{\Gamma}}^{disp}(\mathrm{T})\mathbf{F}\mathbf{\Gamma}^{disp}(\mathrm{T})$, from which Equation (7.24) follows because $\mathbf{\Gamma}^{disp}(\mathrm{T})$ is real and unitary.

(*d*) *Partial diagonalization of the force matrix* $\mathbf{F}$

In general, the displacement representation $\mathbf{\Gamma}^{disp}$ of $\mathcal{G}$ is reducible (and, as it is unitary, it is completely reducible). Suppose that the unitary irreducible representation $\mathbf{\Gamma}^p$ with dimension d_p appears n_p times in the reduction of $\mathbf{\Gamma}^{disp}$. Then there exists a $3A \times 3A$ unitary matrix $\mathbf{U}$ such that for all $\mathrm{T} \in \mathcal{G}$

$$\mathbf{U}^{-1}\mathbf{\Gamma}^{disp}(\mathrm{T})\mathbf{U} = \mathbf{\Gamma}'(\mathrm{T}) \equiv \sum_p \oplus\, n_p \mathbf{\Gamma}^p(\mathrm{T}). \tag{7.25}$$

That is, $\mathbf{U}$ induces a similarity transformation that transforms $\mathbf{\Gamma}^{disp}$ into the *direct* sum, $\mathbf{\Gamma}'$, of unitary irreducible representations. (In general, some of these unitary irreducible representations may not be real, so $\mathbf{U}$ need not be real.)

Let $\mathbf{F}'$ be the $3A \times 3A$ matrix defined by

$$\mathbf{F}' = \mathbf{U}^{-1}\mathbf{F}\mathbf{U}. \tag{7.26}$$

Then Equation (7.24) can be written as

$$\mathbf{\Gamma}'(\mathrm{T})\mathbf{F}' = \mathbf{F}'\mathbf{\Gamma}'(\mathrm{T}) \tag{7.27}$$

for all $\mathrm{T} \in \mathcal{G}$. Now let $\mathbf{F}'$ be partitioned in the same way as $\mathbf{\Gamma}'$, that is, as

$$\mathbf{\Gamma}' = \begin{bmatrix} \mathbf{\Gamma}^1 & \mathbf{0} & \ldots & \mathbf{0} & \mathbf{0} & \ldots \\ \mathbf{0} & \mathbf{\Gamma}^1 & \ldots & \mathbf{0} & \mathbf{0} & \ldots \\ \vdots & \vdots & & \vdots & \vdots & \\ \mathbf{0} & \mathbf{0} & & \mathbf{\Gamma}^2 & \mathbf{0} & \ldots \\ \mathbf{0} & \mathbf{0} & & \mathbf{0} & \mathbf{\Gamma}^2 & \\ \vdots & \vdots & & & & \end{bmatrix}$$

This implies

$$\mathbf{F}' = \begin{bmatrix} \mathbf{F}'^{11}_{11} & \ldots & \mathbf{F}'^{11}_{1n_1} & \mathbf{F}'^{12}_{11} & \ldots & \mathbf{F}'^{12}_{1n_2} & \ldots \\ \vdots & & \vdots & \vdots & & \vdots & \\ \mathbf{F}'^{11}_{n_1 1} & \ldots & \mathbf{F}'^{11}_{n_1 n_1} & \mathbf{F}'^{12}_{n_1 1} & \ldots & \mathbf{F}'^{12}_{n_1 n_2} & \ldots \\ \mathbf{F}'^{21}_{11} & \ldots & \mathbf{F}'^{21}_{1n_1} & \mathbf{F}'^{22}_{11} & \ldots & \mathbf{F}'^{22}_{1n_2} & \ldots \\ \vdots & & \vdots & \vdots & & \vdots & \\ \mathbf{F}'^{21}_{n_2 1} & \ldots & \mathbf{F}'^{21}_{n_2 n_1} & \mathbf{F}'^{22}_{n_2 1} & \ldots & \mathbf{F}'^{22}_{n_2 n_2} & \ldots \\ \vdots & & \vdots & \vdots & & \vdots & \end{bmatrix}, \tag{7.28}$$

where $\mathbf{F}'^{pq}_{\alpha\beta}$ is a submatrix of dimension $d_p \times d_p$ for $\alpha = 1, 2, \ldots, n_p$ and $\beta = 1, 2, \ldots, n_q$. Then Equation (7.27) gives

$$\mathbf{\Gamma}^p(\mathrm{T})\mathbf{F}'^{pq}_{\alpha\beta} = \mathbf{F}'^{pq}_{\alpha\beta}\mathbf{\Gamma}^q(\mathrm{T})$$

for all p and q such that $n_p \neq 0$ and $n_q \neq 0$ and for all $\mathrm{T} \in \mathcal{G}$. Thus, as $\mathbf{\Gamma}^p$ is not equivalent to $\mathbf{\Gamma}^q$ for $p \neq q$, by Schur's Lemma (see Chapter 4, Section 5),

$$\mathbf{F}'^{pq}_{\alpha\beta} = \begin{cases} \mathbf{0}, & \text{if } \; p \neq q, \\ f^p_{\alpha\beta}\mathbf{1}_{d_p}, & \text{if } \; p = q, \end{cases} \tag{7.29}$$

where $f^p_{\alpha\beta}$ are a set of $(n_p)^2$ complex numbers. As $\mathbf{F}$ is Hermitian, Equation (7.25) implies that $\mathbf{F}'$ is also Hermitian, so that $f^p_{\alpha\beta} = (f^p_{\beta\alpha})^*$ for $\alpha, \beta = 1, 2, \ldots, n_p$. The set of complex numbers $f^p_{\alpha\beta}$ can then be regarded as forming an $n_p \times n_p$ Hermitian matrix $\mathbf{f}^p$.

To clarify this general analysis, consider a particular example with

$n_1 = 2$, $d_1 = 3$ and $n_p = 0$ for $p \neq 1$. In this case $\mathbf{F}'$ has the form

$$\mathbf{F}' = \begin{bmatrix} f^1_{11} & 0 & 0 & f^1_{12} & 0 & 0 \\ 0 & f^1_{11} & 0 & 0 & f^1_{12} & 0 \\ 0 & 0 & f^1_{11} & 0 & 0 & f^1_{12} \\ f^1_{21} & 0 & 0 & f^1_{22} & 0 & 0 \\ 0 & f^1_{21} & 0 & 0 & f^1_{22} & 0 \\ 0 & 0 & f^1_{21} & 0 & 0 & f^1_{22} \end{bmatrix}.$$

A further similarity transformation by an orthogonal matrix $\mathbf{V}$ can be applied to rearrange the rows and columns to give the form

$$\mathbf{F}'' = \mathbf{V}^{-1}\mathbf{F}'\mathbf{V} = \begin{bmatrix} f^1_{11} & f^1_{12} & 0 & 0 & 0 & 0 \\ f^1_{21} & f^1_{22} & 0 & 0 & 0 & 0 \\ 0 & 0 & f^1_{11} & f^1_{12} & 0 & 0 \\ 0 & 0 & f^1_{21} & f^1_{22} & 0 & 0 \\ 0 & 0 & 0 & 0 & f^1_{11} & f^1_{12} \\ 0 & 0 & 0 & 0 & f^1_{21} & f^1_{22} \end{bmatrix} = \begin{bmatrix} \mathbf{f}^1 & \mathbf{0} & \mathbf{0} \\ \mathbf{0} & \mathbf{f}^1 & \mathbf{0} \\ \mathbf{0} & \mathbf{0} & \mathbf{f}^1 \end{bmatrix}.$$

Obviously a similar procedure can be carried through in every case, giving $\mathbf{F}'' = \mathbf{V}^{-1}\mathbf{F}'\mathbf{V}$ as a direct sum in which, for each p such that $n_p \neq 0$, the $n_p \times n_p$ Hermitian matrix $\mathbf{f}^p$ appears d_p times. As $\mathbf{F}$ and $\mathbf{F}''$ are related by a similarity transformation they have the same set of eigenvalues. But the eigenvalue equation for $\mathbf{F}''$ is

$$\det \begin{bmatrix} \mathbf{f}^1 - \lambda\mathbf{1} & \mathbf{0} & \cdots & \mathbf{0} & \mathbf{0} & \cdots & \mathbf{0} & \cdots \\ \mathbf{0} & \mathbf{f}^1 - \lambda\mathbf{1} & \cdots & \mathbf{0} & \mathbf{0} & \cdots & \mathbf{0} & \cdots \\ \vdots & \vdots & & \vdots & \vdots & & \vdots & \\ \mathbf{0} & \mathbf{0} & \cdots & \mathbf{f}^1 - \lambda\mathbf{1} & \mathbf{0} & \cdots & \mathbf{0} & \cdots \\ \mathbf{0} & \mathbf{0} & \cdots & \mathbf{0} & \mathbf{f}^2 - \lambda\mathbf{1} & \cdots & \mathbf{0} & \cdots \\ \vdots & \vdots & & & & & \vdots & \\ \mathbf{0} & \mathbf{0} & \cdots & \mathbf{0} & \mathbf{0} & \cdots & \mathbf{f}^2 - \lambda\mathbf{1} & \cdots \\ \vdots & \vdots & & \vdots & \vdots & & \vdots & \end{bmatrix} = 0.$$

Thus, for each p such that $n_p \neq 0$, the set of eigenvalues λ^p_α of $\mathbf{f}^p$ ($\alpha = 1, 2, \ldots, n_p$) occur d_p times.

To summarize, the above analysis shows that:

(i) each eigenvalue of $\mathbf{F}$ may be associated with an irreducible representation Γ^p of $\mathscr{G}$;

(ii) every eigenvalue λ^p_α of $\mathbf{F}$ associated with Γ^p (of dimension d_p) is automatically d_p-fold degenerate;

(iii) the eigenvalues λ_α^p of $\mathbf{F}$ associated with Γ^p may be found by solving the $n_p \times n_p$ secular equation

$$\det(\mathbf{f}^p - \lambda \mathbf{1}) = 0. \qquad (7.30)$$

In addition to the automatic d_p-fold degeneracy mentioned in (ii), there may exist "accidental degeneracies", in which two or more eigenvalues of the matrices $\mathbf{f}^p$ are equal. (Degeneracies created by the reality of $\mathbf{F}$ are classed as accidental degeneracies in this sense. They will be considered further in subsection (*f*).)

Precise numerical values for the frequencies can only be obtained by the explicit solution of Equations (7.30), but conclusions (i) and (ii) provide very useful information with minimal effort, for the determination of which irreducible representations appear in Γ^{disp} (and how often they appear) needs only the characters. Indeed, for the case in which $\mathscr{G}$ is a finite group, Theorem V of Chapter 4, Section 6 shows that

$$n_p = (1/g) \sum_{T \in \mathscr{G}} \chi^{disp}(T)\chi^p(T)^*, \qquad (7.31)$$

g being the order of $\mathscr{G}$, with the corresponding expression holding for a compact Lie group.

Example V *Eigenvalue spectrum for the ammonium molecule* NH_3

From the character table (Table D.21) of $\mathscr{G} = C_{3v}$ and the expressions for $\chi^{disp}(T)$ given in Example III, it follows from Equation (7.31) that

$$\Gamma^{disp} \approx 3\Gamma^1 \oplus \Gamma^2 \oplus 4\Gamma^3,$$

i.e. $n_1 = 3$, $n_2 = 1$, $n_3 = 4$. (Alternatively, expressed in the A, B, E, T notation for irreducible representations, $\Gamma^{disp} \sim 3A_1 \oplus A_2 \oplus 4E$.) Thus there are three non-degenerate eigenvalues associated with Γ^1 (i.e. with A_1), found by solving a 3×3 secular equation, one non-degenerate eigenvalue associated with Γ^2 (i.e. with A_2) (the relevant part of $\mathbf{F}''$ being completely diagonalized), and four two-fold degenerate eigenvalues associated with Γ^3 (i.e. with E), found from a 4×4 secular equation. Not all of these eigenvalues correspond to vibrational modes. (See Example VIII below for the identification and separation of the translational and rotational modes.)

Example VI *Eigenvalue spectrum for the acetylene molecule* C_2H_2

The expressions given for $\chi^{disp}(T)$ in Example IV and the character table (Table D.46) for $\mathscr{G} = D_{\infty h}$ imply that

$$\Gamma^{disp} \approx 2A_{1g} \oplus 2A_{2u} \oplus 2E_{1g} \oplus 2E_{1u}.$$

Thus these are two non-degenerate eigenvalues associated with both A_{1g} and A_{2u} and two two-fold degenerate eigenvalues associated with both E_{1g} and E_{1u}. In each case the eigenvalues are found from a 2×2 secular equation. Again, not all of these eigenvalues correspond to vibrational modes. (The identification and separation of the translational and rotational modes is carried out in Example IX below.)

(e) *Analysis in terms of basis vectors*

The striking similarity between the results of the previous subsection and those on degeneracies of the eigenvalues of the Schrödinger equation will now be made even closer.

The first stage is to introduce a carrier space V for the displacement representation $\boldsymbol{\Gamma}^{disp}$ of $\mathscr{G}$ within the general formulation of Chapter 4, Section 1. This V may be defined as the inner product space consisting of all $3A\times 1$ complex column matrices $\mathbf{v}$, with inner product given by

$$(\mathbf{u}, \mathbf{v}) = \mathbf{u}^{\dagger}\mathbf{v}\left(= \sum_{m=1}^{3A} u_m^* v_m\right)$$

(V may be recognized as the complex $3A$-dimensional space $\mathbb{C}^{3A}$ (see Appendix B, Section 2, Example I)).

The next stage is to *define* for every $T\in\mathscr{G}$ an *operator* $\Phi(T)$ acting on V by

$$\Phi(T)\mathbf{v} = \boldsymbol{\Gamma}^{disp}(T)\mathbf{v} \qquad (7.32)$$

for all $\mathbf{v}\in V$. As $\boldsymbol{\Gamma}^{disp}(T_1T_2) = \boldsymbol{\Gamma}^{disp}(T_1)\boldsymbol{\Gamma}^{disp}(T_2)$ for all $T_1, T_2\in\mathscr{G}$, it follows immediately that

$$\Phi(T_1T_2) = \Phi(t_1)\Phi(T_2),$$

that is, Equation (4.2) is satisfied.

Let $\mathbf{v}_m$ $(m = 1, 2, \ldots, 3A)$ be the ortho-normal basis of V, defined by

$$(\mathbf{v}_m)_n = \delta_{mn}, \qquad m, n = 1, 2, \ldots, 3A. \qquad (7.33)$$

Then, as noted in Chapter 4, Section 1, there exists a $3A$-dimensional representation $\boldsymbol{\Gamma}$ of $\mathscr{G}$ defined by

$$\Phi(T)\mathbf{v}_n = \sum_{m=1}^{3A} \Gamma(T)_{mn}\mathbf{v}_m, \qquad n = 1, 2, \ldots, 3A, \qquad (7.34)$$

for all $T\in\mathscr{G}$, the elements of $\boldsymbol{\Gamma}(T)$ being given explicitly by

$$\Gamma(T)_{mn} = (\mathbf{v}_m, \Phi(T)\mathbf{v}_n), \qquad m, n = 1, 2, \ldots, 3A. \qquad (7.35)$$

(Equations (7.34) and (7.35) are essentially Equations (4.1) and (4.3) respectively.) However, Equations (7.32), (7.33) and (7.35) imply that

$$\Gamma(\mathrm{T})_{mn} = \Gamma^{disp}(\mathrm{T})_{mn}, \qquad m, n = 1, 2, \ldots, 3A,$$

so that with the particular choice of basis made here this representation $\boldsymbol{\Gamma}$ is *identical* to $\boldsymbol{\Gamma}^{disp}$. Thus Equation (7.34) becomes

$$\Phi(\mathrm{T})\mathbf{v}_n = \sum_{m=1}^{3A} \Gamma^{disp}(\mathrm{T})_{mn}\mathbf{v}_m, \qquad n = 1, 2, \ldots, 3A. \tag{7.36}$$

It should be emphasized that Equation (7.32) provides an explicit concrete definition of the operators $\Phi(\mathrm{T})$, which will be used later in setting up projection operators. It should also be noted that, with this definition (Equation (7.32)), Equation (7.36) becomes

$$\boldsymbol{\Gamma}^{disp}(\mathrm{T})\mathbf{v}_n = \sum_{m=1}^{3A} \Gamma^{disp}(\mathrm{T})_{mn}\mathbf{v}_m,$$

in which the same quantities appear on both sides of the equation in significantly different ways!

It will now be demonstrated that the operators $\Phi(\mathrm{T})$ play essentially the same role in vibration theory that the operators $P(\mathrm{T})$ play in the theory of the Schrödinger equation. The first step is to translate the definition of basis functions given in Chapter 1, Section 4, to the present context.

Definition *Basis vectors of a representation of the invariance group $\mathscr{G}$*

A set of d linearly independent $3A \times 1$ complex column matrices $\boldsymbol{\psi}_1, \boldsymbol{\psi}_2, \ldots, \boldsymbol{\psi}_d$ forms a basis for a d-dimensional representation $\boldsymbol{\Gamma}$ of $\mathscr{G}$ if, for every $\mathrm{T} \in \mathscr{G}$,

$$\Phi(\mathrm{T})\boldsymbol{\psi}_n = \sum_{m=1}^{d} \Gamma(\mathrm{T})_{mn}\boldsymbol{\psi}_m, \qquad n = 1, 2, \ldots, d. \tag{7.37}$$

The matrix $\boldsymbol{\psi}_m$ is then said to "transform as the mth row of $\boldsymbol{\Gamma}$".

Here $\boldsymbol{\Gamma}$ may be *any* representation of $\mathscr{G}$, reducible or irreducible. The basis vectors are simply column matrices with components that are constants (in the sense of *not* being functions of $\mathbf{r}$). In particular, as Equation (7.36) has the form of Equation (7.37), the content of Equation (7.36) may be expressed by saying that $\mathbf{v}_1, \mathbf{v}_2, \ldots, \mathbf{v}_{3A}$ are basis vectors for the displacement representation $\boldsymbol{\Gamma}^{disp}$.

The theorem on energy eigenfunctions given in Chapter 1, Section 4, has the following direct analogue in vibration theory.

Theorem II The eigenvectors of the force matrix $\mathbf{F}$ corresponding to a d-fold degenerate eigenvalue form a set of basis vectors for a d-dimensional representation of the invariance group $\mathcal{G}$.

Proof Let $\mathbf{c}_1, \mathbf{c}_2, \ldots, \mathbf{c}_d$ be a set of linearly independent eigenvectors of $\mathbf{F}$ corresponding to a d-fold degenerate eigenvalue λ, that is

$$\mathbf{F}\mathbf{c}_n = \lambda\mathbf{c}_n, \qquad n = 1, 2, \ldots, d \tag{7.38}$$

Then for every $\mathrm{T} \in \mathcal{G}$ and each $n = 1, 2, \ldots, d$, Equations (7.32), (7.24) and (7.38) imply

$$\begin{aligned}\mathbf{F}\{\mathbf{\Phi}(\mathrm{T})\mathbf{c}_n\} &= \mathbf{F}\{\mathbf{\Gamma}^{disp}(\mathrm{T})\mathbf{c}_n\} = \mathbf{\Gamma}^{disp}(\mathrm{T})\{\mathbf{F}\mathbf{c}_n\} \\ &= \mathbf{\Gamma}^{disp}(\mathrm{T})\{\lambda\mathbf{c}_n\} \\ &= \lambda\{\mathbf{\Phi}(\mathrm{T})\mathbf{c}_n\},\end{aligned}$$

so that $\mathbf{\Phi}(\mathrm{T})\mathbf{c}_n$ is also an eigenvector of $\mathbf{F}$ with the same eigenvalue λ. Thus one can write

$$\mathbf{\Phi}(\mathrm{T})\mathbf{c}_n = \sum_{m=1}^{d} \Gamma(\mathrm{T})_{mn}\mathbf{c}_m,$$

thereby defining a $d \times d$ matrix $\mathbf{\Gamma}(\mathrm{T})$ for each $\mathrm{T} \in \mathcal{G}$. These matrices $\mathbf{\Gamma}(\mathrm{T})$ form a d-dimensional representation of $\mathcal{G}$, the proof being essentially the same as that of the theorem of Chapter 1, Section 4.

As in Theorem III of Chapter 4, Section 2, if $\mathbf{U}$ is a $3A \times 3A$ unitary matrix and $\mathbf{v}'_1, \mathbf{v}'_2, \ldots, \mathbf{v}'_{3A}$ are a set of $3A \times 1$ complex column vectors defined by

$$\mathbf{v}'_n = \sum_{m=1}^{3A} \mathrm{U}_{mn}\mathbf{v}_m$$

then $\mathbf{v}'_1, \mathbf{v}'_2, \ldots, \mathbf{v}'_n$ form a basis for the representation $\mathbf{\Gamma}'$ of $\mathcal{G}$, defined by $\mathbf{\Gamma}'(\mathrm{T}) = \mathbf{U}^{-1}\mathbf{\Gamma}^{disp}(\mathrm{T})\mathbf{U}$. In particular, with $\mathbf{U}$ chosen as in Equation (7.25) so that $\mathbf{\Gamma}'$ is completely reduced, then subsets of $\mathbf{v}'_1, \mathbf{v}'_2, \ldots, \mathbf{v}'_{3A}$ form bases for each of the irreducible representations appearing in $\mathbf{\Gamma}'$.

This clarifies the observation made in subsection (d) that, in general, each eigenvalue of $\mathbf{F}$ is associated with an irreducible representation of $\mathcal{G}$ and, apart from accidental degeneracies, the degeneracy of the eigenvalue is equal to the dimension of the representation.

The eigenvectors of $\mathbf{F}$ may all be chosen to be real (and indeed they have to be real in order to produce real normal coordinates by Equation (7.9)). On the other hand, the basis vectors of an irreducible representation cannot be taken to be real unless the irreducible representation is real. For potentially real irreducible representations this is achieved simply by applying an appropriate similarity transformation (see Chapter 5, Section 8). For pseudo-real and essentially complex representations the resolution of this problem is discussed in the next subsection.

Suppose for the moment that the irreducible representation $\boldsymbol{\Gamma}^p$ is real and occurs n_p times in $\boldsymbol{\Gamma}^{disp}$. Then there exist n_p linearly independent real eigenvectors of $\mathbf{F}$ that transform as each row m of $\boldsymbol{\Gamma}^p$. These may be denoted by $\mathbf{c}^p_{\alpha m}$, $\alpha = 1, 2, \ldots, n_p$. The corresponding normal coordinates may be denoted by $Q^p_{\alpha m}$ and defined, as in Equation (7.9), by

$$Q^p_{\alpha m} = \tilde{\mathbf{c}}^p_{\alpha m}\mathbf{q}. \tag{7.39}$$

Thus, in the case in which $\boldsymbol{\Gamma}^p$ is real, the relevant normal coordinates may be identified with the rows of $\boldsymbol{\Gamma}^p$.

The arbitrariness in the choice of eigenvectors when degeneracies occur that was mentioned in Section 2(a) (which in this case occurs when $d_p > 1$) is a reflection of the fact that any similarity transformation with a $d_p \times d_p$ real orthogonal matrix applied to the representation $\boldsymbol{\Gamma}^p$ will produce an equivalent representation that is also real which could be used in place of $\boldsymbol{\Gamma}^p$.

Returning to the general case in which the irreducible representations need not all be real, consider the matrix $\mathbf{U}$ of subsection (d). As $\mathbf{U}$ relates $\boldsymbol{\Gamma}^{disp}$ to $\sum_p \oplus n_p \boldsymbol{\Gamma}^p$ it must be possible to determine its elements from these representations alone. One method is to invoke the projection operator technique of Chapter 5, Section 1, merely replacing P(T) by Φ(T). Thus the projection operator $\mathcal{P}^p_{mn}$ may be defined by

$$\mathcal{P}^p_{mn} = (d_p/g) \sum_{T \in \mathcal{G}} \Gamma^p(T)^*_{mn} \Phi(T)$$

for a finite group of order g, and by

$$\mathcal{P}^p_{mn} = d_p \int_{\mathcal{G}} dT \Gamma^p(T)^*_{mn} \Phi(T)$$

for a compact Lie group. These operators have essentially the same properties as those given in the second theorem of Chapter 5, Section 1, except that the inner product space L^2 is now replaced by $V = \mathbb{C}^{3A}$. In particular, $\mathcal{P}^p_{nn}$ projects out of any $3A \times 1$ column matrix $\mathbf{v}$ the part transforming as the nth row of $\boldsymbol{\Gamma}^p$ and $\mathcal{P}^p_{mn} \boldsymbol{\Psi}^q_j = \delta_{pq}\delta_{nj}\boldsymbol{\Psi}^p_m$.

Assuming that $\boldsymbol{\Gamma}^p$ appears n_p times in $\boldsymbol{\Gamma}^{disp}$, the technique is to apply $\mathcal{P}^p_{11}$ to a subset of the ortho-normal set $\mathbf{v}_1, \mathbf{v}_2, \ldots, \mathbf{v}_{3A}$ to obtain n_p normalized, linearly independent vectors transforming as the first row of Γ^p. Their partners transforming as the other rows of Γ^p can then be obtained by applying $\mathcal{P}^p_{m1}$ to each of these for $m = 2, 3, \ldots, d_p$. As the resulting vectors are linear combinations of $\mathbf{v}_1, \mathbf{v}_2, \ldots, \mathbf{v}_{3A}$, the elements of $\mathbf{U}$ can be read off. The calculations are simple but tedious. They are helped by the observation that Equations (7.32) and (7.33) imply that $\Phi(T)\mathbf{v}_m$ is merely the mth row of $\Gamma^{\text{disp}}(T)$.

Example VII *The matrix* **U** *for the ammonium molecule* NH_3
To illustrate the technique, just a part of the calculation of **U** will be given. From the expressions given for **p**(T) and **R**(T) in Example III and Table D.1, it follows that

$$\left.\begin{aligned}\Phi(E)\mathbf{v}_1 &= \Phi(IC_{2y})\mathbf{v}_1 = \mathbf{v}_1,\\ \Phi(C_{3z})\mathbf{v}_1 &= \Phi(IC_{2D})\mathbf{v}_1 = -\tfrac{1}{2}\mathbf{v}_7 - \tfrac{1}{2}\sqrt{3}\mathbf{v}_8,\\ \Phi(C_{3z}^{-1})\mathbf{v}_1 &= \Phi(IC_{2C})\mathbf{v}_1 = -\tfrac{1}{2}\mathbf{v}_4 + \tfrac{1}{2}\sqrt{3}\mathbf{v}_5.\end{aligned}\right\}$$

Thus, from the character table (Table D.21), for $\mathcal{G} = C_{3v}$

$$\mathcal{P}^1_{11}\mathbf{v}_1 = \tfrac{1}{3}\{\mathbf{v}_1 - \tfrac{1}{2}\mathbf{v}_4 + \tfrac{1}{2}\sqrt{3}\mathbf{v}_5 - \tfrac{1}{2}\mathbf{v}_7 - \tfrac{1}{2}\sqrt{3}\mathbf{v}_8\}.$$

Thus $(1/\sqrt{3})\{\mathbf{v}_1 - \tfrac{1}{2}\mathbf{v}_4 + \tfrac{1}{2}\sqrt{3}\mathbf{v}_5 - \tfrac{1}{2}\mathbf{v}_7 - \tfrac{1}{2}\sqrt{3}\mathbf{v}_8\}$ is a normalized basis vector of the first (and only) row of Γ^1. Labelling the rows of Γ' so that the $4\,\Gamma^1$ representations occur first, this vector may be taken to transform as the first row of Γ'. Thus

$$U_{11} = 1/\sqrt{3}, \quad U_{41} = -1/2\sqrt{3}, \quad U_{51} = \tfrac{1}{2}, \quad U_{71} = -1/2\sqrt{3}, \quad U_{81} = -\tfrac{1}{2}$$

with $U_{m1} = 0$ for $m = 2, 3, 6, 9, 10, 11, 12$.

(*f*) *Extra degeneracies caused by the reality of* **F**

As the force matrix **F** and all its eigenvalues are real, if **c** is an eigenvector of **F** with eigenvalue λ, then $\mathbf{c}^*$ is also an eigenvector with the same eigenvalue λ. The question of whether this implies any additional degeneracies is resolved by the following theorem:

Theorem III Suppose that $\mathbf{c}^p_1, \mathbf{c}^p_2, \ldots$ are a set of d_p linearly independent eigenvectors of **F** corresponding to the eigenvalue λ and transforming as basis vectors for the d_p-dimensional unitary irreducible representation Γ^p of $\mathcal{G}$. Then

(*a*) if Γ^p is pseudo-real or essentially complex, then the vectors $(\mathbf{c}^p_1)^*, (\mathbf{c}^p_2)^*, \ldots$ are linearly independent of $\mathbf{c}^p_1, \mathbf{c}^p_2, \ldots$ and so the eigenvalue λ is $2d_p$-fold degenerate, but (*b*) if Γ^p is potentially real, there is *no* requirement that $(\mathbf{c}_1^p)^*, (\mathbf{c}_2^p)^*, \ldots$ be linearly independent of $\mathbf{c}_1^p, \mathbf{c}_2^p, \ldots$, and hence no requirement that λ be more than d_p-fold degenerate.

Proof The proof is essentially the same as for Theorem I of Chapter 6, Section 5, which deals with time-reversal symmetry in the spin-independent Schrödinger equation, the essential observation being that, as $\Gamma^{disp}(T)$ is real for all T, it follows from Equations (7.29) that

$$\{\Phi(T)\mathbf{v}\}^* = \Phi(T)\mathbf{v}^* \tag{7.40}$$

for any $T \in \mathcal{G}$ and any $\mathbf{v} \in V(=\mathbb{C}^{3A})$.

Suppose that $(\mathbf{c}_1^p)^*, (\mathbf{c}_2^p)^*, \ldots$ depend linearly on $\mathbf{c}_1^p, \mathbf{c}_2^p, \ldots$. Then there must exist a $d_p \times d_p$ matrix $\mathbf{Z}$ such that

$$(\mathbf{c}_n^p)^* = \sum_{m=1}^{d} Z_{mn}\mathbf{c}_m, \qquad n = 1, 2, \ldots, d_p. \tag{7.41}$$

Equation (7.40) implies that $(\mathbf{c}_1^p)^*, (\mathbf{c}_2^p)^*, \ldots$ are basis vectors of the complex conjugate representation Γ^{p*} defined by $\Gamma^{p*}(T) = \{\Gamma^p(T)\}^*$. By virtue of Equation (7.41), Γ^{p*} is equivalent to Γ^p, with $\Gamma^{p*}(T) = \mathbf{Z}^{-1}\Gamma^p(T)\mathbf{Z}$ for all $T \in \mathcal{G}$. Thus Γ^p cannot be essentially complex. Moreover, taking the complex conjugate of Equation (7.41), and substituting back into Equation (7.41), gives $\mathbf{Z}^*\mathbf{Z} = \mathbf{1}$. Consequently, from Theorem II of Chapter 5, Section 8, Γ^p must be essentially real. Conversely, if Γ^p is potentially real, a similarity transformation will make Γ^p real. When this is so the basis vectors of Γ^p can be taken to be purely real, so that $(\mathbf{c}_m^p)^* = \mathbf{c}_m^p$ for $m = 1, 2, \ldots, d_p$. Hence there is no extra degeneracy in this case.

The doubling of degeneracies in the cases in which Γ^p is pseudo-real or essentially complex is consistent with the observation (see Chapter 5, Section 6) that, as Γ^{disp} is real, any pseudo-real irreducible representation can only appear an even number of times, and if Γ^{disp} contains an essentially complex representation Γ^p, then Γ^p and its complex conjugate Γ^{p*} occur the same number of times. In the latter case the extra degeneracy follows from the identity $(\mathbf{f}^p)^* = \mathbf{f}^{p*}$.

If Γ^p is pseudo-real or essentially complex and $\mathbf{c}_{\alpha 1}^p, \mathbf{c}_{\alpha 2}^p, \ldots$ is a set of d_p eigenvectors of $\mathbf{F}$ that transform as basis vectors of Γ^p, then, as shown in the above theorem, $(\mathbf{c}_{\alpha 1}^p)^*, (\mathbf{c}_{\alpha 2}^p)^*, \ldots$ is a linearly independent set with the same eigenvalue. From these two sets, a set of $2d_p$ *real* normalized eigenvectors of $\mathbf{F}$ with the same eigenvalue can be constructed from the prescription:

$$\left.\begin{aligned} \mathbf{a}_{\alpha m}^p &= (1/\sqrt{2})\{\mathbf{c}_{\alpha m}^p + (\mathbf{c}_{\alpha m}^p)^*\}, \\ \mathbf{b}_{\alpha m}^p &= (1/\sqrt{2}i)\{\mathbf{c}_{\alpha m}^p - (\mathbf{c}_{\alpha m}^p)^*\}, \end{aligned}\right\} \quad m = 1, 2, \ldots, d_p. \tag{7.42}$$

If Γ^p is pseudo-real, the linearly independent eigenvectors may initially be chosen to be $\mathbf{c}^p_{\alpha m}$ and $(\mathbf{c}^p_{\alpha m})^*$ for $m = 1, 2, \ldots, d_p$ and $\alpha = 1, 2, \ldots, \frac{1}{2}n_p$. As $(\mathbf{c}^p_{\alpha m})^* = \sum_{m'=1}^{d_p} Z_{mn}\mathbf{c}^p_{\alpha m}$, with $\mathbf{Z}$ defined as in Chapter 5, Section 8, $(\mathbf{c}^p_{\alpha m})^*$ does not transform as the mth row of Γ^p. Consequently $\mathbf{a}^p_{\alpha m}$ and $\mathbf{b}^p_{\alpha m}$ do not transform as the mth row of Γ^p, but rather they are basis functions of the reducible real representation

$$\begin{bmatrix} 2^{-1/2}\mathbf{1} & -i2^{-1/2}\mathbf{1} \\ 2^{-1/2}\mathbf{1} & i2^{-1/2}\mathbf{1} \end{bmatrix}^{-1} \begin{bmatrix} \Gamma^p(T) & 0 \\ 0 & \Gamma^{p*}(T) \end{bmatrix} \begin{bmatrix} 2^{-1/2}\mathbf{1} & -i2^{-1/2}\mathbf{1} \\ 2^{-1/2}\mathbf{1} & i2^{-1/2}\mathbf{1} \end{bmatrix} \quad (7.43)$$

which appears as a subrepresentation of Γ^{disp}.

If Γ^p is essentially complex, $(\mathbf{c}^p_{\alpha m})^*$ transforms as the mth row of Γ^{p*}. In this case $\mathbf{a}^p_{\alpha m}$ and $\mathbf{b}^p_{\alpha m}$ do not transform as basis vectors for just a single irreducible representation, but they are basis functions for the reducible real representation of Expression (7.43).

In both cases the corresponding normal coordinates are given according to Equation (7.9) by $\tilde{\mathbf{a}}^p_{\alpha m}\mathbf{q}$ and $\tilde{\mathbf{b}}^p_{\alpha m}\mathbf{q}$, but they *cannot* be identified with particular rows of particular irreducible representations.

(g) *Symmetry coordinates*

It is helpful, though not essential, to re-express some of the foregoing results in terms of the "symmetry coordinates" $\hat{Q}_1, \hat{Q}_2, \ldots, \hat{Q}_{3A}$. These may be regarded as forming a $3A \times 1$ column matrix $\hat{\mathbf{Q}}$, the definition then being

$$\hat{\mathbf{Q}} = \mathbf{U}^\dagger \mathbf{q}, \quad (7.44)$$

$\mathbf{q}$ being the column matrix of generalized coordinates, and $\mathbf{U}$ the unitary matrix defined in Equation (7.25), whose evaluation was described at the end of subsection (*e*). In general $\mathbf{U}$ is complex, in which case the symmetry coordinates are also complex.

As $\mathbf{q}$ is real, Equation (7.4) can be rewritten as $V = \frac{1}{2}\mathbf{q}^\dagger\mathbf{F}\mathbf{q}$, from which it follows by Equations (7.26) and (7.44) that $V = \frac{1}{2}\hat{\mathbf{Q}}^\dagger\mathbf{F}'\hat{\mathbf{Q}}$. It is convenient to relabel the components of $\hat{\mathbf{Q}}$ so that they correspond to the partitioning of Γ' and $\mathbf{F}'$ of subsection (*d*). To this end let $\hat{Q}^p_{\alpha m}$ be the αth component of $\hat{\mathbf{Q}}$ that belongs to the mth row of the irreducible representation $\Gamma^p (\alpha = 1, 2, \ldots, n_p)$. Then, by Equation (7.28),

$$V = \tfrac{1}{2} \sum_{p,q,\alpha,\beta,m,n} \hat{Q}^{p*}_{\alpha m}(\mathbf{F}'^{pq}_{\alpha\beta})_{mn}\hat{Q}^q_{\beta n},$$

and so, by Equations (7.29),

$$V = \tfrac{1}{2} \sum_{p, n_p \neq 0} \sum_{m=1}^{d_p} \sum_{\alpha,\beta=1}^{n_p} \hat{Q}^{p*}_{\alpha m} f^p_{\alpha\beta} \hat{Q}^p_{\alpha m}. \quad (7.45)$$

Here Equation (7.45) is a sum of quadratic forms in which each sum involves *only* symmetry coordinates belonging to the *same* row of the *same* irreducible representation.

The relationship between these symmetry coordinates and the normal coordinates will be considered in detail only for the case in which each $\boldsymbol{\Gamma}^p$ is real. As in subsection (*e*), let $\mathbf{c}^p_{\alpha m}$ ($\alpha = 1, 2, \ldots, n_p$; $m = 1, 2, \ldots, d_p$) be the real eigenvectors of $\mathbf{F}$ corresponding to the mth row of $\boldsymbol{\Gamma}^p$, the associated eigenvalue being λ^p_α. With the normal coordinates defined by Equation (7.39), Equation (7.10) can be written as

$$V = \tfrac{1}{2} \sum_{p, n_p \neq 0} \sum_{m=1}^{d_p} \sum_{\alpha=1}^{n_p} \lambda^p_\alpha (Q^p_{\alpha m})^2. \tag{7.46}$$

It is obvious that to go from the form of Equation (7.45) to the form of Equation (7.46) requires only the diagonalization of $\mathbf{f}^p$, exactly as in subsection (*d*). Let $\mathbf{Q}^p_m$ and $\hat{\mathbf{Q}}^p_m$ be $n_p \times 1$ real column matrices with elements $Q^p_{\alpha m}$ and $\hat{Q}^p_{\alpha m}$ respectively (for $\alpha = 1, 2, \ldots, n_p$). Then the quadratic forms appearing in Equations (7.45) and (7.46) can be written as $(\hat{\mathbf{Q}}^p_m)^\dagger \mathbf{f}^p \hat{\mathbf{Q}}^p_m$ and $(\mathbf{Q}^p_m)^\dagger \operatorname{diag}(\lambda^p_1, \lambda^p_2, \ldots) \mathbf{Q}^p_m$. Thus, if $\mathbf{X}^p$ is the real orthogonal matrix that diagonalizes $\mathbf{f}^p$, that is, if $(\mathbf{X}^p)^{-1} \mathbf{f}^p \mathbf{X}^p = \operatorname{diag}(\lambda^p_1, \lambda^p_2, \ldots)$, then $\mathbf{Q}^p_m = \tilde{\mathbf{X}}^p \hat{\mathbf{Q}}^p_m$. Thus

$$Q^p_{\alpha m} = \sum_{\beta=1}^{n_p} X^p_{\beta\alpha} \hat{Q}^p_{\beta m}. \tag{7.47}$$

This shows that the normal coordinate $Q^p_{\alpha m}$ is a linear combination involving *only* the symmetry coordinates $\hat{Q}^p_{\alpha m}$ transforming as the *same* row m of the *same* irreducible representation $\boldsymbol{\Gamma}^p$. In the special case $n_p = 1$, the symmetry coordinate $\hat{Q}^p_{1m}$ is automatically a normal coordinate.

If $\boldsymbol{\Gamma}^p$ is pseudo-real or essentially complex, the corresponding normal coordinates have to be found from the *real* eigenvectors $\mathbf{a}^p_{\alpha m}$ and $\mathbf{b}^p_{\alpha m}$ of Equations (7.42). In neither case can they be put into one-to-one correspondence with the rows of a single irreducible representation. In both cases the symmetry coordinates $\hat{Q}^p_{\alpha m}$ are complex and the relationship between the normal coordinates and the symmetry coordinates is more complicated than Equation (7.47). It is probably easiest in these cases to find the normal coordinates directly without introducing the symmetry coordinates.

(*h*) *Zero-frequency modes of molecules*

As noted in subsection (*a*), every molecule is expected to have three zero-frequency translational normal modes. There must also be zero-frequency rotational normal modes, three for a non-linear molecule but only two for a linear molecule. These expectations will now be verified in detail, by constructing the appropriate number of eigenvectors of the force matrix **F** corresponding to zero eigenvalue. The representation of $\mathcal{G}$ for which these eigenvectors form bases will then be deduced.

The first stage is to derive certain identities involving the components of the force matrix **F**.

Theorem IV If no external forces act on the molecule, the components of the force matrix **F** satisfy the following conditions:

(A) $\sum_{j=1}^{A} (m_j)^{1/2} F_{m,3j-3+s} = 0$

for all $m = 1, 2, \ldots, 3A$, and $s = 1, 2, 3$;

(B) $\sum_{j=1}^{A} (m_j)^{1/2} \{y_j^e F_{m,3j} - z_j^e F_{m,3j-1}\} = 0,$

(C) $\sum_{j=1}^{A} (m_j)^{1/2} \{z_j^e F_{m,3j-2} - x_j^e F_{m,3j}\} = 0,$

(D) $\sum_{j=1}^{A} (m_j)^{1/2} \{x_j^e F_{m,3j-1} - y_j^e F_{m,3j-2}\} = 0,$

(b), (c) and (d) being valid for all $m = 1, 2, \ldots, 3A$.

Proof Let $\mathbf{F}_j$ be the total force acting on the jth nucleus, $j = 1, 2, \ldots, A$. If there are only internal forces present $\sum_{j=1}^{A} \mathbf{F}_j = \mathbf{0}$. The first component of this equality gives $\sum_{j=1}^{A} \partial V/\partial x_j = 0$ and hence, by Equations (7.14), $\sum_{j=1}^{3A} (m_j)^{1/2} \partial V/\partial q_{3j-2} = 0$. Similar arguments applied to the second and third components give in all the three conditions:

$$\sum_{j=1}^{A} (m_j)^{1/2} \partial V/\partial q_{3j-3+s} = 0, \qquad s = 1, 2, 3.$$

As $V = \frac{1}{2} \sum_{m,n=1}^{3A} F_{mn} q_m q_n$, these give (for $s = 1, 2, 3$)

$$\sum_{m=1}^{3A} \sum_{j=1}^{A} (m_j)^{1/2} F_{m,3j-3+s} q_m = 0,$$

and, as this must be true for all $q_1, q_2, \ldots, q_{3A}$, condition (A) follows immediately.

When the nuclei are in their equilibrium positions there is zero total moment acting on the system, so

$$\sum_{j=1}^{A} \mathbf{r}_j^e \wedge \mathbf{F}_j = \mathbf{0}. \tag{7.48}$$

Consider the first component of this equality, which can be written as

$$\sum_{j=1}^{A} \{y_j^e(\partial V/\partial z_j) - z_j^e(\partial V/\partial y_j)\} = 0.$$

This becomes, by Equations (7.14),

$$\sum_{j=1}^{A} (m_j)^{1/2}\{y_j^e(\partial V/\partial q_{3j}) - z_j^e(\partial V/\partial q_{3j-1})\} = 0$$

and so

$$\sum_{j=1}^{A} \sum_{m=1}^{3A} (m_j)^{1/2}\{y_j^e F_{m,3j} - z_j^e F_{m,3j-1}\} q_m = 0,$$

from which (B) follows immediately. Conditions (C) and (D) follow from consideration of the second and third components of Equation (7.48).

Theorem V Let $\mathbf{c}_1^{tr}, \mathbf{c}_2^{tr}$ and $\mathbf{c}_3^{tr}$ be three "translational" $3A \times 1$ column matrices, defined by

$$\mathbf{c}_1^{tr} = \begin{bmatrix} (m_1)^{1/2} \\ 0 \\ 0 \\ (m_2)^{1/2} \\ 0 \\ 0 \\ \vdots \end{bmatrix}, \quad \mathbf{c}_2^{tr} = \begin{bmatrix} 0 \\ (m_1)^{1/2} \\ 0 \\ 0 \\ (m_2)^{1/2} \\ 0 \\ \vdots \end{bmatrix}, \quad \mathbf{c}_3^{tr} = \begin{bmatrix} 0 \\ 0 \\ (m_1)^{1/2} \\ 0 \\ 0 \\ (m_2)^{1/2} \\ \vdots \end{bmatrix}, \qquad (7.49)$$

and let $\mathbf{c}_1^{rot}$, $\mathbf{c}_2^{rot}$ and $\mathbf{c}_3^{rot}$ be three "rotational" $3A \times 1$ column matrices, defined by

$$\mathbf{c}_1^{rot} = \begin{bmatrix} 0 \\ -(m_1)^{1/2} z_1^e \\ (m_1)^{1/2} y_1^e \\ 0 \\ -(m_2)^{1/2} z_2^e \\ (m_2)^{1/2} y_2^e \\ \vdots \end{bmatrix}, \quad \mathbf{c}_2^{rot} = \begin{bmatrix} (m_1)^{1/2} z_1^e \\ 0 \\ -(m_1)^{1/2} x_1^e \\ (m_2)^{1/2} z_2^e \\ 0 \\ -(m_2)^{1/2} x_2^e \\ \vdots \end{bmatrix}, \quad \mathbf{c}_3^{rot} = \begin{bmatrix} -(m_1)^{1/2} y_1^e \\ (m_1)^{1/2} x_1^e \\ 0 \\ -(m_2)^{1/2} x_2^e \\ (m_2)^{1/2} x_2^e \\ 0 \\ \vdots \end{bmatrix}. \qquad (7.50)$$

(In all cases the dots indicate that the same pattern of elements is repeated.) Then $\mathbf{F}\mathbf{c}_s^{tr} = 0$ and $\mathbf{F}\mathbf{c}_s^{rot} = 0$ for $s = 1, 2, 3$, so that $\mathbf{c}_s^{tr}$ and $\mathbf{c}_s^{rot}$ are eigenvectors of $\mathbf{F}$ with zero eigenvalue (i.e. they correspond to *zero*-frequency modes).

Proof The equations $\mathbf{F}\mathbf{c}_s^{tr} = \mathbf{0}$ follow immediately from Equations (7.49) and condition (A) of Theorem IV. Similarly, $\mathbf{F}\mathbf{c}_s^{rot} = \mathbf{0}$ follows from Definitions (7.50) using conditions (B), (C) and (D).

The association of $\mathbf{c}_1^{tr}$, $\mathbf{c}_2^{tr}$ and $\mathbf{c}_3^{tr}$ with translational modes follows from the fact that (A) is an expression of the condition that the molecule experiences no external forces, which is the condition that implies the existence of the zero-frequency translational modes. Indeed the normal coordinates $Q_1^{tr}, Q_2^{tr}, Q_3^{tr}$ defined by $Q_s^{tr} = \tilde{\mathbf{c}}_s^{tr}\mathbf{q}$ (cf. Equation (7.9)) for $s = 1, 2, 3$ are then

$$\left.\begin{aligned} Q_1^{tr} &= \sum_{j=1}^{A} m_j (x_j - x_j^e), \\ Q_2^{tr} &= \sum_{j=1}^{A} m_j (y_j - y_j^e), \\ Q_3^{tr} &= \sum_{j=1}^{A} m_j (z_j - z_j^e). \end{aligned}\right\}$$

With the origin of the coordinate system chosen so that

$$\sum_{j=1}^{A} m_j x_j^e = \sum_{j=1}^{A} m_j y_j^e = \sum_{j=1}^{A} m_j z_j^e = 0,$$

Q_1^{tr}/M, Q_2^{tr}/M and Q_3^{tr}/M are merely the coordinates of the centre of mass of the molecule.

Similarly, the association of $\mathbf{c}_1^{rot}$, $\mathbf{c}_2^{rot}$ and $\mathbf{c}_3^{rot}$ with rotational modes is consequential on (B), (C) and (D) being expressions of the condition that the system experiences no net moment, which is precisely the condition that gives rise to the zero-frequency rotational modes. This can also be seen by considering the corresponding normal coordinates $Q_1^{rot}, Q_2^{rot}, Q_3^{rot}$ defined by $Q_s^{rot} = \tilde{\mathbf{c}}_s^{rot}\mathbf{q}$ for $s = 1, 2, 3$. Then, from Equations (7.14) and (7.50),

$$(Q_1^{rot}, Q_2^{rot}, Q_3^{rot}) = \sum_{j=1}^{A} m_j \mathbf{r}_j^e \wedge \mathbf{r}_j.$$

Thus $(\dot{Q}_1^{rot}, \dot{Q}_2^{rot}, \dot{Q}_3^{rot}) = \sum_{j=1}^{A} m_j \mathbf{r}_j^e \wedge \dot{\mathbf{r}}_j$, which may be approximately identified with the angular momentum of the whole system $(\sum_{j=1}^{A} m_j \mathbf{r}_j \wedge \dot{\mathbf{r}}_j)$ when the displacements from the equilibrium position are small.

It is easy to see why a linear molecule is exceptional. Consider the acetylene molecule of Example II. As $y_j^e = z_j^e = 0$ for $j = 1, 2, 3, 4$, it follows that $\mathbf{c}_1^{rot} = \mathbf{0}$, so that there are only two rotational eigenvectors, namely $\mathbf{c}_2^{rot}$ and $\mathbf{c}_3^{rot}$.

The symmetry properties of the translational and rotational eigenvectors are embodied in the following three theorems.

Theorem VI For every $T \in \mathscr{G}$

$$\Phi(T)\mathbf{c}_n^{tr} = \sum_{m=1}^{3} \Gamma^{tr}(T)_{mn}\mathbf{c}_m^{tr}, \qquad n = 1, 2, 3,$$

where Γ^{tr} is the three-dimensional representation of $\mathscr{G}$ defined by

$$\Gamma^{tr}(T) = \mathbf{R}(T)$$

for each $T \in \mathscr{G}$. Thus $\mathbf{c}_1^{tr}, \mathbf{c}_2^{tr}, \mathbf{c}_3^{tr}$ are basis vectors for this representation of $\mathscr{G}$.

Proof Let $\mathbf{e}_1, \mathbf{e}_2, \mathbf{e}_3$ be three 3×1 column matrices defined by $(\mathbf{e}_m)_j = \delta_{jm}$, $j, m = 1, 2, 3$. Then

$$\mathbf{R}(T)\mathbf{e}_n = \sum_{m=1}^{3} R(T)_{mn}\mathbf{e}_m \tag{7.51}$$

for any T. By Equations (7.49)

$$\mathbf{c}_n^{tr} = \begin{bmatrix} (m_1)^{1/2}\mathbf{e}_n \\ (m_2)^{1/2}\mathbf{e}_n \\ \vdots \end{bmatrix}.$$

For clarity let $A = 3$ and consider the typical transformation of Equation (7.20). Then, by Equation (7.32),

$$\Phi(T)\mathbf{c}_n^{tr} = \Gamma^{disp}(T)\mathbf{c}_n^{tr} = \begin{bmatrix} \mathbf{0} & \mathbf{0} & \mathbf{R}(T) \\ \mathbf{R}(T) & \mathbf{0} & \mathbf{0} \\ \mathbf{0} & \mathbf{R}(T) & \mathbf{0} \end{bmatrix} \begin{bmatrix} (m_1)^{1/2}\mathbf{e}_n \\ (m_2)^{1/2}\mathbf{e}_n \\ (m_3)^{1/2}\mathbf{e}_n \end{bmatrix}$$

$$= \begin{bmatrix} (m_3)^{1/2}\mathbf{R}(T)\mathbf{e}_n \\ (m_1)^{1/2}\mathbf{R}(T)\mathbf{e}_n \\ (m_2)^{1/2}\mathbf{R}(T)\mathbf{e}_n \end{bmatrix} = \sum_{m=1}^{3} R(T)_{mn}\mathbf{c}_m^{tr}$$

by Equation (7.51), as $m_1 = m_2 = m_3$ because nuclei related by the transformation of Equation (7.13) must be of the same species. This argument can obviously be generalized to any T of any $\mathscr{G}$.

Theorem VII For any non-linear molecule and any $T \in \mathscr{G}$

$$\Phi(T)\mathbf{c}_n^{rot} = \sum_{m=1}^{3} \Gamma^{rot}(T)_{mn}\mathbf{c}_m^{rot}, \qquad n = 1, 2, 3, \tag{7.52}$$

where $\boldsymbol{\Gamma}^{\text{rot}}$ is the three-dimensional representation of $\mathcal{G}$ defined by

$$\boldsymbol{\Gamma}^{\text{rot}}(\text{T}) = \mathbf{R}(\text{T})\{\det \mathbf{R}(\text{T})\}$$

for each $\text{T} \in \mathcal{G}$. Thus $\mathbf{c}_1^{\text{rot}}, \mathbf{c}_2^{\text{rot}}, \mathbf{c}_3^{\text{rot}}$ are basis vectors for this representation of $\mathcal{G}$.

Proof Let $\mathbf{e}_1, \mathbf{e}_2, \mathbf{e}_3$ be defined as in the proof of Theorem VI. Then, from Equations (7.50),

$$\mathbf{c}_n^{\text{rot}} = \begin{bmatrix} (m_1)^{1/2}\mathbf{r}_1^e \wedge \mathbf{e}_n \\ (m_2)^{1/2}\mathbf{r}_2^e \wedge \mathbf{e}_n \\ \vdots \end{bmatrix},$$

where $\mathbf{r}_j^e \wedge \mathbf{e}_n$ is the 3×1 column matrix whose elements are given by the vector product indicated. Again consider the case $A = 3$ and the typical transformation of Equation (7.20). Then, by Equation (7.32),

$$\Phi(\text{T})\mathbf{c}_n^{\text{rot}} = \boldsymbol{\Gamma}^{disp}(\text{T})\mathbf{c}_n^{\text{rot}} = \begin{bmatrix} \mathbf{0} & \mathbf{0} & \mathbf{R}(\text{T}) \\ \mathbf{R}(\text{T}) & \mathbf{0} & \mathbf{0} \\ \mathbf{0} & \mathbf{R}(\text{T}) & \mathbf{0} \end{bmatrix} \begin{bmatrix} (m_1)^{1/2}\mathbf{r}_1^e \wedge \mathbf{e}_n \\ (m_2)^{1/2}\mathbf{r}_2^e \wedge \mathbf{e}_n \\ (m_3)^{1/2}\mathbf{r}_3^e \wedge \mathbf{e}_n \end{bmatrix}$$

$$= \begin{bmatrix} (m_3)^{1/2}\mathbf{R}(\text{T})\{\mathbf{r}_3^e \wedge \mathbf{e}_n\} \\ (m_1)^{1/2}\mathbf{R}(\text{T})\{\mathbf{r}_1^e \wedge \mathbf{e}_n\} \\ (m_2)^{1/2}\mathbf{R}(\text{T})\{\mathbf{r}_2^e \wedge \mathbf{e}_n\} \end{bmatrix} \tag{7.53}$$

However, for any two vectors $\mathbf{a}$ and $\mathbf{b}$ and any 3×3 real orthogonal matrix $\mathbf{R}(\text{T})$ there exists the easily established identity

$$\mathbf{R}(\text{T})\{\mathbf{a} \wedge \mathbf{b}\} = \{\mathbf{R}(\text{T})\mathbf{a} \wedge \mathbf{R}(\text{T})\mathbf{b}\}\{\det \mathbf{R}(\text{T})\}.$$

Then, as Equations (7.13) and (7.20) imply that $\mathbf{R}(\text{T})\mathbf{r}_3^e = \mathbf{r}_1^e$, $\mathbf{R}(\text{T})\mathbf{r}_1^e = \mathbf{r}_2^e$, and $\mathbf{R}(\text{T})\mathbf{r}_2^e = \mathbf{r}_3^e$, and as $m_1 = m_2 = m_3$, being masses of nuclei of the same species, Equation (7.53) gives Equation (7.52). The argument can be generalized in the obvious way for any rotation T of any point group $\mathcal{G}$.

Of course the value of $\det \mathbf{R}(\text{T})$ is simply $+1$ for a proper rotation and -1 for an improper rotation.

Let $\chi^{\text{tr}}(\text{T})$ and $\chi^{\text{rot}}(\text{T})$ be the characters of the representations $\boldsymbol{\Gamma}^{\text{tr}}$ and $\boldsymbol{\Gamma}^{\text{rot}}$. Then, if $\mathbf{R}(\text{T})$ is a proper rotation through an angle θ, by Equation (12.6),

$$\chi^{\text{tr}}(\text{T}) = \chi^{\text{rot}}(\text{T}) = 1 + 2\cos\theta, \tag{7.54}$$

while if $\mathbf{R}(\text{T})$ is an improper rotation that is the product of the spatial inversion with a proper rotation through θ

$$-\chi^{tr}(T) = \chi^{rot}(T) = 1 + 2\cos\theta. \tag{7.55}$$

For a linear molecule the results on the rotational eigenvectors require minor modification.

Theorem VIII For a linear molecule, whose nuclei may be assumed to lie along the Ox axis, for any $T \in \mathcal{G}$

$$\Phi(T)\mathbf{c}_n^{rot} = \sum_{m=2}^{3} \Gamma^{rot}(T)_{mn}\mathbf{c}_m^{rot}, \qquad n = 2, 3,$$

where $\mathbf{\Gamma}^{rot}$ is the two-dimensional representation of $\mathcal{G}$ defined by

$$\Gamma^{rot}(T)_{mn} = R(T)_{mn}\{\det \mathbf{R}(T)\}, \qquad m, n = 2, 3.$$

In this case $\mathbf{c}_1^{rot} = \mathbf{0}$ and $\mathbf{c}_2^{rot}$ and $\mathbf{c}_3^{rot}$ are basis vectors for this representation.

Proof This follows immediately from the proof of Theorem VII, with the additional observation that, if the nuclei all lie along Ox, $\mathbf{c}_1^{rot} = \mathbf{0}$ and $R(T)_{1m} = R(T)_{m1} = 0$ for $m = 2, 3$ and all $T \in \mathcal{G}$ (cf. Example II).

For such a linear molecule

$$\chi^{rot}(T) = \{\det \mathbf{R}(T)\}\left\{\sum_{m=2}^{3} \mathbf{R}(T)_{mm}\right\}. \tag{7.56}$$

(As Example IX will show, in this case $\chi^{rot}(T)$ depends not only on the angle of rotation but also on the axis.)

Example VIII *Separation of vibrational, translational and rotational modes for the ammonium molecule* NH_3

From Equations (7.54) and (7.55) and Example I,

$$\chi^{tr}(\mathscr{C}_1) = 3, \qquad \chi^{tr}(\mathscr{C}_2) = 0, \qquad \chi^{tr}(\mathscr{C}_3) = 1,$$
$$\chi^{rot}(\mathscr{C}_1) = 3, \qquad \chi^{rot}(\mathscr{C}_2) = 0, \qquad \chi^{rot}(\mathscr{C}_3) = -1.$$

Thus $\mathbf{\Gamma}^{tr} \approx \mathbf{\Gamma}^1 \oplus \mathbf{\Gamma}^3$ (i.e. $A_1 \oplus E$) and $\mathbf{\Gamma}^{rot} \approx \mathbf{\Gamma}^2 \oplus \mathbf{\Gamma}^3$ (i.e. $A_2 \oplus E$). This shows that of the three non-degenerate eigenvalues of Example V associated with $\mathbf{\Gamma}^1$ (i.e. with A_1) one has value zero and the remaining two correspond to vibrational modes. The non-degenerate eigenvalue associated with $\mathbf{\Gamma}^2$ (i.e. with A_2) has value zero, while of the four two-fold degenerate eigenvalues associated with $\mathbf{\Gamma}^3$ (i.e. with E) two have value zero leaving only two to correspond to vibrational modes.

Example IX *Separation of vibrational, translational and rotational modes for the acetylene molecule* C_2H_2

From Equations (7.54), (7.55) and (7.56), and Example VI of Chapter 2, Section 7,

$$\left.\begin{array}{lll}\chi^{tr}(\mathrm{E})=3, & \chi^{tr}(\mathrm{C}_\theta)=1+2\cos\theta, & \chi^{tr}(\mathrm{C}_{2\Phi})=-1,\\ \chi^{tr}(\mathrm{I})=-3, & \chi^{tr}(\mathrm{IC}_\theta)=-(1+2\cos\theta), & \chi^{tr}(\mathrm{IC}_{2\Phi})=1,\end{array}\right\}$$

and

$$\left.\begin{array}{l}\chi^{rot}(\mathrm{E})=\chi^{rot}(\mathrm{I})=2, \quad \chi^{rot}(\mathrm{C}_\theta)=\chi^{rot}(\mathrm{IC}_\theta)=2\cos\theta,\\ \qquad\chi^{rot}(\mathrm{C}_{2\Phi})=\chi^{rot}(\mathrm{IC}_{2\Phi})=0.\end{array}\right\}$$

Thus, from Table D.46, $\Gamma^{tr}\approx \mathrm{A}_{2u}\oplus \mathrm{E}_{1u}$ and $\Gamma^{rot}\approx \mathrm{E}_{1g}$. It follows from Example VI that the vibrational spectrum consists of two non-degenerate eigenvalues associated with A_{1g}, one non-degenerate eigenvalue associated with A_{2u}, one two-fold degenerate eigenvalue associated with E_{1g}, and one two-fold degenerate eigenvalue associated with E_{1u}.

By the use of "internal coordinates" it is possible to remove the translational and vibrational modes at an earlier stage of the calculations. Details of this approach have been given by Wilson *et al.* (1955).

8

The Translational Symmetry of Crystalline Solids

1 The Bravais lattices

An infinite three-dimensional lattice may be defined in terms of three linearly independent real "basic lattice vectors" $\mathbf{a}_1$, $\mathbf{a}_2$ and $\mathbf{a}_3$. The set of all lattice vectors of the lattice is then given by

$$\mathbf{t_n} = n_1\mathbf{a}_1 + n_2\mathbf{a}_2 + n_3\mathbf{a}_3,$$

where $\mathbf{n} = (n_1, n_2, n_3)$, and n_1, n_2 and n_3 are *integers* that take all possible values, positive, negative and zero. Points in $\mathbb{R}^3$ having lattice vectors as their position vectors are called "lattice points" and a pure translation through a lattice vector $\mathbf{t_n}$, $\{\mathbf{1} \mid \mathbf{t_n}\}$, is called a "primitive" translation.

Suppose that in a crystalline solid there are S nuclei per lattice point, and that the equilibrium positions of the nuclei associated with the lattice point $\mathbf{r} = \mathbf{0}$ have position vectors $\boldsymbol{\tau}_1, \boldsymbol{\tau}_2, \ldots, \boldsymbol{\tau}_S$. Then the equilibrium positions of the whole set of nuclei are given by

$$\mathbf{r}^e_{\mathbf{n}\gamma} = \mathbf{t_n} + \boldsymbol{\tau}_\gamma, \tag{8.1}$$

where $\gamma = 1, 2, \ldots, S$ and $\mathbf{t_n}$ is any lattice vector. In the special case when $S = 1$, $\boldsymbol{\tau}_1$ may be taken to be $\mathbf{0}$ and the index γ may be omitted, so that $\mathbf{r}^e_{\mathbf{n}} = \mathbf{t_n}$.

The set of all primitive translations of a lattice form a group which will be denoted by $\mathcal{T}^\infty$. $\mathcal{T}^\infty$ is Abelian but of infinite order. In Section 2 the Born cyclic boundary conditions will be introduced. They have the effect of replacing this infinite group by a similar group of large but finite order, so that all the theorems on finite groups of the previous chapters apply.

The "maximal point group" $\mathscr{G}_0^{max}$ of a crystal lattice may be defined as the set of all pure rotations $\{\mathbf{R}(T) \mid \mathbf{0}\}$ such that, for every lattice vector $\mathbf{t_n}$, the quantity $\mathbf{R}(T)\mathbf{t_n}$ is also a lattice vector. Clearly $\mathbf{R}(T) \in \mathscr{G}_0^{max}$ if and only if $\mathbf{R}(T)\mathbf{a}_j$ is a lattice vector for $j = 1, 2, 3$.

There are essentially 14 different types of crystal lattice. They are known as the "Bravais lattices". These will be described briefly, but no attempt will be made to give a logical derivation or to show that there are no others. (In this context two types of lattice are regarded as being different if they have different maximal point groups, even though one type is a special case of the other. For example, as may be seen from Table 8.1, the simple cubic lattice Γ_c is a special case of the simple tetragonal lattice Γ_q with $a = b$, but $\mathscr{G}_0^{max} = O_h$ for Γ_c, whereas $\mathscr{G}_0^{max} = d_{4h}$ for Γ_q.)

Lattices with the same maximal point group are said to belong to the same "symmetry system", there being only seven different symmetry systems. Complete details are given in Table 8.1, in which the notation for point groups is that of Schönfliess (1923). (A full specification of these and the other crystallographic point groups may be found in Appendix D.)

The cubic system is probably the most significant, the body-centred and face-centred lattices occurring for a large number of important solids. The basic lattice vectors of the cubic lattices are shown in Figures 8.1, 8.2 and 8.3. The lattice points of the simple cubic lattice Γ_c merely form a repeated cubic array, and the basic lattice vectors lie along three edges of a cube. For the body-centred cubic lattice Γ_c^v the basic lattice vectors join a point at the centre of a cube to three of the vertices of the cube, so that the lattice points form a repeated cubic array with lattice points also occurring at every cube centre. For the face-centred lattice Γ_c^f the lattice points again form a repeated cubic array with additional points also occurring at the mid-points of every cube face, the basic lattice vectors then joining a cube vertex to the mid-points of the three adjacent cube faces.

A symmetry system α may be regarded as being "subordinate" to a symmetry system β if $\mathscr{G}_0^{max}$ for α is a subgroup of $\mathscr{G}_0^{max}$ for β and at least one lattice of β is a special case of a lattice of α. The complete subordination scheme is then:

triclinic $<$ monoclinic $<$ orthorhombic $<$ tetragonal $<$ cubic;

monoclinic $<$ rhombohedral; orthorhombic $<$ hexagonal;

(Here $\alpha < \beta$ indicates that α is subordinate to β.)

(1) Triclinic symmetry system ($\mathscr{G}_0^{max} = C_i$):
  (i) simple triclinic lattice, Γ_t:
    $\mathbf{a}_1$, $\mathbf{a}_2$ and $\mathbf{a}_3$ arbitrary.

(2) Monoclinic symmetry system ($\mathscr{G}_0^{max} = C_{2h}$):
  (i) simple monoclinic lattice, Γ_m:
    $\mathbf{a}_3$ perpendicular to both $\mathbf{a}_1$ and $\mathbf{a}_2$;
  (ii) base-centred monoclinic lattice, Γ_m^b:
    $\mathbf{a}_1 = (a, b, 0)$, $\mathbf{a}_2 = (a, -b, 0)$, $\mathbf{a}_3 = (c, 0, d)$.

(3) Orthorhombic symmetry system ($\mathscr{G}_0^{max} = D_{2h}$):
  (i) simple orthorhombic lattice, Γ_0:
    $\mathbf{a}_1 = (a, 0, 0)$, $\mathbf{a}_2 = (0, b, 0)$, $\mathbf{a}_3 = (0, 0, c)$;
  (ii) base-centred orthorhombic lattice, Γ_0^b:
    $\mathbf{a}_1 = (a, b, 0)$, $\mathbf{a}_2 = (a, -b, 0)$, $\mathbf{a}_3 = (0, 0, c)$;
  (iii) body-centred orthorhombic lattice, Γ_0^v:
    $\mathbf{a}_1 = (a, b, c)$, $\mathbf{a}_2 = (a, b, -c)$, $\mathbf{a}_3 = (a, -b, -c)$;
  (iv) face-centred orthorhombic lattice Γ_0^f:
    $\mathbf{a}_1 = (a, b, 0)$, $\mathbf{a}_2 = (0, b, c)$, $\mathbf{a}_3 = (a, 0, c)$.

(4) Tetragonal symmetry system ($\mathscr{G}_0^{max} = D_{4h}$):
  (i) simple tetragonal lattice, Γ_q:
    $\mathbf{a}_1 = (a, 0, 0)$, $\mathbf{a}_2 = (0, a, 0)$, $\mathbf{a}_3 = (0, 0, b)$;
  (ii) body-centred tetragonal lattice, Γ_q^v:
    $\mathbf{a}_1 = (a, a, b)$, $\mathbf{a}_2 = (a, a, -b)$, $\mathbf{a}_3 = (a, -a, b)$.

(5) Cubic symmetry system ($\mathscr{G}_0^{max} = O_h$):
  (i) simple cubic lattice, Γ_c:
    $\mathbf{a}_1 = (a, 0, 0)$, $\mathbf{a}_2 = (0, a, 0)$, $\mathbf{a}_3 = (0, 0, a)$;
  (ii) body-centred cubic lattice, Γ_c^v:
    $\mathbf{a}_1 = \frac{1}{2}a(1, 1, 1)$, $\mathbf{a}_2 = \frac{1}{2}a(1, 1, -1)$, $\mathbf{a}_3 = \frac{1}{2}a(1, -1, -1)$;
  (iii) face-centred cubic lattice, Γ_c^f:
    $\mathbf{a}_1 = \frac{1}{2}a(1, 1, 0)$, $\mathbf{a}_2 = \frac{1}{2}a(0, 1, 1)$, $\mathbf{a}_3 = \frac{1}{2}a(1, 0, 1)$.

(6) Rhombohedral (or trigonal) symmetry system ($\mathscr{G}_0^{max} = D_{3d}$):
  (i) simple rhombohedral lattice, Γ_{rh}:
    $\mathbf{a}_1 = (a, 0, b)$, $\mathbf{a}_2 = (\frac{1}{2}a\sqrt{3}, -\frac{1}{2}a, b)$, $\mathbf{a}_3 = (-\frac{1}{2}a\sqrt{3}, -\frac{1}{2}a, b)$.

(7) Hexagonal symmetry system ($\mathscr{G}_0^{max} = D_{6h}$):
  (i) simple hexagonal lattice, Γ_h:
    $\mathbf{a}_1 = (0, 0, c)$, $\mathbf{a}_2 = (a, 0, 0)$, $\mathbf{a}_3 = (-\frac{1}{2}a, -\frac{1}{2}a\sqrt{3}, 0)$.

Table 8.1 The Bravais lattices. (The real parameters a, b, c and d are arbitrary.)

For a perfect crystalline solid the group of the Schrödinger equation (or, in the case of lattice vibrations, the invariance group) is a crystallographic space group, which contains rotations as well as pure primitive translations. The crystallographic space groups will be investigated in detail in Chapter 9. However, it is very enlightening, as a first stage in their study, to limit attention to the subgroup $\mathscr{T}$ of pure primitive translations of the relevant lattice. Only the translational symmetry is then being taken into account.

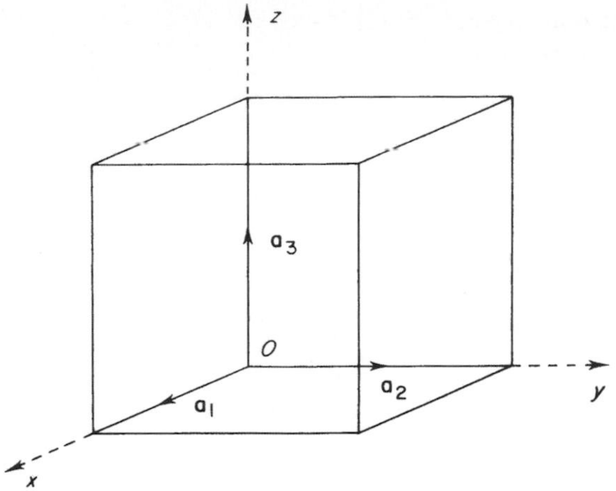

Figure 8.1 Basic lattice vectors of the simple cubic lattice, Γ_c.

In particular, the energy eigenfunctions must transform according to the irreducible representations of this subgroup, which is equivalent to saying that they satisfy Bloch's Theorem, as will be demonstrated in Section 3. Bloch's Theorem has now become so much an essential part of the theory of solids that it is sometimes forgotten that it is basically a group theoretical result. The elementary energy band and lattice vibration theories based upon Bloch's Theorem themselves require no knowledge of group theory and so are presented in most textbooks on solid state theory. However, the neglect

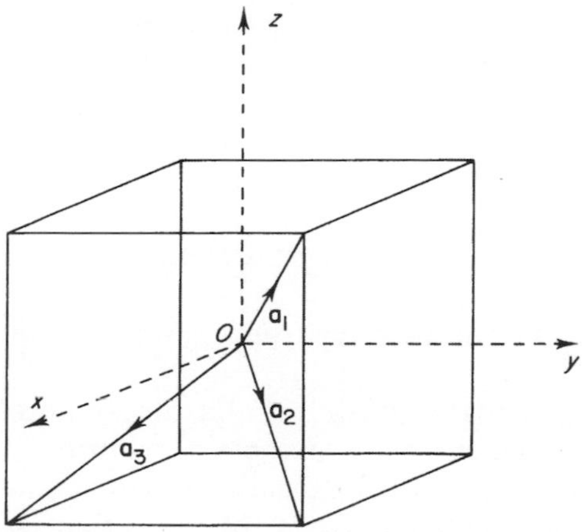

Figure 8.2 Basic lattice vectors of the body-centred cubic lattice, Γ_c^v.

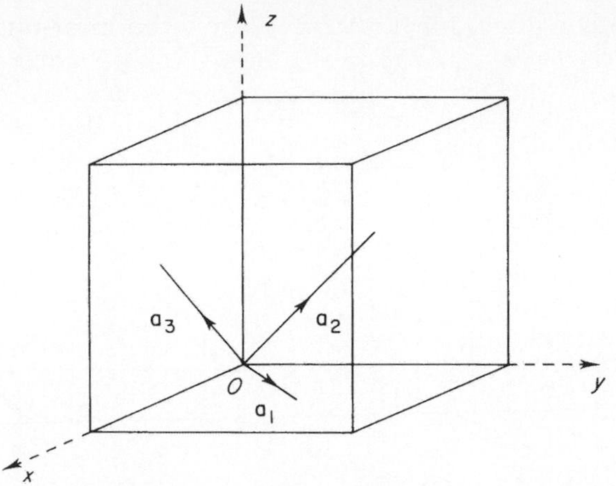

Figure 8.3 Basic lattice vectors of the face-centred cubic lattice, Γ_c^f.

of rotational symmetries in these elementary theories does mean that some phenomena are overlooked, and, in particular, it cannot predict the extra degeneracies which can occur in electronic energy levels and vibrational spectra. Moreover, it is only by taking into account the rotational symmetries that it is possible to reduce the numerical work in energy band calculations to a manageable amount and still produce accurate results.

A proof of Bloch's Theorem that involves only an elementary application of the ideas of the previous chapters is given is Section 3. The rest of the chapter is then devoted to a brief account of the elementary electronic energy band and lattice vibration theories that are based on this theorem. Chapter 9 then describes how the theory is modified when the *full* space group is introduced in place of its translational subgroup $\mathcal{T}$. It will be seen there that the concepts introduced in this chapter still play a fundamental role.

2 The cyclic boundary conditions

(*a*) *Electronic wave functions*

Strictly speaking, a *real* crystalline solid cannot possess any translational symmetry because it is necessarily finite in extent. Consequently any translation will shift some electron or nucleus from just inside some surface to the outside of the body, that is, to a completely different environment.

On the other hand, for a normal sample the inter-nuclear spacing is so much smaller than the dimensions of the sample and the interactions that directly affect each electron or nucleus are of such short range, that for electrons and nuclei well inside the body the situation is almost exactly as if the solid were infinite in extent. Moreover, the evidence of X-ray crystallography is that the nuclei within a solid can be ordered as if they were based on an infinite lattice, except near the surfaces. As most of the properties of a solid depend only on the behaviour of the vast majority of electrons or nuclei that lie in the interior, it is a very reasonable approximation to idealize the situation by working with models based on infinite lattices. The translational symmetry possessed by such models then permits a considerable simplification of the analysis of electronic and vibrational problems.

However, the symmetry groups based on infinite lattices are necessarily of infinite order, and it is easier to work with groups of finite order. This can be achieved by imposing "cyclic" boundary conditions on the infinite lattice.

For electrons it may be assumed that for every energy eigenfunction $\phi(\mathbf{r})$

$$\phi(\mathbf{r}) = \phi(\mathbf{r} + N_1\mathbf{a}_1) = \phi(\mathbf{r} + N_2\mathbf{a}_2) = \phi(\mathbf{r} + N_3\mathbf{a}_3), \tag{8.2}$$

where N_1, N_2 and N_3 are very large positive integers and $\mathbf{a}_1$, $\mathbf{a}_2$ and $\mathbf{a}_3$ are the basic lattice vectors of the lattice. This implies that the infinite crystal is considered to consist of a set of basic blocks in the form of parallelepipeds having edges $N_1\mathbf{a}_1$, $N_2\mathbf{a}_2$ and $N_3\mathbf{a}_3$, and that the physical situation is identical in corresponding points of different blocks. These boundary conditions cannot affect the behaviour of electrons well inside each basic block to any significant extent, so the bulk properties are again unchanged. The integers N_1, N_2 and N_3 may be taken to be as large as desired.

The integration involved in the inner product (ϕ, ψ) defined in Equation (1.19) must now be taken as being over just one basic block of the crystal, B. For any pure primitive translation T the operators P(T) retain the unitary property of Equation (1.20), provided all functions involved satisfy Equation (8.2). This follows because $(P(T)\phi, P(T)\psi)$ is equal to

$$\iiint_B \phi(\mathbf{r} - \mathbf{t}(T))^* \psi(\mathbf{r} - \mathbf{t}(T))\, dx\, dy\, dz = \iiint_{B'} \phi(\mathbf{r})^* \psi(\mathbf{r})\, dx\, dy\, dz,$$

where B' is obtained from B by a translation $-\mathbf{t}(T)$. As every part of B' can be mapped into a part of B by an appropriate combination of translations through $N_1\mathbf{a}_1$, $N_2\mathbf{a}_2$ and $N_3\mathbf{a}_3$, by Equations (8.2) the last integral becomes (ϕ, ψ).

The conditions in Equations (8.2) are often referred to as the "Born cyclic boundary conditions", as they are the analogues for electronic states of the vibrational boundary conditions first proposed by Born and von Karman (1912). They imply that

$$\mathrm{P}(\{\mathbf{1} \mid N_j\mathbf{a}_j\}) = \mathrm{P}(\{\mathbf{1} \mid \mathbf{0}\}) \tag{8.3}$$

for every function of interest and for $j = 1, 2, 3$. (Of course $\mathrm{P}(\{\mathbf{1} \mid \mathbf{0}\})$ is merely the identity operator.) Consequently

$$\mathrm{P}(\{\mathbf{1} \mid \mathbf{t_n} + l_1N_1\mathbf{a}_1 + l_2N_2\mathbf{a}_2 + l_3N_3\mathbf{a}_3\}) = \mathrm{P}(\{\mathbf{1} \mid \mathbf{t_n}\})$$

for any lattice vector $\mathbf{t_n}$ and any set of integers l_1, l_2 and l_3, so that only $N = N_1N_2N_3$ of these operators are *distinct*. The set of distinct operators may be taken to be $\mathrm{P}(\{\mathbf{1} \mid n_1\mathbf{a}_1 + n_2\mathbf{a}_2 + n_3\mathbf{a}_3\})$ with

$$0 \leq n_j < N_j, \qquad j = 1, 2, 3. \tag{8.4}$$

Moreover, as $\mathrm{P}(\{\mathbf{1} \mid N_j\mathbf{a}_j\}) = \mathrm{P}(\{\mathbf{1} \mid \mathbf{a}_j\})^{N_j}$, it follows from Equation (8.3) that

$$\mathrm{P}(\{\mathbf{1} \mid \mathbf{a}_j\})^{N_j} = \mathrm{P}(\{\mathbf{1} \mid \mathbf{0}\}), \qquad j = 1, 2, 3. \tag{8.5}$$

Thus this set of distinct operators forms a *finite* group $\mathcal{T}$ of order $N = N_1N_2N_3$. Henceforth this group $\mathcal{T}$ will be used in place of the infinite group of pure primitive translations $\mathcal{T}^\infty$.

Incidentally, as it remains true that

$$\mathrm{P}(\{\mathbf{1} \mid \mathbf{t_n}\})\mathrm{P}(\{\mathbf{1} \mid \mathbf{t_{n'}}\}) = \mathrm{P}(\{\mathbf{1} \mid \mathbf{t_n}\}\{\mathbf{1} \mid \mathbf{t_{n'}}\})$$

for any two lattice vectors $\mathbf{t_n}$ and $\mathbf{t_{n'}}$ of a lattice, the mapping $\phi(\{\mathbf{1} \mid \mathbf{t_n}\}) = \mathrm{P}(\{\mathbf{1} \mid \mathbf{t_n}\})$ is a homomorphic mapping of $\mathcal{T}^\infty$ onto $\mathcal{T}$. The kernel $\mathcal{K}$ of this mapping is the infinite set of pure primitive translations of the form $\{\mathbf{1} \mid l_1N_1\mathbf{a}_1 + l_2N_2\mathbf{a}_2 + l_3N_3\mathbf{a}_3\}$, where l_1, l_2, l_3 are any set of integers.

(*b*) *Lattice vibrations*

It is easiest to justify the cyclic boundary conditions for lattice vibrations by considering first a model in which all the nuclei are of the same species and lie in a straight line, their equilibrium position vectors being $\mathbf{r}^e_{n'} = n'\mathbf{a}$, where $n' = 0, \pm 1, \pm 2, \ldots$. Then, under a pure primitive translation through $\mathbf{t}_n = n\mathbf{a}$,

$$\mathbf{r}^e_{n'} \to \mathbf{r}^e_{n+n'}, \qquad \text{for all } n' = 0, \pm 1, \pm 2, \ldots \tag{8.6}$$

This infinite system can be approximated by a finite system with N

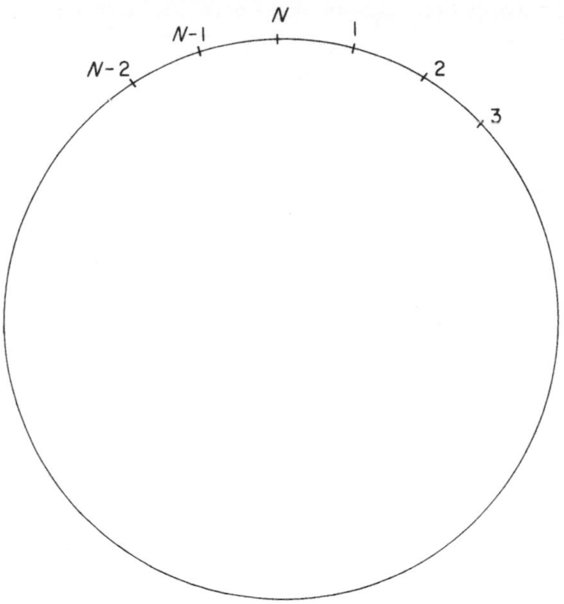

Figure 8.4 Cyclic boundary conditions applied to a linear chain.

nuclei, N being very large, the nuclei lying on a circle of very large radius in such a way that the first nucleus and the Nth nucleus are the same distance apart as all other successive nuclei. Figure 8.4 shows the situation envisaged. Assuming that *every* nucleus experiences the same forces as its neighbours, the system is invariant under the permutations

$$\mathbf{r}^e_{n'} \to \mathbf{r}^e_{n+n'-lN} \qquad \text{for all } n' = 1, 2, \ldots, N, \tag{8.7}$$

where l is such that

$$1 \leqslant n + n' - lN \leqslant N.$$

This is a transformation in which each of the set of N nuclei is mapped into another nucleus of the *same* set. It may be associated with a pure primitive translation through $\mathbf{t}_n = n\mathbf{a}$. Clearly when N is large there will not be any appreciable difference between the vibrations of the infinite system and its finite counterpart. However, Equations (8.6) and (8.7) show that there is a significant difference between the symmetries of the two systems.

For a *three*-dimensional situation it is easy to give a simple algebraic generalization of Equation (8.7) in which each of the set of $N = N_1N_2N_3$ lattice points $\mathbf{t}_{\mathbf{n}'} = n'_1\mathbf{a}_1 + n'_2\mathbf{a}_2 + n'_3\mathbf{a}_3$ of the basic block with $0 \leqslant n'_j < N_j$ $(j = 1, 2, 3)$ is mapped into other lattice point of the *same* block. Let $\{1 \mid \mathbf{t_n}\}$, where $\mathbf{t_n} = n_1\mathbf{a}_1 + n_2\mathbf{a}_2 + n_3\mathbf{a}_3$, be *any* pure primitive translation. Then the generalization of Equation (8.7) is that associated with this translation

$$\mathbf{t_{n'}} \to \mathbf{t_{n''}}$$

where $\mathbf{t_{n''}} = n_1''\mathbf{a}_1 + n_2''\mathbf{a}_2 + n_3''\mathbf{a}_3$, and where

$$n_j'' = n_j + n_j' - l_j N_j, \tag{8.8}$$

l_1, l_2, l_3 being integers chosen so that $0 \leq n_j'' < N_j$ $(j = 1, 2, 3)$. Moreover, if there are S nuclei per lattice site, the corresponding mapping of equilibrium positions is

$$\mathbf{r}^e_{\mathbf{n}'\gamma'} \to \mathbf{r}^e_{\mathbf{n}''\gamma''}, \tag{8.9}$$

where $\mathbf{r}^e_{\mathbf{n}'\gamma'} = \mathbf{t_{n'}} + \boldsymbol{\tau}_{\gamma'}$, $\mathbf{r}^e_{\mathbf{n}''\gamma''} = \mathbf{t_{n''}} + \boldsymbol{\tau}_{\gamma''}$, where

$$\gamma' = \gamma''(=1, 2, \ldots, S),$$

and where n_j'' and n_j' are related by Equation (8.8). The mapping (Expression (8.9)) can be written in the form

$$\mathbf{r}^e_{\mathbf{n}''\gamma''} = \{1 \mid \mathbf{t_n} - l_1N_1\mathbf{a}_1 - l_2N_2\mathbf{a}_2 - l_3N_3\mathbf{a}_3\}\mathbf{r}^e_{\mathbf{n}'\gamma'} \tag{8.10}$$

An alternative way of expressing the mapping (Expression (8.9)) is to invoke the permutation matrix $\mathbf{p}(\{1 \mid \mathbf{t_n}\})$ of Chapter 7, Section 3(*c*). With the equilibrium positions specified (as in Equation (8.1)) by a triple of integers $\mathbf{n}$ and an index γ, $\mathbf{p}(T)$ is an $N_1N_2N_3S \times N_1N_2N_3S$ matrix with rows and columns specified by triples and single indices such that

$$\mathbf{p}(\{1 \mid \mathbf{t_n}\})_{\mathbf{n}''\gamma'',\mathbf{n}'\gamma'} = \begin{cases} 1, & \text{if } \gamma'' = \gamma' \text{ and there exist integers } l_1, l_2, l_3 \\ & \quad \text{such that } n_j + n_j' = n_j'' + l_jN_j, \\ 0, & \text{otherwise.} \end{cases} \tag{8.11}$$

Then

$$\mathbf{p}(\{1 \mid \mathbf{t_n}\})\mathbf{p}(\{1 \mid \mathbf{t_{n'}}\}) = \mathbf{p}(\{1 \mid \mathbf{t_n}\}\{1 \mid \mathbf{t_{n'}}\})$$

for any two lattice vectors $\mathbf{t_n}$ and $\mathbf{t_{n'}}$ and

$$\mathbf{p}(\{1 \mid N_j\mathbf{a}_j\}) = \mathbf{p}(\{1 \mid \mathbf{0}\}) = \mathbf{1}$$

for $j = 1, 2, 3$. Thus the set of matrices $\mathbf{p}(\{1 \mid n_1\mathbf{a}_1 + n_2\mathbf{a}_2 + n_3\mathbf{a}_3\})$ with n_1, n_2 and n_3 satisfying Conditions (8.4) and with

$$\mathbf{p}(\{1 \mid \mathbf{a}_j\})^{N_j} = \mathbf{p}(\{1 \mid \mathbf{0}\}) \tag{8.12}$$

again form a group of order $N = N_1N_2N_3$. Clearly the mapping $\psi(P(\{1 \mid \mathbf{t_n}\})) = \mathbf{p}(\{1 \mid \mathbf{t_n}\})$ is an *isomorphic* mapping of $\mathcal{T}$ onto this group. No confusion will be caused if this group and the corresponding group of transformations in Expression (8.9) are also denoted by $\mathcal{T}$.

This approximation can easily be incorporated into the general formulation of vibrational problems of Chapter 7, Section 3. The mapping (Expression (8.9)) defines a symmetry transformation T of the equilibrium positions of the nuclei in one basic block, that is, of those whose equilibrium position vectors are $\mathbf{r}^e_{\mathbf{n}'\gamma'} = \mathbf{t}_{\mathbf{n}'} + \boldsymbol{\tau}_{\gamma'}$ with $0 \leq n'_j < N_j$ $(j = 1, 2, 3)$. Thus A, the number of relevant nuclei, must be taken to be equal to $N_1N_2N_3S$. The invariance group $\mathcal{G}$ of the system is then $\mathcal{T}$. The details will be examined in Section 7, after the irreducible representations of $\mathcal{T}$ have been found.

3 The irreducible representations of the group $\mathcal{T}$ of pure primitive translations and Bloch's Theorem

As the group $\mathcal{T}$ is a finite Abelian group of order $N = N_1N_2N_3$, it possesses N inequivalent irreducible representations, all of which are one-dimensional (see Chapter 5, Section 6). These are easily found, for $\mathcal{T}$ is isomorphic to the direct product of three cyclic groups.

Consider a particular one-dimensional irreducible representation Γ of $\mathcal{T}$ and suppose that $\Gamma(\{\mathbf{1} \mid \mathbf{a}_j\}) = [c_j]$, for $j = 1, 2, 3$. Then, from Equation (8.5) (and Equation (8.12)), it follows that

$$c_j^{N_j} = 1, \tag{8.13}$$

so that

$$c_j = \exp(-2\pi i p_j/N_j), \qquad j = 1, 2, 3,$$

where p_j is an integer. As $\exp\{-2\pi i(p_j + N_j)/N_j\} = \exp(-2\pi i p_j/N_j)$, there are only N_j *different* values of c_j allowed by Equation (8.13) and each of these, by convention, may be taken to correspond to a p_j having one of the values $0, 1, 2, \ldots, N_j - 1$. Then

$$\Gamma(\{\mathbf{1} \mid n_j\mathbf{a}_j\}) = [\exp(-2\pi i p_j n_j/N_j)]$$

and hence

$$\Gamma(\{\mathbf{1} \mid \mathbf{t_n}\}) = [\exp(-2\pi i\{(p_1n_1/N_1) + (p_2n_2/N_2) + (p_3n_3/N_3)\})], \tag{8.14}$$

where $\mathbf{t_n} = n_1\mathbf{a}_1 + n_2\mathbf{a}_2 + n_3\mathbf{a}_3$. There are $N = N_1N_2N_3$ sets of integers (p_1, p_2, p_3) allowed by the above convention which can be used to label the N different irreducible representations of $\mathcal{T}$.

Equation (8.14) can be simplified and given a simple geometric interpretation by introducing the following notation. Define the "basic lattice vectors of the reciprocal lattice" $\mathbf{b}_1$, $\mathbf{b}_2$ and $\mathbf{b}_3$ by

$$\mathbf{a}_j \,.\, \mathbf{b}_k = 2\pi\delta_{jk}, \qquad j, k = 1, 2, 3, \tag{8.15}$$

so that, explicitly,

$$\mathbf{b}_1 = 2\pi \mathbf{a}_2 \wedge \mathbf{a}_3 / \{\mathbf{a}_1 \,.\, (\mathbf{a}_2 \wedge \mathbf{a}_3)\}, \tag{8.16}$$

with similar expressions for $\mathbf{b}_2$ and $\mathbf{b}_3$. Then define the so-called "allowed $\mathbf{k}$-vectors" by

$$\mathbf{k} = k_1\mathbf{b}_1 + k_2\mathbf{b}_2 + k_3\mathbf{b}_3, \tag{8.17}$$

where $k_j = p_j/N_j$. Thus $\mathbf{k} \,.\, \mathbf{t_n} = 2\pi\{(p_1n_1/N_1) + (p_2n_2/N_2) + (p_3n_3/N_3)\}$, so that Equation (8.14) becomes

$$\Gamma^{\mathbf{k}}(\{1 \mid \mathbf{t_n}\}) = [\exp(-i\mathbf{k} \,.\, \mathbf{t_n})], \tag{8.18}$$

where the N irreducible representations are now labelled by the allowed $\mathbf{k}$-vectors.

Suppose that $\phi_1^{\mathbf{k}}(\mathbf{r})$ is a basis function transforming as the first (and only) row of $\Gamma^{\mathbf{k}}$. Then, by Equations (1.26) and (8.18),

$$P(\{1 \mid \mathbf{t_n}\})\phi_1^{\mathbf{k}}(\mathbf{r}) = \Gamma^{\mathbf{k}}(\{1 \mid \mathbf{t_n}\})_{11}\phi_1^{\mathbf{k}}(\mathbf{r}) = \exp(-i\mathbf{k} \,.\, \mathbf{t_n})\phi_1^{\mathbf{k}}(\mathbf{r}). \tag{8.19}$$

However, by Equation (1.17),

$$P(\{1 \mid \mathbf{t_n}\})\phi_1^{\mathbf{k}}(\mathbf{r}) = \phi_1^{\mathbf{k}}(\{1 \mid \mathbf{t_n}\}^{-1}\mathbf{r}) = \phi_1^{\mathbf{k}}(\mathbf{r} - \mathbf{t_n}),$$

so that

$$\phi_1^{\mathbf{k}}(\mathbf{r} - \mathbf{t_n}) = \exp(-i\mathbf{k} \,.\, \mathbf{t_n})\phi_1^{\mathbf{k}}(\mathbf{r}).$$

Thus

$$\phi_1^{\mathbf{k}}(\mathbf{r}) = \exp(i\mathbf{k} \,.\, \mathbf{r})u_{\mathbf{k}}(\mathbf{r}), \tag{8.20}$$

where $u_{\mathbf{k}}(\mathbf{r})$ is a function that has the perodicity of the lattice, that is, $u_{\mathbf{k}}(\mathbf{r} - \mathbf{t_n}) = u_{\mathbf{k}}(\mathbf{r})$ for any lattice vector $\mathbf{t_n}$.

Equation (8.20) is the statement of the theorem of Bloch (1928) in its usual form, for electronic energy eigenfunctions must be basis functions of the irreducible representations $\Gamma^{\mathbf{k}}$ of $\mathcal{T}$. A function of the form in Equation (8.20) is therefore called a "Bloch function". The corresponding energy eigenvalue may be denoted by $\varepsilon(\mathbf{k})$, so that

$$H(\mathbf{r})\phi_1^{\mathbf{k}}(\mathbf{r}) = \varepsilon(\mathbf{k})\phi_1^{\mathbf{k}}(\mathbf{r}). \tag{8.21}$$

The notation for basis functions here follows the standard practice in which the irreducible representation is specified by a superscript (or set of superscripts) and the rows by a subscript (or set of subscripts). In particular, the wave vector $\mathbf{k}$ appears as a superscript with this convention. However, it should be pointed out that in most of the solid state literature $\mathbf{k}$ is written as a subscript, so that $\phi_1^{\mathbf{k}}(\mathbf{r})$ would be written as $\phi_{\mathbf{k}}(\mathbf{r})$ and Equation (8.20) would become $H(\mathbf{r})\phi_{\mathbf{k}}(\mathbf{r}) = \varepsilon(\mathbf{k})\phi_{\mathbf{k}}(\mathbf{r})$.

4 Brillouin zones

The set of lattice vectors of the reciprocal lattice is defined by

$$\mathbf{K_m} = m_1\mathbf{b}_1 + m_2\mathbf{b}_2 + m_3\mathbf{b}_3, \tag{8.22}$$

where $\mathbf{m} = (m_1, m_2, m_3)$, m_1, m_2 and m_3 are integers, and $\mathbf{b}_1$, $\mathbf{b}_2$ and $\mathbf{b}_3$ are the basic lattice vectors of the reciprocal lattice defined by Equation (8.15). They have the property that

$$\exp(i\mathbf{K_m}\,.\,\mathbf{t_n}) = 1 \tag{8.23}$$

for any $\mathbf{K_m}$ and $\mathbf{t_n}$. It is useful to note that

$$\sum_{\mathbf{t_n}} \exp(i\mathbf{k}\,.\,\mathbf{t_n}) = \begin{cases} N, & \text{if } \mathbf{k} = \mathbf{K_m}, \\ 0, & \text{if } \mathbf{k} \neq \mathbf{K_m}, \end{cases} \tag{8.24}$$

where the sum is over all the lattice vectors of one basic block of Section 2, this result being a consequence of the fact that the left-hand side is a product of three simple geometric series. Similarly,

$$\sum_{\mathbf{k}} \exp(-i\mathbf{k}\,.\,\mathbf{t_n}) = \begin{cases} N, & \text{if } \mathbf{t_n} = \mathbf{0}, \\ 0, & \text{if } \mathbf{t_n} \neq \mathbf{0}, \end{cases}$$

the sum being over all allowed $\mathbf{k}$-vectors.

In Section 3, N irreducible representations of $\mathcal{T}$ were found and described by the allowed $\mathbf{k}$-vectors (Equation (8.17)). These $\mathbf{k}$-vectors can be imagined as being plotted in the so-called "$\mathbf{k}$-space" or "reciprocal space" defined by the reciprocal lattice vectors. The allowed $\mathbf{k}$-vectors lie on a very fine lattice (defined by Equation (8.17)) within and upon three faces of the parallelepiped having edges $\mathbf{b}_1$, $\mathbf{b}_2$ and $\mathbf{b}_3$ that is shown in Figure 8.5.

It is, however, more convenient to replot the allowed $\mathbf{k}$-vectors into a more symmetrical region of $\mathbf{k}$-space surrounding the point $\mathbf{k} = \mathbf{0}$. To do this consider the equation

$$\mathbf{k}' = \mathbf{k} + \mathbf{K_m}, \tag{8.25}$$

where $\mathbf{K_m}$ is a reciprocal lattice vector. A pair of vectors $\mathbf{k}$ and $\mathbf{k}'$ satisfying Equation (8.25) are said to be "equivalent", because

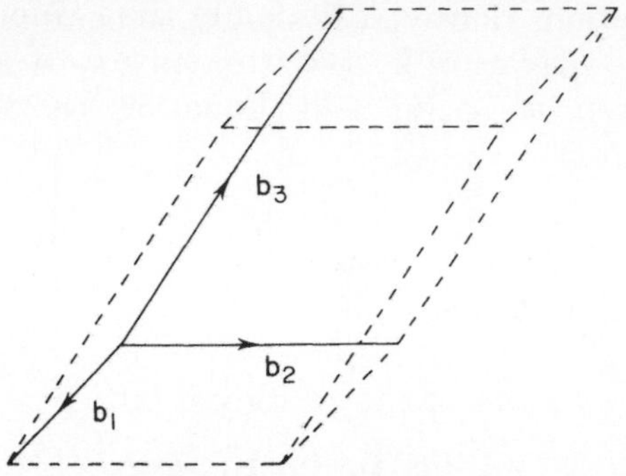

Figure 8.5 The basic parallelepiped of **k**-space.

$\exp(-i\mathbf{k}' . \mathbf{t_n}) = \exp(-i\mathbf{k} . \mathbf{t_n})$ by Equation (8.23), and hence

$$\Gamma^{\mathbf{k}'}(\{1 \mid \mathbf{t_n}\}) = \Gamma^{\mathbf{k}}(\{1 \mid \mathbf{t_n}\})$$

for every $\{1 \mid \mathbf{t_n}\}$ of $\mathcal{T}$. Thus the irreducible representation described by **k** could equally be described by **k**′. The more symmetrical region of **k**-space is called the "Brillouin zone" (or sometimes the "first Brillouin zone"), and it is defined to consist of all those points of **k** space that lie closer to $\mathbf{k} = \mathbf{0}$ than to any other reciprocal lattice points. Its boundaries are therefore the planes that are the perpendicular bisectors of the lines joining the point $\mathbf{k} = \mathbf{0}$ to the nearer reciprocal lattice points, the plane bisecting the line from $\mathbf{k} = \mathbf{0}$ to $\mathbf{k} = \mathbf{K_m}$ having the equation

$$\mathbf{k} . \mathbf{K_m} = \tfrac{1}{2} |\mathbf{K_m}|^2,$$

as is clear from Figure 8.6. For some lattices, such as the body-centred cubic lattice Γ_c^v, only *nearest* neighbour reciprocal lattice

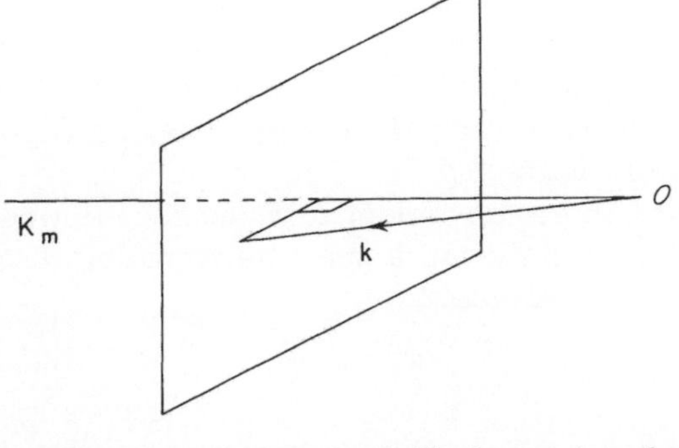

Figure 8.6 Construction of a Brillouin zone boundary.

points are involved in the construction of the Brillouin zone, but for others, such as the face-centred cubic lattice Γ_c^f, ***next-nearest*** neighbours are involved as well. The irreducible representations of $\mathcal{T}$ then correspond to a very fine lattice of points inside the Brillouin zone and on one half of its surface.

The mapping of the parallelepiped of Figure 8.5 into the Brillouin zone can be quite complicated because different regions of the parallelepiped are mapped using different reciprocal lattice vectors. The following *two*-dimensional example shown in Figure 8.7 of a square lattice demonstrates this clearly. In this example the analogue of the three-dimensional parallelepiped of Figure 8.6 is the square with sides $\mathbf{b}_1$ and $\mathbf{b}_2$, which consists of four regions 1, 2, 3 and 4, and the analogue of the Brillouin zone is the square having $\mathbf{k}=\mathbf{0}$ at its centre, which consists of the four regions 1′, 2′, 3′ and 4′. The region 1 is mapped into 1′ by $\mathbf{K}_{(0,0,0)}=\mathbf{0}$, 2 is mapped into 2′ by $\mathbf{K}_{(-1,0,0)}=-\mathbf{b}_1$, 3 is mapped into 3′ by $\mathbf{K}_{(0,-1,0)}=-\mathbf{b}_2$ and 4 is mapped into 4′ by $\mathbf{K}_{(-1,-1,0)}=-\mathbf{b}_1-\mathbf{b}_2$.

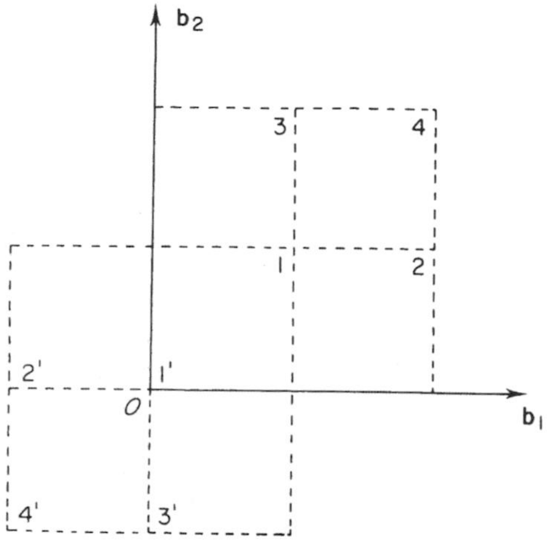

Figure 8.7 Construction of a two-dimensional Brillouin zone.

By construction, the volume of the Brillouin zone is the same as that of the parallelepiped from which it is formed, namely $\mathbf{b}_1 \cdot (\mathbf{b}_2 \wedge \mathbf{b}_3)$. It follows from Equation (8.16) that this is equal to $(2\pi)^3/\{\mathbf{a}_1 \cdot (\mathbf{a}_2 \wedge \mathbf{a}_3)\}$, where $\mathbf{a}_1 \cdot (\mathbf{a}_2 \wedge \mathbf{a}_3)$ is the volume of the parallelepiped whose sides are $\mathbf{a}_1$, $\mathbf{a}_2$ and $\mathbf{a}_3$.

For the simple cubic lattice Γ_c, the basic lattice vectors of the reciprocal lattice obtained from Table 8.1 and Equation (8.16) are

$$\mathbf{b}_1=(2\pi/a)(1,0,0), \qquad \mathbf{b}_2=(2\pi/a)(0,1,0), \qquad \mathbf{b}_3=(2\pi/a)(0,0,1).$$

The Brillouin zone is given in Figure 8.8. The position vectors of the "symmetry points" are as follows: for Γ, $\mathbf{k}=(0,0,0)$; for X, $\mathbf{k}=(\pi/a)(0,0,1)$; for M, $\mathbf{k}=(\pi/a)(0,1,1)$; and for R, $\mathbf{k}=(\pi/a)(1,1,1)$. The significance of the term "symmetry point" will be explained in Chapter 9, Section 2. The notation is that of Bouckaert *et al.* (1936).

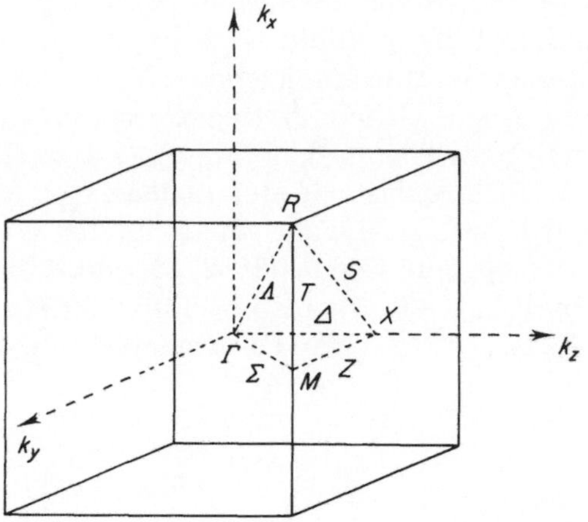

Figure 8.8 Brillouin zone corresponding to the simple cubic lattice Γ_c.

Similarly, for the body-centred cubic lattice Γ_c^v the basic lattice vectors of the reciprocal lattice are

$$\mathbf{b}_1=(2\pi/a)(1,0,1), \qquad \mathbf{b}_2=(2\pi/a)(0,1,-1), \qquad \mathbf{b}_3=(2\pi/a)(1,-1,0),$$

the Brillouin zone being shown in Figure 8.9. The position vectors of the symmetry points are as follows: for Γ, $\mathbf{k}=(0,0,0)$; for H, $\mathbf{k}=(\pi/a)(0,0,2)$; for N, $\mathbf{k}=(\pi/a)(0,1,1)$; and for P, $\mathbf{k}=(\pi/a)(1,1,1)$, the notation being that of Bouckaert *et al.* (1936).

Finally, for the face-centred cubic lattice Γ_c^f the basic lattice vectors of the reciprocal lattice are

$$\mathbf{b}_1=(2\pi/a)(1,1,-1), \qquad \mathbf{b}_2=(2\pi/a)(-1,1,1), \qquad \mathbf{b}_3=(2\pi/a)(1,-1,1).$$

The Brillouin zone is given in Figure 8.10, the position vectors of the points indicated (in the notation of Bouckaert *et al.* (1936)) being: for Γ, $\mathbf{k}=(0,0,0)$; for K, $\mathbf{k}=(\pi/a)(0,\frac{3}{2},\frac{3}{2})$; for L, $\mathbf{k}=(\pi/a)(1,1,1)$; for U, $\mathbf{k}=(\pi/a)(\frac{1}{2},\frac{1}{2},2)$; for W, $\mathbf{k}=(\pi/a)(0,1,2)$; and for X, $\mathbf{k}=(\pi/a)(0,0,2)$.

The Brillouin zones corresponding to the other eleven Bravais lattices may be found in the review article by Koster (1957).

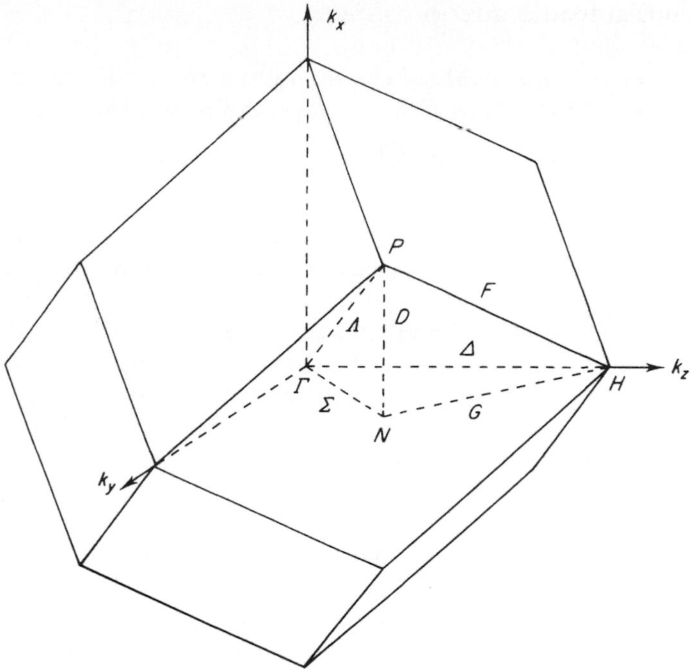

Figure 8.9 Brillouin zone corresponding to the body-centred cubic lattice Γ_c^v.

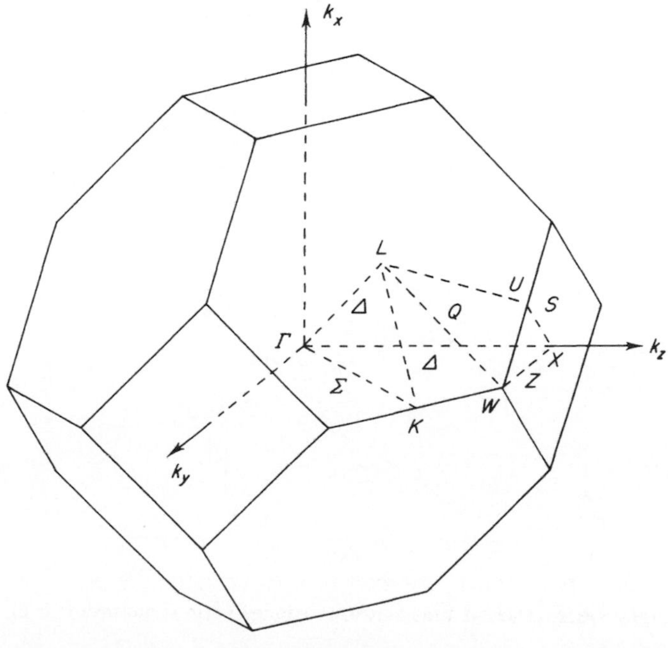

Figure 8.10 Brillouin zone corresponding to the face-centred cubic lattice Γ_c^f.

5 Electronic energy bands

The set of energy eigenvalues corresponding to an allowed **k**-vector may be denoted by $\varepsilon_1(\mathbf{k})$, $\varepsilon_2(\mathbf{k})$, . . . , with the convention that

$$\varepsilon_n(\mathbf{k}) \leqslant \varepsilon_{n+1}(\mathbf{k}) \tag{8.26}$$

for all $n = 1, 2, \ldots$. The set of energy eigenvalues $\varepsilon_n(\mathbf{k})$ corresponding to a particular n are said to form the "nth energy band" and the set of energy bands is said to constitute the electronic energy band "structure".

To visualize the energy band structure, it is convenient to consider, one at a time, the axes of the Brillouin zone that join the symmetry points, and for each allowed **k**-vector on each axis to plot the energy eigenvalues $\varepsilon_n(\mathbf{k})$. Two typical examples are shown in Figures 8.11 and 8.12, which give the energy levels along the axis Δ for iron (as calculated by Wood (1962)) and silicon (as calculated by Chelikowsky and Cohen (1976)), the lattices being the body-centred cubic and face-centred cubic respectively. The number in curved brackets gives the band index n, as defined in Expression (8.26) and the number in square brackets indicates the degeneracy of the corresponding eigenvalue. (The occurrence of degenerate eigenvalues is a consequence of the rotational symmetry, which is being neglected in this chapter but

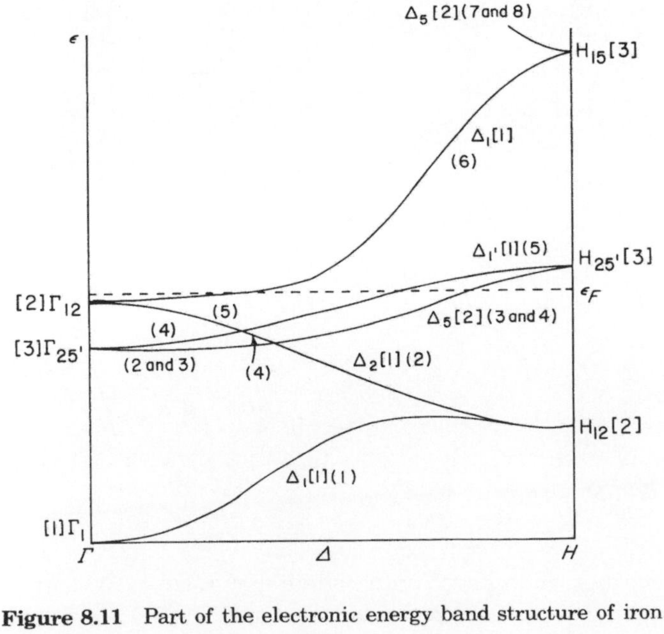

Figure 8.11 Part of the electronic energy band structure of iron.

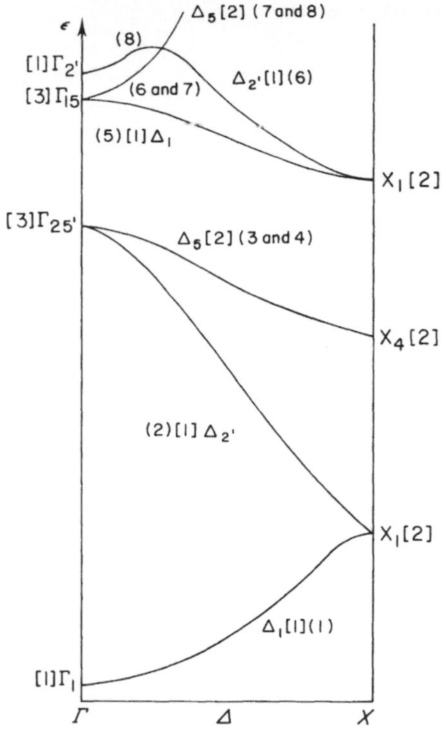

Figure 8.12 Part of the electronic energy band structure of silicon.

which will be investigated in Chapter 9. The other symbols will also be explained in Chapter 9.)

The positive integers N_1, N_2 and N_3 introduced in Equations (8.2) are arbitrarily large, and it is frequently convenient to consider the limiting case in which they tend to infinity. The allowed **k**-vectors can then take *all* values inside the Brillouin zone and on half of its surface, and the $\varepsilon_n(\mathbf{k})$ are continuous functions of **k** for each n. Moreover, $\text{grad}_{\mathbf{k}}\, \varepsilon_n(\mathbf{k})$ are also continuous functions of **k**, except possibly at points where two bands touch. The plots in Figures 8.11 and 8.12 are made for this limiting case.

In the "single-particle" approximation (see Chapter 1, Section 3(*a*)), the Pauli exclusion principle implies that no two electrons can "occupy" the same one-electron state. With the present neglect of spin-dependent terms, such a state is specified by an allowed **k**-vector, a band index n, and a spin quantum number that can take one of two possible values. It follows that each energy level $\varepsilon_n(\mathbf{k})$ can "hold" *two* electrons and hence each energy band can hold $2N$ electrons. If there are V valence or conduction electrons per atom and S atoms per lattice point of the crystal, there will be NVS valence or conduction electrons in each large basic block of the crystal (in the sense of Section 2(a)), which will therefore require the equivalent of $\frac{1}{2}VS$ bands to hold them.

In the ground state of the system all the energy levels will be doubly occupied up to a certain energy ε_F, the "Fermi energy", and all levels above this energy will be unoccupied. The surface in $\mathbf{k}$-space defined by

$$\varepsilon_n(\mathbf{k}) = \varepsilon_F$$

is called the "Fermi surface". The distribution of energy levels near the Fermi energy largely determines the electronic properties of a solid. If one and only one band contains the Fermi energy, and all others are entirely above it or below it, then the Fermi surface merely consists of one sheet. If no band contains the Fermi energy, as happens for insulators and semiconductors, there is no Fermi surface. In all other cases the Fermi surface consists of several sheets, to visualize which one considers a number of identical Brillouin zones, with one zone for each band. A full band corresponds to a full Brillouin zone, but a partially occupied band corresponds to a partially occupied Brillouin zone and hence to a sheet of the Fermi surface in that zone.

As an example, consider body-centred cubic iron for which $S = 1$ and $V = 8$, so that the equivalent of four complete bands are needed to hold the valence electrons. The Fermi energy ε_F is shown in Figure 8.11 and is clearly consistent with this. Bands 3, 4, 5 and 6 are partially occupied, giving rise to four sheets in the Fermi surface.

For silicon $V = 4$ and $S = 2$, so that again four bands (or their equivalent) are required to hold the valence electrons. However, in this case there is no overlap between the 4th and 5th bands, so that the Fermi energy lies between the X_1 and $\Gamma_{25'}$ levels of Figure 8.12. There are therefore four completely filled bands and no Fermi surface.

Equation (8.18) shows that the complex conjugate representation to $\mathbf{\Gamma}^{\mathbf{k}}$ is $\mathbf{\Gamma}^{-\mathbf{k}}$ and that both are essentially complex (except when $\mathbf{k} = \mathbf{0}$ or when $\mathbf{k}$ is one-half of some reciprocal lattice vector, that is, except at a few isolated points of the Brillouin zone). Time-reversal symmetry (see Chapter 6, Section 5(a)) then implies that the energy eigenvalues associated with the irreducible representations $\mathbf{\Gamma}^{\mathbf{k}}$ and $\mathbf{\Gamma}^{-\mathbf{k}}$ must be equal, that is,

$$\varepsilon_n(\mathbf{k}) = \varepsilon_n(-\mathbf{k}). \tag{8.27}$$

(When the extra rotational symmetries of the space group are taken into account, time-reversal symmetry may cause extra degeneracies (cf. Cornwell 1969).)

6 Irreducible representations of the double group $\mathcal{T}^D$ of pure primitive translations

So far spin-dependent terms have been neglected. It will now be shown that their inclusion has virtually no effect as far as the translational symmetry is concerned.

The elements of the double group $\mathcal{T}^D$ may be denoted by $O([\mathbf{1} \mid \mathbf{t_n}])$ and $O([\bar{\mathbf{1}} \mid \mathbf{t_n}])$, where $\mathbf{t_n} = n_1\mathbf{a}_1 + n_2\mathbf{a}_2 + n_3\mathbf{a}_3$ and $0 \leq n_j < N_j$, $(j = 1, 2, 3)$, the Born cyclic boundary conditions implying that

$$O([\mathbf{1} \mid \mathbf{a}_j])^{N_j} = O([\mathbf{1} \mid \mathbf{0}])$$

for $j = 1$, 2 and 3 (cf. Equation (8.5)). It follows from Equations (6.32) and (6.33) that $\mathcal{T}^D$ is an Abelian group of order $2N$ $(=2N_1N_2N_3)$. As $\mathcal{T}^D$ is isomorphic to the direct product of the group of order 2 consisting of $[\mathbf{1} \mid \mathbf{0}]$ and $[\bar{\mathbf{1}} \mid \mathbf{0}]$ and the group $\mathcal{T}$, $\mathcal{T}^D$ possesses N "extra" irreducible representations (in the sense of Chapter 6, Section 4(*e*)). These are given by

$$\mathbf{\Gamma}^{\mathbf{k}}([\mathbf{1} \mid \mathbf{t_n}]) = -\mathbf{\Gamma}^{\mathbf{k}}([\bar{\mathbf{1}} \mid \mathbf{t_n}]) = [\exp(-i\mathbf{k} \cdot \mathbf{t_n})], \tag{8.28}$$

where $\mathbf{k}$ is an allowed $\mathbf{k}$-vector, defined exactly as in Equation (8.17).

It follows from Equations (6.28) that $O([\mathbf{1} \mid \mathbf{t_n}])\boldsymbol{\psi}(\mathbf{r}) = \boldsymbol{\psi}(\mathbf{r} - \mathbf{t_n})$ and $O([\bar{\mathbf{1}} \mid \mathbf{t_n}])\boldsymbol{\psi}(\mathbf{r}) = -\boldsymbol{\psi}(\mathbf{r} - \mathbf{t_n})$, thereby showing again that two-component spinor functions can transform as bases *only* for the *extra* representations of $\mathcal{T}^D$. If $\boldsymbol{\phi}_1^{\mathbf{k}}(\mathbf{r})$ is a two-component spinor basis function transforming as the first (and only) row of $\mathbf{\Gamma}^{\mathbf{k}}$ of Equation (8.28), then Equation (6.43) implies

$$O([\mathbf{1} \mid \mathbf{t_n}) \boldsymbol{\phi}_1^{\mathbf{k}}(\mathbf{r}) (= \boldsymbol{\phi}_1^{\mathbf{k}}(\mathbf{r} - \mathbf{t_n})) = \exp(-i\mathbf{k} \cdot \mathbf{t_n}) \boldsymbol{\phi}_1^{\mathbf{k}}(\mathbf{r}),$$

so that

$$\boldsymbol{\phi}_1^{\mathbf{k}}(\mathbf{r}) = \mathbf{u_k}(\mathbf{r}) \exp(i\mathbf{k} \cdot \mathbf{r}),$$

where $\mathbf{u_k}(\mathbf{r})$ is a *two-component spinor* having the periodicity of the lattice. This is the extension of the theorem of Bloch (1928) for spinors. It shows that a two-component spinor that is an energy eigenfunction has components that are Bloch functions. All the consequences of Bloch's Theorem that are described in Sections 4 and 5 still apply, except that each energy level $\varepsilon_n(\mathbf{k})$ can now "hold" only one electron and so each energy band can hold only N electrons. The concept of the Brillouin zone remains valid.

Time-reversal symmetry again implies the degeneracy expressed in Equation (8.27), but when the extra rotational symmetries of the space group are taken into account the effect is more profound than in the spin-independent situation (cf. Cornwell 1969).

7 Lattice vibrations

As noted in Section 2(*b*), the vibrations of the nuclei of a crystalline solid with Born cyclic boundary conditions imposed can be incorporated into the general formulation of vibrational problems given in Chapter 7, Section 3, provided that the invariance group $\mathscr{G}$ is taken to be $\mathscr{T}$, and that A, the number of relevant nuclei, is taken to be $N_1N_2N_3S$, S being the number of nuclei per lattice site.

The detailed investigation of lattice vibrations will begin with a study of the displacement representation Γ^{disp}. The number of nuclei that are unmoved by the transformation associated with $\{\mathbf{1}\mid \mathbf{t_n}\}$, $u(\{\mathbf{1}\mid \mathbf{t_n}\})$, is related to the corresponding permutation matrix $\mathbf{p}(\{\mathbf{1}\mid \mathbf{t_n}\})$ by $u(\{\mathbf{1}\mid \mathbf{t_n}\}) = \mathrm{tr}\,\mathbf{p}(\{\mathbf{1}\mid \mathbf{t_n}\})$. Equations (8.11) then show that, for $\{\mathbf{1}\mid \mathbf{t_n}\}\in\mathscr{T}$,

$$u(\{\mathbf{1}\mid \mathbf{t_n}\}) = \begin{cases} N_1N_2N_2S, & \text{if } \mathbf{t_n}=\mathbf{0},\\ 0, & \text{if } \mathbf{t_n}\neq\mathbf{0}.\end{cases}$$

By Equation (7.22) the character of Γ^{disp} for $\{\mathbf{1}\mid \mathbf{t_n}\}\in\mathscr{T}$ is then

$$\chi^{disp}(\{\mathbf{1}\mid \mathbf{t_n}\}) = \begin{cases} 3N_1N_2N_3S, & \text{if } \mathbf{t_n}=\mathbf{0},\\ 0, & \text{if } \mathbf{t_n}\neq\mathbf{0}.\end{cases}$$

Equations (7.31) and (8.18) then show that *every* irreducible representation $\Gamma^{\mathbf{k}}$ of $\mathscr{T}$ occurs exactly $3S$ times in the reduction of Γ^{disp}. Thus, to every allowed $\mathbf{k}$-vector there correspond $3S$ eigenvalues $\lambda_n(\mathbf{k})$ ($n = 1, 2, \ldots, 3S$) of the force matrix $\mathbf{F}$, which may be found by solving a $3S\times 3S$ secular equation of the form of Equation (7.30) (with $\Gamma^{p} = \Gamma^{\mathbf{k}}$).

The concept of the Brillouin zone may be introduced exactly as in Section 4. *To each allowed* $\mathbf{k}$*-vector of the Brillouin zone there then corresponds* $3S$ *eigenvalues* $\lambda_n(\mathbf{k})$ *of* $\mathbf{F}$ *and hence* $3S$ *frequencies* $\omega_n(\mathbf{k})$, given by $\omega_n(\mathbf{k}) = \{\lambda_n(\mathbf{k})\}^{1/2}$, $n = 1, 2, \ldots, 3S$. These frequencies may be plotted in the same way as that described for electronic energy bands in Section 5.

As examples consider body-centred cubic iron and silicon, for which S is 1 and 2 respectively. The lattice vibrational spectra along the Δ axes as calculated by Brescansin *et al.* (1976) for iron and Martin (1969) for silicon are given in Figures 8.13 and 8.14. (The

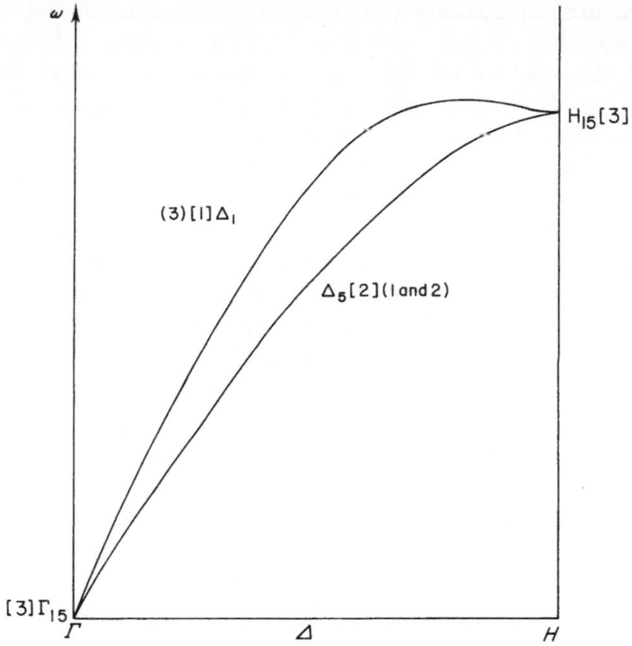

Figure 8.13 Part of the lattice vibrational spectrum of iron.

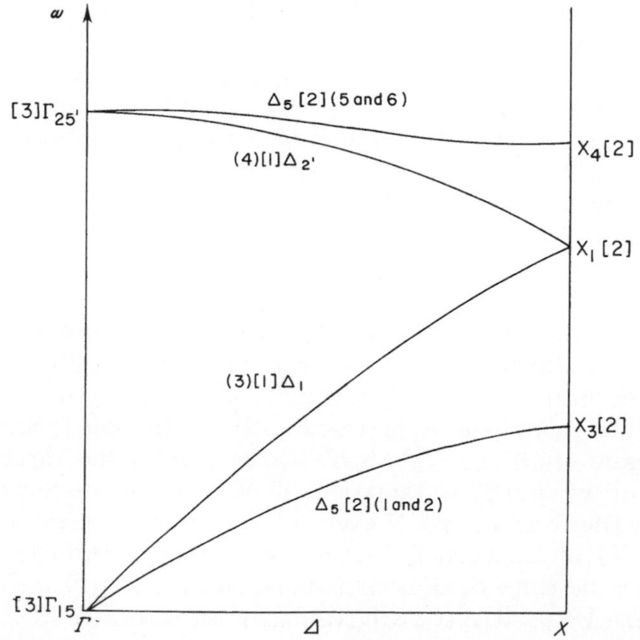

Figure 8.14 Part of the lattice vibrational spectrum of silicon.

number in curved brackets gives the band index n, defined by $\omega_n(\mathbf{k}) \leq \omega_{n+1}(\mathbf{k})$ ($n = 1, 2, \ldots, 3S$), and the number in square brackets indicates the degeneracy of the corresponding frequency. Again, the occurrence of degenerate eigenvalues is a consequence of the rotational symmetry, which is at present being neglected but which will be studied in Chapter 9. The other symbols will also be explained in Chapter 9.)

These figures exhibit the general property (which will be proved shortly) that there *always* exist three bands of frequencies $\omega_n(\mathbf{k})$ ($n = 1, 2, 3$) such that $\omega_n(\mathbf{k}) \to 0$ as $\mathbf{k} \to \mathbf{0}$. These correspond to the so-called "acoustic" modes of vibration. When $S > 1$, as it is for silicon, there exist other bands associated with the "optical" modes of vibration.

In the quantum mechanical treatment (cf. Chapter 7, Section 2(b)) consider the excitation of the nuclear system from one state to another state in which all the occupation numbers are the same, except for that associated with wave vector $\mathbf{k}$ and frequency $\omega_n(\mathbf{k})$, which may be assumed to be increased by one. This excitation may be looked upon as the creation of a pseudo-particle of energy $\hbar\omega_n(\mathbf{k})$ called a "phonon". Accordingly, the lattice vibrational spectra, such as those displayed in Figures 8.13 and 8.14, are often called "phonon dispersion curves". (For a much more detailed treatment of these and related matters the reader is referred to the works of Born and Huang (1954), Ziman (1960), Maradudin *et al.* (1963), Joshi and Rajagopal (1968) and Cochrane (1973).)

As noted in Chapter 7, Section 3(f), the reality of $\mathbf{F}$ can cause extra degeneracies. Equation (8.18) shows that the complex conjugate representation to $\mathbf{\Gamma}^{\mathbf{k}}$ is $\mathbf{\Gamma}^{-\mathbf{k}}$, and that both are essentially complex (except when $\mathbf{k} = \mathbf{0}$ or is one-half of some reciprocal lattice vector, that is, except at a few isolated points of the Brillouin zone). Consequently, as in Equation (8.27),

$$\omega_n(\mathbf{k}) = \omega_n(-\mathbf{k}), \qquad n = 1, 2, \ldots, 3S.$$

It remains only to consider the zero-frequency modes of vibration, the theory of which for molecules was given in Chapter 7, Section 3(h). When Born cyclic boundary conditions are imposed on a crystalline solid the conditions (A) of Theorem IV of Chapter 7, Section 3(h) remain valid (with $A = N_1N_2N_3S$). Consequently the three column matrices $\mathbf{c}_1^{tr}$, $\mathbf{c}_2^{tr}$ and $\mathbf{c}_3^{tr}$ of Equations (7.49) are zero-frequency eigenvectors of the force matrix $\mathbf{F}$ even for the crystalline solid case. By Theorem VI of Chapter 7, Section 3(h), they transform as basis vectors for the three-dimensional representation $\mathbf{\Gamma}^{tr}$ of $\mathcal{T}$ for which $\mathbf{\Gamma}^{tr}(\{\mathbf{1} \mid \mathbf{t_n}\}) = \mathbf{1}$ for all $\{\mathbf{1} \mid \mathbf{t_n}\} \in \mathcal{T}$. Thus, by Equation (8.18), $\mathbf{\Gamma}^{tr}$ is the direct sum in which the irreducible representation corresponding to $\mathbf{k}=\mathbf{0}$ occurs three times. Hence *for* $\mathbf{k}=\mathbf{0}$ *there are always three zero-frequency "translational" modes.*

The situation for the "rotational" modes is completely different, for the Born cyclic boundary conditions destroy the identities (B), (C) and (D) of Theorem IV of Chapter 7, Section 3(h). This is most easily seen by an example. Consider the "linear" model set up in Section 2(b) with $\mathbf{a}=(a,0,0)$ and the potential energy

$$V=\alpha[\{\mathbf{r}_1-\mathbf{r}_1^e)-(\mathbf{r}_2-\mathbf{r}_2^e)\}^2+\{(\mathbf{r}_2-\mathbf{r}_2^e)-(\mathbf{r}_3-\mathbf{r}_3^e)\}^2+\dots$$
$$\dots+\{(\mathbf{r}_{N-1}-\mathbf{r}_{N-1}^e)-(\mathbf{r}_N-\mathbf{r}_N^e)\}^2+\{(\mathbf{r}_N-\mathbf{r}_N^3)-(\mathbf{r}_1-\mathbf{r}_1^e)\}^2].$$

This is clearly invariant under the transformation in Expression (8.7) for every $n=1,2,\dots,N$. With all the nuclei assumed to have the same mass m, in terms of the generalized coordinates of Equations (7.14)

$$V=(\alpha/m)\left[\sum_{s=1}^{3}\sum_{j=1}^{N-1}(q_{3j-3+s}^2-2q_{3j-3+s}q_{3j+s}+q_{3j+s}^2)\right.$$
$$\left.+\sum_{s=1}^{3}(q_{3N-3+s}^2-2q_{3N-3+s}q_s+q_s^2)\right],$$

Comparison with the form of Equation (7.2) shows that

$$F_{mm}=4\alpha/m, \qquad \text{for } m=1,2,\dots,3N,$$

$$F_{m,m+3}=F_{m+3,m}=-2\alpha/m, \qquad \text{for } m=1,2,\dots,3N-3,$$

and

$$F_{1,3N-2}=F_{3N-2,1}=F_{2,3N-1}=F_{3N-1,2}=F_{3,3N}=F_{3N,3}=-2\alpha/m,$$

all other matrix elements of $\mathbf{F}$ being zero. Direct substitution shows that the conditions (A) of Theorem IV of Chapter 7, Section 3(h) remain valid. However the left-hand side of (C) with $m=3$ becomes

$$\sum_{j=1}^{N}m(-ja)F_{3,3j}=-ma[F_{33}+2F_{36}+NF_{3,3N}]=2a\alpha N\neq 0,$$

so that the condition (C) is no longer valid. The same is true of condition (D). Although (B) is valid, every term on its left-hand side being zero, it leads to a trivial $\mathbf{c}^{\text{rot}}$ that is identically zero. Thus there are *no* "rotational" zero-frequency modes after the Born cyclic boundary conditions have been applied.

9

The Crystallographic Space Groups

1 A survey of the crystallographic space groups

Consider an infinite crystalline solid for which the equilibrium positions of the nuclei are given by Equation (8.1). The set of all coordinate transformations that map the set of equilibrium positions into itself forms an infinite group $\mathcal{G}^{\infty}$ that is known as a "crystallographic space group". Clearly $\mathcal{G}^{\infty}$ contains as a subgroup the infinite group of pure primitive translations $\mathcal{T}^{\infty}$ of the relevant lattice, but contains no other pure translations.

If $\{\mathbf{R}\,|\,\mathbf{t}\}$ is a member of the space group $\mathcal{G}^{\infty}$ and $\mathbf{t_n}$ is any lattice vector of its lattice, then $\mathbf{Rt_n}$ must also be a lattice vector. (This follows because (by Equations (1.7) and (1.8)) $\{\mathbf{R}\,|\,\mathbf{t}\}\{\mathbf{1}\,|\,\mathbf{t_n}\}\{\mathbf{R}\,|\,\mathbf{t}\}^{-1} = \{\mathbf{1}\,|\,\mathbf{Rt_n}\}$, so that $\{\mathbf{1}\,|\,\mathbf{Rt_n}\}$ must be a member of $\mathcal{G}^{\infty}$ and, being a pure translation, must be a primitive translation. Thus the set of all rotational parts $\mathbf{R}$ of the space group operations $\{\mathbf{R}\,|\,\mathbf{t}\}$ form a subgroup $\mathcal{G}_0$ of the maximal point group $\mathcal{G}_0^{max}$ of its crystal lattice (though $\mathcal{G}_0$ need not be a proper subgroup of $\mathcal{G}_0^{max}$). $\mathcal{G}_0$ is known as the "point group of the space group".

Detailed investigations show that the only possible proper rotations of $\mathcal{G}_0$ are through multiples of $2\pi/6$ or $2\pi/4$ and the only possible improper rotations are products of these proper rotations with the spatial inversion operator. Consequently $\mathcal{G}_0$ is always a finite group, and must be one of the 32 crystallographic point groups that are specified in detail in Appendix D.

Space groups having the same point group $\mathcal{G}_0$ are said to belong to the same "crystal class", so there are 32 different crystal classes. In the classification of space groups by Schönfliess (1923), each space group is denoted by the Schönfliess symbol for its point group $\mathscr{G}_0$, together with a superscript. Thus, for example, the space groups for crystals possessing the cubic Bravais lattices Γ_c, Γ_c^v and Γ_c^f, and having only nuclei at the lattice points, all have O_h as point group, and are denoted by O_h^1, O_h^9 and O_h^5 respectively. Similarly, the space group of the diamond structure, which also has the Bravais lattice Γ_c^f and point group O_h but which has two nuclei per lattice site, is denoted by O_h^7. As will be seen, the assignment of superscripts by Schönfliess is rather arbitrary. An alternative that is more explicit but more complicated is the "international notation" (see Henry and Lonsdale 1965, Shubnikov and Koptsik 1974, Burns and Glazer 1978).

To every rotation $\{\mathbf{R} \mid \mathbf{0}\}$ of $\mathscr{G}_0$ there exists a vector $\boldsymbol{\tau}_{\mathbf{R}}$ such that $\{\mathbf{R} \mid \boldsymbol{\tau}_{\mathbf{R}} + \mathbf{t}_{\mathbf{n}}\}$ is a member of $\mathscr{G}^\infty$ for every lattice vector $\mathbf{t}_{\mathbf{n}}$ of the lattice. Moreover, $\boldsymbol{\tau}_{\mathbf{R}}$ is unique (up to a lattice vector), as if $\{\mathbf{R} \mid \boldsymbol{\tau}_{\mathbf{R}}\}$ and $\{\mathbf{R} \mid \boldsymbol{\tau}'_{\mathbf{R}}\}$ are both members of $\mathscr{G}^\infty$ then so must be $\{\mathbf{R} \mid \boldsymbol{\tau}_{\mathbf{R}}\}\{\mathbf{R} \mid \boldsymbol{\tau}'_{\mathbf{R}}\}^{-1} = \{\mathbf{1} \mid \boldsymbol{\tau}_{\mathbf{R}} - \boldsymbol{\tau}'_{\mathbf{R}}\}$, which, being a pure translation, is bound to be a primitive translation, so that $\boldsymbol{\tau}'_{\mathbf{R}} = \boldsymbol{\tau}_{\mathbf{R}} + \mathbf{t}_{\mathbf{n}}$ for some $\mathbf{t}_{\mathbf{n}}$. For definiteness it will be assumed henceforth that $\boldsymbol{\tau}_{\mathbf{R}}$ is always chosen so that $\boldsymbol{\tau}_{\mathbf{R}} = q_1\mathbf{a}_1 + q_2\mathbf{a}_2 + q_3\mathbf{a}_3$ with $0 \leq q_j < 1$, $j = 1, 2, 3$.

If $\boldsymbol{\tau}_{\mathbf{R}} = \mathbf{0}$ for *every* $\{\mathbf{R} \mid \mathbf{0}\} \in \mathscr{G}_0$, then $\mathscr{G}^\infty$ is said to be a "symmorphic" space group. That is, for a symmorphic space group *every* transformation is of the form $\{\mathbf{R} \mid \mathbf{t}_{\mathbf{n}}\}$, the translational part *always* being a *lattice vector*. Obviously if $\mathscr{G}^\infty$ is symmorphic then $\mathscr{G}_0$ is a subgroup of $\mathscr{G}^\infty$. Only 73 of the 230 crystallographic space groups are symmorphic. Important examples include the cubic space groups O_h^1, O_h^5 and O_h^9. As symmorphic space groups are the easiest type to study, their representations will be considered first (in Section 2).

If $\mathscr{G}^\infty$ is non-symmorphic, then for some $\{\mathbf{R} \mid \mathbf{0}\}$ of $\mathscr{G}_0$ there exists a *non-zero* $\boldsymbol{\tau}_{\mathbf{R}}$. Thus, if $\mathscr{G}^\infty$ is non-symmorphic, $\mathscr{G}_0$ is *not* a subgroup of $\mathscr{G}^\infty$. The representations of non-symmorphic space groups will be considered in Section 3, the space group O_h^7 of the diamond structure being treated as an example.

If a crystal has nuclei of only one species and their equilibrium positions lie only at the lattice points of a Bravais lattice, then the corresponding space group is symmorphic. The same is true if the crystal has more than one species of nuclei and if the arrays of nuclei of each species each form a Bravais lattice. (This is only possible if there is only one nucleus of each species per lattice site.) By contrast, non-symmorphic space groups are associated with crystals in which there are more than one nuclei of a given species per lattice site.

The general relationship between $\mathcal{G}^\infty$, $\mathcal{T}^\infty$ and $\mathcal{G}_0$ is summarized by the following theorem.

Theorem I $\mathcal{T}^\infty$ is an invariant subgroup of $\mathcal{G}^\infty$, and the factor group $\mathcal{G}^\infty/\mathcal{T}^\infty$ is isomorphic to $\mathcal{G}_0$.

Proof If $\{\mathbf{R}\mid\mathbf{t}\}$ is any member of $\mathcal{G}^\infty$ and $\{\mathbf{1}\mid\mathbf{t_n}\}$ is any member of $\mathcal{T}^\infty$, then $\{\mathbf{R}\mid\mathbf{t}\}\{\mathbf{1}\mid\mathbf{t_n}\}\{\mathbf{R}\mid\mathbf{t}\}^{-1}$ is a pure translation, and so it must be a member of $\mathcal{T}^\infty$. Thus $\mathcal{T}^\infty$ is an invariant subgroup of $\mathcal{G}^\infty$.

The right coset $\mathcal{T}^\infty\{\mathbf{R}\mid\boldsymbol{\tau}_\mathbf{R}\}$ comprises all transformations of the form $\{\mathbf{R}\mid\mathbf{t_n}+\boldsymbol{\tau}_\mathbf{R}\}$, that is, all the transformations of $\mathcal{G}^\infty$ having rotational part $\mathbf{R}$. Let $\Psi(\mathcal{T}^\infty\{\mathbf{R}\mid\boldsymbol{\tau}_\mathbf{R}\})=\{\mathbf{R}\mid\mathbf{0}\}$ be the one-to-one mapping of $\mathcal{G}^\infty/\mathcal{T}^\infty$ onto $\mathcal{G}_0$. Then, for any $\{\mathbf{R}\mid\boldsymbol{\tau}_\mathbf{R}\}$ and $\{\mathbf{R}'\mid\boldsymbol{\tau}_{\mathbf{R}'}\}$ of $\mathcal{G}^\infty$,

$$\Psi(\mathcal{T}^\infty\{\mathbf{R}\mid\boldsymbol{\tau}_\mathbf{R}\})\Psi(\mathcal{T}^\infty\{\mathbf{R}'\mid\boldsymbol{\tau}_{\mathbf{R}'}\})=\{\mathbf{R}\mid\mathbf{0}\}\{\mathbf{R}'\mid\mathbf{0}\}=\{\mathbf{RR}'\mid\mathbf{0}\}.$$

However, by Equation (2.6), $\mathcal{T}^\infty\{\mathbf{R}\mid\boldsymbol{\tau}_\mathbf{R}\}\mathcal{T}^\infty\{\mathbf{R}'\mid\boldsymbol{\tau}_{\mathbf{R}'}\}=\mathcal{T}^\infty\{\mathbf{RR}'\mid\mathbf{R}\boldsymbol{\tau}_{\mathbf{R}'}+\boldsymbol{\tau}_\mathbf{R}\}$, so that

$$\Psi(\mathcal{T}^\infty\{\mathbf{R}\mid\boldsymbol{\tau}_\mathbf{R}\}\mathcal{T}^\infty\{\mathbf{R}'\mid\boldsymbol{\tau}_{\mathbf{R}'}\})=\{\mathbf{RR}'\mid\mathbf{0}\},$$

and hence Ψ is an *isomorphic* mapping of $\mathcal{G}^\infty/\mathcal{T}^\infty$ onto $\mathcal{G}_0$.

A complete description of all 230 space groups may be found in the "International Tables for X-ray Crystallography" (Henry and Lonsdale 1965), which employ both the Schönfliess and international notations. (A simple prescription for determining the symmetry elements of a space group from the "general position" listed in the International Tables has been given by Wondratschek and Neubüser (1967).) Another complete specification that is particularly clear and thorough has been given by Shubnikov and Koptsik (1974). A further clear and comprehensive description of the crystallographic space groups has been given by Burns and Glazer (1978). Lists of the space groups to which elements and compounds belong have been compiled by Wyckoff (1963, 1964, 1965) and by Donnay and Nowacki (1954).

The effect of imposing Born cyclic boundary conditions is to replace the infinite space group $\mathcal{G}^\infty$ by a *finite* group $\mathcal{G}$. As a preliminary, let $N_1=N_2=N_3$, where N_1, N_2, N_3 are as defined in Equations (8.2). (This ensures for every $\{\mathbf{R}\mid\mathbf{0}\}\in\mathcal{G}_0$ that $\mathbf{R}(N_j\mathbf{a}_j)=l_1N_1\mathbf{a}_1+l_2N_2\mathbf{a}_2+l_3N_3\mathbf{a}_3$ for some *integers* l_1, l_2 and l_3 and each $j=1,2,3$.)

Consider first the case of electronic wave functions. In this case $\mathcal{G}$ may be defined to be the group of operators $P(\{\mathbf{R}\mid\boldsymbol{\tau}_\mathbf{R}+\mathbf{t_n}\})$ for all $\{\mathbf{R}\mid\mathbf{0}\}\in\mathcal{G}_0$ and all lattice vectors $\mathbf{t_n}$ of the finite group $\mathcal{T}$ with Equation (8.5) applied. Then $\mathcal{G}$ has order g_0N, where g_0 and N are the orders of $\mathcal{G}_0$ and $\mathcal{T}$ respectively.

As it remains true that $P(\{\mathbf{R}\mid\mathbf{t}\})P(\{\mathbf{R}'\mid\mathbf{t}'\})=P(\{\mathbf{R}\mid\mathbf{t}\}\{\mathbf{R}'\mid\mathbf{t}'\})$ for every $\{\mathbf{R}\mid\mathbf{t}\}$ and $\{\mathbf{R}'\mid\mathbf{t}'\}$ of $\mathscr{G}$, the mapping $\Phi(\{\mathbf{R}\mid\mathbf{t}\})=P(\{\mathbf{R}\mid\mathbf{t}\})$ is a homomorphic mapping of $\mathscr{G}^{\infty}$ onto $\mathscr{G}$. The kernel $\mathscr{K}$ of this mapping is again the infinite set of pure primitive translations of the form $\{\mathbf{1}\mid l_1N_1\mathbf{a}_1+l_2N_2\mathbf{a}_2+l_3N_3\mathbf{a}_3\}$, where l_1, l_2, l_3 are any set of integers.

With the inner product defined as in Chapter 8, Section 2(a), i.e. to involve an integral over just one basic block of the crystal, B, the operators $P(\{\mathbf{R}\mid\mathbf{t}\})$ retain the unitary property of Equation (1.20), provided all the functions on which they act satisfy the Born cyclic boundary conditions (Equations (8.2)).

For the study of lattice vibrations the finite space group may be taken to be the set of mappings associated with $\{\mathbf{R}\mid\boldsymbol{\tau}_{\mathbf{R}}+\mathbf{t}_{\mathbf{n}}\}$ for all $\{\mathbf{R}\mid\mathbf{0}\}\in\mathscr{G}_0$ and $\mathbf{t}_{\mathbf{n}}\in\mathscr{T}$. The generalization of Equation (8.10) is

$$\mathbf{r}^{e}_{\mathbf{n}''\gamma''}=\{\mathbf{R}\mid\boldsymbol{\tau}_{\mathbf{R}}+\mathbf{t}_{\mathbf{n}}-l_1N_1\mathbf{a}_1-l_2N_2\mathbf{a}_2-l_3N_3\mathbf{a}_3\}\mathbf{r}^{e}_{\mathbf{n}'\gamma'}\tag{9.1}$$

where $\mathbf{r}^{e}_{\mathbf{n}'\gamma'}=\mathbf{t}_{\mathbf{n}'}+\boldsymbol{\tau}_{\gamma'}$, $\mathbf{r}^{e}_{\mathbf{n}''\gamma''}=\mathbf{t}_{\mathbf{n}''}+\boldsymbol{\tau}_{\gamma''}$, and l_1, l_2 and l_3 are integers chosen so that $\mathbf{t}_{\mathbf{n}''}$ is within the basic block whenever $\mathbf{t}_{\mathbf{n}'}$ is in that block. This group is isomorphic to the group $\mathscr{G}$ just defined above and no confusion will be caused if it too is denoted by $\mathscr{G}$.

By construction, the finite space group $\mathscr{G}$ contains the same point group $\mathscr{G}_0$ as its infinite counterpart $\mathscr{G}^{\infty}$. This gives a finite analogue for the previous theorem (the proof being essentially identical).

Theorem II $\mathscr{T}$ is an invariant subgroup of $\mathscr{G}$ and the factor group $\mathscr{G}/\mathscr{T}$ is isomorphic to $\mathscr{G}_0$.

The following sections will be devoted to the study of the representations of the finite space groups $\mathscr{G}$ and of their consequences. In this context no confusion will be caused if the Schönfliess or international notations are applied to the finite space groups as well as to the corresponding infinite groups.

An identity that is worth noting is

$$\mathbf{t}\,.\,(\mathbf{R}\mathbf{k})=(\mathbf{R}^{-1}\mathbf{t})\,.\,\mathbf{k},\tag{9.2}$$

which is valid for *any* rotation $\mathbf{R}$ and any vectors $\mathbf{t}$ and $\mathbf{k}$. (As $\mathbf{R}$ is a 3×3 orthogonal matrix (see Chapter 1, Section 2(a)), $\mathbf{t}\,.\,(\mathbf{R}\mathbf{k})=\sum_{i=1}^{3}t_i(\mathbf{R}\mathbf{k})_i=\sum_{i,j=1}^{3}t_iR_{ij}k_j=\sum_{i,j=1}^{3}(\mathbf{R}^{-1})_{ji}t_ik_j=(\mathbf{R}^{-1}\mathbf{t})\,.\,\mathbf{k}$).

This implies that the reciprocal lattice (as defined in Chapter 8, Section 4) has the *same* symmetry as the crystal lattice to which it belongs. This is shown by the following theorem.

Theorem III If $\{\mathbf{R}\mid\mathbf{0}\}\in\mathscr{G}_0$ and $\mathbf{K}_{\mathbf{m}}$ is a lattice vector of the reciprocal lattice, then $\mathbf{R}\mathbf{K}_{\mathbf{m}}$ is also a lattice vector of the reciprocal lattice.

Proof Suppose that $\mathbf{RK_m}$ is *not* a lattice vector of the reciprocal lattice. Then there exists a lattice vector $\mathbf{t_n}$ of the crystal lattice such that $\exp\{i\mathbf{t_n}\cdot(\mathbf{RK_m})\} \neq 1$. Thus, by Equation (9.2), $\exp\{i(\mathbf{R}^{-1}\mathbf{t_n})\cdot\mathbf{K_m}\} \neq 1$. However, $\{\mathbf{R}^{-1}\,|\,\mathbf{0}\}$ is a member of $\mathcal{G}_0$, so $\mathbf{R}^{-1}\mathbf{t_n}$ must be a lattice vector of the crystal lattice. Equation (8.23) then provides a contradiction.

2 Irreducible representations of symmorphic space groups

(a) *The fundamental theorem on irreducible representations of symmorphic space groups*

The first stage in the analysis of symmorphic space groups is to observe that they possess a particularly straightforward structure.

Theorem I If $\mathcal{G}$ is a *symmorphic* space group, then $\mathcal{G}$ is isomorphic to the semi-direct product $\mathcal{T}$Ⓢ$\mathcal{G}_0$.

Proof All that has to be verified is that the three requirements of the definition of Chapter 2, Section 7, are satisfied. Firstly, $\mathcal{T}$ is certainly an invariant subgroup of $\mathcal{G}$ (by Theorem II of Section 1 above). Clearly $\mathcal{T}$ and $\mathcal{G}_0$ have only the identity in common. Finally, if $\mathcal{G}$ is symmorphic, every element of $\mathcal{G}$ is the product of a pure primitive translation of $\mathcal{T}$ with a pure rotation of $\mathcal{G}_0$.

As $\mathcal{T}$ is Abelian, the theory of induced representations given in Chapter 5, Section 7, can be applied to produce *all* the irreducible representations of $\mathcal{G}$. Moreover, all the "little groups" from which the irreducible representations of $\mathcal{G}$ are induced are subgroups of $\mathcal{G}_0$, and hence every "little group" is a crystallographic point group. As all the irreducible representations of the crystallographic point groups are known (and are listed in Appendix D), the irreducible representations of the space group $\mathcal{G}$ follow immediately.

In applying the results of the induced representation theory it is very helpful to re-cast some of the concepts in terms of the geometric picture of the Brillouin zone and its allowed **k**-vectors developed in Chapter 8. The quantities that will now be introduced will be identified in the proof of the fundamental theorem with certain of the entities of the induced representation theory.

Definition $\mathcal{G}_0(\mathbf{k})$, *the point group of the allowed wave vector* **k**
The point group $\mathcal{G}_0(\mathbf{k})$ is the *subgroup* of the point group $\mathcal{G}_0$ of the

space group $\mathscr{G}$ that consists of all the rotations $\{\mathbf{R}\,|\,\mathbf{0}\}$ of $\mathscr{G}_0$ that rotate $\mathbf{k}$ into itself or an "equivalent" vector (in the sense of Equation (8.25)). That is, $\{\mathbf{R}\,|\,\mathbf{0}\}$ of $\mathscr{G}_0$ is a member of $\mathscr{G}_0(\mathbf{k})$ if there exists a lattice vector $\mathbf{K}_\mathbf{m}$ of the reciprocal lattice (as defined in Equation (8.22)), which may be zero, such that

$$\mathbf{Rk} = \mathbf{k} + \mathbf{K}_\mathbf{m}, \tag{9.3}$$

Let g_0 and $g_0(\mathbf{k})$ be the orders of $\mathscr{G}_0$ and $\mathscr{G}_0(\mathbf{k})$ respectively, and let

$$M(\mathbf{k}) = g_0/g_0(\mathbf{k}). \tag{9.4}$$

Then (see Chapter 2, Section 4) $M(\mathbf{k})$ is always an *integer*.

Definition *General points, symmetry points, symmetry axes and symmetry planes of the Brillouin zone*

If $\mathscr{G}_0(\mathbf{k})$ is the trivial group consisting only of the identity transformation $\{\mathbf{1}\,|\,\mathbf{0}\}$, then $\mathbf{k}$ is said to be a "general point" of the Brillouin zone. If $\mathbf{k}$ is such that $\mathscr{G}_0(\mathrm{k})$ is a larger group than those corresponding to *all* neighbouring points of the Brillouin zone, then $\mathbf{k}$ is known as a "symmetry point". If all the points on a line or plane have the same *non-trivial* $\mathscr{G}_0(\mathbf{k})$, then this line or plane is said to be a "symmetry axis" or "symmetry plane".

It is easily shown that this list of definitions exhausts all the possible situations.

Example I *Symmetry points, axes and planes for the space group* O_h^9

The Brillouin zone of the body-centred cubic lattice Γ_c^v is shown in Figure 8.9 in the previous chapter. For the space group O_h^9 with this lattice the symmetry points are Γ, H, N and P, and the symmetry axes are Δ, Λ, Σ, D, F and G, and the symmetry planes are the planes containing two symmetry axes.

Example II *Symmetry points, axes and planes for the space group* O_h^5

The Brillouin zone of the face-centred cubic lattice Γ_c^f is given in Figure 8.10. For the space group O_h^5 with this lattice the symmetry points are Γ, L, W and X, the symmetry axes are Δ, Λ, Σ, Q, S and Z, and the symmetry planes are ΓKWX, ΓKL, ΓLUX and UWX. The point K is *not* a symmetry point because its $\mathscr{G}_0(\mathbf{k})$ is the same as that for the axis Σ that ends at K. For the same reason U is *not* a symmetry point. It should be noted also that $KLUW$ is not a symmetry plane. Finally, the lines LK, KW, LU and UW have only the symmetry of the symmetry planes to which they belong, so they are not regarded as symmetry axes. (These considerations apply equally to the space group O_h^7: see Example I of Section 3 below.)

Definition *The "star" of* $\mathbf{k}$
Let $\{\mathbf{R}_j \mid \mathbf{0}\}$, $j = 1, 2, \ldots, M(\mathbf{k})$, be a set of coset representatives for the decomposition of $\mathcal{G}_0$ into *left* cosets with respect to $\mathcal{G}_0(\mathbf{k})$ (see Chapter 2, Section 4). Then the set of $M(\mathbf{k})$ vectors $\mathbf{k}_j$ defined by $\mathbf{k}_j = \mathbf{R}_j\mathbf{k}$ ($j = 1, 2, \ldots, M(\mathbf{k})$) is called the "star" of $\mathbf{k}$.

Of course *any* member of a left coset can be chosen to be the coset representative, but once a choice is made it should be adhered to. If $\{\mathbf{R}'_j \mid \mathbf{0}\} \in \{\mathbf{R}_j \mid \mathbf{0}\}\mathcal{G}_0(\mathbf{k})$, then there exists an element $\{\mathbf{R} \mid \mathbf{0}\}$ of $\mathcal{G}_0(\mathbf{k})$ such that $\mathbf{R}'_j = \mathbf{R}_j\mathbf{R}$. Then $\mathbf{R}'_j\mathbf{k} = \mathbf{R}_j(\mathbf{R}\mathbf{k}) = \mathbf{R}_j\mathbf{k} + \mathbf{R}_j\mathbf{K}_\mathbf{m}$ (by Equation (9.3)), so that $\mathbf{R}'_j\mathbf{k}$ is equivalent to $\mathbf{R}_j\mathbf{k}$, as $\mathbf{R}_j\mathbf{K}_\mathbf{m}$ is a reciprocal lattice vector. Thus a different choice of the coset representatives merely results in a set of vectors that are equivalent to those of the set $\mathbf{k}_1, \mathbf{k}_2, \ldots$ Consequently the star of $\mathbf{k}$ is unique up to equivalence. It is always convenient to choose $\mathbf{R}_1 = \mathbf{1}$, so that $\mathbf{k}_1 = \mathbf{k}$.

The fundamental theorem on the irreducible representations of a symmorphic space group can now be presented.

Theorem II Let $\mathbf{k}$ be any allowed $\mathbf{k}$-vector of the Brillouin zone and let $\{\mathbf{R}_j \mid \mathbf{0}\}$ ($j = 1, 2, \ldots, M(\mathbf{k})$) be a set of coset representatives for the decomposition of $\mathcal{G}_0$ into left cosets with respect to $\mathcal{G}_0(\mathbf{k})$. Let $\Gamma^p_{\mathcal{G}_0(\mathbf{k})}$ be a unitary irreducible representation of $\mathcal{G}_0(\mathbf{k})$, assumed to be of dimension d_p. Then there exists a corresponding unitary irreducible representation of the space group $\mathcal{G}$ of dimension $d_pM(\mathbf{k})$, which may be denoted by $\Gamma^{\mathbf{k}p}$, such that

$$\Gamma^{\mathbf{k}p}(\{\mathbf{R} \mid \mathbf{t_n}\})_{kt,jr} = \begin{cases} \exp\{-i(\mathbf{R}_k\mathbf{k}) \cdot \mathbf{t_n}\}\Gamma^p_{\mathcal{G}_0(\mathbf{k})}(\{\mathbf{R}_k^{-1}\mathbf{R}\mathbf{R}_j \mid \mathbf{0}\})_{tr}, \\ \qquad \text{if } \{\mathbf{R}_k^{-1}\mathbf{R}\mathbf{R}_j \mid \mathbf{0}\} \in \mathcal{G}_0(\mathbf{k}), \\ 0, \qquad \text{otherwise}, \end{cases} \tag{9.5}$$

for $j, k = 1, 2, \ldots, M(\mathbf{k})$, and $r, t = 1, 2, \ldots, d_p$. (Here each row and column is specified by a *pair* of indices.) Moreover, *all* the inequivalent irreducible representations of $\mathcal{G}$ may be obtained in this way by working through all the inequivalent irreducible representations of $\mathcal{G}_0(\mathbf{k})$ for all allowed $\mathbf{k}$-vectors that are in different stars.

Proof All that is required is to identify the concepts introduced above with those developed for induced representations in Chapter 5, Section 7. The required result is then an immediate consequence of Theorem II of Chapter 5, Section 7.

Theorem I above showed that $\mathcal{G}$ is isomorphic to $\mathcal{T} \circledS \mathcal{G}_0$, so the groups $\mathcal{A}$ and $\mathcal{B}$ can be identified with $\mathcal{T}$ and $\mathcal{G}_0$ respectively. As the characters of $\mathcal{T}$ are specified by the allowed $\mathbf{k}$-vectors and are given by Equation (8.18), the label q may be identified with $\mathbf{k}$. Thus with $\mathrm{A} = \{\mathbf{1} \mid \mathbf{t_n}\}$

$$\chi^q_{\mathscr{A}}(\mathrm{A}) = \exp(-i\mathbf{k} \cdot \mathbf{t_n}). \tag{9.6}$$

With $\mathrm{B} = \{\mathbf{R} \mid \mathbf{0}\}$, $\mathrm{BAB}^{-1} = \{\mathbf{1} \mid \mathbf{Rt_n}\}$, so $\chi^q_{\mathscr{A}}(\mathrm{BAB}^{-1}) = \exp\{-i\mathbf{k} \cdot (\mathbf{Rt_n})\} = \exp\{-i(\mathbf{R}^{-1}\mathbf{k}) \cdot \mathbf{t_n}\}$ (by Equation (9.2)). Thus Equation (5.51) implies that the subgroup $\mathscr{B}(q)$ of $\mathscr{B}$ is merely $\mathscr{G}_0(\mathbf{k})$, that is, *the* $\mathscr{G}_0(\mathbf{k})$ *are the "little groups"*. Clearly $b = g_0$, $b(q) = g_0(\mathbf{k})$, and $M(q) = M(\mathbf{k})$.

If $\{\mathbf{R}_j \mid \mathbf{0}\}$ $(j = 1, 2, \ldots, M(\mathbf{k}))$ are the coset representatives for the decomposition of $\mathscr{G}_0$ into *left* cosets with respect to $\mathscr{G}_0(\mathbf{k})$, then $\{\mathbf{R}_j \mid \mathbf{0}\}^{-1} (j = 1, 2, \ldots, M(\mathbf{k}))$ may be taken as the coset representatives for the decomposition of $\mathscr{G}_0$ into *right* cosets with respect to $\mathscr{G}_0(\mathbf{k})$. (This follows because, if $\{\mathbf{R}_j \mid \mathbf{0}\}^{-1}$ and $\{\mathbf{R}_k \mid \mathbf{0}\}^{-1}$ belong to the same right coset, there exists an $\{\mathbf{R} \mid \mathbf{0}\}^{-1}$ of $\mathscr{G}_0$ such that $\{\mathbf{R}_k \mid \mathbf{0}\}^{-1} = \{\mathbf{R} \mid \mathbf{0}\}\{\mathbf{R}_j \mid \mathbf{0}\}^{-1}$. Then $\{\mathbf{R}_k \mid \mathbf{0}\} = \{\mathbf{R}_j \mid \mathbf{0}\}\{\mathbf{R} \mid \mathbf{0}\}^{-1}$, which implies that $\{\mathbf{R}_j \mid \mathbf{0}\}$ and $\{\mathbf{R}_k \mid \mathbf{0}\}$ belong to the same left coset.) Thus the coset representatives B_j $(j = 1, 2, \ldots, M(q))$ for the decomposition of $\mathscr{B}$ into right cosets with respect to $\mathscr{B}(q)$ may be identified with $\{\mathbf{R}_j \mid \mathbf{0}\}^{-1}$ $(j = 1, 2, \ldots, M(\mathbf{k}))$.

Then, by Equations (5.52), (9.2) and (9.6), with $\mathrm{A} = \{\mathbf{1} \mid \mathbf{t_n}\}$,

$$\chi^{\mathrm{B}_j(q)}_{\mathscr{A}}(\mathrm{A}) = \exp\{-i(\mathbf{R}_j\mathbf{k}) \cdot \mathbf{t_n}\}, \tag{9.7}$$

and Equations (5.55) reduce to Equations (9.5). Finally, the *orbit* of q is obviously merely the *star* of $\mathbf{k}$.

This theorem removes the need for an explicit display of the character table for $\mathscr{G}$, which would in any case be very difficult because of the vast number of irreducible representations that $\mathscr{G}$ possesses. All the information required about $\mathscr{G}$ on such things as basis functions and degeneracies is immediately provided by the theorem in terms of the corresponding quantities for crystallographic point groups, which are very easily obtained.

The character $\chi^{\mathbf{k}p}(\{\mathbf{R} \mid \mathbf{t_n}\})$ of $\{\mathbf{R} \mid \mathbf{t_n}\}$ in the irreducible representation $\boldsymbol{\Gamma}^{\mathbf{k}p}$ of $\mathscr{G}$ is given by

$$\chi^{\mathbf{k}p}(\{\mathbf{R} \mid \mathbf{t_n}\}) = \sum_j \exp\{-i(\mathbf{R}_j\mathbf{k}) \cdot \mathbf{t_n}\}\chi^p_{\mathscr{G}_0(\mathbf{k})}(\{\mathbf{R}_j^{-1}\mathbf{R}\mathbf{R}_j \mid \mathbf{0}\}), \tag{9.8}$$

where the sum is over all the coset representatives $\mathbf{R}_j$ such that $\{\mathbf{R}_j^{-1}\mathbf{R}\mathbf{R}_j \mid \mathbf{0}\} \in \mathscr{G}_0(\mathbf{k})$, and where $\chi^p_{\mathscr{G}_0(\mathbf{k})}$ denotes the character of the irreducible representation $\boldsymbol{\Gamma}^p_{\mathscr{G}_0(\mathbf{k})}$ of $\mathscr{G}_0(\mathbf{k})$. (This follows immediately from Equation (5.56) using Equation (9.7).)

In applications the following properties of the basis functions of $\boldsymbol{\Gamma}^{\mathbf{k}p}$ of $\mathscr{G}$ are useful.

Theorem III Let $\psi_{jr}^{\mathbf{k}p}(\mathbf{r})$ ($j=1,2,\ldots,M(\mathbf{k})$, and $r=1,2,\ldots,d_p$) be a set of basis functions of the unitary irreducible representation $\Gamma^{\mathbf{k}p}$ of the space group $\mathscr{G}$ defined by Equations (9.5). Then

(*a*) $\psi_{jr}^{\mathbf{k}p}(\mathbf{r})$ is a Bloch function with wave vector $\mathbf{R}_j\mathbf{k}$,
(*b*) the functions $\psi_{1r}^{\mathbf{k}p}$ ($r=1,2,\ldots,d_p$) form a basis for the unitary irreducible representation $\Gamma^{p}_{\mathscr{G}_0(\mathbf{k})}$ of $\mathscr{G}_0(\mathbf{k})$, and
(*c*) $\psi_{jr}^{\mathbf{k}p}(\mathbf{r})=\mathrm{P}(\{\mathbf{R}_j\mid\mathbf{0}\})\psi_{1r}^{\mathbf{k}p}(\mathbf{r})$, $j=1,2,\ldots,M(\mathbf{k})$.

Proof In the double subscript notation and the present context, Equation (1.26) becomes

$$\mathrm{P}(\{\mathbf{R}\mid\mathbf{t_n}\})\psi_{jr}^{\mathbf{k}p}(\mathbf{r})=\sum_{k=1}^{M(\mathbf{k})}\sum_{t=1}^{d_p}\Gamma^{\mathbf{k}p}(\{\mathbf{R}\mid\mathbf{t_n}\})_{kt,jr}\psi_{kt}^{\mathbf{k}p}(\mathbf{r}) \tag{9.9}$$

for all $\{\mathbf{R}\mid\mathbf{t_n}\}\in\mathscr{G}$.

(*a*) It follows from Equations (9.5) that

$$\Gamma^{\mathbf{k}p}(\{\mathbf{1}\mid\mathbf{t_n}\})_{kt,jr}=\delta_{jk}\delta_{tr}\exp\{-i(\mathbf{R}_j\mathbf{k})\cdot\mathbf{t_n}\},$$

so, by Equation (9.9),

$$\mathrm{P}(\{\mathbf{1}\mid\mathbf{t_n}\})\psi_{jr}^{\mathbf{k}p}(\mathbf{r})=\exp\{-i(\mathbf{R}_j\mathbf{k})\cdot\mathbf{t_n}\}\psi_{jr}^{\mathbf{k}p}(\mathbf{r}).$$

Comparison with Equations (8.19) shows that $\psi_{jr}^{\mathbf{k}p}(\mathbf{r})$ is a Bloch function with wave vector $\mathbf{R}_j\mathbf{k}$.

(*b*) If $\{\mathbf{R}\mid\mathbf{0}\}\in\mathscr{G}_0(\mathbf{k})$, Equations (9.5) imply that

$$\Gamma^{\mathbf{k}p}(\{\mathbf{R}\mid\mathbf{0}\})_{kt,1r}=\delta_{k1}\Gamma^{p}_{\mathscr{G}_0(\mathbf{k})}(\{\mathbf{R}\mid\mathbf{0}\})_{tr}.$$

Then, by Equation (9.9), for $\{\mathbf{R}\mid\mathbf{0}\}\in\mathscr{G}_0(\mathbf{k})$

$$\mathrm{P}(\{\mathbf{R}\mid\mathbf{0}\})\psi_{1r}^{\mathbf{k}p}(\mathbf{r})=\sum_{t=1}^{d_p}\Gamma^{p}_{\mathscr{G}_0(\mathbf{k})}(\{\mathbf{R}\mid\mathbf{0}\})_{tr}\psi_{1t}^{\mathbf{k}p}(\mathbf{r}).$$

(*c*) From Equations (9.5)

$$\Gamma^{\mathbf{k}p}(\{\mathbf{R}_j\mid\mathbf{0}\})_{kt,1r}=\delta_{jk}\delta_{tr}$$

so that, by Equation (9.9), $\mathrm{P}(\{\mathbf{R}_j\mid\mathbf{0}\})\psi_{1r}^{\mathbf{k}p}(\mathbf{r})=\psi_{jr}^{\mathbf{k}p}(\mathbf{r})$.

This latter theorem has a converse.

Theorem IV Suppose that $\phi_{r}^{\mathbf{k}p}(\mathbf{r})$, $r=1,2,\ldots,d_p$, are Bloch functions, with wave vector $\mathbf{k}$, that are also basis functions of the unitary irreducible representation $\Gamma^{p}_{\mathscr{G}_0(\mathbf{k})}$ of $\mathscr{G}_0(\mathbf{k})$. Let

$$\psi_{jr}^{\mathbf{k}p}(\mathbf{r})=\mathrm{P}(\{\mathbf{R}_j\mid\mathbf{0}\})\phi_{r}^{\mathbf{k}p}(\mathbf{r}),\ j=1,2,\ldots,M(\mathbf{k}). \tag{9.10}$$

Then the set of $d_p M(\mathbf{k})$ functions $\psi^{\mathbf{k}p}_{jr}(\mathbf{r})$ form a basis for the irreducible representation $\mathbf{\Gamma}^{\mathbf{k}p}$ of $\mathscr{G}$ defined by Equations (9.5).

Proof All that has to be shown is that the functions $\psi^{\mathbf{k}p}_{jr}(\mathbf{r})$, as defined in Equation (9.10), satisfy Equation (9.9). This follows immediately from the identity

$$P(\{\mathbf{R} \mid \mathbf{t}_n\})P(\{\mathbf{R}_j \mid \mathbf{0}\}) = P(\{\mathbf{R}_k \mid \mathbf{0}\})P(\{\mathbf{1} \mid \mathbf{R}_k^{-1}\mathbf{t}_n\})P(\{\mathbf{R}_k^{-1}\mathbf{R}\mathbf{R}_j \mid \mathbf{0}\}),$$

in which $\mathbf{R}_k$ can be chosen so that $\{\mathbf{R}_k^{-1}\mathbf{R}\mathbf{R}_j \mid \mathbf{0}\} \in \mathscr{G}_0(\mathbf{k})$.

The analysis of Chapter 6, Section 1, shows that considerable simplification of the calculation of electronic energy eigenvalues can be achieved if the energy eigenfunctions corresponding to an irreducible representation of the group of the Schrödinger equation are expanded in terms of basis functions of that representation. Moreover, it is sufficient to restrict attention just to functions transforming as one row of the irreducible representation. In the present context, the group of the Schrödinger equation is a symmorphic space group $\mathscr{G}$, and so for an energy eigenvalue corresponding to $\mathbf{\Gamma}^{\mathbf{k}p}$ it is sufficient to restrict attention to functions transforming as the $j = 1$, $r = 1$ row of $\mathbf{\Gamma}^{\mathbf{k}p}$, that is, to Bloch functions with wave vector $\mathbf{k}$ that transform under the rotations of $\mathscr{G}_0(\mathbf{k})$ as the first row of $\mathbf{\Gamma}^{p}_{\mathscr{G}_0(\mathbf{k})}$.

Such "symmetrized wave functions" can be constructed using the projection operator technique of Chapter 5, Section 1. Applied to spherical harmonics, plane waves or atomic orbitals this technique produces the "symmetrized spherical harmonics", "symmetrized plane waves" or "symmetrized atomic orbitals" that are needed for the various methods of electronic energy band calculation. (Good accounts of these methods exist in the reviews of Callaway (1958, 1964), Fletcher (1971), Pincherle (1960, 1971) and Reitz (1955).) Details of these applications of the projection operator technique may be found elsewhere (e.g. Cornwell (1969)). Tabulations of the symmetrized spherical harmonics for the space groups O_h^1, O_h^5 and O_h^9, the so-called "kubic harmonics" have been given by von der Lage and Bethe (1947), Howarth and Jones (1952), Bell (1954), Altmann (1957) and Altmann and Cracknell (1965). The construction of symmetrized plane waves for all 73 symmorphic space groups may be simplified by the use of tables given by Luehrmann (1968), which supplement those given previously for O_h^1, O_h^9 and T_d^2 by Schlosser (1962).

(*b*) *The irreducible representations of the cubic space groups* O_h^1, O_h^5 *and* O_h^9

The simple cubic space group O_h^1, the face-centred cubic space group O_h^5 and the body-centred cubic space group O_h^9 provide examples of symmorphic space groups of great importance. The following description of them will serve to introduce the standard notations and conventions that are used for all symmorphic space groups.

As shown by Theorem II of subsection (*a*), an irreducible representation of the space group $\mathscr{G}$ is specified by an allowed **k** vector and a label *p* of the irreducible representation of $\mathscr{G}_0(\mathbf{k})$ to which it corresponds. The convention is that the symmetry points and axes of the Brillouin zone are denoted by capital letters, as for example in Figures 8.8, 8.9 and 8.10. The irreducible representations of the corresponding point groups $\mathscr{G}_0(\mathbf{k})$ are then labelled by assigning a subscript or set of subscripts to the appropriate capital letter. For example, the centre of the Brillouin zone for the space groups O_h^1, O_h^5 and O_h^9 is the point Γ, so that the ten irreducible representations of the point group $\mathscr{G}_0(\mathbf{k})$ for Γ (which is actually O_h) are called $\Gamma_1, \Gamma_2, \Gamma_{12}, \Gamma_{15}, \Gamma_{25}, \Gamma_{1'}, \Gamma_{2'}, \Gamma_{12'}, \Gamma_{15'}$ and $\Gamma_{25'}$ in the most commonly used notation of Bouckaert *et al.* (1936). (This assignment of subscripts is almost entirely arbitrary and unfortunately conveys no direct information about the nature of the corresponding irreducible representation. More informative notations have been proposed by Howarth and Jones (1952) and by Bell (1954), but the notation of Bouckaert *et al.* (1936) is so widely used that to employ any other notation now would just cause confusion.)

The point groups $\mathscr{G}_0(\mathbf{k})$ for the symmetry points, axes and planes of the space groups O_h^1, O_h^5 and O_h^9 are given in Tables 9.1, 9.2 and 9.3

Point	Coordinates	$\mathscr{G}_0(\mathbf{k})$
Γ	$(0,0,0)$	O_h
M	$(\pi/a)(0,1,1)$	D_{4h}
R	$(\pi/a)(1,1,1)$	O_h
X	$(\pi/a)(0,0,1)$	D_{4h}
Axis	**Coordinates, $0<\kappa<1$**	$\mathscr{G}_0(\mathbf{k})$
Δ	$(\pi/a)(0,0,\kappa)$	C_{4v}
Λ	$(\pi/a)(\kappa,\kappa,\kappa)$	C_{3v}
Σ	$(\pi/a)(0,\kappa,\kappa)$	C_{2v}
S	$(\pi/a)(\kappa,\kappa,1)$	C_{2v}
T	$(\pi/a)(\kappa,1,1)$	C_{4v}
Z	$(\pi/a)(0,\kappa,1)$	C_{2v}
Plane	**Equation**	$\mathscr{G}_0(\mathbf{k})$
ΓMX	$k_x=0$	$C_s(IC_{2x})$
ΓRM	$k_y=k_z>k_x$	$C_s(IC_{2f})$
ΓRX	$k_x=k_y<k_z$	$C_s(IC_{2b})$
MRX	$k_z=\pi/a$	$C_s(IC_{2z})$

Table 9.1 The point groups $\mathscr{G}_0(\mathbf{k})$ for the symmetry points, axes and planes of the simple cubic space group O_h^1.

Point	Coordinates	$\mathscr{G}_0(\mathbf{k})$
Γ	$(0,0,0)$	O_h
L	$(\pi/a)(1,1,1)$	D_{3d}
W	$(\pi/a)(0,1,2)$	D_{2d}
X	$(\pi/a)(0,0,2)$	D_{4h}
Axis	**Coordinates**	$\mathscr{G}_0(\mathbf{k})$
Δ	$(\pi/a)(0,0,2\kappa)$, $0<\kappa<1$	C_{4v}
Λ	$(\pi/a)(\kappa,\kappa,\kappa)$, $0<\kappa<1$	C_{3v}
Σ	$(\pi/a)(0,\frac{3}{2}\kappa,\frac{3}{2}\kappa)$, $0<\kappa\leqslant 1$	C_{2v}
Q	$(\pi/a)(1-\kappa,1,1+\kappa)$, $0<\kappa<1$	C_2
S	$(\pi/a)(\frac{1}{2}\kappa,\frac{1}{2}\kappa,2)$, $0<\kappa\leqslant 1$	C_{2v}
Z	$(\pi/a)(0,\kappa,2)$, $0<\kappa<1$	C_{2v}
Plane	**Equation**	$\mathscr{G}_0(\mathbf{k})$
ΓKWX	$k_x=0$	$C_s(IC_{2x})$
ΓKL	$k_y=k_z>k_x$	$C_s(IC_{2f})$
ΓLUX	$k_x=k_y<k_z$	$C_s(IC_{2b})$
UWX	$k_z=2\pi/a$	$C_s(IC_{2z})$

Table 9.2 The point groups $\mathscr{G}_0(\mathbf{k})$ for the symmetry points, axes and planes of the face-centred cubic space group O_h^5.

respectively. The notation for the point groups is that of Schönfliess (1923). This notation is used again in Appendix D, where the character tables for all 32 crystallographic point groups are listed, together with explicit sets of matrices for all irreducible representations of

Point	Coordinates	$\mathscr{G}_0(\mathbf{k})$
Γ	$(0,0,0)$	O_h
H	$(\pi/a)(0,0,2)$	O_h
N	$(\pi/a)(0,1,1)$	D_{2h}
P	$(\pi/a)(1,1,1)$	T_d
Axis	**Coordinates, $0<\kappa<1$**	$\mathscr{G}_0(\mathbf{k})$
Δ	$(\pi/a)(0,0,2\kappa)$	C_{4v}
Λ	$(\pi/a)(\kappa,\kappa,\kappa)$	C_{3v}
Σ	$(\pi/a)(0,\kappa,\kappa)$	C_{2v}
D	$(\pi/a)(\kappa,1,1)$	C_{2v}
F	$(\pi/a)(1-\kappa,1-\kappa,1+\kappa)$	C_{3v}
G	$(\pi/a)(0,1-\kappa,1+\kappa)$	C_{2v}
Plane	**Equation**	$\mathscr{G}_0(\mathbf{k})$
ΓHN	$k_x=0$	$C_s(IC_{2x})$
ΓNP	$k_y=k_z>k_x$	$C_s(IC_{2f})$
ΓHP	$k_x=k_y<k_z$	$C_s(IC_{2b})$
HNP	$k_y+k_z=2\pi/a$	$C_s(IC_{2e})$

Table 9.3 The point groups $\mathscr{G}_0(\mathbf{k})$ for the symmetry points, axes and planes of the body-centred cubic lattice O_h^9.

dimension greater than one. The tables of Appendix D also give, when appropriate, the notation of Bouckaert *et al.* (1936) for the irreducible representations of $\mathscr{G}_0(\mathbf{k})$. (The rotations of each $\mathscr{G}_0(\mathbf{k})$ listed in Appendix D are for the value of $\mathbf{k}$ specified in Tables 9.1, 9.2 or 9.3.)

It is possible for two or more points in different stars to have point groups $\mathscr{G}_0(\mathbf{k})$ that are isomorphic. In such a situation the actual group elements of the $\mathscr{G}_0(\mathbf{k})$ may be different, the isomorphic groups merely differing in the orientation of their defining axes. For example, for the body-centred cubic space group O_h^9, the points on the axes Σ and D correspond to a $\mathscr{G}_0(\mathbf{k})$ that is C_{2v}. However, with the coordinates of Σ and D given in Table 9.3, the group elements of $\mathscr{G}_0(\mathbf{k})$ for Σ (arranged in classes) are

$$\mathscr{C}_1 = E, \qquad \mathscr{C}_2 = C_{2e}, \qquad \mathscr{C}_3 = IC_{2x}, \qquad \mathscr{C}_4 = IC_{2f},$$

where the group elements of $\mathscr{G}_0(\mathbf{k})$ for D are

$$\mathscr{C}_1 = E, \qquad \mathscr{C}_2 = C_{2x}, \qquad \mathscr{C}_3 = IC_{2e}, \qquad \mathscr{C}_4 = IC_{2f}.$$

The irreducible representations of $\mathscr{G}_0(\mathbf{k})$ for Σ are denoted by Σ_1, Σ_2, Σ_3 and Σ_4, while those for D are denoted by D_1, D_2, D_3 and D_4. The different sets of group elements occurring in this way are all listed in the description of the corresponding point group in Appendix D.

For every point $\mathbf{k}$ on a symmetry plane, the group $\mathscr{G}_0(\mathbf{k})$ is C_s, which contains just the identity transformation and a reflection. The appropriate reflection for each plane is indicated in parentheses in the third column of Tables 9.1, 9.2 and 9.3. The irreducible representation of C_s for which the character of the reflection is $+1$ is described as being "even" and is denoted by a $+$, and the other irreducible representation is described as being "odd" and is denoted by a $-$.

Figures 8.11 and 8.13 are examples of energy bands and lattice vibrational spectra employing the notation described above.

As noted in Chapter 8, Section 7, the three zero-frequency *translational* modes of lattice vibration correspond to $\mathbf{k} = \mathbf{0}$. As the irreducible representation Γ_{15} of $\mathscr{G}_0(\mathbf{0})$ $(= O_h)$ can be chosen (see Appendix D) so that $\mathbf{\Gamma}_{15}(\mathbf{R}(T)) = \mathbf{R}(T)$ for all $T \in \mathscr{G}_0(\mathbf{0})$, Theorem VI of Chapter 7, Section 3, shows that these modes correspond to Γ_{15} for O_h^1, O_h^5 and O_h^9. For each allowed value of $\mathbf{k}$ the irreducible representations $\mathbf{\Gamma}^{\mathbf{k}p}$ corresponding to lattice vibrations with wave vector $\mathbf{k}$ may be found by substituting $\chi^{\mathbf{k}p}(T)$ of Equation (9.8) into Equation (7.31) in the place of $\chi^p(T)$. The quantities $u(T)$ in the expressions for $\chi^{\text{disp}}(T)$ in Equations (7.22) and (7.23) may be obtained from the formulae given by de Angelis *et al.* (1972).

3 Irreducible representations of non-symmorphic space groups

(*a*) *The fundamental theorem on the irreducible representations of non-symmorphic space groups*

For a *non-symmorphic* space group $\mathcal{G}$ it is no longer true that $\mathcal{G}$ is the semi-direct product of $\mathcal{T}$ and $\mathcal{G}_0$. Consequently that part of the induced representation theory of Chapter 5, Section 7, cannot be invoked. However, it is still possible to construct certain subgroups of $\mathcal{G}$ and induce representations of $\mathcal{G}$ from them using Theorem I of Chapter 5, Section 7. Rather remarkably, although this theorem does not guarantee that the resulting representations are irreducible, by a judicious choice of subgroups and their representations not only are the induced representations irreducible, but *all* the inequivalent irreducible representations of $\mathcal{G}$ can be produced. The resulting theory then has many features in common with that for symmorphic space groups.

The subgroup $\mathcal{G}_0(\mathbf{k})$ of the point group $\mathcal{G}_0$ of $\mathcal{G}$ may be defined exactly as in Section 2(*a*), but when $\mathcal{G}$ is non-symmorphic, $\mathcal{G}_0$ is not a subgroup of $\mathcal{G}$, so that in general $\mathcal{G}_0(\mathbf{k})$ is not a subgroup of $\mathcal{G}$ either. A more useful quantity is provided by the following definition:

Definition $\mathcal{G}(\mathbf{k})$, *the group of the allowed wave vector* $\mathbf{k}$
The group $\mathcal{G}(\mathbf{k})$ is the *subgroup* of the *space group* $\mathcal{G}$ that consists of all transformations $\{\mathbf{R} \mid \boldsymbol{\tau}_\mathbf{R} + \mathbf{t}_\mathbf{n}\}$ of $\mathcal{G}$ having the property that

$$\mathbf{R}\mathbf{k} = \mathbf{k} + \mathbf{K}_\mathbf{m}, \qquad (9.11)$$

where $\mathbf{K}_\mathbf{m}$ is some lattice vector of the reciprocal lattice (as defined in Equation (8.22)), which may be zero.

$\mathcal{G}(\mathbf{k})$ always contains the group $\mathcal{T}$ of pure primitive translations as a subgroup. Consequently, for any $\{\mathbf{R} \mid \boldsymbol{\tau}_\mathbf{R}\} \in \mathcal{G}$, *every* transformation of the form $\{\mathbf{R} \mid \boldsymbol{\tau}_\mathbf{R} + \mathbf{t}_\mathbf{n}\}$ is contained in the left coset $\{\mathbf{R} \mid \boldsymbol{\tau}_\mathbf{R}\}\mathcal{G}(\mathbf{k})$. Thus the coset representatives for the decomposition of $\mathcal{G}$ into left cosets with respect to $\mathcal{G}(\mathbf{k})$ can always be chosen to be of the form $\{\mathbf{R} \mid \boldsymbol{\tau}_\mathbf{R}\}$. For typographical convenience these coset representatives will be denoted by $\{\mathbf{R}_j \mid \boldsymbol{\tau}_j\}$, $j = 1, 2, \ldots$.

Clearly $\mathcal{G}_0(\mathbf{k})$ consists of the set of distinct rotational parts of the elements of $\mathcal{G}(\mathbf{k})$. If $g, g_0, g(\mathbf{k}), g_0(\mathbf{k})$ and N are the orders of $\mathcal{G}, \mathcal{G}_0, \mathcal{G}(\mathbf{k}), \mathcal{G}_0(\mathbf{k})$ and $\mathcal{T}$ respectively, then $g = g_0 N$ and $g(\mathbf{k}) = g_0(\mathbf{k})N$, so that for the integer $M(\mathbf{k})$ defined in Equation (9.4):

$$M(\mathbf{k}) = g_0/g_0(\mathbf{k}) = g/g(\mathbf{k}). \qquad (9.12)$$

The concepts of general points and symmetry points, axes and planes may be defined in terms of the point groups $\mathscr{G}_0(\mathbf{k})$ exactly as for symmorphic space groups. Clearly a non-symmorphic space group having the same Bravais lattice and the same point group $\mathscr{G}_0$ as a symmorphic space group has exactly the same symmetry points, axes and planes as that symmorphic space group.

Example I *Symmetry points, axes and planes for the space group* O_h^7 *of the diamond structure*

The space group O_h^7 of the diamond structure will be described in detail in subsection (c). Its Bravais lattice is the face-centred cubic lattice Γ_c^f, whose Brillouin zone is given in Figure 8.10 in the previous chapter, and its point group $\mathscr{G}_0$ is O_h. As these are the same as for the space group O_h^5, O_h^7 has exactly the same symmetry points, axes and planes as those listed for O_h^5 in Example II of Section 2(a).

Definition *The "star" of* $\mathbf{k}$

Let $\{\mathbf{R}_j \mid \boldsymbol{\tau}_j\}$, $j = 1, 2, \ldots, M(\mathbf{k})$, be a set of coset representatives for the decomposition of $\mathscr{G}$ into *left* cosets with respect to $\mathscr{G}(\mathbf{k})$. Then the set of $M(\mathbf{k})$ vectors $\mathbf{k}_j$ defined by $\mathbf{k}_j = \mathbf{R}_j\mathbf{k}$ ($j = 1, 2, \ldots, M(\mathbf{k})$) is called the "star" of $\mathbf{k}$.

As in the case of symmorphic space groups, the star of $\mathbf{k}$ is unique up to equivalence. Again it is convenient to choose $\{\mathbf{R}_1 \mid \boldsymbol{\tau}_{\mathbf{R}_1}\} = \{\mathbf{1} \mid \mathbf{0}\}$, so that $\mathbf{k}_1 = \mathbf{k}$.

The first fundamental theorem on the irreducible representations of a non-symmorphic space group is very similar to that for symmorphic space groups.

Theorem I Let $\mathbf{k}$ be any allowed $\mathbf{k}$-vector of the Brillouin zone and let $\{\mathbf{R}_j \mid \boldsymbol{\tau}_j\}$ ($j = 1, 2, \ldots, M(\mathbf{k})$) be a set of coset representatives for the decomposition of $\mathscr{G}$ into left cosets with respect to $\mathscr{G}(\mathbf{k})$. Let $\boldsymbol{\Gamma}^p_{\mathscr{G}(\mathbf{k})}$ be a unitary irreducible representation of $\mathscr{G}(\mathbf{k})$ of dimension d_p such that

$$\boldsymbol{\Gamma}^p_{\mathscr{G}(\mathbf{k})}(\{\mathbf{1} \mid \mathbf{t}_n\}) = \exp(-i\mathbf{k} \cdot \mathbf{t}_n)\boldsymbol{\Gamma}^p_{\mathscr{G}(\mathbf{k})}(\{\mathbf{1} \mid \mathbf{0}\}) \tag{9.13}$$

for every $\{\mathbf{1} \mid \mathbf{t}_n\}$ of $\mathscr{T}$. Then there exists a corresponding unitary irreducible representation of the space group $\mathscr{G}$ of dimension $d_pM(\mathbf{k})$, which may be denoted by $\boldsymbol{\Gamma}^{\mathbf{k}p}$, such that

$$\Gamma^{\mathbf{k}p}(\{\mathbf{R} \mid \boldsymbol{\tau}_{\mathbf{R}} + \mathbf{t}_n\})_{kt,jr} = \begin{cases} \exp\{-i(\mathbf{R}_k\mathbf{k}) \cdot \mathbf{t}_n\}\Gamma^p_{\mathscr{G}(\mathbf{k})}(\{\mathbf{R}_k \mid \boldsymbol{\tau}_k\}^{-1}\{\mathbf{R} \mid \boldsymbol{\tau}_{\mathbf{R}}\}\{\mathbf{R}_j \mid \boldsymbol{\tau}_j\})_{tr}, \\ \quad \text{if } \{\mathbf{R}_k \mid \boldsymbol{\tau}_k\}^{-1}\{\mathbf{R} \mid \boldsymbol{\tau}_{\mathbf{R}}\}\{\mathbf{R}_j \mid \boldsymbol{\tau}_j\} \in \mathscr{G}(\mathbf{k}), \\ 0, \text{ otherwise,} \end{cases} \tag{9.14}$$

for $j, k = 1, 2, \ldots, M(\mathbf{k})$, and $r, t = 1, 2, \ldots, d_p$. (Here each row and each column is specified by a *pair* of indices.) Moreover, *all* the inequivalent irreducible representations of $\mathcal{G}$ may be obtained in this way by working through all the inequivalent irreducible representations of $\mathcal{G}(\mathbf{k})$ *that satisfy* Equation (9.13) for all the allowed values of $\mathbf{k}$ that are in different stars.

Proof It is very easy to establish that the matrices of Equations (9.14) do form a representation of $\mathcal{G}$ and that this representation is unitary, for these are immediate consequences of Theorem I of Chapter 5, Section 7. In that theorem $\mathcal{S}$ may be identified with $\mathcal{G}(\mathbf{k})$, s with $g(\mathbf{k})$, M with $M(\mathbf{k})$, $\mathbf{\Delta}$ with $\mathbf{\Gamma}^{p}_{\mathcal{G}(\mathbf{k})}$, d with d_p, and the coset representatives T_j with $\{\mathbf{R}_j \mid \boldsymbol{\tau}_j\}^{-1}$, $j = 1, 2, \ldots, M(\mathbf{k})$. (As shown in a similar situation in the proof of Theorem II of Section 2, if $\{\mathbf{R}_1 \mid \boldsymbol{\tau}_1\}, \{\mathbf{R}_2 \mid \boldsymbol{\tau}_2\}, \ldots$ are coset representatives for the decomposition of $\mathcal{G}$ into *left* cosets with respect to $\mathcal{G}(\mathbf{k})$, then the coset representatives for the corresponding decomposition into *right* cosets may be taken to be $\{\mathbf{R}_1 \mid \boldsymbol{\tau}_1\}^{-1}$, $\{\mathbf{R}_2 \mid \boldsymbol{\tau}_2\}^{-1}, \ldots$). Then with $T = \{\mathbf{R} \mid \boldsymbol{\tau}_{\mathbf{R}} + \mathbf{t}_n\}$ in Equations (5.49),

$$
\begin{aligned}
\Delta(T_k T T_j^{-1})_{tr} &= \Gamma^{p}_{\mathcal{G}(\mathbf{k})}(\{\mathbf{R}_k \mid \boldsymbol{\tau}_k\}^{-1}\{\mathbf{R} \mid \boldsymbol{\tau}_{\mathbf{R}} + \mathbf{t}_n\}\{\mathbf{R}_j \mid \boldsymbol{\tau}_j\})_{tr} \\
&= \Gamma^{p}_{\mathcal{G}(\mathbf{k})}(\{\mathbf{1} \mid \mathbf{R}_k^{-1}\mathbf{t}_n\}\{\mathbf{R}_k \mid \boldsymbol{\tau}_k\}^{-1}\{\mathbf{R} \mid \boldsymbol{\tau}_{\mathbf{R}}\}\{\mathbf{R}_j \mid \boldsymbol{\tau}_j\})_{tr} \\
&= \exp\{-i(\mathbf{R}_k\mathbf{k}) \cdot \mathbf{t}_n\}\Gamma^{p}_{\mathcal{G}(\mathbf{k})}(\{\mathbf{R}_k \mid \boldsymbol{\tau}_k\}^{-1}\{\mathbf{R} \mid \boldsymbol{\tau}_{\mathbf{R}}\}\{\mathbf{R}_j \mid \boldsymbol{\tau}_j\})_{tr}
\end{aligned}
$$

by Equations (9.13) and (9.2), which gives Equations (9.14) immediately.

It will now be shown that the representation $\mathbf{\Gamma}^{\mathbf{k}p}$ of $\mathcal{G}$ is *irreducible*. From Equations (9.14) the corresponding characters are given by

$$
\begin{aligned}
&\chi^{\mathbf{k}p}(\{\mathbf{R} \mid \boldsymbol{\tau}_{\mathbf{R}} + \mathbf{t}_n\}) \\
&\qquad = \sum_j \exp\{-i(\mathbf{R}_j\mathbf{k}) \cdot \mathbf{t}_n\}\chi^{p}_{\mathcal{G}(\mathbf{k})}(\{\mathbf{R}_j \mid \boldsymbol{\tau}_j\}^{-1}\{\mathbf{R} \mid \boldsymbol{\tau}_{\mathbf{R}}\}\{\mathbf{R}_j \mid \boldsymbol{\tau}_j\}), \qquad (9.15)
\end{aligned}
$$

where the summation is over all the coset representatives $\{\mathbf{R}_j \mid \boldsymbol{\tau}_j\}$ such that $\{\mathbf{R}_j \mid \boldsymbol{\tau}_j\}^{-1}\{\mathbf{R} \mid \boldsymbol{\tau}_{\mathbf{R}}\}\{\mathbf{R}_j \mid \boldsymbol{\tau}_j\} \in \mathcal{G}(\mathbf{k})$, and $\chi^{p}_{\mathcal{G}(\mathbf{k})}$ denotes the characters of $\mathbf{\Gamma}^{p}_{\mathcal{G}(\mathbf{k})}$. Then, by Equations (8.24),

$$
\sum_{\mathbf{t}_n \in \mathcal{T}} |\chi^{\mathbf{k}p}(\{\mathbf{R} \mid \boldsymbol{\tau}_{\mathbf{R}} + \mathbf{t}_n\})|^2 = N \sum_j |\chi^{p}_{\mathcal{G}(\mathbf{k})}(\{\mathbf{R}_j \mid \boldsymbol{\tau}_j\}^{-1}\{\mathbf{R} \mid \boldsymbol{\tau}_{\mathbf{R}}\}\{\mathbf{R}_j \mid \boldsymbol{\tau}_j\})|^2,
$$

so that

$$
\begin{aligned}
&\sum_{\{\mathbf{R} \mid \boldsymbol{\tau}_{\mathbf{R}} + \mathbf{t}_n\} \in \mathcal{G}} |\chi^{\mathbf{k}p}(\{\mathbf{R} \mid \boldsymbol{\tau}_{\mathbf{R}} + \mathbf{t}_n\})|^2 \\
&\qquad = \sum_{\{\mathbf{R} \mid \mathbf{0}\} \in \mathcal{G}_0} \sum_{\mathbf{t}_n \in \mathcal{T}} \sum_j |\chi^{p}_{\mathcal{G}(\mathbf{k})}(\{\mathbf{R}_j \mid \boldsymbol{\tau}_j\}^{-1}\{\mathbf{R} \mid \boldsymbol{\tau}_{\mathbf{R}} + \mathbf{t}_n\}\{\mathbf{R}_j \mid \boldsymbol{\tau}_j\})|^2,
\end{aligned}
$$

by Equation (9.13). Interchanging the order of summation, for each coset representative $\{\mathbf{R}_j \mid \boldsymbol{\tau}_j\}$ the sum over $\{\mathbf{R} \mid \mathbf{0}\} \in \mathcal{G}_0$ contains only contributions from those $\{\mathbf{R} \mid \boldsymbol{\tau}_\mathbf{R}+\mathbf{t}_n\}$ such that $\{\mathbf{R}_j \mid \boldsymbol{\tau}_j\}^{-1}\{\mathbf{R} \mid \boldsymbol{\tau}_\mathbf{R}+\mathbf{t}_n\}$ $\{\mathbf{R}_j \mid \boldsymbol{\tau}_j\} \in \mathcal{G}(\mathbf{k})$. However, for any $\{\mathbf{R}' \mid \boldsymbol{\tau}_{\mathbf{R}'}+\mathbf{t}_{n'}\} \in \mathcal{G}(\mathbf{k})$, $\{\mathbf{R}_j \mid \boldsymbol{\tau}_j\}$ $\{\mathbf{R}' \mid \boldsymbol{\tau}_{\mathbf{R}'}+\mathbf{t}_{n'}\}\{\mathbf{R}_j \mid \boldsymbol{\tau}_j\}^{-1}$ satisfies this condition. Hence

$$\sum_{\mathcal{G}} |\chi^{\mathbf{k}p}(\{\mathbf{R} \mid \boldsymbol{\tau}_\mathbf{R}+\mathbf{t}_n\})|^2 = \sum_{j=1}^{M(\mathbf{k})} \sum_{\mathcal{G}(\mathbf{k})} |\chi^p_{\mathcal{G}(\mathbf{k})}(\{\mathbf{R}' \mid \boldsymbol{\tau}_{\mathbf{R}'}+\mathbf{t}_{n'}\})|^2$$
$$= M(\mathbf{k})g(\mathbf{k})$$

by Theorem VI of Chapter 4, Section 6. As $M(\mathbf{k})g(\mathbf{k}) = g$, a further application of this theorem shows that $\Gamma^{\mathbf{k}p}$ is irreducible.

It follows immediately from Equation (9.15) that two representations $\Gamma^{\mathbf{k}'p'}$ and $\Gamma^{\mathbf{k}p}$ are not equivalent if $\mathbf{k}$ and $\mathbf{k}'$ are in different stars, nor if $\mathbf{k} = \mathbf{k}'$ but $\Gamma^p_{\mathcal{G}(\mathbf{k})}$ and $\Gamma^{p'}_{\mathcal{G}(\mathbf{k})}$ are inequivalent. Thus, to show that the process produces *all* the inequivalent irreducible representations of $\mathcal{G}$, it is only necessary to show that the sum of the squares of the dimensions of all the representations $\Gamma^{\mathbf{k}p}$ for all inequivalent irreducible representations $\Gamma^p_{\mathcal{G}(\mathbf{k})}$ satisfying Equation (9.13) and all allowed $\mathbf{k}$ vectors in different stars is equal to g (cf. Theorem VII of Chapter 4, Section 6). However, it will now be shown that for each allowed vector $\mathbf{k}$ the sum of the squares of the dimensions of the inequivalent irreducible representations $\Gamma^p_{\mathcal{G}(\mathbf{k})}$ satisfying Equation (9.13) is $g_0(\mathbf{k})$. (Then the required sum is $g_0(\mathbf{k})\{M(\mathbf{k})\}^2$ summed over all $\mathbf{k}$ in different stars. As $g_0(\mathbf{k})M(\mathbf{k}) = g_0$, and the sum of $M(\mathbf{k})$ over $\mathbf{k}$ in different stars is equal to N, the required sum is indeed $g_0N = g$.)

Let $\phi(\mathbf{r})$ be a function chosen so that the $g(\mathbf{k})$ functions $P(\{\mathbf{R} \mid \boldsymbol{\tau}_\mathbf{R}+\mathbf{t}_n\})\phi(\mathbf{r})$ for all $\{\mathbf{R} \mid \boldsymbol{\tau}_\mathbf{R}+\mathbf{t}_n\} \in \mathcal{G}(\mathbf{k})$ are linearly independent. This set of functions forms a basis for the "regular" representation Γ^{reg} of $\mathcal{G}(\mathbf{k})$ (see Appendix C, Section 4). The projection operator $\mathcal{P}^\mathbf{k}_{11} = (1/N)\Sigma_{\mathbf{t}_{n'} \in \mathcal{T}} \exp(i\mathbf{k} \cdot \mathbf{t}_{n'})\mathcal{P}(\{\mathbf{1} \mid \mathbf{t}_{n'}\})$ (see Chapter 5, Section 1) applied to any function of this set produces a Bloch function with wave vector $\mathbf{k}$. However,

$$\mathcal{P}^\mathbf{k}_{11}[P(\{\mathbf{R} \mid \boldsymbol{\tau}_\mathbf{R}+\mathbf{t}_n\})\phi(\mathbf{r})] = \exp(-i\mathbf{k} \cdot \mathbf{t}_n)\mathcal{P}^\mathbf{k}_{11}[P(\{\mathbf{R} \mid \boldsymbol{\tau}_\mathbf{R}\})\phi(\mathbf{r})],$$

so there are only $g_0(\mathbf{k})$ linearly independent combinations of functions of this set that are Bloch functions with wave vector $\mathbf{k}$. However, every basis function of an irreducible representation $\Gamma^p_{\mathcal{G}(\mathbf{k})}$ of $\mathcal{G}(\mathbf{k})$ is a Bloch function with wave vector $\mathbf{k}$ if and only if $\Gamma^p_{\mathcal{G}(\mathbf{k})}$ satisfies Equation (9.13). Thus $\Sigma_p n_p d_p = g_0(\mathbf{k})$, where the sum is over the $\Gamma^p_{\mathcal{G}(\mathbf{k})}$ satisfying Equation (9.13) and n_p is the number of times $\Gamma^p_{\mathcal{G}(\mathbf{k})}$ appears in the reduction of Γ^{reg}. But, by Equations (C.11) and (C.17), $n_p = d_p$, so $\Sigma_p d_p^2 = g_0(\mathbf{k})$.

The representations $\Gamma^p_{\mathcal{G}(\mathbf{k})}$ that satisfy Equation (9.13) will be called the "relevant" representations of $\mathcal{G}(\mathbf{k})$. It should be noted that, for each allowed $\mathbf{k}$, the sum of the squares of the dimensions of the inequivalent "relevant" irreducible representation of $\mathcal{G}(\mathbf{k})$ is equal to $g_0(\mathbf{k})$, the order of $\mathcal{G}_0(\mathbf{k})$. (This result was derived in the course of the above proof.) Thus not every irreducible representation of $\mathcal{G}(\mathbf{k})$ is "relevant".

Theorems III and IV of Section 2 have immediate generalizations to the case of non-symmorphic space groups, the only changes being that $\mathcal{G}_0(\mathbf{k})$ must be replaced by $\mathcal{G}(\mathbf{k})$, $\Gamma^p_{\mathcal{G}_0(\mathbf{k})}$ by the "relevant" representation $\Gamma^p_{\mathcal{G}(\mathbf{k})}$, and $P(\{\mathbf{R}_j \mid \mathbf{0}\})$ by $P(\{\mathbf{R}_j \mid \boldsymbol{\tau}_j\})$.

(*b*) *The "relevant" irreducible representations of $\mathcal{G}(\mathbf{k})$*

Theorem I of the previous subsection reduced the problem of finding the unitary irreducible representations of the space group $\mathcal{G}$ to that of finding the "relevant" irreducible representations of the groups $\mathcal{G}(\mathbf{k})$. However, each $\mathcal{G}(\mathbf{k})$ is a very large group, with a quite complicated structure in general. Fortunately, it is possible in many cases to derive all the "relevant" irreducible representations of $\mathcal{G}(\mathbf{k})$ from the irreducible representations of the corresponding point group $\mathcal{G}_0(\mathbf{k})$, and in the remaining cases it is possible to reduce the problem to the study of certain much smaller groups.

Theorem II If

$$\exp\{-i\mathbf{k}\,.\,(\mathbf{R}\boldsymbol{\tau}_{\mathbf{R}'}-\boldsymbol{\tau}_{\mathbf{R}'})\}=1 \tag{9.16}$$

for every pair of transformations $\{\mathbf{R} \mid \boldsymbol{\tau}_{\mathbf{R}}\}$ and $\{\mathbf{R}' \mid \boldsymbol{\tau}_{\mathbf{R}'}\}$ of $\mathcal{G}(\mathbf{k})$, then the set of matrices defined for all $\{\mathbf{R} \mid \boldsymbol{\tau}_{\mathbf{R}}+\mathbf{t}_{\mathbf{n}}\}\in\mathcal{G}(\mathbf{k})$ by

$$\Gamma^p_{\mathcal{G}(\mathbf{k})}(\{\mathbf{R} \mid \boldsymbol{\tau}_{\mathbf{R}}+\mathbf{t}_{\mathbf{n}}\})=\exp\{-i\mathbf{k}\,.\,(\boldsymbol{\tau}_{\mathbf{R}}+\mathbf{t}_{\mathbf{n}})\}\Gamma^p_{\mathcal{G}_0(\mathbf{k})}(\{\mathbf{R} \mid \mathbf{0}\}), \tag{9.17}$$

where $\Gamma^p_{\mathcal{G}_0(\mathbf{k})}$ is a unitary irreducible representation of $\mathcal{G}_0(\mathbf{k})$, form a "relevant" unitary irreducible representation of $\mathcal{G}(\mathbf{k})$. Moreover, *all* the inequivalent "relevant" irreducible representations of $\mathcal{G}(\mathbf{k})$ may be obtained in this way by working through all the inequivalent irreducible representations of $\mathcal{G}_0(\mathbf{k})$.

Proof To prove that Equation (9.17) defines a representation of $\mathcal{G}(\mathbf{k})$ when Equation (9.16) is valid, let $\{\mathbf{R} \mid \boldsymbol{\tau}_{\mathbf{R}}+\mathbf{t}_{\mathbf{n}}\}$ and $\{\mathbf{R} \mid \boldsymbol{\tau}_{\mathbf{R}'}+\mathbf{t}_{\mathbf{n}'}\}$ be any two members of $\mathcal{G}(\mathbf{k})$. Then, by Equation (9.17),

$$\Gamma^p_{\mathcal{G}(\mathbf{k})}(\{\mathbf{R} \mid \boldsymbol{\tau}_{\mathbf{R}}+\mathbf{t}_{\mathbf{n}}\})\Gamma^p_{\mathcal{G}(\mathbf{k})}(\{\mathbf{R}' \mid \boldsymbol{\tau}_{\mathbf{R}'}+\mathbf{t}_{\mathbf{n}'}\})$$
$$=\exp\{-i\mathbf{k}\,.\,(\boldsymbol{\tau}_{\mathbf{R}}+\mathbf{t}_{\mathbf{n}}+\boldsymbol{\tau}_{\mathbf{R}'}+\mathbf{t}_{\mathbf{n}'})\}\Gamma^p_{\mathcal{G}_0(\mathbf{k})}(\{\mathbf{R}\mathbf{R}' \mid \mathbf{0}\}), \tag{9.18}$$

while

$$\Gamma^p_{\mathscr{G}(\mathbf{k})}(\{\mathbf{R} \mid \boldsymbol{\tau}_\mathbf{R}+\mathbf{t_n}\}\{\mathbf{R}' \mid \boldsymbol{\tau}_{\mathbf{R}'}+\mathbf{t}_{\mathbf{n}'}\})$$
$$= \exp\{-i\mathbf{k}\,.\,(\boldsymbol{\tau}_\mathbf{R}+\mathbf{t_n}+\mathbf{R}\boldsymbol{\tau}_{\mathbf{R}'}+\mathbf{R}\mathbf{t}_{\mathbf{n}'})\}\Gamma^p_{\mathscr{G}_0(\mathbf{k})}(\{\mathbf{RR}' \mid \mathbf{0}\}) \quad (9.19)$$

By virtue of Equations (9.2), (9.11) and (8.23), $\exp\{-i\mathbf{k}\,.(\mathbf{R}\mathbf{t}_{\mathbf{n}'})\} = \exp(-i\mathbf{k}\,.\,\mathbf{t}_{\mathbf{n}'})$, so that the right-hand sides of Equations (9.18) and (9.19) are equal if Equation (9.16) is true.

It is obvious that $\Gamma^p_{\mathscr{G}(\mathbf{k})}$ is unitary and irreducible if $\Gamma^p_{\mathscr{G}_0(\mathbf{k})}$ is unitary and irreducible, and that $\Gamma^p_{\mathscr{G}(\mathbf{k})}$ satisfies Equation (9.13). As the sum of the squares of the dimensions of the inequivalent irreducible representations of $\mathscr{G}_0(\mathbf{k})$ is equal to $g_0(\mathbf{k})$, this must be true of the sum of the squares of the inequivalent "relevant" irreducible representations defined by Equation (9.17), and hence there can be no other such representations.

A symmorphic space group can be considered to be a special case for which $\boldsymbol{\tau}_\mathbf{R} = \mathbf{0}$ for all $\{\mathbf{R} \mid \mathbf{0}\} \in \mathscr{G}_0$. Then Equation (9.16) holds for *every* allowed value of $\mathbf{k}$, so Equation (9.17) gives *every* "relevant" irreducible representation of $\mathscr{G}(\mathbf{k})$. Hence Equations (9.14) reduce to Equations (9.5) and Theorem I of Section 3 reduces in this case to Theorem II of Section 2.

This theorem has the following particularly important consequence for *non*-symmorphic space groups.

Theorem III If $\mathbf{k}$ is any *internal* allowed vector of the Brillouin zone (that is, one that is not on the surface of the Brillouin zone), then *every* "relevant" unitary irreducible representation of $\mathscr{G}(\mathbf{k})$ is given in terms of a unitary irreducible representation of $\mathscr{G}_0(\mathbf{k})$ by Equation (9.17).

Proof By Equation (9.2), Equation (9.16) can be rewritten as $\exp\{-i(\mathbf{R}^{-1}\mathbf{k})\,.\,\boldsymbol{\tau}_{\mathbf{R}'}\} = \exp(-i\mathbf{k}\,.\,\boldsymbol{\tau}_{\mathbf{R}'})$, as $\{\mathbf{R}^{-1} \mid \mathbf{0}\} \in \mathscr{G}_0(\mathbf{k})$, $\mathbf{R}^{-1}\mathbf{k} = \mathbf{k}+\mathbf{K_m}$ by Equation (9.3), and, if $\mathbf{k}$ is an internal point, $\mathbf{K_m} = \mathbf{0}$, so Equation (9.16) is satisfied.

As all the irreducible representations of the crystallographic point groups are known (and are listed in Appendix D), it remains only to consider allowed values of $\mathbf{k}$ on the *surface* of the Brillouin zone. In some such cases (including all general points on the surface) Equation (9.16) still applies, but for others the following development is useful.

Definition *The translation group $\mathscr{T}(\mathbf{k})$ of the allowed wave vector* $\mathbf{k}$

$\mathscr{T}(\mathbf{k})$ consists of all $\{\mathbf{1} \mid \mathbf{t_n}\}$ of $\mathscr{T}$ such that

$$\exp(-i\mathbf{k}\,.\,\mathbf{t_n}) = 1 \tag{9.20}$$

Lemma $\mathscr{T}(\mathbf{k})$ is an *invariant* subgroup of $\mathscr{G}(\mathbf{k})$.

Proof As $\mathscr{T}(\mathbf{k})$ is a subgroup of $\mathscr{T}$, and $\mathscr{T}$ is a subgroup of $\mathscr{G}(\mathbf{k})$, $\mathscr{T}(\mathbf{k})$ must be a subgroup of $\mathscr{G}(\mathbf{k})$. Suppose $\{\mathbf{R} \mid \boldsymbol{\tau}_\mathbf{R} + \mathbf{t_n}\} \in \mathscr{G}(\mathbf{k})$ and $\{\mathbf{1} \mid \mathbf{t_{n'}}\} \in \mathscr{T}(\mathbf{k})$. Then $\{\mathbf{R} \mid \boldsymbol{\tau}_\mathbf{R} + \mathbf{t_n}\}\{\mathbf{1} \mid \mathbf{t_{n'}}\}\{\mathbf{R} \mid \boldsymbol{\tau}_\mathbf{R} + \mathbf{t_n}\}^{-1} = \{\mathbf{1} \mid \mathbf{R}\mathbf{t_{n'}}\}$, by Equations (1.7) and (1.8). However, $\{\mathbf{1} \mid \mathbf{R}\mathbf{t_{n'}}\}$ is also a member of $\mathscr{T}(\mathbf{k})$, as $\exp(-i\mathbf{k}\,.\,(\mathbf{R}\mathbf{t_{n'}})) = 1$ by Equations (9.2), (9.11) and (9.20).

It is then possible to construct the factor group $\mathscr{G}(\mathbf{k})/\mathscr{T}(\mathbf{k})$ (see Chapter 2, Section 5). The elements of $\mathscr{G}(\mathbf{k})/\mathscr{T}(\mathbf{k})$ will be taken to be left cosets (of the form $\{\mathbf{R} \mid \mathbf{t}\}\mathscr{T}(\mathbf{k})$), although they could equally well be taken as right cosets. The identity of $\mathscr{G}(\mathbf{k})/\mathscr{T}(\mathbf{k})$ will be taken to be $\{\mathbf{1} \mid \mathbf{0}\}\mathscr{T}(\mathbf{k})$.

Theorem IV Let $\boldsymbol{\Gamma}^p_{\mathscr{G}(\mathbf{k})/\mathscr{T}(\mathbf{k})}$ be a unitary irreducible representation of $\mathscr{G}(\mathbf{k})/\mathscr{T}(\mathbf{k})$ that satisfies

$$\boldsymbol{\Gamma}^p_{\mathscr{G}(\mathbf{k})/\mathscr{T}(\mathbf{k})}(\{\mathbf{1} \mid \mathbf{t_n}\}\mathscr{T}(\mathbf{k})) = \exp(-i\mathbf{k}\,.\,\mathbf{t_n})\boldsymbol{\Gamma}^p_{\mathscr{G}(\mathbf{k})/\mathscr{T}(\mathbf{k})}(\{\mathbf{1} \mid \mathbf{0}\}\mathscr{T}(\mathbf{k})) \tag{9.21}$$

for every coset $\{\mathbf{1} \mid \mathbf{t_n}\}\mathscr{T}(\mathbf{k})$ formed from primitive translations. Then the set of matrices $\boldsymbol{\Gamma}^p_{\mathscr{G}(\mathbf{k})}(\{\mathbf{R} \mid \boldsymbol{\tau}_\mathbf{R} + \mathbf{t_n}\})$ defined by

$$\boldsymbol{\Gamma}^p_{\mathscr{G}(\mathbf{k})}(\{\mathbf{R} \mid \boldsymbol{\tau}_\mathbf{R} + \mathbf{t_n}\}) = \boldsymbol{\Gamma}^p_{\mathscr{G}(\mathbf{k})/\mathscr{T}(\mathbf{k})}(\{\mathbf{R} \mid \boldsymbol{\tau}_\mathbf{R} + \mathbf{t_n}\}\mathscr{T}(\mathbf{k})) \tag{9.22}$$

for every $\{\mathbf{R} \mid \boldsymbol{\tau}_\mathbf{R} + \mathbf{t_n}\}$ of $\mathscr{G}(\mathbf{k})$ forms a unitary irreducible representation of $\mathscr{G}(\mathbf{k})$ that satisfies Equation (9.13). Moreover, every such "relevant" irreducible representation of $\mathscr{G}(\mathbf{k})$ can be constructed in this way.

Proof It will first be shown that Equation (9.22) defines a representation of $\mathscr{G}(\mathbf{k})$. For any $\{\mathbf{R} \mid \boldsymbol{\tau}_\mathbf{R} + \mathbf{t_n}\}$ and $\{\mathbf{R}' \mid \boldsymbol{\tau}_{\mathbf{R}'} + \mathbf{t_{n'}}\}$ of $\mathscr{G}(\mathbf{k})$, by Equation (9.22),

$$\begin{aligned}
&\boldsymbol{\Gamma}^p_{\mathscr{G}(\mathbf{k})}(\{\mathbf{R} \mid \boldsymbol{\tau}_\mathbf{R} + \mathbf{t_n}\})\boldsymbol{\Gamma}^p_{\mathscr{G}(\mathbf{k})}(\{\mathbf{R}' \mid \boldsymbol{\tau}_{\mathbf{R}'} + \mathbf{t_{n'}}\})\\
&= \boldsymbol{\Gamma}^p_{\mathscr{G}(\mathbf{k})/\mathscr{T}(\mathbf{k})}(\{\mathbf{R} \mid \boldsymbol{\tau}_\mathbf{R} + \mathbf{t_n}\}\mathscr{T}(\mathbf{k})\,.\,\{\mathbf{R}' \mid \boldsymbol{\tau}_{\mathbf{R}'} + \mathbf{t_{n'}}\}\mathscr{T}(\mathbf{k}))\\
&= \boldsymbol{\Gamma}^p_{\mathscr{G}(\mathbf{k})/\mathscr{T}(\mathbf{k})}((\{\mathbf{R} \mid \boldsymbol{\tau}_\mathbf{R} + \mathbf{t_n}\}\{\mathbf{R}' \mid \boldsymbol{\tau}_{\mathbf{R}'} + \mathbf{t_{n'}}\})\mathscr{T}(\mathbf{k}), \text{ by Equation (2.6),}\\
&= \boldsymbol{\Gamma}^p_{\mathscr{G}(\mathbf{k})}(\{\mathbf{R} \mid \boldsymbol{\tau}_\mathbf{R} + \mathbf{t_n}\}\{\mathbf{R}' \mid \boldsymbol{\tau}_{\mathbf{R}'} + \mathbf{t_{n'}}\}), \text{ by Equation (9.22) again.}
\end{aligned}$$

That $\Gamma^p_{\mathscr{G}(\mathbf{k})}$ is irreducible follows from applying Theorem VI of Chapter 4, Section 6, for, by Equation (9.22),

$$\sum |\chi^p_{\mathscr{G}(\mathbf{k})}(\{\mathbf{R} \mid \boldsymbol{\tau}_\mathbf{R}+\mathbf{t}_n\})|^2 = t(\mathbf{k}) \sum |\chi^p_{\mathscr{G}(\mathbf{k})/\mathscr{T}(\mathbf{k})}(\{\mathbf{R} \mid \boldsymbol{\tau}_\mathbf{R}+\mathbf{t}_n\}\mathscr{T}(\mathbf{k}))|^2,$$

the summations on the left- and right-hand sides being over all the elements of $\mathscr{G}(\mathbf{k})$ and $\mathscr{G}(\mathbf{k})/\mathscr{T}(\mathbf{k})$ respectively and $t(\mathbf{k})$ being the order of $\mathscr{T}(\mathbf{k})$. As $\Gamma^p_{\mathscr{G}(\mathbf{k})/\mathscr{T}(\mathbf{k})}$ is assumed irreducible, the sum on the right-hand side is equal to $g(\mathbf{k})/t(\mathbf{k})$, the order of $\mathscr{G}(\mathbf{k})/\mathscr{T}(\mathbf{k})$, so the sum on the left-hand side must be equal to $g(\mathbf{k})$.

It follows immediately from Equations (9.21) and (9.22) that $\Gamma^p_{\mathscr{G}(\mathbf{k})}$ satisfies Equation (9.13). Conversely, if $\Gamma^p_{\mathscr{G}(\mathbf{k})}$ satisfies Equation (9.13), then every member of $\mathscr{T}(\mathbf{k})$ is represented in $\Gamma^p_{\mathscr{G}(\mathbf{k})}$ by a unit matrix, so that all members of a coset of $\mathscr{T}(\mathbf{k})$ are represented by the same matrix and hence such a $\Gamma^p_{\mathscr{G}(\mathbf{k})}$ *must* have the form of Equation (9.22).

A representation $\Gamma^p_{\mathscr{G}(\mathbf{k})/\mathscr{T}(\mathbf{k})}$ that satisfies Equation (9.21) will be referred to as a "relevant" representation of $\mathscr{G}(\mathbf{k})/\mathscr{T}(\mathbf{k})$. In general, not every irreducible representation of $\mathscr{G}(\mathbf{k})/\mathscr{T}(\mathbf{k})$ is "relevant", so the character table of "relevant" representations need not be square. For all symmetry points, axes and planes, the order of $\mathscr{G}(\mathbf{k})/\mathscr{T}(\mathbf{k})$ is less than that of $\mathscr{G}(\mathbf{k})$ and is often much less, so that the irreducible representations of $\mathscr{G}(\mathbf{k})/\mathscr{T}(\mathbf{k})$ can be found more easily. Nevertheless, when Equation (9.16) does not hold, the structure of $\mathscr{G}(\mathbf{k})/\mathscr{T}(\mathbf{k})$ can be quite complicated. The standard method for the construction of character tables for these groups may be found in the paper by Herring (1942). Alternative methods have recently been described by Zak (1960), Raghavacharyulu (1961), Raghavacharyulu and Bhavanacharyulu (1962), McIntosh (1963), Olbrychski (1963), Kitz (1965), Rudra (1965), Slechta (1966), Glück *et al.* (1967) and Streitwolf (1971). The calculations have now been performed for *every* non-symmorphic space group. Complete character tables may be found in the works of Kovalev (1965), Miller and Love (1967), Bradley and Cracknell (1972) and Cracknell *et al.* (1979). (A further comprehensive list of individual calculations may be found in Appendix 3 of Cornwell (1969).)

(c) *The irreducible representations of the space group* O_h^7 *of the diamond structure*

The diamond structure consists of *two* face-centred cubic lattices Γ_c^f of identical nuclei, one lattice being displaced relative to the other by a translation through $\boldsymbol{\tau}$, where $\boldsymbol{\tau}=\frac{1}{4}a(1,1,1)$. (Here a is the quantity appearing in the definition of the basic lattice vectors of Γ_c^f in Table 8.1, so that the translation $\boldsymbol{\tau}$ is along one-quarter of one of the cube diagonals shown in Figure 8.3.) The technologically important semiconductors silicon and germanium possess this structure.

As indicated by the notation, the point group $\mathscr{G}_0$ of O_h^7 is O_h, whose rotations may be taken to be as in Appendix D, Section 1(1). For the 24 rotations E, $C_{3\alpha}$, $C_{3\beta}$, $C_{3\gamma}$, $C_{3\delta}$, $C_{3\alpha}^{-1}$, $C_{3\beta}^{-1}$, $C_{3\gamma}^{-1}$, $C_{3\delta}^{-1}$, C_{2x}, C_{2y}, C_{2z}, IC_{4x}, IC_{4y}, IC_{4z}, IC_{4x}^{-1}, IC_{4y}^{-1}, IC_{4z}^{-1}, IC_{2a}, IC_{2b}, IC_{2c}, IC_{2d}, IC_{2e} and IC_{2f}, $\boldsymbol{\tau}_{\mathbf{R}}=\mathbf{0}$, whereas for the remaining 24 rotations of O_h, $\boldsymbol{\tau}_{\mathbf{R}}=\boldsymbol{\tau}$. Of course the Bravais lattice of O_h^7 is Γ_c^f, so Figure 8.10 gives the Brillouin zone.

The irreducible representations of O_h^7 were first deduced by Herring (1942), who employed a similar convention for labelling as that described for O_h^1, O_h^5 and O_h^9 in Section 2(*b*). That is, the "relevant" irreducible representations of $\mathscr{G}(\mathbf{k})$ for a symmetry point or axis are labelled by the capital letter denoting the point or axis with a subscript or set of subscripts attached.

As noted in the previous subsection, if $\mathbf{k}$ is an internal point of the Brillouin zone, the "relevant" irreducible representations of $\mathscr{G}(\mathbf{k})$ are given by Equation (9.17) in terms of the irreducible representations of $\mathscr{G}_0(\mathbf{k})$. Thus the characters of the "relevant" irreducible representations of $\mathscr{G}(\mathbf{k})$ corresponding to the symmetry point Γ and the symmetry axes Δ, Λ and Σ follow (by Table 9.2) from Tables D.2, D.17, D.21 and D.22 respectively. A similar situation exists for the symmetry planes ΓKWX, ΓKL and ΓLUX, for which the appropriate reflections are indicated in Table 9.2 and for which $\mathscr{G}_0=C_s$ in each case. For the symmetry point L and the axis Q Equation (9.16) is also valid, so the "relevant" representations for L and Q may be found, using Equation (9.17), from Tables D.13 and D.25 respectively. Equation (9.16) is also valid for the point K and the lines LK, KW and LU, so K can be considered to be an ordinary point of the axis Σ and the points of LK, KW and LU as ordinary points of the planes ΓKL, ΓKWX and ΓLUX respectively. In all of these cases the label of the irreducible representation of $\mathscr{G}_0(\mathbf{k})$ (with the conventions of Bouckaert *et al.* (1936)) is used as the label of the corresponding "relevant" representation of $\mathscr{G}(\mathbf{k})$.

For the axis S Equation (9.16) is not valid but a slight generalization of Equation (9.19) holds (Streitwolf 1971), namely

$$\boldsymbol{\Gamma}^p_{\mathscr{G}(\mathbf{k})}(\{\mathbf{R}\mid\boldsymbol{\tau}_{\mathbf{R}}+\mathbf{t}_n\})=\exp\{-i\mathbf{k}\,.\,(\mathbf{t}_{\mathbf{R}}+\mathbf{t}_n)\}\boldsymbol{\Gamma}^p_{\mathscr{G}_0(\mathbf{k})}(\{\mathbf{R}\mid\mathbf{0}\}),$$

where $\mathbf{t}_{\mathbf{R}}=\mathbf{0}$ for E and IC_{2b} but $\mathbf{t}_{\mathbf{R}}=\frac{1}{2}\mathbf{a}_1$ for C_{2a} and IC_{2z}.

For the remaining symmetry points W and X and axis Z, the characters of the "relevant" irreducible representations of $\mathscr{G}(\mathbf{k})/\mathscr{T}(\mathbf{k})$ are given in Tables 9.4, 9.5 and 9.6 respectively. It is convenient to divide the coset representatives of $\mathscr{G}(\mathbf{k})/\mathscr{T}(\mathbf{k})$ into three sets, namely

	W_1	W_2
$\{E\,\vert\,\mathbf{0}\}\mathcal{T}(\mathbf{k})$	2	2
$\{C_{2y}\,\vert\,\mathbf{0}\}\mathcal{T}(\mathbf{k})$	0	0
$\{IC_{4y}\,\vert\,\mathbf{0}\}\mathcal{T}(\mathbf{k})$, $\{IC_{4y}^{-1}\,\vert\,\mathbf{0}\}\mathcal{T}(\mathbf{k})$	$(1-i)$	$-(1-i)$
$\{IC_{2x}\,\vert\,\boldsymbol{\tau}\}\mathcal{T}(\mathbf{k})$, $\{IC_{2z}\,\vert\,\boldsymbol{\tau}\}\mathcal{T}(\mathbf{k})$	0	0
$\{C_{2c}\,\vert\,\boldsymbol{\tau}\}\mathcal{T}(\mathbf{k})$, $\{C_{2d}\,\vert\,\boldsymbol{\tau}\}\mathcal{T}(\mathbf{k})$	0	0

Table 9.4 Characters of the "relevant" irreducible representations of $\mathcal{G}(\mathbf{k})/\mathcal{T}(\mathbf{k})$ at the point W for the space group O_h^7.

	X_1	X_2	X_3	X_4
$\{E\,\vert\,\mathbf{0}\}\mathcal{T}(\mathbf{k})$	2	2	2	2
$\{C_{2x}\,\vert\,\mathbf{0}\}\mathcal{T}(\mathbf{k})$, $\{C_{2y}\,\vert\,\mathbf{0}\}\mathcal{T}(\mathbf{k})$	0	0	0	0
$\{C_{2z}\,\vert\,\mathbf{0}\}\mathcal{T}(\mathbf{k})$	2	2	-2	-2
$\{C_{4z}\,\vert\,\boldsymbol{\tau}\}\mathcal{T}(\mathbf{k})$, $\{C_{4z}^{-1}\,\vert\,\boldsymbol{\tau}\}\mathcal{T}(\mathbf{k})$	0	0	0	0
$\{C_{2a}\,\vert\,\boldsymbol{\tau}\}\mathcal{T}(\mathbf{k})$, $\{C_{2b}\,\vert\,\boldsymbol{\tau}\}\mathcal{T}(\mathbf{k})$	0	0	2	-2
$\{I\,\vert\,\boldsymbol{\tau}\}\mathcal{T}(\mathbf{k})$	0	0	0	0
$\{IC_{2x}\,\vert\,\boldsymbol{\tau}\}\mathcal{T}(\mathbf{k})$, $\{IC_{2y}\,\vert\,\boldsymbol{\tau}\}\mathcal{T}(\mathbf{k})$	0	0	0	0
$\{IC_{2z}\,\vert\,\boldsymbol{\tau}\}\mathcal{T}(\mathbf{k})$	0	0	0	0
$\{IC_{4z}\,\vert\,\mathbf{0}\}\mathcal{T}(\mathbf{k})$, $\{IC_{4z}^{-1}\,\vert\,\mathbf{0}\}\mathcal{T}(\mathbf{k})$	0	0	0	0
$\{IC_{2a}\,\vert\,\mathbf{0}\}\mathcal{T}(\mathbf{k})$, $\{IC_{2b}\,\vert\,\mathbf{0}\}\mathcal{T}(\mathbf{k})$	2	-2	0	0

Table 9.5 Characters of the "relevant" irreducible representations of $\mathcal{G}(\mathbf{k})/\mathcal{T}(\mathbf{k})$ at the point X for the space group O_h^7.

	Z_1
$\{E\,\vert\,\mathbf{0}\}\mathcal{T}(\mathbf{k})$	2
$\{C_{2y}\,\vert\,\mathbf{0}\}\mathcal{T}(\mathbf{k})$	0
$\{IC_{2z}\,\vert\,\boldsymbol{\tau}\}\mathcal{T}(\mathbf{k})$	0
$\{IC_{2x}\,\vert\,\boldsymbol{\tau}\}\mathcal{T}(\mathbf{k})$	0

Table 9.6 Characters of the "relevant" irreducible representation of $\mathcal{G}(\mathbf{k})/\mathcal{T}(\mathbf{k})$ along the axis Z for the space group O_h^7.

pure primitive translations, transformations of the form $\{\mathbf{R}\,|\,\boldsymbol{\tau}_\mathbf{R}\}$, and products of members of these two sets. In any "relevant" irreducible representation of $\mathcal{G}(\mathbf{k})/\mathcal{T}(\mathbf{k})$, Equation (9.21) gives the matrices representing coset representatives of the first set, so it is clearly sufficient to specify *only* the matrices for coset representatives of the form $\{\mathbf{R}\,|\,\boldsymbol{\tau}_\mathbf{R}\}$. Consequently the left-hand columns of these tables do *not* give a complete listing of the classes of $\mathcal{G}(\mathbf{k})/\mathcal{T}(\mathbf{k})$. (Tables 9.5 and 9.6

were calculated by Herring (1942). Table 9.4 is the corrected version by Elliott (1954) of Herring's original table.)

For an allowed vector $\mathbf{k}$ on the symmetry plane UWX, $g_0(\mathbf{k})=2$, so there are only two inequivalent "relevant" irreducible representations of $\mathscr{G}(\mathbf{k})/\mathscr{T}(\mathbf{k})$, both of which must be one-dimensional. For these representations

$$\begin{aligned}[\Gamma^p_{\mathscr{G}(\mathbf{k})/\mathscr{T}(\mathbf{k})}(\{R(IC_{2z})\mid\boldsymbol{\tau}\}\mathscr{T}(\mathbf{k}))]^2 &= \Gamma^p_{\mathscr{G}(\mathbf{k})/\mathscr{T}(\mathbf{k})}(\{R(IC_{2z})\mid\boldsymbol{\tau}\}^2\mathscr{T}(\mathbf{k}))\\ &= \Gamma^p_{\mathscr{G}(\mathbf{k})/\mathscr{T}(\mathbf{k})}(\{1\mid\mathbf{a}_1\}\mathscr{T}(\mathbf{k}))\\ &= \exp(-i\mathbf{k}\cdot\mathbf{a}_1),\end{aligned}$$

by Equation (9.21), so

$$\Gamma^p_{\mathscr{G}(\mathbf{k})/\mathscr{T}(\mathbf{k})}(\{R(IC_{2z})\mid\boldsymbol{\tau}\}\mathscr{T}(\mathbf{k}))=[\pm\exp(-\tfrac{1}{2}i\mathbf{k}\cdot\mathbf{a}_1)],$$

the two possible signs giving the two irreducible representations.

Finally, if $\mathbf{k}$ is a general point either within the Brillouin zone or on the plane $KLUW$, $g_0(\mathbf{k})=1$, so there is only one "relevant" irreducible representation of $\mathscr{G}(\mathbf{k})$ and this is necessarily one-dimensional. As Equation (9.16) holds, Equation (9.17) implies that this representation is

$$\Gamma^1_{\mathscr{G}(\mathbf{k})}(\{1\mid\mathbf{t}_n\})=[\exp(-i\mathbf{k}\cdot\mathbf{t}_n)]$$

for all $\{1\mid\mathbf{t}_n\}\in\mathscr{T}\ (=\mathscr{G}(\mathbf{k}))$.

Figures 8.12 and 8.14 give examples of electronic energy bands and lattice vibrational spectra using the notation described above. Exactly as in Section 2(*b*), the zero-frequency *translational* modes of lattice vibration always belong to the irreducible representation Γ_{15} at $\mathbf{k}=\mathbf{0}$. The determination of the other irreducible representations that appear proceeds exactly as in Section 2(*b*), except that Equation (9.15) should be used in place of Equation (9.8).

4 Consequences of the fundamental theorems

(*a*) *Degeneracies of eigenvalues and the symmetry of* $\varepsilon(\mathbf{k})$ *and* $\omega(\mathbf{k})$

For convenience, the arguments will be given for electronic energy bands, but the results apply equally to lattice vibrational spectra. The discussion will be presented first for *symmorphic* space groups.

Suppose first that $\mathbf{k}$ is a *general point* of the Brillouin zone. Then $\mathscr{G}_0(\mathbf{k})$ consists only of the identity transformation $\{1\mid\mathbf{0}\}$ and so has only one irreducible representation, namely the one-dimensional representation for which $\Gamma^1(\{1\mid\mathbf{0}\})=[1]$. As $g_0(\mathbf{k})=1$, it follows from

Equation (9.4) that $M(\mathbf{k}) = g_0$. There is therefore only one irreducible representation of $\mathscr{G}$ corresponding to this $\mathbf{k}$, and its dimension is $M(\mathbf{k})d_1 = g_0 \,.\, 1 = g_0$, so that the corresponding energy eigenvalue is g_0-fold degenerate.

Any Bloch function $\exp(i\mathbf{k}\,.\,\mathbf{r})u_{\mathbf{k}}(\mathbf{r})$ with wave vector $\mathbf{k}$ is a basis function for the irreducible representation of this $\mathscr{G}_0(\mathbf{k})$, and the set of basis functions of the corresponding irreducible representation of $\mathscr{G}$ formed from this function are $\exp(i\mathbf{k}\,.\,\mathbf{r})u_{\mathbf{k}}(\mathbf{r})$, $\exp(i\mathbf{k}_2\,.\,\mathbf{r})u_{\mathbf{k}}(\mathbf{R}_2^{-1}\mathbf{r}), \ldots$. Now suppose that these Bloch functions are energy eigenfunctions. As they correspond to wave vectors $\mathbf{k}$ $(=\mathbf{k}_1)$, $\mathbf{k}_2, \ldots$, they correspond, according to Equation (8.21), to energy eigenvalues $\varepsilon(\mathbf{k})$, $\varepsilon(\mathbf{k}_2), \ldots$. As they are degenerate, being a basis for an irreducible representation of $\mathscr{G}$, it follows that

$$\varepsilon(\mathbf{R}_j\mathbf{k}) \equiv \varepsilon(\mathbf{k}_j) = \varepsilon(\mathbf{k}) \tag{9.23}$$

for $j = 1, 2, \ldots, g_0$, the $\mathbf{R}_j$ being the set of rotations of $\mathscr{G}_0$. Thus the g_0 wave vectors in the star of $\mathbf{k}$ have the same energies, and $\varepsilon(\mathbf{k})$ has the symmetry of the point group $\mathscr{G}_0$ of the space group $\mathscr{G}$. This means that if the band structure is known in one basic section of the Brillouin zone containing only $(1/g_0)$ of the volume of the Brillouin zone and no two wave vectors in the same star, then it can be obtained immediately throughout the *whole* Brillouin zone. For example, for the body-centred cubic space group O_h^9, whose Brillouin zone is shown in Figure 8.9, $g_0 = 48$ and the basic section is the wedge-shaped region ΓHNP (or, more precisely, the region bounded by the planes containing three of the four points Γ, H, N and P).

The symmetry of $\varepsilon(\mathbf{k})$ is widely employed in calculations of energy band structures and in determinations of the Fermi surface from experimental measurements, such as those of the de Haas–van Alphen effect. (Incidentally, it has been shown (Cornwell 1964) that the Fermi surface of the "many-electron" theory of Luttinger (1960) continues to have the symmetry of $\mathscr{G}_0$ even though the "single-particle" energies $\varepsilon(\mathbf{k})$ lose their usual meaning.) For a general point of the Brillouin zone the inclusion of the rotational parts of $\mathscr{G}$ in addition to the translational parts of $\mathscr{T}$ already taken into account in Bloch's Theorem is of *no* assistance in simplifying the numerical task of actually finding the energy eigenvalues by the technique described in Chapter 6, Section 1. For this reason, relatively few accurate calculations of $\varepsilon(\mathbf{k})$ have been performed for general points of the Brillouin zone.

For lattice vibrational spectra the result corresponding to Equation (9.23) is

$$\omega(\mathbf{R}_j\mathbf{k}) \equiv \omega(\mathbf{k}_j) = \omega(\mathbf{k})$$

for $j = 1, 2, \ldots, g_0$.

The second case that will be considered is at the other extreme. For the point $\mathbf{k}=\mathbf{0}$, $\mathbf{R}\mathbf{k}=\mathbf{0}$ for every $\{\mathbf{R}\,|\,\mathbf{0}\}$ of $\mathcal{G}_0$, so that $\mathcal{G}_0(\mathbf{0})=\mathcal{G}_0$. The star of $\mathbf{k}$ then consists only of $\mathbf{k}=\mathbf{0}$ itself and so $M(\mathbf{0})=1$. The basis functions of $\mathcal{G}$ corresponding to $\mathbf{k}=\mathbf{0}$ are merely the periodic basis functions of $\mathcal{G}_0$, and the corresponding degeneracies of energy eigenvalues are those of the dimensions of the irreducible representations of $\mathcal{G}_0$. In this case the technique of Chapter 6, Section 1, allows an appreciable simplification of the numerical work involved in finding the energy eigenvalues, even beyond that already brought about by the consideration of $\mathcal{T}$ alone.

Some space groups possess other symmetry points for which $\mathcal{G}_0(\mathbf{k})=\mathcal{G}_0$. These do not require further examination, as the comments made about the point $\mathbf{k}=\mathbf{0}$ also apply to these points. The point H of the Brillouin zone of the body-centred cubic lattice is an example.

The third and final case is that of the intermediate situation in which $\mathcal{G}_0(\mathbf{k})$ is not trivial but is a proper subgroup of $\mathcal{G}_0$. Included in this case are all symmetry points other than those of the second case and all points on symmetry axes and planes. For such a point $\mathcal{G}_0(\mathbf{k})$ will have more than one irreducible representation, some of these possibly being of more than one dimension. As $g_0(\mathbf{k})<g_0$, it follows from Equation (9.4) that $M(\mathbf{k})>1$.

Consider the energy eigenvalue corresponding to a d_p-dimensional irreducible representation of $\mathcal{G}_0(\mathbf{k})$. This will be $d_pM(\mathbf{k})$-fold degenerate by the theorem, and this degeneracy is made up as follows:

(*a*) By a similar argument to that used in the case of a general point, it follows that

$$\varepsilon(\mathbf{k}_j)=\varepsilon(\mathbf{k}),\qquad j=1,2,\ldots,M(\mathbf{k}),$$

so that again $\varepsilon(\mathbf{k})$ exhibits the symmetry of the point group $\mathcal{G}_0$.

(*b*) In addition, each $\varepsilon(\mathbf{k}_j)$ is "d_p-fold degenerate", in the sense that there are d_p linearly independent Bloch eigenfunctions of $\mathrm{H}(\mathbf{r})$ corresponding to this eigenvalue *and* to this particular wave vector $\mathbf{k}_j$. The degree of simplification of numerical work depends on the order of $g_0(\mathbf{k})$.

It is very easy to generalize these arguments to *non-symmorphic* space groups. The symmetry relation (Equation (9.23)) remains true, so that $\varepsilon(\mathbf{k})$ continues to have the symmetry of the point group $\mathcal{G}_0$. Also, the energy eigenvalue corresponding to a d_p-dimensional "relevant" irreducible representation of $\mathcal{G}(\mathbf{k})$ (or equivalently of $\mathcal{G}(\mathbf{k})/\mathcal{T}(\mathbf{k})$) will be $d_pM(\mathbf{k})$-fold degenerate, this degeneracy being made up in exactly the same way as above.

These arguments show that, although the concept of a star appears in the fundamental theorems for both symmorphic and non-symmorphic space groups, it is in some contexts possible and convenient to revert to the *description* that appeared in Chapter 8 with Bloch's Theorem, in which there corresponds a set of energy levels to *every* allowed **k**-vector of the Brillouin zone and not merely to those lying in different stars. In this description, a d_p-fold degeneracy of $\varepsilon(\mathbf{k})$ means, as above, that there are d_p linearly independent Bloch eigenfunctions of $H(\mathbf{r})$ corresponding to this eigenvalue *and* to the particular wave vector **k**. A d_p-fold degeneracy of $\varepsilon(\mathbf{k})$ then corresponds to a d_p-dimensional irreducible representation of $\mathcal{G}_0(\mathbf{k})$ for a symmorphic space group (or of $\mathcal{G}(\mathbf{k})$ for a non-symmorphic space group). This degeneracy was indicated in Figures 8.11 to 8.14 by $[d_p]$. Furthermore each energy band then has the symmetry of $\mathcal{G}_0$. This is the description that is commonly used in the solid state literature.

It is worthwhile pointing out here that even if $\mathcal{G}_0$ does not contain the spatial inversion operator I, the symmetry $\varepsilon(\mathbf{k}) = \varepsilon(-\mathbf{k})$ always remains because of time-reversal symmetry, as was shown in Equation (8.27).

(b) *Continuity and compatibility of the irreducible representations of $\mathcal{G}_0(\mathbf{k})$ and $\mathcal{G}(\mathbf{k})$*

Again the arguments will be presented in the context of electronic energy bands, but the conclusions still apply to lattice vibrational spectra. For convenience, the discussion will be given for a symmorphic space group, but everything applies equally to non-symmorphic space groups, provided only that references to irreducible representations of $\mathcal{G}_0(\mathbf{k})$ are replaced by references to "relevant" irreducible representations of $\mathcal{G}(\mathbf{k})$.

A typical section of an energy band diagram for a symmetry axis is shown in Figure 9.1. The axis displayed there is the Δ axis of the cubic space groups O_h^1, O_h^5 and O_h^9. The numbers in parentheses are the band labels, as defined in Expression (8.26), so that the bands 1 and 2 "touch" at one point $\mathbf{k}_0$. This figure exhibits two general characteristics of energy bands.

The first feature is that, along any part of a band which does not touch another band, the corresponding irreducible representation of $\mathcal{G}_0(\mathbf{k})$ remains the same. Thus, for example, the whole of the left-hand part of band 1 corresponds to Δ_2 and the whole of the right-hand part to $\Delta_{1'}$. It is therefore possible to talk of *the* symmetry of a band, or part of a band, along an axis. The reason for this behaviour is essentially that the energy eigenvalues corresponding to a particular irreducible representation of $\mathcal{G}_0(\mathbf{k})$ can be obtained from a secular

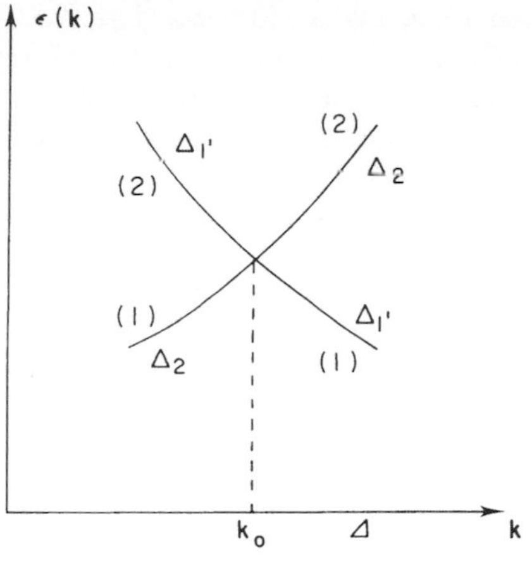

Figure 9.1 "Touching" of energy bands along the Δ axis of the cubic space groups.

equation involving only basis functions of that representation, as was noted in Chapter 6, Section 1. A small change in **k** will then only produce a small change in the energy eigenvalues produced from this secular equation.

The second general characteristic is that the symmetries of the bands are *interchanged* at a point where the bands touch. For example, in Figure 9.1 band 1 changes from Δ_2 to $\Delta_{1'}$ on moving from left to right, while band 2 changes from $\Delta_{1'}$ to Δ_2. The reason for this is that the secular equation corresponding to an irreducible representation of $\mathscr{G}_0(\mathbf{k})$ produces energy eigenvalues that are analytic functions of **k**, the degeneracy corresponding to the touching of two bands having no effect on this as it is "accidental". This also implies that $\text{grad}_{\mathbf{k}}\, \varepsilon(\mathbf{k})$ is continuous for bands, or parts of bands, belonging to the *same* irreducible representation, and this continuity is not affected by the interchange of band labels when bands touch. Thus, for example, in Figure 9.1 the limit as $\mathbf{k}\rightarrow\mathbf{k}_0$ from the left of $\text{grad}_{\mathbf{k}}\, \varepsilon(\mathbf{k})$ for band 1 is equal to the limit as $\mathbf{k}\rightarrow\mathbf{k}_0$ from the right of $\text{grad}_{\mathbf{k}}\, \varepsilon(\mathbf{k})$ for band 2. The "touching" of energy bands has been investigated in detail by Herring (1937b).

The concept of "compatibility" is best described by an example. Consider therefore a Γ_{12} energy level at the point Γ of the Brillouin zone for the body-centred cubic space group O_h^9. This level, being two-fold degenerate, belongs to two energy bands. What then are the symmetries of these bands near along the symmetry axis Δ? That is, what irreducible representations of the group $\mathscr{G}_0(\mathbf{k})$ for Δ are "compatible" with Γ_{12}?

The investigation proceeds as follows. Let $\mathscr{G}_0(\Gamma)$ and $\mathscr{G}_0(\Delta)$ be the point groups $\mathscr{G}_0(\mathbf{k})$ for $\mathbf{k}$ at Γ and on Δ respectively. Then $\mathscr{G}_0(\Delta)$ is a subgroup of $\mathscr{G}_0(\Gamma)$ and so the irreducible representations of $\mathscr{G}_0(\Gamma)$ are representations of $\mathscr{G}_0(\Delta)$ that are, in general, reducible. The actual reduction can be determined immediately from the characters, for if $\mathbf{\Gamma}$ and $\mathbf{\Gamma}^p$ are irreducible representations of $\mathscr{G}_0(\Gamma)$ and $\mathscr{G}_0(\Delta)$ respectively, with characters denoted by χ and χ^p, then the number of times n_p that $\mathbf{\Gamma}^p$ appears in the reduction of $\mathbf{\Gamma}$ is given by

$$n_p = (1/g_0(\Delta)) \sum_{\{\mathbf{R}|\mathbf{0}\}\in\mathscr{G}_0(\Delta)} \chi(\{\mathbf{R}\,|\,\mathbf{0}\})\chi^p(\{\mathbf{R}\,|\,\mathbf{0}\})^*,$$

where $g_0(\Delta)$ is the order of $\mathscr{G}_0(\Delta)$. (This is just an immediate application of Theorem V of Chapter 4, Section 6.) Thus, for example, the reduction of Γ_{12} in this context is given by

$$\Gamma_{12} = \Delta_1 \oplus \Delta_2.$$

The point Γ could be considered as an ordinary point of Δ by ignoring the elements of $\mathscr{G}_0(\Gamma)$ that are not in $\mathscr{G}_0(\Delta)$, and then the energy levels at Γ could be classified in terms of the irreducible representations of $\mathscr{G}_0(\Delta)$. The Γ_{12} level would then be regarded as a non-degenerate Δ_1 level and a non-degenerate Δ_2 level that happen to have the same value. Because of the continuity of irreducible representations along that was mentioned above, the two bands that touch at Γ in the Γ_{12} level will along the Δ axis near Γ have symmetries Δ_1 and Δ_2. The degeneracy that exists at Γ is split on moving away from Γ. This situation is shown in Figure 9.2.

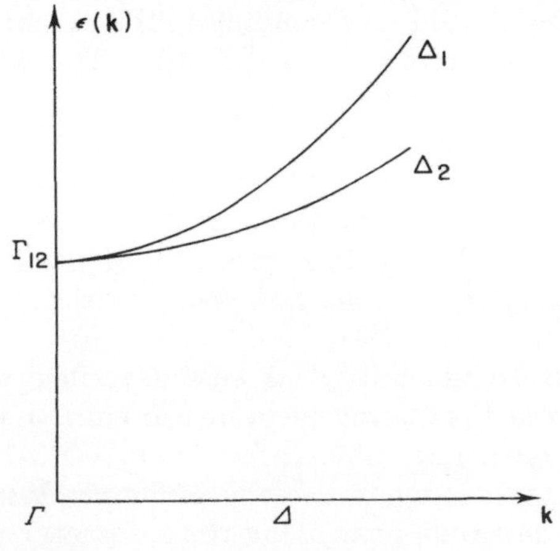

Figure 9.2 "Compatibility" of the Γ_{12} energy level with the Δ_1 and Δ_2 energy levels along the Δ axis.

Although this argument shows that the Δ_1 and Δ_2 irreducible representations are compatible with the Γ_{12} level, it cannot predict whether the band having Δ_1 symmetry has lower or higher than the band having Δ_2 symmetry. This can only be determined by direct calculation.

The same analysis can be applied to all irreducible representations at every symmetry point for all the symmetry axes going through that point. A similar analysis can also be used to determine the compatibility of the irreducible representations corresponding to a symmetry axis with those of the symmetry planes containing the axis. These results can be expressed in "compatibility tables". A typical example is exhibited in Table 9.7. A comprehensive set of such tables for the space groups O_h^1, O_h^5 and O_h^9 was given by Bouckaert *et al.* (1936) (see also Cornwell (1969)).

Γ_1	Γ_2	Γ_{12}	Γ_{15}	Γ_{25}	$\Gamma_{1'}$	$\Gamma_{2'}$	$\Gamma_{12'}$	$\Gamma_{15'}$	$\Gamma_{25'}$
Δ_1	Δ_2	$\Delta_1\Delta_2$	$\Delta_1\Delta_5$	$\Delta_2\Delta_5$	$\Delta_{1'}$	$\Delta_{2'}$	$\Delta_{1'}\Delta_{2'}$	$\Delta_{1'}\Delta_5$	$\Delta_{2'}\Delta_5$
Σ_1	Σ_4	$\Sigma_1\Sigma_4$	$\Sigma_1\Sigma_3\Sigma_4$	$\Sigma_1\Sigma_2\Sigma_4$	Σ_2	Σ_3	$\Sigma_2\Sigma_3$	$\Sigma_2\Sigma_3\Sigma_4$	$\Sigma_1\Sigma_2\Sigma_3$
Λ_1	Λ_2	Λ_3	$\Lambda_1\Lambda_3$	$\Lambda_2\Lambda_3$	Λ_2	Λ_1	Λ_3	$\Lambda_2\Lambda_3$	$\Lambda_1\Lambda_3$

Table 9.7 Compatibility relations between the symmetry point Γ and the symmetry axes Δ, Σ and Λ for the cubic space groups O_h^1, O_h^5, O_h^7 and O_h^9.

It is possible to develop this type of analysis to investigate certain finer features of the electronic energy bands, particularly the vanishing of components of $\mathrm{grad}_{\mathbf{k}}\, \varepsilon(\mathbf{k})$ and the form of the intersection of constant energy contours with symmetry axes. This allows the location of the "critical points" (van Hove 1953, Phillips 1956) to be located. For details see Cornwell (1969), and also Rashba (1959), Sheka (1960) and Kudryavtseva (1967).

(c) *Origin and orientation dependence of the symmetry labelling of electronic and lattice vibrational states*

Consider, as an example, the NaCl structure, in which the Na nuclei occupy the lattice sites of one Γ_c^f lattice and the Cl nuclei occupy the lattice sites of another Γ_c^f lattice, one lattice being obtained from the other by a pure translation through $\mathbf{t}_0 = \frac{1}{2}a(1,0,0)$. (Here a is the quantity appearing in the definition of the basic lattice vectors of Γ_c^f in Table 8.1). With the origin of the coordinate system taken at one of the Na nuclei, Equation (1.13) is satisfied for every coordinate transformation T of the space group O_h^5. Similarly, with the origin at a Cl nucleus, Equation (1.13) is again satisfied for every T of O_h^5. Let $H(\mathbf{r})$ and $H'(\mathbf{r})$ be the Hamiltonian operators corresponding to these two choices of origin. Although $H(\mathbf{r})$ and $H'(\mathbf{r})$ are related, they are obviously *not* identical, that is, $H(\mathbf{r}) \neq H'(\mathbf{r})$.

It can be shown (Cornwell 1971e) in a situation such as this that, while the actual *values* of the electronic energy levels and lattice vibrational frequencies are naturally independent of the choice of origin of the coordinate system, the *labelling* of the states in terms of the irreducible representations of the space group can be origin dependent. In fact the wave vector label $\mathbf{k}$ is always unaffected by the translation, but for a symmorphic space group the translation $\{\mathbf{1} \mid \mathbf{t}_0\}$ changes the state at $\mathbf{k}$ corresponding to the irreducible representation $\Gamma^p_{\mathscr{G}_0(\mathbf{k})}$ of $\mathscr{G}_0(\mathbf{k})$ into a state corresponding to $\Gamma^q_{\mathscr{G}_0(\mathbf{k})}$, where

$$\Gamma^q_{\mathscr{G}_0(\mathbf{k})}(\{\mathbf{R} \mid \mathbf{0}\}) = \exp\{-i\mathbf{k}\,.\,(\mathbf{R}\mathbf{t}_0 - \mathbf{t}_0)\}\Gamma^p_{\mathscr{G}_0(\mathbf{k})}(\{\mathbf{R} \mid \mathbf{0}\}) \tag{9.24}$$

for all $\{\mathbf{R} \mid \mathbf{0}\} \in \mathscr{G}_0(\mathbf{k})$. The corresponding relation for a non-symmorphic space group is

$$\Gamma^q_{\mathscr{G}(\mathbf{k})}(\{\mathbf{R} \mid \boldsymbol{\tau}_\mathbf{R} + \mathbf{t}_n\}) = \exp\{-i\mathbf{k}\,.\,(\mathbf{R}\mathbf{t}_0 - \mathbf{t}_0)\}\Gamma^p_{\mathscr{G}(\mathbf{k})}(\{\mathbf{R} \mid \boldsymbol{\tau}_\mathbf{R} + \mathbf{t}_n\})$$

for all $\{\mathbf{R} \mid \boldsymbol{\tau}_\mathbf{R} + \mathbf{t}_n\} \in \mathscr{G}(\mathbf{k})$. Thus in all cases a necessary and sufficient condition that the translation $\{\mathbf{1} \mid \mathbf{t}_0\}$ does *not* change the labelling of states with wave vector $\mathbf{k}$ is that

$$\exp\{-i\mathbf{k}\,.\,(\mathbf{R}\mathbf{t}_0 - \mathbf{t}_0)\} = 1 \tag{9.25}$$

for all $\{\mathbf{R} \mid \mathbf{0}\} \in \mathscr{G}_0(\mathbf{k})$. This condition is always satisfied for every internal point $\mathbf{k}$ of the Brillouin zone (as for such a point $\mathbf{R}\mathbf{k} = \mathbf{k}$ for all $\{\mathbf{R} \mid \mathbf{0}\} \in \mathscr{G}_0(\mathbf{k})$), but need not be satisfied for points on the surface of the Brillouin zone. (The condition is satisfied if $\mathbf{t}_0$ is a lattice vector of the crystal lattice, so that changing the origin from one lattice site to another on the *same* lattice does not change the labelling of states, which is exactly what one would expect.)

Example I *Label changes at the point L for the space group* O_h^5
With $\mathbf{t}_0 = \frac{1}{2}a(1, 0, 0)$ (as in the NaCl example mentioned above) and $\mathbf{k} = (\pi/a)(1, 1, 1)$,

$$\exp\{-i\mathbf{k}\,.\,(\mathbf{R}\mathbf{t}_0 - \mathbf{t}_0)\} = \begin{cases} 1, \text{ for } E, C_{3\delta}, C_{3\delta}^{-1}, IC_{2b}, IC_{2d}, IC_{2f}, \\ -1, \text{ for } I, IC_{3\delta}, IC_{3\delta}^{-1}, C_{2b}, C_{2d}, C_{2f}, \end{cases}$$

so that by Equation (9.24) the translation $\{\mathbf{1} \mid \mathbf{t}_0\}$ causes the following changes in state labels:

$$L_1 \to L_{2'},\ L_2 \to L_{1'},\ L_3 \to L_{3'},\ L_{1'} \to L_2,\ L_{2'} \to L_1,\ L_{3'} \to L_3.$$

Clearly when this situation occurs it is essential to state precisely which origin has been chosen for a calculation, although this is rarely done. In practice the problem only arises for crystals containing nuclei of more than one species, for in other cases there is little temptation to choose the origin other than at a lattice site.

The more complicated dependence of the symmetry labels of states on the choice of the *orientation* of the coordinate axes has been discussed in detail elsewhere (Cornwell 1972a) and will not be repeated here.

5 Irreducible representations of double space groups

The theorems of Sections 2 and 3 for *single* space groups have immediate generalizations to the case of *double* space groups. For simplicity, consider first the *symmorphic* space groups. In Theorems I and II of Section 2 the only change that is necessary is to replace the single groups $\mathcal{G}, \mathcal{G}_0$ and $\mathcal{G}_0(\mathbf{k})$ by their corresponding double groups $\mathcal{G}^D, \mathcal{G}_0^D$, and $\mathcal{G}_0^D(\mathbf{k})$. In particular, $\mathcal{G}^D$ is isomorphic to the semi-direct product $\mathcal{T} \circledS \mathcal{G}_0^D$ and the irreducible representations of $\mathcal{G}^D$ are induced from the "little groups" $\mathcal{G}_0^D(\mathbf{k})$, the generalization of Equations (9.5) being

$$\Gamma^{\mathbf{k}p}([\mathbf{R} \mid \mathbf{t_n}])_{kt,jr} = \begin{cases} \exp\{-i(\mathbf{R}_k\mathbf{k}) \,.\, \mathbf{t_n}\}\Gamma^p_{\mathcal{G}_0^D(\mathbf{k})}([\mathbf{R}_k \mid \mathbf{0}]^{-1}[\mathbf{R} \mid \mathbf{0}][\mathbf{R}_j \mid \mathbf{0}])_{tr}, \\ \quad \text{if } [\mathbf{R}_k \mid \mathbf{0}]^{-1}[\mathbf{R} \mid \mathbf{0}][\mathbf{R}_j \mid \mathbf{0}] \in \mathcal{G}_0^D(\mathbf{k}), \\ 0, \text{ otherwise,} \end{cases}$$

and

$$\Gamma^{\mathbf{k}p}([\bar{\mathbf{R}} \mid \mathbf{t_n}])_{kt,jr}, = \begin{cases} \exp\{-i(\mathbf{R}_k\mathbf{k}) \,.\, \mathbf{t_n}\}\Gamma^p_{\mathcal{G}_0^D(\mathbf{k})}([\mathbf{R}_k \mid \mathbf{0}]^{-1}[\bar{\mathbf{R}} \mid \mathbf{0}][\mathbf{R}_j \mid \mathbf{0}])_{tr}, \\ \quad \text{if } [\mathbf{R}_k \mid \mathbf{0}]^{-1}[\bar{\mathbf{R}} \mid \mathbf{0}][\mathbf{R}_j \mid \mathbf{0}] \in \mathcal{G}_0^D(\mathbf{k}), \\ 0, \text{ otherwise} \end{cases}$$

Clearly the "extra" irreducible representations of $\mathcal{G}^D$ are induced from the "extra" irreducible representations of the $\mathcal{G}_0^D(\mathbf{k})$. (As $\mathcal{G}_0^D(\mathbf{k})$ contains both $[\mathbf{1} \mid \mathbf{0}]$ and $[\bar{\mathbf{1}} \mid \mathbf{0}]$, if $[\mathbf{R}_j \mid \mathbf{0}]$ is a member of a left coset of $\mathcal{G}_0^D$ with respect to $\mathcal{G}_0^D(\mathbf{k})$, then so is $[\bar{\mathbf{R}}_j \mid \mathbf{0}]$. Consequently the set of coset representatives may all be taken to be "unbarred rotations" $[\mathbf{R}_j \mid \mathbf{0}]$, and the "star" of $\mathbf{k}$ may be defined to be the corresponding set of vectors $\mathbf{k}_j = \{\mathbf{R}_j \mid \mathbf{0}\}\mathbf{k}, j = 1, 2, \ldots, M(\mathbf{k})$). Similarly, in the generalization of Theorems III and IV of Section 2, the only additional changes are that the $\psi_{jr}^{\mathbf{k}p}(\mathbf{r})$ and $\phi_r^{\mathbf{k}p}(\mathbf{r})$ must be replaced by two-component spinors $\boldsymbol{\psi}_{jr}^{\mathbf{k}p}(\mathbf{r})$ and $\boldsymbol{\phi}_r^{\mathbf{k}p}(\mathbf{r})$, and $P(\{\mathbf{R}_j \mid \mathbf{0}\})$ must be replaced by $O([\mathbf{R}_j \mid \mathbf{0}])$.

All the characters of the "extra" irreducible representations of the double crystallographic point groups are given in Appendix D, Section 2.

The situation is very similar for *double non-symmorphic* space groups. $\mathcal{G}(\mathbf{k})$ must be replaced by its double group $\mathcal{G}^D(\mathbf{k})$, and again the coset representations for the decomposition of $\mathcal{G}^D$ into left cosets with respect to $\mathcal{G}^D(\mathbf{k})$ may be taken to be "unbarred" transformations $[\mathbf{R}_j \mid \boldsymbol{\tau}_j]$. Then Equations (9.13), (9.14) and (9.15) generalize in the obvious way. With the group $\mathcal{T}(\mathbf{k})$ of Section 3(*b*) redefined as the set of generalized translations $[\mathbf{1} \mid \mathbf{t}_\mathbf{n}]$ such that $\exp(-i\mathbf{k}\,.\,\mathbf{t}_\mathbf{n})=1$, so that $\mathcal{T}(\mathbf{k})$ is an invariant subgroup of $\mathcal{G}^D(\mathbf{k})$. Theorem IV of Section 3 has an equally obvious generalization in which $\mathcal{G}(\mathbf{k})/\mathcal{T}(\mathbf{k})$ is replaced by $\mathcal{G}^D(\mathbf{k})/\mathcal{T}(\mathbf{k})$.

The consequences of these theorems on double groups for degeneracies of energy eigenvalues and the symmetry of $\varepsilon(\mathbf{k})$ are exactly as given for single groups in Section 4(*a*), except that scalar functions have to be replaced by two-component spinors. In particular, $\varepsilon(\mathbf{k})$ retains the symmetry of the point group $\mathcal{G}_0$. Similarly, continuity and compatibility may be discussed in the same way as in Section 4(*b*).

The double groups corresponding to O_h^1, O_h^5, O_h^7 and O_h^9 were first discussed by Elliott (1954), whose notation appears in Appendix D, Section 2, where relevant. (The labelling of the "extra" irreducible representations is related to the choice of the sign of the matrices $\mathbf{u}(\mathbf{R}_\mathbf{p})$. The particular choice made in Table D.26 is consistent with the labelling conventions of Elliott (1954), Parmenter (1955) and Dresselhaus (1955) (cf. Chapter 6, Section 4(*b*), and Cornwell (1969)).

It will now be shown that for *symmorphic* space groups the splitting of energy levels caused by spin-orbit coupling can be determined directly from the *point* groups $\mathcal{G}_0(\mathbf{k})$. (The general theory of this splitting was given in Chapter 6, Section 4(*f*).) Theorem III of Section 2 and its generalization to double groups and Theorem VIII of Chapter 6, Section 4 (*f*) together imply that if $\Gamma^{\mathbf{k}p}$ is an irreducible representation of the single space group $\mathcal{G}$ induced from the irreducible representation $\Gamma^p_{\mathcal{G}_0(\mathbf{k})}$ of $\mathcal{G}_0(\mathbf{k})$, then $\Gamma^{\mathbf{k}p}\otimes\mathbf{u}$ is the representation of the double space group $\mathcal{G}^D$ induced from the representation $\Gamma^p_{\mathcal{G}_0(\mathbf{k})}\otimes\mathbf{u}$ of $\mathcal{G}_0^D(\mathbf{k})$. Moreover, if $\Gamma^q_{\mathcal{G}_0^D(\mathbf{k})}$, $\Gamma^r_{\mathcal{G}_0^D(\mathbf{k})}, \ldots$ are "extra" irreducible representations of $\mathcal{G}_0^D(\mathbf{k})$ such that

$$\Gamma^p_{\mathcal{G}_0(\mathbf{k})}\otimes\mathbf{u}\approx n_q\Gamma^q_{\mathcal{G}_0^D(\mathbf{k})}\oplus n_r\Gamma^r_{\mathcal{G}_0^D(\mathbf{k})}\oplus\ldots \tag{9.26}$$

then

$$\Gamma^{\mathbf{k}p}\otimes\mathbf{u}\approx n_q\Gamma^{\mathbf{k}q}\oplus n_r\Gamma^{\mathbf{k}r}\oplus\ldots, \tag{9.27}$$

where $\Gamma^{\mathbf{t}q}, \Gamma^{\mathbf{k}r}, \ldots$ are the "extra" irreducible representations of $\mathscr{G}^D$ induced from $\Gamma^q_{\mathscr{G}_0^D(\mathbf{k})}, \Gamma^r_{\mathscr{G}_0^D(\mathbf{k})}, \ldots$. As Equations (6.50) take the form of Equation (9.27) in the present context, the reduction series of Equation (9.26) for the *point* groups $\mathscr{G}_0^D(\mathbf{k})$ completely determines how the splitting takes place.

The situation is very similar for ***non-symmorphic*** space groups, the only changes being that $\mathscr{G}_0(\mathbf{k})$ and $\mathscr{G}_0^D(\mathbf{k})$ in the above discussion have to be replaced by $\mathscr{G}(\mathbf{k})$ and $\mathscr{G}^D(\mathbf{k})$ respectively. Indeed, for an internal point $\mathbf{k}$ of the Brillouin zone, Equation (9.17) and its generalization imply that the analysis reduces again to the point groups $\mathscr{G}_0(\mathbf{k})$ and $\mathscr{G}_0^D(\mathbf{k})$.

Table 9.8 gives the reduction series of Equation (9.27) for the point Γ of the cubic space groups O_h^1, O_h^5, O_h^7 and O_h^9. As an example of its

Γ_j	Γ_1	Γ_2	Γ_{12}	$\Gamma_{15'}$	$\Gamma_{25'}$	$\Gamma_{1'}$	$\Gamma_{2'}$	$\Gamma_{12'}$	Γ_{15}	Γ_{25}
$\Gamma_j \otimes \mathbf{u}$	Γ_6^+	Γ_7^+	Γ_8^+	$\Gamma_6^+ \oplus \Gamma_8^+$	$\Gamma_7^+ \oplus \Gamma_8^+$	Γ_6^-	Γ_7^-	Γ_8^-	$\Gamma_6^- \oplus \Gamma_8^-$	$\Gamma_7^- \oplus \Gamma_8^-$

Table 9.8 Reduction series for the point Γ for the cubic space groups O_h^1, O_h^5, O_h^7 and O_h^9.

use, consider the energy levels at Γ, which, in the absence of spin-orbit coupling, correspond to the three-dimensional representation $\Gamma_{15'}$. When the two possible linearly independent spin functions that can be attached to every orbital eigenfunction are taken into account, these levels are six-fold degenerate. As $\Gamma_{15'} \otimes \mathbf{u} \approx \Gamma_6^+ \oplus \Gamma_8^+$, and Γ_6^+ and Γ_8^+ have dimensions 2 and 4 respectively (see Table D.27), the introduction of spin-orbit coupling splits the six-fold degenerate level into a two-fold degenerate level and a four-fold degenerate level. (This argument cannot predict which of these is the highest.)

6 Selection rules

As shown in the general treatment of Chapter 6, Section 2, a list of forbidden transitions can be determined from the Clebsch–Gordan series alone. In the present context the invariance group of the unperturbed system is a crystallographic space group $\mathscr{G}$, whose irreducible representations $\Gamma^{\mathbf{k}p}$ are specified by an allowed vector $\mathbf{k}$ and a label p of an irreducible representation of $\mathscr{G}_0(\mathbf{k})$ (if $\mathscr{G}$ is symmorphic) or of $\mathscr{G}(\mathbf{k})$ (if $\mathscr{G}$ is non-symmorphic). If the wave functions $\phi_i(\mathbf{r})$ and $\phi_f(\mathbf{r})$ transform as some row of $\Gamma^{\mathbf{k}p}$ and $\Gamma^{\mathbf{k}''p''}$ respectively, and if the operator Q governing the process under consideration is an irreducible tensor operator of $\Gamma^{\mathbf{k}'p'}$, the transition from $\phi_i(\mathbf{r})$ to $\phi_f(\mathbf{r})$ is forbidden (to first order) if $n^{\mathbf{k}''\mathbf{p}''}_{\mathbf{kp},\mathbf{k}'\mathbf{p}'}=0$, where $n^{\mathbf{k}''\mathbf{p}''}_{\mathbf{kp},\mathbf{k}'\mathbf{p}'}$ is the number of times that $\Gamma^{\mathbf{k}''\mathbf{p}''}$ occurs in the reduction of $\Gamma^{\mathbf{kp}}\otimes\Gamma^{\mathbf{k}'\mathbf{p}'}$.

Consider first the case in which $\mathscr{G}$ is symmorphic. Substitution of Equation (9.8) in Equation (5.17) shows that $n^{\mathbf{k}''\mathbf{p}''}_{\mathbf{kp},\mathbf{k}'\mathbf{p}'}\neq 0$ only if there exist vectors $\mathbf{k}_i$, $\mathbf{k}'_j$ and $\mathbf{k}''_k$ from the stars of $\mathbf{k}$, $\mathbf{k}'$ and $\mathbf{k}''$ respectively, such that

$$\mathbf{k}_i+\mathbf{k}'_j=\mathbf{k}''_k+\mathbf{K}_\mathbf{m}, \tag{9.28}$$

where $\mathbf{K}_\mathbf{m}$ is some reciprocal lattice vector (which may be zero). When this is so, the labelling of vectors within each star may be chosen so that

$$\mathbf{k}+\mathbf{k}'=\mathbf{k}''+\mathbf{K}_\mathbf{m}, \tag{9.29}$$

where the reciprocal lattice vector $\mathbf{K}_\mathbf{m}$ may be zero. An example of such a relabelling will be given shortly. Then, after some algebra (see Cornwell (1969) for details),

$$n^{\mathbf{k}''\mathbf{p}''}_{\mathbf{kp},\mathbf{k}'\mathbf{p}'}=(1/g')\sum \chi^{\mathbf{p}}_{\mathscr{G}_0(\mathbf{k})}(\{\mathbf{R}\mid\mathbf{0}\})\chi^{\mathbf{p}'}_{\mathscr{G}_0(\mathbf{k}')}(\{\mathbf{R}\mid\mathbf{0}\})\chi^{\mathbf{p}''}_{\mathscr{G}_0(\mathbf{k}'')}(\{\mathbf{R}\mid\mathbf{0}\})^*, \tag{9.30}$$

where the sum is over all $\{\mathbf{R}\mid\mathbf{0}\}$ of $\mathscr{G}_0(\mathbf{k},\mathbf{k}',\mathbf{k}'')$, the common subgroup of $\mathscr{G}_0(\mathbf{k})$, $\mathscr{G}_0(\mathbf{k}')$, and $\mathscr{G}_0(\mathbf{k}'')$, and where g' is the order of $\mathscr{G}_0(\mathbf{k},\mathbf{k}',\mathbf{k}'')$.

The situation is very similar for a non-symmorphic space group. In this case it is Equation (9.15) that has to be substituted into Equation (5.17). Again $n^{\mathbf{k}''\mathbf{p}''}_{\mathbf{kp},\mathbf{k}'\mathbf{p}'}\neq 0$ only if Equation (9.28) holds. When this is so, and with the relabelling that gives Equation (9.29),

$$n^{\mathbf{k}''\mathbf{p}''}_{\mathbf{kp},\mathbf{k}'\mathbf{p}'}=(1/g')\sum \chi^{\mathbf{p}}_{\mathscr{G}(\mathbf{k})}(\{\mathbf{R}\mid\boldsymbol{\tau}_\mathbf{R}\})\chi^{\mathbf{p}'}_{\mathscr{G}(\mathbf{k}')}(\{\mathbf{R}\mid\boldsymbol{\tau}_\mathbf{R}\})\chi^{\mathbf{p}''}_{\mathscr{G}(\mathbf{k}'')}(\{\mathbf{R}\mid\boldsymbol{\tau}_\mathbf{R}\})^*,$$

where again the sum is over all $\{\mathbf{R}\mid\mathbf{0}\}$ of $\mathscr{G}_0(\mathbf{k},\mathbf{k}',\mathbf{k}'')$. (Direct proofs of this result have been given by Zak (1962) and Bradley (1966). The relationships with alternative methods of calculating selection rules have been studied by Lax (1965) and Birman (1966).)

As an example of the application of Equation (9.30), consider the body-centred cubic space group O_h^9 and the case where $\mathbf{k}=(0,0,k)$, $\mathbf{k}'=(0,k,0)$ and $\mathbf{k}''=(0,k,k)$. Clearly here $\mathbf{k}+\mathbf{k}'=\mathbf{k}''$. The point $\mathbf{k}''$ is on a Σ axis, while $\mathbf{k}$ and $\mathbf{k}'$ are on two different Δ axes. In fact $\mathbf{k}'$ is in the star of $\mathbf{k}$, and $\mathbf{k}'=\mathbf{R}(C_{4x})\mathbf{k}$. This is an example of the relabelling that is referred to above. The groups $\mathscr{G}_0(\mathbf{k})$ and $\mathscr{G}_0(\mathbf{k}')$ are here isomorphic, and if $\mathbf{R}(T)$ is a rotation of $\mathscr{G}_0(\mathbf{k})$, then the corresponding rotation of $\mathscr{G}_0(\mathbf{k}')$ is $\mathbf{R}(C_{4x})\mathbf{R}(T)\mathbf{R}(C_{4x}^{-1})$. The character table of $\mathscr{G}_0(\mathbf{k}')$ is then easily constructed from that given for $\mathscr{G}_0(\mathbf{k})$ in Table D.17, for the classes of $\mathscr{G}_0(\mathbf{k}')$ are $\mathscr{C}_1=E$; $\mathscr{C}_2=C_{2y}$; $\mathscr{C}_3=C_{4y}$, C_{4y}^{-1}; $\mathscr{C}_4=IC_{2x}$, IC_{2z}; $\mathscr{C}_5=IC_{2d}$, IC_{2c}. (The above correspondence is a consequence of the following argument. If $\phi_s^{\mathbf{kp}}(\mathbf{r})$ are a set of basis functions of the pth irreducible representation of $\mathcal{G}_0(\mathbf{k})$, then $P(\{\mathbf{R}(C_{4x})\,|\,\mathbf{0}\})\phi_s^{\mathbf{k}p}(\mathbf{r})$ have wave vector $\mathbf{k}'=\mathbf{R}(C_{4x})\mathbf{k}$ and therefore transform as a set of basis functions for an irreducible representation of $\mathcal{G}_0(\mathbf{k}')$, which may naturally be called the pth irreducible representation. Then, if $\{\mathbf{R}'\,|\,\mathbf{0}\}$ is a member of $\mathcal{G}_0(\mathbf{k}')$,

$$
\begin{aligned}
&P(\{\mathbf{R}'\,|\,\mathbf{0}\})[P(\{\mathbf{R}(C_{4x})\,|\,\mathbf{0}\})\phi_s^{\mathbf{k}p}(\mathbf{r})\\
&\quad=P(\{\mathbf{R}(C_{4x})\,|\,\mathbf{0}\})P(\{\mathbf{R}(C_{4x}^{-1})\mathbf{R}'\mathbf{R}(C_{4x})\,|\,\mathbf{0}\})\,\phi_s^{\mathbf{k}p}(\mathbf{r})\\
&\quad=\sum_t \Gamma^p(\mathbf{R}(C_{4x}^{-1})\mathbf{R}'\mathbf{R}(C_{4x}))_{ts}[P(\{\mathbf{R}(C_{4x})\,|\,\mathbf{0}\})\phi_t^{\mathbf{k}p}(\mathbf{r})],
\end{aligned}
$$

which demonstrates the correspondence.) In this particular example the only elements of $\mathcal{G}_0(\mathbf{k},\mathbf{k}',\mathbf{k}'')$ are E and IC_{2x}. Then, for instance, if Q transforms as Δ_2 and $\phi_i(\mathrm{r})$ as $\Delta_{2'}$, the only allowed transitions are to a final state $\phi_f(\mathrm{r})$ transforming as Σ_2 or Σ_3. In the literature this is sometimes written compactly but misleadingly as

$$\Delta_2\otimes\Delta_{2'}=\Sigma_2\oplus\Sigma_3, \tag{9.31}$$

but this equation should not be interpreted literally, for the irreducible representations involved belong to different groups. Indeed, the dimension $d_p d_{p'}$ of the "direct product representation" on the left-hand side of Equation (9.31) is, in general, not equal to the sum of the dimensions $\sum d_{p''}$ of the representations on the right-hand side. The reason for this is, of course, that this result is true for the corresponding irreducible representations of the *space group* $\mathcal{G}$, which have dimensions $M(\mathbf{k})d_p$, $M(\mathbf{k}')d_p$ and $M(\mathbf{k}'')d_{p''}$ respectively, $M(\mathbf{k})$, $M(\mathbf{k}')$ and $M(\mathbf{k}'')$ being the number of wave vectors in the stars of $\mathbf{k}$, $\mathbf{k}'$ and $\mathbf{k}''$ respectively, so that $M(\mathbf{k})M(\mathbf{k}')d_p d_{p'}=M(\mathbf{k}'')\sum d_{p''}$ and, in general, $M(\mathbf{k})M(\mathbf{k}')\neq M(\mathbf{k}'')$. In fact, in the above example $M(\mathbf{k})=6$, $M(\mathbf{k}')=6$ and $M(\mathbf{k}'')=12$.

Cracknell and Davies (1979) and Davies and Cracknell (1979, 1980) have calculated $n_{\mathbf{k}p,\mathbf{k}'p'}^{\mathbf{k}''p''}$ for *all* combinations of $\mathbf{k}p$, $\mathbf{k}'p'$ and $\mathbf{k}''p''$ for *every* crystallographic space group. A general discussion has also been given by Dirl (1979b). For the diamond structure space group O_h^7 selection rules have also been tabulated by Elliott and Loudon (1960), Lax and Hopfield (1961), Birman (1962, 1963) and Balkonski and Nusimovici (1964).

As noted in Chapter 6, Section 2, *more* information can be obtained from the Clebsch–Gordan *coefficients*. If $n_{\mathbf{k}p,\mathbf{k}'p'}^{\mathbf{k}''p''}=1$ it is possible to substitute Equations (9.5) or Equations (9.14) (as appropriate) into Equation (5.31) to obtain an explicit expression for these coefficients. For the case when $n_{\mathbf{k}p,\mathbf{k}'p'}^{\mathbf{k}''p''}\geqslant 2$ the reader is referred to the detailed discussions of van den Broek and Cornwell (1978), van den

Broek (1979) and Dirl (1979c, d). (See also Koster (1958), Litvin and Zak (1968), Novosadov *et al.* (1969), Cornwell (1970) and Berenson and Birman (1975).) The actual calculations are very tedious, the most extensive tabulations at present being those of Berenson *et al.* (1975) for O_h^5 and O_h^7.

For absorption or induced emission of electromagnetic radiation the appropriate operator is $Q = \mathbf{A}_0 \,.\, \mathbf{grad}$, and for spontaneous emission it is $Q = \mathbf{n} \,.\, \mathbf{grad}$ (see Chapter 6, Section 2). Example III of Chapter 5, Section 3, implies that $\partial/\partial x$, $\partial/\partial y$ and $\partial/\partial z$ transform as the first, second and third rows respectively of the three-dimensional representation $\boldsymbol{\Gamma}$ of the space group $\mathcal{G}$, defined by $\boldsymbol{\Gamma}(\{\mathbf{R} \mid \boldsymbol{\tau}_{\mathbf{R}} + \mathbf{t}_{\mathbf{n}}\}) = \mathbf{R}$ for all $\{\mathbf{R} \mid \boldsymbol{\tau}_{\mathbf{R}} + \mathbf{t}_{\mathbf{n}}\} \in \mathcal{G}$. As the basis functions of $\boldsymbol{\Gamma}$ are Bloch functions with zero wave vector, $\boldsymbol{\Gamma}$ is an irreducible representation of $\mathcal{G}$ with wave vector zero or is equivalent to a direct sum of such irreducible representations. Thus, in the notation of the above analysis, $\mathbf{k}' = \mathbf{0}$, so Equation (9.29) becomes

$$\mathbf{k} = \mathbf{k}''. \tag{9.32}$$

Consequently, on an electronic energy band diagram of $\varepsilon(\mathbf{k})$ plotted against $\mathbf{k}$, the transitions are "vertical". Moreover, $\mathcal{G}_0(\mathbf{k}') = \mathcal{G}_0$ and $\mathcal{G}_0(\mathbf{k}) = \mathcal{G}_0(\mathbf{k}'')$, so $\mathcal{G}_0(\mathbf{k}, \mathbf{k}', \mathbf{k}'') = \mathcal{G}_0(\mathbf{k})$. For the space groups O_h^1, O_h^5, O_h^7 and O_h^9, as $\boldsymbol{\Gamma}_{15}(\{\mathbf{R} \mid \mathbf{0}\}) = \mathbf{R}$ for all $\{\mathbf{R} \mid \mathbf{0}\} \in \mathcal{G}_0$ (see Appendix D, Section 1(1)), Equations (9.5) and (9.14) imply that $\boldsymbol{\Gamma}$ is actually the three-dimensional irreducible representation Γ_{15}.

The transitions just considered are often called "direct" optical transitions. If a phonon is created or absorbed at the same time, Equation (9.32) is no longer valid, and the transition is said to be "indirect". For a detailed discussion of this case see Streitwolf (1971).

Appendices

Appendix A

Matrices

The object of this appendix is to give the definitions, notations and terminology for matrices that are used in this book, together with a brief but coherent account of their relevant properties.

1 Definitions

An $m \times n$ "matrix" $\mathbf{A}$ is defined as a rectangular array of mn elements A_{jk} $(1 \leq j \leq m, 1 \leq k \leq n)$, each of which is a real or complex number, arranged in m rows and n columns. That is

$$\mathbf{A} = \begin{bmatrix} A_{11} & A_{12} \dots A_{1n} \\ A_{21} & A_{22} \dots A_{2n} \\ \vdots & \vdots \\ A_{m1} & A_{m2} \dots A_{mn} \end{bmatrix}.$$

A matrix whose elements are all zero is called a "null matrix" and is denoted by $\mathbf{0}$.

When $m = n$, as is the case for most matrices encountered in this book, the matrix is said to be "square". In this case the elements A_{jk} with $j = k$ are called the "diagonal" elements, while those with $j < k$ are referred to as being in the "upper off-diagonal" positions. If $A_{jk} = 0$ for $j \neq k$ then $\mathbf{A}$ is said to be a "diagonal matrix". The most important example is the "unit matrix" $\mathbf{1}$, which is defined by

$$(\mathbf{1})_{jk} = \delta_{jk} = \begin{cases} 1, & \text{if } j = k, \\ 0, & \text{if } j \neq k, \end{cases}$$

δ_{jk} being the Kronecker delta symbol. (The dimension of $\mathbf{1}$ is usually clear from its context, but when this is not so the $m \times m$ unit matrix will be denoted by $\mathbf{1}_m$.)

The "sum" of two $m \times n$ matrices $\mathbf{A}$ and $\mathbf{B}$ is defined to be another $m \times n$ matrix $\mathbf{A}+\mathbf{B}$ such that $(\mathbf{A}+\mathbf{B})_{jk} = \mathrm{A}_{jk} + \mathrm{B}_{jk}$ $(1 \leq j \leq m, 1 \leq k \leq n)$. Similarly, the "scalar product" of an $m \times n$ matrix $\mathbf{A}$ with a real or complex number λ is an $m \times n$ matrix $\lambda\mathbf{A}$ defined such that $(\lambda\mathbf{A})_{jk} = \lambda \mathrm{A}_{jk}$ $(1 \leq j \leq m, 1 \leq k \leq n)$. The "matrix product" of an $m \times n$ matrix $\mathbf{A}$ and an $n \times p$ matrix $\mathbf{B}$ is defined as an $m \times p$ matrix $\mathbf{AB}$ whose elements are given by

$$(\mathbf{AB})_{jl} = \sum_{k=1}^{n} \mathrm{A}_{jk}\mathrm{B}_{kl} \tag{A.1}$$

When $m = n = p$, both $\mathbf{AB}$ and $\mathbf{BA}$ exist and have the same dimensions, but even so, in general, $\mathbf{AB} \neq \mathbf{BA}$. However, if $\mathbf{A}$ and $\mathbf{B}$ are both diagonal matrices, then necessarily $\mathbf{AB} = \mathbf{BA}$. Of course, for any $m \times n$ matrix $\mathbf{A}$, $\mathbf{A1} = \mathbf{1A} = \mathbf{A}$.

The "transpose" $\tilde{\mathbf{A}}$ of an $m \times n$ matrix $\mathbf{A}$ is defined to be the $n \times m$ matrix whose elements are given by $(\tilde{\mathrm{A}})_{jk} = \mathrm{A}_{kj}$, so that if $\mathbf{C} = \mathbf{AB}$ then $\tilde{\mathbf{C}} = \tilde{\mathbf{B}}\tilde{\mathbf{A}}$. The "complex conjugate" $\mathbf{A}^*$ of $\mathbf{A}$ is defined as the $m \times n$ matrix such that $(\mathbf{A}^*)_{jk} = \mathrm{A}_{jk}^*$, $*$ denoting complex conjugation. Combining these concepts gives the "Hermitian adjoint" $\mathbf{A}^\dagger$ of $\mathbf{A}$, which is the $n \times m$ matrix defined by $\mathbf{A}^\dagger = (\tilde{\mathbf{A}})^*$. (In the mathematical literature this $\mathbf{A}^\dagger$ is often referred to as the "associate" of $\mathbf{A}$, the term "adjoint" being reserved for another matrix.)

The "determinant" $\det \mathbf{A}$ of an $m \times m$ matrix $\mathbf{A}$ is the real or complex number defined by $\det \mathbf{A} = \sum (-1)^p \mathrm{A}_{1c_1} \mathrm{A}_{2c_2} \ldots \mathrm{A}_{mc_m}$, where $(c_1, c_2, \ldots, c_m)$ is a permutation of $(1, 2, \ldots, m)$, p being the number of transpositions required to bring $(c_1, c_2, \ldots, c_m)$ to the "natural" order $(1, 2, \ldots, n)$, and the sum is over all such permutations. Then $\det \mathbf{1} = 1$,

$$\det \tilde{\mathbf{A}} = \det \mathbf{A}, \tag{A.2}$$

$$\det \mathbf{A}^* = (\det \mathbf{A})^*, \tag{A.3}$$

$$\det(\mathbf{AB}) = (\det \mathbf{A})(\det \mathbf{B}), \tag{A.4}$$

and, if $\mathbf{B}$ is a matrix obtained from $\mathbf{A}$ by interchanging all the elements of a pair of rows (or a pair of columns), then $\det \mathbf{B} = -\det \mathbf{A}$.

The "inverse" $\mathbf{A}^{-1}$ of an $m \times m$ matrix $\mathbf{A}$ is defined as the $m \times m$ matrix such that $\mathbf{A}^{-1}\mathbf{A} = \mathbf{A}\mathbf{A}^{-1} = \mathbf{1}$. $\mathbf{A}^{-1}$ exists if and only if $\det \mathbf{A} \neq 0$, in which case $\mathbf{A}$ is described as being "non-singular". If $\mathbf{A}$ and $\mathbf{B}$ are two non-singular $m \times m$ matrices, then $(\mathbf{AB})^{-1} = \mathbf{B}^{-1}\mathbf{A}^{-1}$.

Name	Defining property
symmetric	$\tilde{\mathbf{A}} = \mathbf{A}$
antisymmetric (or skew-symmetric)	$\tilde{\mathbf{A}} = -\mathbf{A}$
real	$\mathbf{A}^* = \mathbf{A}$
orthogonal	$\tilde{\mathbf{A}} = \mathbf{A}^{-1}$
Hermitian (or self-adjoint)	$\mathbf{A}^\dagger = \mathbf{A}$
anti-Hermitian	$\mathbf{A}^\dagger = -\mathbf{A}$
unitary	$\mathbf{A}^\dagger = \mathbf{A}^{-1}$
anti-unitary	$\mathbf{A}^\dagger = -\mathbf{A}^{-1}$

Table A.1 Definitions of special types of matrix.

Table A.1 gives the definitions of a number of important special types of matrix. It should be noted that a matrix that is both real and symmetric is necessarily Hermitian, and a matrix that is both real and orthogonal is necessarily unitary. For an orthogonal matrix $\mathbf{A}$, as $\tilde{\mathbf{A}}\mathbf{A} = \mathbf{1}$, Equations (A.2) and (A.4) imply that $(\det \mathbf{A})^2 = 1$, so that

$$\det \mathbf{A} = +1 \quad \text{or} \quad -1, \tag{A.5}$$

Similarly, for a unitary matrix $\mathbf{A}$, as $\mathbf{A}^\dagger\mathbf{A} = \mathbf{1}$, Equations (A.2), (A.3) and (A.4) imply that $|\det \mathbf{A}|^2 = 1$, so that

$$\det \mathbf{A} = e^{i\alpha}, \tag{A.6}$$

where α is some real number.

The "trace" $\operatorname{tr}\mathbf{A}$ of an $m \times m$ matrix $\mathbf{A}$ is defined by $\operatorname{tr}\mathbf{A} = \sum_{j=1}^{m} A_{jj}$, that is, it is the sum of the diagonal elements of $\mathbf{A}$. If $\mathbf{A}$ and $\mathbf{B}$ are *any* two $m \times m$ matrices, it follows immediately from Equation (A.1) that $\operatorname{tr}(\mathbf{AB}) = \operatorname{tr}(\mathbf{BA})$. Also $\operatorname{tr}(\mathbf{A}+\mathbf{B}) = \operatorname{tr}\mathbf{A} + \operatorname{tr}\mathbf{B}$, and, for any complex number α, $\operatorname{tr}(\alpha\mathbf{A}) = \alpha \operatorname{tr}\mathbf{A}$. Moreover, if $\mathbf{A}, \mathbf{B}, \mathbf{C}$ are any three $m \times m$ matrices, Equation (A.1) gives $\operatorname{tr}(\mathbf{ABC}) = \operatorname{tr}(\mathbf{BCA}) = \operatorname{tr}(\mathbf{CAB})$. If $\operatorname{tr}\mathbf{A} = 0$, then $\mathbf{A}$ is said to be "traceless". If $\mathbf{A}$ and $\mathbf{A}'$ are $m \times m$ matrices related by a so-called "similarity transformation" $\mathbf{A}' = \mathbf{S}^{-1}\mathbf{AS}$, where $\mathbf{S}$ is any $m \times m$ non-singular matrix, then $\operatorname{tr}\mathbf{A}' = \operatorname{tr}\mathbf{A}$.

A matrix may be "partitioned" into submatrices by inserting dividing lines between arbitrarily chosen adjacent pairs of rows and columns. For example,

$$\mathbf{A} = \left[\begin{array}{c|cc|c} A_{11} & A_{12} & A_{13} & A_{14} \\ A_{21} & A_{22} & A_{23} & A_{24} \\ \hline A_{31} & A_{32} & A_{33} & A_{34} \end{array}\right] = \left[\begin{array}{c|c|c} \mathbf{A}^{11} & \mathbf{A}^{12} & \mathbf{A}^{13} \\ \hline \mathbf{A}^{21} & \mathbf{A}^{22} & \mathbf{A}^{23} \end{array}\right]$$

is a partitioning of a 3×4 matrix $\mathbf{A}$ into six submatrices $\mathbf{A}^{jk}$, defined by

$$\mathbf{A}^{11}=\begin{bmatrix}A_{11}\\A_{21}\end{bmatrix},\qquad \mathbf{A}^{12}=\begin{bmatrix}A_{12} & A_{13}\\A_{22} & A_{23}\end{bmatrix},\qquad \mathbf{A}^{13}=\begin{bmatrix}A_{14}\\A_{24}\end{bmatrix},$$
$$\mathbf{A}^{21}=[A_{31}],\qquad \mathbf{A}^{22}=[A_{32}\quad A_{33}],\qquad \mathbf{A}^{23}=[A_{34}].$$

In terms of submatrices, the matrix product $\mathbf{C}=\mathbf{AB}$ of an $m\times n$ matrix $\mathbf{A}$ with an $n\times p$ matrix $\mathbf{B}$ has a remarkably simple property, provided that the "column" partitioning of $\mathbf{A}$ is chosen to be the same as the "row" partitioning of $\mathbf{B}$. Explicitly, if $\mathbf{A}, \mathbf{B}$ and $\mathbf{C}$ are partitioned into st submatrices $\mathbf{A}^{jk}$, tu submatrices $\mathbf{B}^{kl}$ and su submatrices $\mathbf{C}^{jl}$ respectively, where $1\leq j\leq s\leq m, 1\leq k\leq t\leq n, 1\leq l\leq u\leq p$, and where $\mathbf{A}^{jk}, \mathbf{B}^{kl}$ and $\mathbf{C}^{jl}$ have dimensions $m_j\times n_k$, $n_k\times p_l$ and $m_j\times p_l$ respectively (where $\sum_{j=1}^{s} m_j=m$, $\sum_{k=1}^{t} n_k=n$, and $\sum_{l=1}^{u} p_l=p$), then

$$\mathbf{C}^{jl}=\sum_{k=1}^{t}\mathbf{A}^{jk}\mathbf{B}^{kl} \tag{A.7}$$

There is a striking similarity of form with Equation (A.1). It is as though the submatrices $\mathbf{A}^{jk}, \mathbf{B}^{kl}$ and $\mathbf{C}^{jl}$ can be regarded as being matrix elements, but with the product of $\mathbf{A}^{jk}$ and $\mathbf{B}^{kl}$ being given by Equation (A.1). For clarity, the various submatrices have been distinguished here by superscripts. However, it is sometimes convenient to use subscripts instead, as, for example, in Chapter 4, Section 4. Moreover, the dividing lines will be omitted when there is no possibility of confusion, so that for the above example

$$\mathbf{A}=\begin{bmatrix}\mathbf{A}^{11} & \mathbf{A}^{12} & \mathbf{A}^{13}\\ \mathbf{A}^{21} & \mathbf{A}^{22} & \mathbf{A}^{23}\end{bmatrix}.$$

The "direct product" (or "Kronecker product") of an $m\times m$ matrix $\mathbf{A}$ and an $n\times n$ matrix $\mathbf{B}$ is defined to be an $mn\times mn$ matrix $\mathbf{A}\otimes\mathbf{B}$, whose rows and columns are each labelled by a *pair* of indices in such a way that

$$(\mathbf{A}\otimes\mathbf{B})_{js,kt}=A_{jk}B_{st} \qquad (1\leq j,k\leq m;\ 1\leq s,t\leq n). \tag{A.8}$$

In order to express such matrices in the usual form in which rows and columns are each labelled by a *single* index it is necessary to put the set of mn pairs (j,s) $(1\leq j\leq m, 1\leq s\leq n)$ into one-to-one correspondence with a set of mn integers p $(1\leq p\leq mn)$, with an identical correspondence between the pairs (k,t) and a set of integers q. The most convenient choice is $p=n(j-1)+s$, $q=n(k-1)+t$. With this prescription, the matrix $\mathbf{A}\otimes\mathbf{B}$ for $m=2$ and $n=2$ would be displayed as

$$\begin{bmatrix} (\mathbf{A}\otimes\mathbf{B})_{11,11} & (\mathbf{A}\otimes\mathbf{B})_{11,12} & (\mathbf{A}\otimes\mathbf{B})_{11,21} & (\mathbf{A}\otimes\mathbf{B})_{11,22} \\ (\mathbf{A}\otimes\mathbf{B})_{12,11} & (\mathbf{A}\otimes\mathbf{B})_{12,12} & (\mathbf{A}\otimes\mathbf{B})_{12,21} & (\mathbf{A}\otimes\mathbf{B})_{12,22} \\ (\mathbf{A}\otimes\mathbf{B})_{21,11} & (\mathbf{A}\otimes\mathbf{B})_{21,12} & (\mathbf{A}\otimes\mathbf{B})_{21,21} & (\mathbf{A}\otimes\mathbf{B})_{21,22} \\ (\mathbf{A}\otimes\mathbf{B})_{22,11} & (\mathbf{A}\otimes\mathbf{B})_{22,12} & (\mathbf{A}\otimes\mathbf{B})_{22,21} & (\mathbf{A}\otimes\mathbf{B})_{22,22} \end{bmatrix}.$$

Clearly, the diagonal elements of $\mathbf{A}\otimes\mathbf{B}$ in the pair-labelling scheme are those for which $j=k$ *and* $s=t$. If $\mathbf{A}$ and $\mathbf{B}$ are both diagonal, then $\mathbf{A}\otimes\mathbf{B}$ is also diagonal (for if $A_{jk}=a_j\delta_{jk}$ and $B_{st}=b_s\delta_{st}$, then $(\mathbf{A}\otimes\mathbf{B})_{js,kt}=a_jb_s\delta_{jk}\delta_{st}$).

If $\mathbf{A}$ and $\mathbf{A}'$ are both $m\times m$ matrices and $\mathbf{B}$ and $\mathbf{B}'$ are both $n\times n$ matrices, then

$$(\mathbf{A}\otimes\mathbf{B})(\mathbf{A}'\otimes\mathbf{B}')=(\mathbf{AA}')\otimes(\mathbf{BB}'), \tag{A.9}$$

where all products other than those indicated by the symbol $\otimes$ are ordinary matrix products (as defined in Equation (A.1)). The proof of Equation (A.9) is straightforward, for the (js, kt) element of the right-hand side is

$$(\mathbf{AA}')_{jk}(\mathbf{BB}')_{st}=\sum_{l=1}^{m}\sum_{u=1}^{n}A_{jl}A'_{lk}B_{su}B'_{ut}$$

while the (js, kt) element of the left-hand side is

$$\sum_{l=1}^{m}\sum_{u=1}^{n}(\mathbf{A}\otimes\mathbf{B})_{js,lu}(\mathbf{A}'\otimes\mathbf{B}')_{lu,kt}=\sum_{l=1}^{m}\sum_{u=1}^{n}A_{jl}B_{su}A'_{lk}B'_{ut}.$$

Finally, if $\mathbf{A}$ and $\mathbf{B}$ are both unitary, then $\mathbf{A}\otimes\mathbf{B}$ is also unitary. (It follows directly from Equation (A.9) that $(\mathbf{A}\otimes\mathbf{B})^\dagger=\mathbf{A}^\dagger\otimes\mathbf{B}^\dagger$, and, as $\mathbf{A}^\dagger\mathbf{A}=\mathbf{AA}^\dagger=\mathbf{1}_m$, $\mathbf{B}^\dagger\mathbf{B}=\mathbf{BB}^\dagger=\mathbf{1}_n$ and $\mathbf{1}_m\otimes\mathbf{1}_n=\mathbf{1}_{mn}$, the unitary property of $\mathbf{A}\otimes\mathbf{B}$ is an immediate consequence of Equation (A.9).)

2 Eigenvalues and eigenvectors

If $\mathbf{A}$ is an $m\times m$ matrix and λ is a real or complex number which, together with an $m\times 1$ "column" matrix $\mathbf{c}$ ($\neq\mathbf{0}$), satisfies the equation

$$\mathbf{Ac}=\lambda\mathbf{c}, \tag{A.10}$$

then λ is said to be an "eigenvalue" of $\mathbf{A}$ and $\mathbf{c}$ is said to be an "eigenvector" corresponding to λ. Equation (A.10) has a non-trivial solution if and only if

$$\det(\mathbf{A}-\lambda\mathbf{1})=0, \tag{A.11}$$

which is often referred to in the mathematical physics literature as a "secular equation". The left-hand side of Equation (A.11) is a polynomial $P(\lambda)$ of degree m, known as the "characteristic polynomial", whose coefficients are determined by explicit evaluation of the determinant. The eigenvalues are given therefore by the "characteristic equation"

$$P(\lambda)=0, \tag{A.12}$$

that is, they are the roots of $P(\lambda)$. Suppose that $P(\lambda)$ has R distinct roots $\lambda_1, \lambda_2, \ldots, \lambda_R$, and that λ_j has multiplicity r_j, $j=1,2,\ldots,R$, so that

$$P(\lambda)=(\lambda-\lambda_1)^{r_1}(\lambda-\lambda_2)^{r_2}\ldots.(\lambda-\lambda_R)^{r_R}. \tag{A.13}$$

Then the Cayley-Hamilton Theorem states that the matrix **A** also satisfies the characteristic equation (Equation (A.12)), that is,

$$P(\mathbf{A})=(\mathbf{A}-\lambda_1\mathbf{1})^{r_1}(\mathbf{A}-\lambda_2\mathbf{1})^{r_2}\ldots.(\mathbf{A}-\lambda_R\mathbf{1})^{r_R}=\mathbf{0}. \tag{A.14}$$

If $\mathbf{A}'$ is related to $\mathbf{A}$ by a similarity transformation $\mathbf{A}'=\mathbf{S}^{-1}\mathbf{AS}$, where $\mathbf{S}$ is any non-singular $m\times m$ matrix, then Equation (A.10) can be written as

$$\mathbf{A}'(\mathbf{S}^{-1}\mathbf{c})=\lambda(\mathbf{S}^{-1}\mathbf{c}). \tag{A.15}$$

Thus $\mathbf{A}$ and $\mathbf{A}'$ have the same set of eigenvalues, with identical multiplicities and, if $\mathbf{c}$ is an eigenvector of $\mathbf{A}$ corresponding to λ then $\mathbf{c}'=\mathbf{S}^{-1}\mathbf{c}$ is an eigenvector of $\mathbf{A}'$ corresponding to λ and vice versa.

The question now arises as to whether $\mathbf{A}'$ can be made diagonal by an appropriate choice of $\mathbf{S}$. If this is so then $\mathbf{A}$ is said to be "diagonalizable". This is certainly true if $\mathbf{A}$ is Hermitian or unitary, and in both of these cases $\mathbf{S}$ can be chosen to be unitary (Gantmacher 1959). As the operators corresponding to physical observables in quantum mechanics are self-adjoint, whenever the corresponding operator eigenvalue equation is cast in matrix form (as in Appendix B, Section 4) the matrix involved is Hermitian. Thus all the matrices occurring in such a context are automatically diagonalizable. However, non-diagonalizable matrices do occur in various contexts, so it is worthwhile analysing the question of diagonalizability in more detail. The most important result is embodied in the following theorem.

Theorem I An $m\times m$ matrix $\mathbf{A}$ is diagonalizable if and only if $\mathbf{A}$ possesses m linearly independent eigenvectors.

Proof Suppose first that $\mathbf{A}$ is diagonalizable and that $\mathbf{A}'=\mathbf{S}^{-1}\mathbf{AS}$ is the diagonal form. Then each diagonal element of $\mathbf{A}'$ is an eigenvalue of $\mathbf{A}'$ (and of $\mathbf{A}$) and the eigenvector of $\mathbf{A}'$ corresponding to the eigenvalue A'_{jj} can be taken to be $\mathbf{c}'_j$, where

$$(\mathbf{c}'_j)_{k1} = \begin{cases} 1, & j = k, \\ 0, & j \neq k. \end{cases} \tag{A.16}$$

Thus $\mathbf{A}'$ possesses m linearly independent eigenvectors $\mathbf{c}'_j, j = 1, 2, \ldots, m$, and hence $\mathbf{A}$ possesses m linearly independent eigenvectors $\mathbf{S}\mathbf{c}'_j, j = 1, 2, \ldots, m$.

Conversely, suppose that $\mathbf{A}$ possesses m linearly independent eigenvectors $\mathbf{c}_j, j = 1, 2, \ldots, m$. Define the $m \times m$ matrix $\mathbf{S}$ by

$$\mathbf{S} = [\mathbf{c}_1 \mid \mathbf{c}_2 \mid \ldots \mid \mathbf{c}_m],$$

so that $\det \mathbf{S} \neq 0$. By virtue of Equation (A.7), if $\mathbf{c}'_j$ is defined by Equations (A.16) then $\mathbf{S}\mathbf{c}'_j = \mathbf{c}_j$, so $\mathbf{c}'_j = \mathbf{S}^{-1}\mathbf{c}_j$ for $j = 1, 2, \ldots, m$. But the set $\mathbf{c}'_j$ $(j = 1, 2, \ldots, m)$ can be eigenvectors of $\mathbf{A}'$ (as required by Equation (A.15)) only if $\mathbf{A}'$ is diagonal, so $\mathbf{A}$ must be diagonalizable.

This theorem implies that if $\mathbf{A}$ is diagonalizable there are r_j linearly independent eigenvectors corresponding to each eigenvalue of multiplicity r_j, whereas if $\mathbf{A}$ is non-diagonalizable at least one eigenvalue has less linearly independent eigenvectors than its multiplicity.

The following example illustrates the two possible situations. Suppose first that

$$\mathbf{A} = \begin{bmatrix} 0 & 1 & 0 \\ 1 & 0 & 0 \\ 0 & 0 & 1 \end{bmatrix}, \tag{A.17}$$

which is Hermitian and hence diagonalizable. The characteristic polynomial is $P(\lambda) = (\lambda - 1)^2(\lambda + 1)$. For the eigenvalue $\lambda = 1$, the $(1, 1)$ and $(2, 1)$ components of Equation (A.10) both give $c_{21} = c_{11}$, while the $(3, 1)$ component is trivially satisfied. Thus

$$\begin{bmatrix} 1 \\ 1 \\ 0 \end{bmatrix} \quad \text{and} \quad \begin{bmatrix} 0 \\ 0 \\ 1 \end{bmatrix}$$

are two linearly independent eigenvectors corresponding to eigenvalue $\lambda = 1$. Similarly, for the eigenvalue $\lambda = -1$, the $(1, 1)$ and $(2, 1)$ components of Equation (A.10) give $c_{21} = -c_{11}$, while the $(3, 1)$ component gives $c_{31} = 0$. Thus

$$\begin{bmatrix} 1 \\ -1 \\ 0 \end{bmatrix}$$

is the only linearly independent eigenvector corresponding to eigenvalue $\lambda = -1$.

As an example of a non-diagonalizable matrix, consider

$$\mathbf{A} = \begin{bmatrix} 0 & 1 \\ 0 & 0 \end{bmatrix}, \tag{A.18}$$

for which $P(\lambda) = \lambda^2$. For the eigenvalue $\lambda = 0$, the (1, 1) component of Equation (A.10) gives $c_{21} = 0$, while the (2, 1) component is trivially satisfied. Thus although $\lambda = 0$ is an eigenvalue of multiplicity 2,

$$\begin{bmatrix} 1 \\ 0 \end{bmatrix}$$

is the *only* linearly independent eigenvector.

There exists a useful criterion for diagonalizability involving the "minimal polynomial" $M(\mathbf{A})$ of $\mathbf{A}$, which is defined as the polynomial of *lowest* degree in $\mathbf{A}$ such that $M(\mathbf{A}) = \mathbf{0}$. In some cases the minimal polynomial is identical to the characteristic polynomial, but otherwise its degree is less than m. It can be shown (Gantmacher 1959) that $M(\mathbf{A})$ is unique and has the form

$$M(\mathbf{A}) = (\mathbf{A} - \lambda_1 \mathbf{1})^{s_1} (\mathbf{A} - \lambda_2 \mathbf{1})^{s_2} \dots (\mathbf{A} - \lambda_R \mathbf{1})^{s_R},$$

where $1 \leq s_j \leq r_j$, $j = 1, 2, \dots, R$, and where λ_j and r_j are as defined in Equation (A.13). (In particular this implies that *every* distinct eigenvalue of $\mathbf{A}$ appears in a factor of $M(\mathbf{A})$.) Moreover, $\mathbf{A}$ is diagonalizable if and only if $s_j = 1$ for *every* $j = 1, 2, \dots, R$, that is, if and only if $M(\mathbf{A})$ consists only of *linear* factors. This has the corollary that $\mathbf{A}$ is necessarily diagonalizable if every eigenvalue of $\mathbf{A}$ has multiplicity 1.

The examples that were considered above demonstrate this criterion very neatly. For the matrix $\mathbf{A}$ of Equation (A.17), $M(\mathbf{A}) = \mathbf{A}^2 - \mathbf{1} = (\mathbf{A} - \mathbf{1})(\mathbf{A} + \mathbf{1})$, which has only linear factors, so that $\mathbf{A}$ must be diagonalizable. By contrast, for the matrix $\mathbf{A}$ of Equation (A.18), $M(\mathbf{A}) = \mathbf{A}^2$, so this $\mathbf{A}$ is non-diagonalizable.

Even when $\mathbf{A}$ is not diagonalizable, it can be transformed by an appropriate similarity transformation into a standard form, the "Jordan canonical form". More precisely, for any $m \times m$ matrix $\mathbf{A}$ there exists an $m \times m$ matrix $\mathbf{S}$ such that all the elements of $\mathbf{A}' = \mathbf{S}^{-1}\mathbf{A}\mathbf{S}$ are zero except possibly the A'_{rr} elements (for $r = 1, 2, \dots, m$) and the $A'_{r,r+1}$ elements (for $r = 1, 2, \dots, m-1$). Moreover, $A'_{r,r+1} = 0$ or 1 for all $r = 1, 2, \dots, m-1$, and

$$A'_{r,r+1} = \begin{cases} 0 \text{ or } 1, & \text{if } A'_{rr} = A'_{r+1,r+1}, \\ 0 & , \text{if } A'_{rr} \neq A'_{r+1,r+1}. \end{cases}$$

For example,

$$\mathbf{A}' = \begin{bmatrix} 2 & 1 & 0 & 0 & 0 \\ 0 & 2 & 0 & 0 & 0 \\ 0 & 0 & 5 & 1 & 0 \\ 0 & 0 & 0 & 5 & 1 \\ 0 & 0 & 0 & 0 & 5 \end{bmatrix}$$

is in Jordan canonical form. So too is the matrix **A** of Equation (1.18). (Clearly the case in which **A**′ is diagonal is merely a special case in which $A'_{r,r+1} = 0$ for all $r = 1, 2, \ldots, m-1$.) Obviously the eigenvalues of **A**′ are just the set of diagonal elements A'_{rr}, $r = 1, 2, \ldots, m$.

Appendix B

Vector Spaces

This appendix is intended both to provide an introduction to vector spaces and to give the various notations and conventions that are used throughout this book.

1 The concept of a vector space

A general vector space is obtained by selectively abstracting certain properties of vectors of the three-dimensional Euclidean space $\mathbb{R}^3$.

A vector ψ of $\mathbb{R}^3$ may be specified by a triple of real numbers x_1, x_2 and x_3, that is, $\psi = (x_1, x_2, x_3)$. (In elementary treatments of $\mathbb{R}^3$ it is conventional to indicate a vector by using bold type, but for treatments of higher-dimensional spaces this convention is discontinued. To help avoid confusion, as far as possible vectors in this appendix will be denoted by the Greek letters $\psi, \phi, \chi, \ldots$ and scalars (that is, real or complex numbers) by $a, b, c, \ldots$.) The product of a vector ψ of $\mathbb{R}^3$ with a real number a is defined to be another vector $a\psi$ such that

$$a\psi = (ax_1, ax_2, ax_3), \tag{B.1}$$

from which it follows that if b is any other real number

$$b(a\psi) = (ba)\psi. \tag{B.2}$$

The sum of two vectors $\psi = (x_1, x_2, x_3)$ and $\phi = (y_1, y_2, y_3)$ of $\mathbb{R}^3$ is defined by

$$\psi + \phi = (x_1 + y_1, x_2 + y_2, x_3 + y_3), \tag{B.3}$$

so that

$$\psi + \phi = \phi + \psi. \tag{B.4}$$

Similarly, if $\chi = (z_1, z_2, z_3)$ is any other vector of $\mathbb{R}^3$,

$$\psi + (\phi + \chi) = (\psi + \phi) + \chi. \tag{B.5}$$

With these definitions it is easily verified that

$$a(\psi + \phi) = a\psi + a\phi \tag{B.6}$$

for any real number a and any two vectors ψ and ϕ. Similarly,

$$(a + b)\psi = a\psi + b\psi \tag{B.7}$$

for any real numbers and any ψ. Finally, there exists a vector $0 = (0, 0, 0)$ such that

$$\psi + 0 = \psi \tag{B.8}$$

for every ψ of $\mathbb{R}^3$.

In a *general* vector space V it is assumed that multiplication of vectors by scalar and vector addition can always be defined (though *not* necessarily by Equations (B.1) and (B.3)) in such a way that the properties in Equations (B.2), (B.4), (B.5), (B.6), (B.7) and (B.8) are retained. The precise definition is as follows:

Definition *Vector space*
A "vector space" V is a collection of elements $\psi, \phi, \chi, \ldots$ (called vectors) for which "scalar multiplication" $a\psi$ is defined for any "scalar" from a certain set and for any vector ψ, and for which "vector addition" $\psi + \phi$ is defined for all vectors ψ and ϕ, such that

$$\left.\begin{aligned} a(b\psi) &= (ba)\psi, \\ \psi + \phi &= \phi + \psi, \\ \psi + (\phi + \chi) &= (\psi + \phi) + \chi, \\ a(\phi + \psi) &= a\phi + a\psi, \\ (a + b)\psi &= a\psi + b\psi. \end{aligned}\right\}$$

Moreover, there must exist in V a "zero vector" 0 such that

$$\psi + 0 = \psi$$

for all $\psi \in \mathrm{V}$. If the set of scalars consists of all real numbers then V is said to be a "real vector space". Similarly, if the set of scalars comprises of complex numbers, V is called a "complex vector space". The set of scalars is often referred to as the "field".

A set of vectors $\psi_1, \psi_2, \ldots, \psi_d$ of V is described as being ***linearly dependent*** if there exists a set of non-zero scalars (of the appropriate set) $a_1, a_2, \ldots, a_d$ such that

$$a_1\psi_1 + a_2\psi_2 + \ldots + a_d\psi_d = 0 \tag{B.9}$$

Thus the set $\psi_1, \psi_2, \ldots, \psi_d$ is ***linearly independent*** if the only solution of Equation (B.9) (in the appropriate set of scalars) is $a_1 = a_2 = \ldots = a_d = 0$. If V contains a set of d linearly independent vectors, but every set of $(d+1)$ vectors is linearly dependent, then V is known as a "d-dimensional space". (For example, the vectors of $\mathbb{R}^3$ form a three-dimensional real vector space.) If there is no limit on the number of linearly independent vectors then V is said to be an "infinite-dimensional space". Finite-dimensional spaces are much easier to deal with and fortunately most of the vector spaces encountered in this book will be of this type.

Let V be a vector space of finite dimension d and let $\psi_1, \psi_2, \ldots, \psi_d$ be any set of linearly independent vectors of V. Then *any* $\psi \in \mathrm{V}$ can be uniquely expressed in terms of $\psi_1, \psi_2, \ldots, \psi_d$ by

$$\psi = a_1\psi_1 + a_2\psi_2 + \ldots + a_d\psi_d, \tag{B.10}$$

where $a_1, a_2, \ldots, a_d$ are a set of scalars that depend on ψ. (This follows from the definition just given, for $\psi, \psi_1, \psi_2, \ldots, \psi_d$ must be linearly dependent, so there exists a set of scalars $b, b_1, \ldots, b_d$ such that $b\psi + b_1\psi_1 + b_2\psi_2 + \ldots + b_d\psi_d = 0$, and not all of the set $b, b_1, \ldots, b_d$ are zero. As b must be non-zero, dividing by b and putting $a_j = -b_j/b$ gives Equation (B.10) immediately. The decomposition (Equation (B.10)) is *unique* because if $\psi = \sum_{j=1}^{d} a_j\psi_j$ and $\psi = \sum_{j=1}^{d} a_j'\psi_j$ then $\sum_{j=1}^{d} (a_j - a_j')\psi_j = 0$, for which the only solution is $a_j = a_j'$ for $j = 1, 2, \ldots, d$, as the vectors $\psi_1, \psi_2, \ldots, \psi_d$ are assumed to be linearly independent.) The set $\psi_1, \psi_2, \ldots, \psi_d$ is therefore said to provide a "basis" for V. In $\mathbb{R}^3$ a very convenient basis is provided by $\psi_1 = (1, 0, 0)$, $\psi_2 = (0, 1, 0)$ and $\psi_3 = (0, 0, 1)$ for, if $\psi = (x_1, x_2, x_3)$, then $\psi = x_1\psi_1 + x_2\psi_2 + x_3\psi_3$.

The set of d vectors $\psi_1', \psi_2', \ldots, \psi_d'$ defined in terms of the basis $\psi_1, \psi_2, \ldots, \psi_d$ by

$$\psi_n' = \sum_{m=1}^{d} \mathrm{S}_{mn}\psi_m$$

for $n = 1, 2, \ldots, d$ form a linearly independent set if and only if $\mathbf{S}$ is non-singular. Thus when $\mathbf{S}$ is non-singular $\psi_1', \psi_2', \ldots, \psi_d'$ provides an alternative basis for V.

It is occasionally convenient to regard a d-dimensional complex vector space as a $2d$-dimensional real vector space. If $\psi_1, \psi_2, \ldots, \psi_d$ is a basis for the complex space, then $\psi_1, \psi_2, \ldots, \psi_d$ together with $i\psi_1, i\psi_2, \ldots, i\psi_d$, form a basis for the real space. (It should be noted that ψ_j and $i\psi_j$ are linearly independent elements of the real space (although they are linearly dependent in the complex space) as $a\psi_j + bi\psi_j = 0$ has no solution for *real* numbers a and b, other than $a = b = 0$.)

The following examples show some of the widely differing forms of vector space that are encompassed by the definition.

Example I *The three-dimensional complex vector space* $\mathbb{C}^3$
$\mathbb{C}^3$ consists of the set of triples $\psi = (x_1, x_2, x_3)$, where x_1, x_2 and x_3 are *complex* numbers. Scalar multiplication by an arbitrary complex number a is defined by $a\psi = (ax_1, ax_2, ax_3)$ and vector addition is defined by Equation (B.4). The set of vectors $(1, 0, 0), (0, 1, 0)$ and $(0, 0, 1)$ again provides a convenient basis.

Example II *The set of all $N \times N$ traceless anti-Hermitian matrices ($N > 1$)*
Let $\mathbf{A}$ and $\mathbf{B}$ be any two $N \times N$ traceless anti-Hermitian matrices (see Table A.1). Then the scalar product $a\mathbf{A}$ and vector sum $\mathbf{A}+\mathbf{B}$ may be taken to be the scalar product and matrix sum defined in Appendix A, Section 1. Then $a\mathbf{A}$ and $\mathbf{A}+\mathbf{B}$ are $N \times N$ traceless anti-Hermitian matrices, provided that a is real. Thus the set of such matrices form a *real* vector space, the $N \times N$ matrix $\mathbf{0}$ (all of whose elements are zero) providing the zero element. It should be noted that even though this vector space is real the elements of matrices involved may be complex! As will be seen in Example II of Chapter 10, Section 5, this vector space has an additional structure and forms the Lie algebra su(N) of the linear Lie group SU(N). It is shown there that the dimension of this space is $(N^2 - 1)$.

Example III *Set of all functions defined in* $\mathbb{R}^3$
Let $\psi(\mathbf{r})$ and $\phi(\mathbf{r})$ be any two complex-valued functions defined for all $\mathbf{r} \in \mathbb{R}^3$. Then $\psi + \phi$ is defined in the natural way by $(\psi + \phi)(\mathbf{r}) = \psi(\mathbf{r}) + \phi(\mathbf{r})$ for all $\mathbf{r} \in \mathbb{R}^3$ and, for any complex number a, $a\psi$ is defined by $(a\psi)(\mathbf{r}) = a(\psi(\mathbf{r}))$ for all $\mathbf{r} \in \mathbb{R}^3$. The set of all such functions then forms an infinite-dimensional complex vector space, the zero vector being defined to be the function that is zero for all $\mathbf{r} \in \mathbb{R}^3$.

A "subspace" of a vector space V is a subset of V that is itself a vector space. The subspace is said to be "proper" if its dimension is less than that of V. V is said to be the "direct sum" of two subspaces V_1 and V_2 if every $\phi \in V$ can be written *uniquely* in the form $\phi = \phi_1 + \phi_2$, where $\phi_1 \in V_1$ and $\phi_2 \in V_2$. This implies that V_1 and V_2

have only the zero element of V in common. If $\psi_1, \psi_2, \ldots, \psi_d$ is a basis for V and $1 \leq d' < d$, then $\psi_1, \psi_2, \ldots, \psi_{d'}$ and $\psi_{d'+1}, \ldots, \psi_d$ are bases for two subspaces of V of dimensions d' and $(d-d')$ respectively. Moreover, V is the direct sum of these two subspaces, because if $\phi = \sum_{j=1}^{d} a_j\psi_j$, then $\phi = \phi_1 + \phi_2$, where $\phi_1 = \sum_{j=1}^{d'} a_j\psi_j$ and $\phi_2 = \sum_{j=d'+1}^{d} a_j\psi_j$, this decomposition being unique because the set $a_1, a_2, \ldots, a_d$ depends uniquely on ϕ. The concept of a direct sum can be generalized to more than two subspaces in the obvious way.

2 Inner product spaces

Many vector spaces have the additional attribute of being endowed with an "inner product". Consider first the example of vectors of $\mathbb{R}^3$, in which the inner product is the familiar scalar product. Thus, if $\psi = (x_1, x_2, x_3)$ and $\phi = (y_1, y_2, y_3)$ are any two vectors of $\mathbb{R}^3$, their inner product (ψ, ϕ) is the real number defined by

$$(\psi, \phi) = x_1y_1 + x_2y_2 + x_3y_3.$$

The "length" of $\psi = (x_1, x_2, x_3)$ is given by $\{x_1^2 + x_2^2 + x_3^2\}^{1/2}$, which is real and non-negative. Indeed it is only zero when $\psi = 0$, the zero vector. It may be denoted by $\|\psi\|$ and will be called the "norm" of ψ. Clearly $\|\psi\| = \{(\psi, \psi)\}^{1/2}$.

In $\mathbb{C}^3$ (see Example I of the previous section) it is natural to again require that the norm $\|\psi\|$ be always real and non-negative and also that $\|\psi\| = 0$ only when $\psi = 0$. This is achieved by the definition $\|\psi\| = \{|x_1|^2 + |x_2|^2 + |x_3|^2\}^{1/2}$. The identity $\|\psi\| = \{(\psi, \psi)\}^{1/2}$ can be retained if the inner product of any two vectors $\psi = (x_1, x_2, x_3)$ and $\phi = (y_1, y_2, y_3)$ (where the components are now complex numbers) is defined by

$$(\psi, \phi) = x_1^*y_1 + x_2^*y_2 + x_3^*y_3.$$

With this definition

$$(\psi, \phi) = (\phi, \psi)^*,$$

and for any two complex numbers a and b

$$(a\psi, b\phi) = a^*b(\psi, \phi).$$

Also, if $\chi = (z_1, z_2, z_3)$, then

$$(\psi + \phi, \chi) = (\psi, \chi) + (\phi, \chi).$$

A general "inner product space" is a vector space possessing an inner product that has the properties exhibited by these examples (even though the definition of this inner product may be quite different). The precise requirements are as follows.

Definition *Inner product space*
A complex vector space V is said to be an "inner product" space if to every pair of vectors ψ and ϕ of V there corresponds a complex number (ψ, ϕ) (called the inner product of ψ with ϕ) such that:

(*a*) $(\psi, \phi) = (\phi, \psi)^*$;
(*b*) $(a\psi, b\phi) - a^*b(\psi, \phi)$ for any two complex numbers a and b;
(*c*) $(\psi + \phi, \chi) = (\psi, \chi) + (\phi, \chi)$ for any $\chi \in$ V;
(*d*) $(\psi, \psi) \geqslant 0$ for all ψ; and
(*e*) $(\psi, \psi) = 0$ if and only if $\psi = 0$, the zero vector.

If V is a real vector space the inner product is required to be a real number, and in (*b*) a and b are restricted to being real numbers, but otherwise the requirements (*a*) to (*e*) are the same as for a complex space.

An "abstract" inner product space is a space that satisfies all the axioms without possessing a "concrete" realization for the inner product. It should be noted that (*a*) and (*c*) imply that

$$(\chi, \psi + \phi) = (\chi, \psi) + (\chi, \phi),$$

so that, by (*b*),

$$(\chi, a\psi + b\phi) = a(\chi, \psi) + b(\chi, \phi),$$

whereas (*b*) and (*c*) imply

$$(a\psi + b\phi, \chi) = a^*(\psi, \chi) + b^*(\phi, \chi).$$

Also, (*a*) implies that (ψ, ψ) is necessarily real (which is implicit in the requirement (*d*)).

For any inner product space the "norm" $\|\psi\|$ may be defined by

$$\|\psi\| = \{(\psi, \psi)\}^{1/2}.$$

It follows from (*b*) that

$$\|a\psi\| = |a|\,\|\psi\|. \tag{B.11}$$

Two other properties that are easily proved (Akhiezer and Glazman 1961) are the "Schwarz inequality" (with strict inequality applying if ϕ and ψ are linearly independent),

$$|(\psi, \phi)| \leqslant \|\psi\|\,\|\phi\|,$$

and the "triangle inequality"

$$\|\psi + \phi\| \leqslant \|\psi\| + \|\phi\|,$$

both valid for any ψ and ϕ of an inner product space.

By analogy with the situation in $\mathbb{R}^3$, the "distance" $d(\psi, \phi)$ between two vectors ψ and ϕ in a general inner product space may be defined by

$$d(\psi, \phi) = \|\psi - \phi\|. \tag{B.12}$$

Then it follows immediately that

(i) $d(\psi, \phi) = d(\phi, \psi)$;
(ii) $d(\psi, \psi) = 0$;
(iii) $d(\psi, \phi) > 0$ if $\phi \neq \psi$; and
(iv) $d(\psi, \phi) \leq d(\psi, \chi) + d(\chi, \phi)$ for any $\psi, \phi, \chi \in V$,

all of which are essential for the interpretation of $d(\psi, \phi)$ as a distance. The distance function $d(\psi, \phi)$ is often called the "metric".

Example I *The d-dimensional complex vector space* $\mathbb{C}^d$
$\mathbb{C}^d$ is the set of d-component quantities $\psi = (x_1, x_2, \ldots, x_d)$, where $x_1, x_2, \ldots, x_d$ are complex numbers. It is a complex vector space of dimension d. The inner product of $\mathbb{C}^d$ may be defined by

$$(\psi, \phi) = \sum_{j=1}^{d} x_j^* y_j, \tag{B.13}$$

where $\psi = (x_1, x_2, \ldots, x_d)$ and $\phi = (y_1, y_2, \ldots, y_d)$, which satisfies all the requirements for $\mathbb{C}^d$ to form an inner product space. From Equations (B.12) and (B.13) it follows that

$$d(\psi, \phi) = \left\{ \sum_{j=1}^{d} |x_j - y_j|^2 \right\}^{1/2}.$$

Example II *The set of all $m \times m$ matrices*
The set of all $m \times m$ matrices with complex elements form a complex vector space of dimension m^2, provided that the scalar product and vector sum are taken to be the scalar product and matrix sum defined in Appendix A, Section 1. The inner product of two such matrices $\mathbf{A}$ and $\mathbf{B}$ may be defined by

$$(\mathbf{A}, \mathbf{B}) = \sum_{j=1}^{m} \sum_{k=1}^{m} A_{jk}^* B_{jk}, \tag{B.14}$$

which again satisfies all the requirements for the vector space to form an inner product space. Moreover, Equations (B.12) and (B.14) imply that

$$d(\mathbf{A}, \mathbf{B}) = \left\{ \sum_{j=1}^{m} \sum_{k=1}^{m} |A_{jk} - B_{jk}|^2 \right\}^{1/2}.$$

This explains the origin of the metric of Equation (3.1). Comparison of Equations (B.12) and (B.13) shows that this inner product space is essentially just $\mathbb{C}^{m^2}$.

Two elements ψ and ϕ of an inner product space are said to be "orthogonal" if $(\psi, \phi) = 0$. (In $\mathbb{R}^3$ this coincides with the usual geometric notion of orthogonality.) A vector ψ is described as being "normalized" if $\|\psi\| = 1$. An "ortho-normal" set is then a set of vectors $\psi_1, \psi_2, \ldots$ such that $(\psi_j, \psi_k) = \delta_{jk}$ for $j, k = 1, 2, \ldots$. From any set of linearly independent vectors $\phi_1, \phi_2, \ldots$ an ortho-normal set $\psi_1, \psi_2, \ldots$ can be constructed by taking appropriate linear combinations. The procedure, often called the "Schmidt orthogonalization process", is as follows. First let

$$\theta_1 = \phi_1;$$

$$\theta_2 = \phi_2 - \{(\theta_1, \phi_2)/(\theta_1, \theta_1)\}\theta_1;$$

$$\theta_3 = \phi_3 - \{(\theta_1, \phi_3)/(\theta_1, \theta_1)\}\theta_1 - \{(\theta_2, \phi_3)/(\theta_2, \theta_2)\}\theta_2;$$

and so on. The vectors $\theta_1, \theta_2, \ldots$ are then mutually orthogonal. Finally let $\psi_d = \{\|\theta_j\|^{-1}\}\theta_d$ for $j = 1, 2, \ldots$, so that, by Equation (B.11), $\|\psi_j\| = \|\theta_j\|^{-1}\|\theta_j\| = 1$. Then $\psi_1, \psi_2, \ldots$ form an ortho-normal set.

Ortho-normal sets are particularly useful as bases. If V is an inner product space of dimension d and the basis $\psi_1, \psi_2, \ldots, \psi_d$ of Equation (B.10) is an ortho-normal set, then forming the inner product of both sides of Equation (B.10) with ψ_j gives

$$(\psi_j, \psi) = \sum_{k=1}^{d} a_k(\psi_j, \psi_k) = \sum_{k=1}^{d} a_k\delta_{jk} = a_j.$$

Thus Equation (B.10) can be rewritten as

$$\psi = \sum_{j=1}^{d} (\psi_j, \psi)\psi_j. \tag{B.15}$$

If $\psi'_1, \psi'_2, \ldots, \psi'_d$ is another basis for V and the $d \times d$ matrix $\mathbf{S}$ is defined by

$$\psi'_n = \sum_{m=1}^{d} \mathrm{S}_{mn}\psi_m$$

for $n = 1, 2, \ldots, d$, then $\psi'_1, \psi'_2, \ldots, \psi'_d$ also form an ortho-normal set if and only if $\mathbf{S}$ is a unitary matrix. (This follows from the fact that

$$(\psi'_j, \psi'_k) = \sum_{m=1}^{d} \sum_{n=1}^{d} \mathrm{S}^*_{mj}\mathrm{S}_{nk}(\psi_m, \psi_n) = (\mathbf{S}^\dagger\mathbf{S})_{jk}.)$$

3 Hilbert spaces

For an infinite-dimensional inner product space it is natural to enquire whether the expansion in Equation (B.15) is valid with the finite sum replaced by an infinite sum. This immediately poses questions of convergence for such infinite series. With the metric introduced in Section 2 one may say that the infinite *sequence* $\phi_1, \phi_2, \ldots$ of vectors in an inner product space V tends to a limit ϕ of V (i.e. $\phi_n \to \phi$ if $n \to \infty$) if and only if $d(\phi_n, \phi) \to 0$ as $n \to \infty$. Then for an infinite *series* one may say that $\sum_{j=1}^{\infty} \psi_j$ converges to ϕ if the sequence of partial sums defined for $n = 1, 2, \ldots$ by $\phi_n = \sum_{j=1}^{n} \psi_j$ converges to ϕ, so that all such questions are reduced to questions about sequences.

A sequence $\phi_1, \phi_2, \ldots$ for which

$$\lim_{m,n\to\infty} d(\phi_n, \phi_m) = 0$$

(where m and n tend to infinity independently) is called a "Cauchy sequence". It follows immediately from property (iv) of the metric d that, if $\phi_1, \phi_2, \ldots$ tends to some limit ϕ, then $\phi_1, \phi_2, \ldots$ must be a Cauchy sequence. Unfortunately, examples can be constructed which demonstrate that, in general, the converse is not true. This makes the general investigation of convergence very difficult, for while it is easy to test whether a sequence is a Cauchy sequence or not, direct examination of the definition of convergence requires some presupposition about the possible limit ϕ. This problem can be completely avoided by confining attention to those spaces for which every Cauchy sequence converges, that is, to "Hilbert spaces". The definition will be given for complex inner product spaces, as the only infinite-dimensional spaces that will be met in this book are of this type.

Definition *Hilbert space*
A "Hilbert space" is a complex inner product space in which every Cauchy sequence converges to an element of the space.

The following further restriction is required in order that Equation (B.15) may be generalized to the desired form.

Definition *Separable Hilbert space*
A Hilbert space V is said to be "separable" if there exists a *countable* set of elements $\mathscr{S}$ contained in V such that every vector $\psi \in V$ has some point $\phi \in \mathscr{S}$ arbitrarily close to it. That is, for any $\psi \in V$ and any

$\varepsilon > 0$ there must exist a $\phi \in \mathcal{S}$ such that $d(\phi, \psi) < \varepsilon$. The set $\mathcal{S}$ is then said to be "dense" in V.

It is easily shown that every *finite*-dimensional complex inner product space is a separable Hilbert space.

Definition *Complete ortho-normal system*
An ortho-normal set of vectors $\psi_1, \psi_2, \ldots$ of a Hilbert space is said to be "complete" if there is no non-zero vector that is orthogonal to every ψ_j, $j = 1, 2, \ldots$.

Obviously in an infinite-dimensional Hilbert space a complete set of vectors necessarily contains an infinite number of elements. The following two theorems then provide the required extension of Equation (B.15).

Theorem I If an infinite-dimensional Hilbert space is *separable*, then the space contains a complete ortho-normal system, and every complete ortho-normal system in the space consists of a *countable* number of vectors.

Theorem II If $\psi_1, \psi_2, \ldots$ form a complete ortho-normal system of an infinite-dimensional Hilbert space, then any vector ψ of the space can be written as

$$\psi = \sum_{j=1}^{\infty} (\psi_j, \psi)\psi_j. \tag{B.16}$$

Moreover,

$$\|\psi\|^2 = \sum_{j=1}^{\infty} |(\psi_j, \psi)|^2. \tag{B.17}$$

Equation (B.17) is often called "Parseval's Relation". Proofs of both theorems may be found in the book of Akhiezer and Glazman (1961).

Example I *The separable Hilbert space* L^2
L^2 is defined to be the set of all complex-valued functions $\phi(\mathbf{r})$ (defined for all $\mathbf{r} \in \mathbb{R}^3$) such that

$$\int_{-\infty}^{\infty} \int_{-\infty}^{\infty} \int_{-\infty}^{\infty} |\phi(\mathbf{r})|^2 \, dx \, dy \, dz$$

exists and is finite, the integral here being the Lebesgue integral (see below). The inner product of L^2 may be defined by

$$(\phi, \psi) = \int_{-\infty}^{\infty} \int_{-\infty}^{\infty} \int_{-\infty}^{\infty} \phi(\mathbf{r})^* \psi(\mathbf{r})\, dx\, dy\, dz, \tag{B.18}$$

where the integral is again the Lebesgue integral. With addition and scalar multiplication defined as in Example III of Section 1, it can be shown that L^2 is an infinite-dimensional separable Hilbert space (cf. Akhiezer and Glazman 1961). Equation (B.18) implies that

$$\|\phi\|^2 = \int_{-\infty}^{\infty} \int_{-\infty}^{\infty} \int_{-\infty}^{\infty} |\phi(\mathbf{r})|^2\, dx\, dy\, dz.$$

For a proper development of the concept of the Lebesgue integral, the reader is referred to specialized texts such as that of Riesz and Sz.-Nagy (1956). However, for the understanding of the present book no detailed knowledge is required. It is sufficient to be aware that the definition of the Lebesgue integral is more general than that of the more familiar Riemann integral, so that functions that are not Riemann-integrable may still be Lebesgue-integrable. Nevertheless, the generalization is such that every Riemann-integrable function is Lebesgue-integrable and the values of the two integrals coincide. Also, if $f(\mathbf{r}) = 0$ except on a "set of measure zero" then

$$\int_{-\infty}^{\infty} \int_{-\infty}^{\infty} \int_{-\infty}^{\infty} f(\mathbf{r})\, dx\, dy\, dz = 0.$$

(It is difficult to give a concise characterization of sets of measure zero, but two important facts are easily stated. Firstly, the set of points $\mathbf{r}$ in any sphere $|\mathbf{r} - \mathbf{r}_0|^2 < \delta$ of $\mathbb{R}^3$ has non-zero measure provided $\delta > 0$. Secondly, a set consisting of a finite or a countable number of points has measure zero.) Two functions $f(\mathbf{r})$ and $g(\mathbf{r})$ that are equal except on a set of measure zero are said to be equal "almost everywhere". For such functions

$$\int_{-\infty}^{\infty} \int_{-\infty}^{\infty} \int_{-\infty}^{\infty} f(\mathbf{r})\, dx\, dy\, dz = \int_{-\infty}^{\infty} \int_{-\infty}^{\infty} \int_{-\infty}^{\infty} g(\mathbf{r})\, dx\, dy\, dz.$$

Consequently two functions $\psi(\mathbf{r})$ and $\phi(\mathbf{r})$ that are equal almost everywhere are to be regarded as being identical members of L^2.

4 Linear operators

Let D be a subset of a separable Hilbert space V. If to every $\psi \in D$ there exists a unique element $\phi \in V$, one can write $\phi = A\psi$, thereby defining the "operator" A. D is called the "domain" of A, and the set Δ consisting of all $\phi = A\psi$, where ψ runs through all of D, is known as the "range" of A. Two operators A and B are then said to be "equal" if they have the same domain D, and $A\psi = B\psi$ for all $\psi \in D$.

If the mapping $\phi = A\psi$ is one-to-one, the inverse operator A^{-1} may be defined by $A^{-1}\phi = \psi$ if and only if $\phi = A\psi$. Clearly the domain and range of A^{-1} are Δ and D respectively.

Definition *Linear operator*
An operator A is said to be "linear" if its domain D is a linear manifold (a set D such that if $\phi, \psi \in D$ then $(a\phi + b\psi) \in D$ for all complex numbers a and b) and if

$$A(a\phi + b\psi) = aA\phi + bA\psi$$

for all $\phi, \psi \in D$ and any two complex numbers a and b.

There is no requirement in general that D be the whole Hilbert space, so the definition accommodates such operators as $\partial/\partial x$ acting in $V = L^2$, for which D is the set of functions of L^2 that are differentiable with respect to x.

Definition *Bounded linear operator*
A linear operator A is said to be "bounded" if there exists a positive constant K such that $\|A\psi\| \leqslant K\,\|\psi\|$ for all $\psi \in D$.

Theorem I If A is a linear operator acting in a *finite*-dimensional inner product space V and D = V, then A is necessarily bounded.

Definition *Unitary operator*
An operator U is said to be "unitary" if $D = \Delta = V$ and

$$(U\phi, U\psi) = (\phi, \psi)$$

for all $\phi, \psi \in V$.

It is easily shown that every unitary operator is a bounded linear operator. It is obvious that if $\psi_1, \psi_2, \ldots$ are a complete ortho-normal set then $\psi'_j = U\psi_j$, $j = 1, 2, \ldots$ also form a complete ortho-normal set. Conversely, if $\psi_1, \psi_2, \ldots$ and $\psi'_1, \psi'_2, \ldots$ are two complete ortho-normal sets in a Hilbert space V, then there exists a unitary operator U such that $\psi'_j = U\psi_j$, $j = 1, 2, \ldots$

For a general treatment of linear operators the reader is referred to the books of Akhiezer and Glazman (1961), Simmons (1963) and Riesz and Sz. Nagy (1956). However, as all the operators associated with finite-dimensional representations of groups and Lie algebras are either unitary or act on finite-dimensional spaces, attention here will henceforth be concentrated exclusively on *bounded* linear operators whose domain is the whole Hilbert space V.

If A is such an operator there exists an "adjoint" operator $A^\dagger$ whose domain is also V such that

$$(A^\dagger\phi, \psi) = (\phi, A\psi)$$

for all $\phi, \psi \in V$. It is easily shown that $(AB)^\dagger = B^\dagger A^\dagger$, $(A^\dagger)^\dagger = A$, and $U^\dagger = U^{-1}$ for a unitary operator U.

Definition *Self-adjoint operator*
A bounded linear operator A whose domain is the whole Hilbert space V is said to be "self-adjoint" if $A = A^\dagger$, that is, if

$$(A\phi, \psi) = (\phi, A\psi) \tag{B.19}$$

for all $\phi, \psi \in V$.

If for a bounded linear operator A there exists a non-zero vector ψ and a complex number λ such that

$$A\psi = \lambda\psi, \tag{B.20}$$

then ψ is said to be an "eigenvector" of A and λ is referred to as the corresponding "eigenvalue". If there exist d linearly independent eigenvectors $\psi_1, \psi_2, \ldots$ of A with the same eigenvalue λ, then λ is said to have "multiplicity d" or to be "d-fold degenerate". In that case any linear combination $(b_1\psi_1 + b_2\psi_2 + \ldots + b_d\psi_d)$ is also an eigenvector with the same eigenvalue λ.

For the special case of self-adjoint operators there are three important theorems:

Theorem II The eigenvalues of a *self-adjoint* operator are all *real*.

Proof Suppose that $A\psi = \lambda\psi$, where $\psi \neq 0$. Then, if A is self-adjoint, $\lambda(\psi, \psi) = (\psi, A\psi) = (A\psi, \psi) = \lambda^*(\psi, \psi)$, so $\lambda = \lambda^*$.

Theorem III Eigenvectors of a *self-adjoint* operator belonging to *different* eigenvalues are *orthogonal*.

Proof Suppose that $A\psi_1 = \lambda_1\psi_1$ and $A\psi_2 = \lambda_2\psi_2$, where $\lambda_1 \neq \lambda_2$. Then, if A is self-adjoint, $\lambda_1(\psi_2, \psi_1) = (\psi_2, A\psi_1) = (A\psi_2, \psi_1) = \lambda_2^*(\psi_2, \psi_1) = \lambda_2(\psi_2, \psi_1)$, so that $(\lambda_1 - \lambda_2)(\psi_2, \psi_1) = 0$. As $\lambda_1 - \lambda_2 \neq 0$, it follows that $(\psi_2, \psi_1) = 0$.

Theorem IV If A is a self-adjoint operator and U is a unitary operator, then $A' = U^{-1}AU$ is also self-adjoint and possesses exactly the same eigenvalues as A.

Proof A′ is self-adjoint because

$$(A')^{\dagger} = (U^{-1}AU)^{\dagger} = U^{\dagger}A^{\dagger}(U^{-1})^{\dagger} = U^{-1}AU = A'.$$

Now suppose that ψ' is an eigenvector of A′ with eigenvalue λ', so that $A'\psi' = \lambda'\psi'$. Then $U^{-1}AU\psi' = \lambda'\psi'$, so that $A(U\psi') = \lambda'(U\psi')$, showing that $U\psi'$ is an eigenvector of A with the same eigenvalue λ'.

Every bounded operator has a matrix representation. Indeed, the operator eigenvalue equation (Equation (B.20)) can be re-cast in the form of the matrix eigenvalue equation (Equation (A.10)). For convenience, the argument will be presented for a finite-dimensional inner product space V of dimension d, but the results generalize in the obvious way to bounded operators acting on a separable infinite-dimensional Hilbert space, although in that case all the matrices involved are infinite-dimensional.

Let $\psi_1, \psi_2, \dots, \psi_d$ be an ortho-normal set of V. Taking the inner product of both sides of Equation (B.20) with any ψ_k and invoking Equation (B.15) gives

$$\sum_{j=1}^{d} (\psi_k, A\psi_j)(\psi_j, \psi) = \lambda \sum_{j=1}^{d} (\psi_k, \psi_j)(\psi_j, \psi)(= \lambda(\psi_k, \psi)) \tag{B.21}$$

for $k = 1, 2, \dots, d$. Let $\mathbf{A}$ be the $d \times d$ matrix defined by

$$\mathbf{A}_{kj} = (\psi_k, A\psi_j) \tag{B.22}$$

for $j, k = 1, 2, \dots, d$, and let $\mathbf{c}$ be the $d \times 1$ column matrix whose elements are specified by $\mathbf{c}_{j1} = (\psi_j, \psi)$, $j = 1, 2, \dots, d$. Then Equation (B.21) can be rewritten as $\mathbf{Ac} = \lambda\mathbf{c}$, that is, as Equation (A.10).

It should be noted that if $A\psi_n$ is expanded in terms of the ortho-normal set, then, by Equation (B.15),

$$A\psi_n = \sum_{m=1}^{d} (\psi_m, A\psi_n)\psi_m = \sum_{m=1}^{d} \mathbf{A}_{mn}\psi_m. \tag{B.23}$$

It will be observed that the ordering of indices is *exactly* as in Equations (4.1) and (4.3).

If A is a *self-adjoint* operator then its corresponding matrix $\mathbf{A}$ is *Hermitian*, as, by Equations (B.19) and (B.22),

$$\mathbf{A}_{kj} = (\psi_k, A\psi_j) = (A\psi_k, \psi_j) = (\psi_j, A\psi_k)^* = \mathbf{A}^*_{jk}.$$

Similarly, if U is a *unitary* operator and **U** is its corresponding matrix, then **U** is a *unitary* matrix. (This follows as $(\psi_k, U\psi_j) = (U\psi_j, \psi_k)^* = (\psi_j, U^{-1}\psi_k)^*$, so that $U_{kj} = (\mathbf{U}^{-1})^*_{jk}$.)

Finally, if A, B and C are three bounded operators such that C = AB, and if **A**, **B** and **C** are their corresponding matrices, then **C** = **AB**. (As $AB\psi_j = C\psi_j$ for each $\psi_j, j = 1, 2, \ldots, d$, $C_{kj} = (\psi_k, C\psi_j) = (\psi_k, AB\psi_j)$. But, from Equation (B.23), $B\psi_j = \sum_{m=1}^{d} (\psi_m, B\psi_j)\psi_m$, so $C_{kj} = \sum_{m=1}^{d} (\psi_k, A\psi_m)(\psi_m, B\psi_j) = \sum_{m=1}^{d} A_{km}B_{mj}$). This is the origin of the duality between operators and matrices that is used repeatedly, particularly in Chapter 1, Section 4, and Chapter 4, Section 1.

5 Bilinear forms

Even when a vector space does not possess an inner product it may possess a symmetric non-degenerate bilinear form which gives rise to rather similar properties. In particular this is true of semi-simple Lie algebras (see Chapter 13).

Definition *Symmetric bilinear form*
A *complex* vector space V possesses a bilinear form B if to every pair of vectors ψ and ϕ of V there corresponds a complex number $B(\psi, \phi)$ such that

(*a*) $B(\psi, \phi) = B(\phi, \psi)$,
(*b*) $B(a\psi, b\phi) = abB(\psi, \phi)$, for any two complex numbers a and b,
(*c*) $B(\psi + \phi, \chi) = B(\psi, \chi) + B(\phi, \chi)$, for any $\chi \in$ V.

If V is a *real* vector space the bilinear form $B(\psi, \phi)$ is required to be real for all $\psi, \phi \in$ V, and in (*b*) a and b are restricted to being real numbers, but otherwise the conditions (*a*) to (*c*) are the same as for a complex space.

It should be noted that (*a*), (*b*) and (*c*) imply that

$$B(\chi, a\psi + b\phi) = aB(\chi, \psi) + bB(\chi, \phi)$$

and

$$B(a\psi + b\phi, \chi) = aB(\psi, \chi) + bB(\phi, \chi).$$

There is no requirement that $B(\psi, \psi)$ be real (unless V is a real vector space), and even then $B(\psi, \psi)$ could be negative, or could be zero with $\psi \neq 0$.

Thus a symmetric bilinear form does *not* in general have the properties of an inner product (as defined in Section 2). Conversely, if

V is a *complex* inner product space then the inner product is not a symmetric bilinear form (because the right-hand sides of parts (a) and (b) of the definition in Section 2 of an inner product involve complex conjugation, whereas the corresponding parts (a) and (b) of the definition of a symmetric bilinear form do not do so). However, if V is a *real* inner product space these particular distinctions disappear, so in this case an inner product is also a symmetric bilinear form.

Example I *Symmetric bilinear form of Minkowski space–time*

Consider the four-dimensional real vector space of four-component vectors (x_1, x_2, x_3, x_4) with scalar multiplication and vector addition being defined by the obvious generalizations of Equations (B.1) and (B.3). Let $\psi = (x_1, x_2, x_3, x_4)$ and $\phi = (y_1, y_2, y_3, y_4)$ be any two vectors of this space and define $B(\psi, \phi)$ by

$$B(\psi, \phi) = x_1 y_1 + x_2 y_2 + x_3 y_3 - x_4 y_4. \tag{B.24}$$

Then it is easily verified that B is a symmetric bilinear form. Minkowski space–time provides an example of such a space. (The first three components of a vector give the spatial coordinates of an event, the fourth component being ct, where t is the time of the event and c the speed of light (Pauli 1958).) It will be shown below (Equation (B.27)) that *every* symmetric bilinear form of *every* real vector space has a form similar to Equation (B.24) if the basis is chosen appropriately.

Let $\psi_1, \psi_2, \ldots, \psi_d$ be a basis for V, and let $\mathbf{B}$ be the $d \times d$ matrix defined by

$$\mathrm{B}_{pq} = B(\psi_p, \psi_q), \qquad p, q = 1, 2, \ldots, d. \tag{B.25}$$

Then, if $\psi = \sum_{j=1}^{d} a_j \psi_j$ and $\phi = \sum_{k=1}^{d} b_k \psi_k$,

$$B(\psi, \phi) = \sum_{j=1}^{d} \sum_{k=1}^{d} B_{jk} a_j b_k \tag{B.26}$$

Suppose that $\psi_1', \psi_2', \ldots, \psi_d'$ is another basis for V, with $\psi_n' = \sum_{m=1}^{d} \mathrm{S}_{mn} \psi_m$, $(n = 1, 2, \ldots, d)$, so that (as noted in Section 1) $\mathbf{S}$ is a $d \times d$ non-singular matrix. Let $\mathbf{B}'$ be the corresponding matrix for the bilinear form defined for this basis, that is, let

$$\mathrm{B}_{pq}' = B(\psi_p', \psi_q'), \qquad p, q = 1, 2, \ldots, d.$$

Then a very straightforward argument shows that

$$\mathbf{B}' = \tilde{\mathbf{S}} \mathbf{B} \mathbf{S}.$$

This implies that $\det \mathbf{B}' = (\det \mathbf{S})^2 \det \mathbf{B}$. Consequently $\det \mathbf{B}' = 0$ if and only if $\det \mathbf{B} = 0$.

If V is a *real* vector space it can be shown (Gantmacher 1959) that $\mathbf{S}$ may be chosen so that $\mathbf{B}'$ is diagonal with diagonal elements $1, -1$, or 0 only. Then, if the basis $\psi'_1, \psi'_2, \ldots, \psi'_d$ is ordered so that the first d_+ ($\geqslant 0$) members correspond to 1, the next d_-($\geqslant 0$) to -1, and the remaining d_0 ($\geqslant 0$) to 0, and if $\psi = \sum_{j=1}^{d} a'_j \psi'_j$ and $\phi = \sum_{k=1}^{d} b'_k \psi'_k$, then

$$B(\psi, \phi) = \sum_{j=1}^{d_+} a'_j b'_j - \sum_{j=d_++1}^{d_++d_-} a'_j b'_j. \tag{B.27}$$

Matrices $\mathbf{S}$ with this property can be chosen in an infinite number of ways, but *all* choices give the same values of the dimensions d_+, d_- and d_0 (Gantmacher 1959). The invariant quantity $\sigma = d_+ - d_-$ is called the "signature" of the bilinear form.

Definition *Degenerate and non-degenerate symmetric bilinear forms*

A symmetric bilinear form B is said to be "degenerate" if there exists in V some $\psi \neq 0$ such that $B(\psi, \phi) = 0$ for *all* $\phi \in \mathrm{V}$. Conversely, a symmetric bilinear form is "non-degenerate" if, for each $\psi \in \mathrm{V}$, the condition

$$B(\psi, \phi) = 0 \quad \text{for all} \quad \phi \in \mathrm{V}$$

implies that $\psi = 0$.

Theorem I The symmetric bilinear form B is non-degenerate if and only if $\det \mathbf{B} \neq 0$, where $\mathbf{B}$ is the $d \times d$ matrix defined in Equation (B.25).

Proof It should be noted that as $\det \mathbf{B}' = 0$ if and only if $\det \mathbf{B} = 0$, this condition for non-degeneracy is actually independent of the choice of basis, as is to be expected.

Suppose that there exists a $\psi \in \mathrm{V}$ such that $B(\psi, \phi) = 0$ for *all* $\phi \in \mathrm{V}$. By Equation (B.26) this is so if and only if $\sum_{j,k=1}^{d} \mathrm{B}_{jk} a_j b_k = 0$ for *all* sets $b_1, b_2, \ldots, b_d$, that is, if and only if $\sum_{j=1}^{d} \mathrm{B}_{jk} a_j = 0$ for each $k = 1, 2, \ldots, d$. As this set of d simultaneous linear equations for $a_1, a_2, \ldots, a_d$ has a non-trivial solution (i.e. a solution other than $\psi = 0$) if and only if $\det \mathbf{B} = 0$, the quoted result follows.

6 Linear functionals

The theory of linear functionals will be considered here only for *finite*-dimensional vector spaces and inner product spaces. The results will be needed in the discussions of Lie algebras in Chapter 13. The generalization to infinite-dimensional Hilbert spaces may be found in the books of Akhiezer and Glazman (1961) and Riesz and Sz. Nagy (1956).

Definition *Linear functional*
If to every member ψ of a *complex* finite-dimensional vector space V a *complex number* $\Phi(\psi)$ is assigned in such a way that

$$\Phi(a\phi + b\psi) = a\Phi(\phi) + b\Phi(\psi) \tag{B.28}$$

for every $\phi, \psi \in V$ and any two complex numbers a and b, then Φ is said to be a "linear functional" on V. Likewise, a linear functional on a *real* finite-dimensional vector space V is an assignment of a *real number* $\Phi(\psi)$ to every $\psi \in V$ such that Equation (B.28) holds for every $\phi, \psi \in V$ and any two real numbers a and b.

If Φ and Ψ are any two linear functionals defined on a finite-dimensional vector space V, then $(\Phi+\Psi)$ may be defined by $(\Phi+\Psi)(\psi) = \Phi(\psi)+\Psi(\psi)$ for all $\psi \in V$. Similarly, $a\Phi$ may be defined by $(a\Phi)(\psi) = a\Phi(\psi)$ for all $\psi \in V$, a being any real or complex number as appropriate. Then the set of linear functionals on V themselves form a vector space V*, called the "dual" of V. (The zero of V* is the functional whose value is 0 for all $\psi \in V$.) V* is real when V is real and is complex when V is complex.

Suppose that V has dimension d and $\psi_1, \psi_2, \ldots, \psi_d$ is a basis for V. Then each linear functional Φ on V is *completely* specified by the d numbers $\Phi(\psi_j), j = 1, 2, \ldots, d$. (*Any* $\psi \in V$ can be written in the form of Equation (B.10) as $\psi = \sum_{j=1}^{d} a_j\psi_j$, so, by Equation (B.28), $\Phi(\psi) = \sum_{j=1}^{d} a_j\Phi(\psi_j)$.) Let $\Phi_k, k = 1, 2, \ldots, d$, be a set of linear functionals defined by

$$\Phi_k(\psi_j) = \delta_{kj} \tag{B.29}$$

for all $j, k = 1, 2, \ldots, d$. The functionals of this set are obviously linearly independent. Moreover, if Φ is *any* linear functional on V then Equation (B.29) implies that $\Phi = \sum_{k=1}^{d} \Phi(\psi_k)\Phi_k$, that is, Φ depends linearly on $\Phi_1, \Phi_2, \ldots, \Phi_d$. Thus the dual space V* has the same dimension d as V, and $\Phi_1, \Phi_2, \ldots, \Phi_d$ provide a basis for V*.

If V is equipped with a symmetric non-degenerate bilinear form, or is an inner product space, the following theorems show that every linear functional is given by a remarkably simple expression.

Theorem I Each linear functional Φ on a finite-dimensional vector space equipped with a symmetric non-degenerate bilinear form can be expressed in the form

$$\Phi(\psi) = B(\psi_\Phi, \psi) \tag{B.30}$$

for all $\psi \in \mathrm{V}$, where $B(\psi_\Phi, \psi)$ is the bilinear form, and ψ_Φ is an element of V which is uniquely determined by the functional Φ.

Proof Suppose that $\psi_\Phi = \sum_{j=1}^{d} a_j \psi_j$ has the required property, $\psi_1, \psi_2, \ldots, \psi_d$ being a basis for V. Then Equation (B.30) can be written in the form $\Phi(\psi) = \sum_{j=1}^{d} B(\psi_j, \psi) a_j$, so that for each $k = 1, 2, \ldots, d$.

$$\Phi(\psi_k) = \sum_{j=1}^{d} B(\psi_j, \psi_k) a_j.$$

Thus if $\mathbf{\Phi}$ and $\mathbf{a}$ are the $d \times 1$ matrices with elements $\Phi(\psi_k)$ and a_k respectively ($k = 1, 2, \ldots, d$), and $\mathbf{B}$ is defined by Equation (B.25), as $\mathbf{B}$ is symmetric these equations can be written as $\mathbf{\Phi} = \mathbf{Ba}$. The linear functional Φ fixes $\mathbf{\Phi}$. This equation has a *unique* solution $\mathbf{a}$ when $\det \mathbf{B} \neq 0$, namely $\mathbf{a} = \mathbf{B}^{-1}\mathbf{\Phi}$, which then determines ψ_Φ uniquely.

Theorem II Each linear functional Φ on a Hilbert space V can be expressed in the form

$$\Phi(\psi) = (\psi_\Phi, \psi)$$

for all $\psi \in \mathrm{V}$, where (ψ_Φ, ψ) is the inner product of V, and ψ_Φ is an element of V which is uniquely determined by the functional Φ.

Proof If V is finite-dimensional, a proof can be given along the lines of that of the previous theorem. For the infinite-dimensional case see Akhiezer and Glazman (1961) or Riesz and Sz. Nagy (1956).

This latter theorem is often called the "Riesz Representation Theorem". It is easily verified that if ψ_Φ is as specified in this theorem and $\psi_1, \psi_2, \ldots$ is an ortho-normal basis for V, then

$$\psi_\Phi = \sum_j \Phi(\psi_j)^* \psi_j.$$

7 Direct product spaces

Let V_1 and V_2 be two complex inner product spaces of dimensions d_1 and d_2. Let ψ_j ($j = 1, 2, \ldots, d_1$) and ϕ_s ($s = 1, 2, \ldots, d_2$) be ortho-normal bases for V_1 and V_2 respectively. Then the "direct product" or "tensor product" space $\mathrm{V}_1 \otimes \mathrm{V}_2$ may be defined as the complex vector space having the set of $d_1 d_2$ "products" $\psi_j \otimes \phi_s$ as its basis, so that

$V_1 \otimes V_2$ is the set of all quantities θ of the form

$$\theta = \sum_{j=1}^{d_1} \sum_{s=1}^{d_2} a_{js} \psi_j \otimes \phi_s, \tag{B.31}$$

where the a_{js} are a set of complex numbers. The direct product of any two elements $\psi = \sum_{j=1}^{d_1} b_j \psi_j$ and $\phi = \sum_{s=1}^{d_2} c_s \phi_s$ of V_1 and V_2 is defined to be

$$\psi \otimes \phi = \sum_{j=1}^{d_1} \sum_{s=1}^{d_2} b_j c_s \psi_j \otimes \phi_s, \tag{B.32}$$

so that the set of such products is a subset of $V_1 \otimes V_2$. It is easily verified that all the requirements for $V_1 \otimes V_2$ to be a vector space are satisfied. (The zero vector of $V_1 \otimes V_2$ corresponds to $a_{js} = 0$ for *all* $j = 1, 2, \ldots, d_1$, and $s = 1, 2, \ldots, d_2$.) The products $\psi_j \otimes \phi_s$ are assumed to be linearly independent, so that $V_1 \otimes V_2$ has dimension $d_1 d_2$. (This is assumed to be the case even when V_1 and V_2 are identical, when one could take $\psi_j = \phi_j$ for $j = 1, 2, \ldots, d_1$ $(= d_2)$, implying that the products $\psi_j \otimes \psi_s$ and $\psi_s \otimes \psi_j$ are linearly independent.)

An inner product can be defined on $V_1 \otimes V_2$ by assuming that the basis elements $\psi_j \otimes \phi_s$ are ortho-normal, i.e. that

$$(\psi_j \otimes \phi_s, \psi_k \otimes \phi_t) = \delta_{jk} \delta_{st}. \tag{B.33}$$

Then if θ is defined as in Equation (B.31), and

$$\chi = \sum_{j=1}^{d_1} \sum_{s=1}^{d_2} d_{js} \psi_j \otimes \phi_s, \tag{B.34}$$

it follows that

$$(\theta, \chi) = \sum_{j=1}^{d_1} \sum_{s=1}^{d_2} a_{js}^* d_{js} \tag{B.35}$$

This inner product has all the required properties of an inner product space.

The definition of $V_1 \otimes V_2$ and its inner product of Equation (B.35) is actually ***independent*** of the choice of the ortho-normal bases of V_1 and V_2. To see this let ψ'_k $(k = 1, 2, \ldots, d_1)$ and ϕ'_t $(t = 1, 2, \ldots, d_2)$ be another pair of ortho-normal bases for V_1 and V_2 respectively. Then (see Section 2) there exists a $d_1 \times d_1$ unitary matrix $\mathbf{F}$ and a $d_2 \times d_2$ unitary matrix $\mathbf{G}$ such that

$$\psi_j = \sum_{k=1}^{d_1} \mathbf{F}_{kj} \psi'_k, \qquad j = 1, 2, \ldots, d_1.$$

and

$$\phi_s = \sum_{t=1}^{d_2} \mathbf{G}_{ts} \phi'_t, \qquad s = 1, 2, \ldots, d_2.$$

Then, for any θ of $V_1 \otimes V_2$, defined as in Equation (B.31),

$$\theta = \sum_{k=1}^{d_1} \sum_{t=1}^{d_2} a'_{kt} \psi'_k \otimes \phi'_t,$$

where

$$a'_{kt} = \sum_{l=1}^{d_1} \sum_{u=1}^{d_2} F_{kl} a_{lu} G_{tu},$$

thereby demonstrating that the set $\psi'_k \otimes \phi'_t$ forms an alternative basis for $V_1 \otimes V_2$. Moreover, as the vector χ of Equation (B.34) can similarly be rewritten as

$$\chi = \sum_{k=1}^{d_1} \sum_{t=1}^{d_2} d'_{kt} \psi'_k \otimes \phi'_t$$

with

$$d'_{kt} = \sum_{l=1}^{d_1} \sum_{u=1}^{d_2} F_{kl} d_{lu} G_{tu},$$

and as **F** and **G** are unitary, it follows that

$$(\theta, \chi) = \sum_{k=1}^{d_1} \sum_{t=1}^{d_2} a'^{*}_{kt} d'_{kt},$$

showing that the inner product is independent of the choice of basis (see Equation B.35).

In the physics literature the $\otimes$ sign is often omitted in products such as $\psi_j \otimes \phi_s$, but it will be retained throughout this book as a warning that the product is *not* ordinary multiplication.

For abstract inner product spaces V_1 and V_2, the product $\otimes$ in $\psi_j \otimes \phi_s$ neither requires nor is amenable to any further specification. However, in concrete examples this product can be defined quite naturally.

As a first example, suppose $V_1 = \mathbb{C}^3$ and $V_2 = \mathbb{C}^2$, with $\psi_1 = (1, 0, 0)$, $\psi_2 = (0, 1, 0)$, $\psi_3 = (0, 0, 1)$ and $\phi_1 = (1, 0)$, $\phi_2 = (0, 1)$. Then the products $\psi_j \otimes \phi_s$ can be *defined* as the six-component quantities:

$$\left.\begin{array}{ll} \psi_1 \otimes \phi_1 = (1,0,0,0,0,0), & \psi_1 \otimes \phi_2 = (0,0,0,1,0,0); \\ \psi_2 \otimes \phi_1 = (0,1,0,0,0,0), & \psi_2 \otimes \phi_2 = (0,0,0,0,1,0); \\ \psi_3 \otimes \phi_1 = (0,0,1,0,0,0), & \psi_3 \otimes \phi_2 = (0,0,0,0,0,1); \end{array}\right\}$$

so that Equation (B.33) is satisfied. Clearly $\mathbb{C}^3 \otimes \mathbb{C}^2$ can be identified with $\mathbb{C}^6$.

As a second example, let V_1 and V_2 be subspaces of L^2, the elements of V_1 being functions $\psi(\mathbf{r}_1)$ of $\mathbf{r}_1$ and the elements of V_2 being functions $\phi(\mathbf{r}_2)$ of $\mathbf{r}_2$. Then the elements of $V_1 \otimes V_2$ are linear combinations of products $\psi(\mathbf{r}_1)\phi(\mathbf{r}_2)$, the inner product for which may be defined by

$$(\psi'(\mathbf{r}_1)\phi'(\mathbf{r}_2), \psi(\mathbf{r}_1)\phi(\mathbf{r}_2)) = \int\int\int\int\int\int dx_1\, dy_1\, dz_1\, dx_2\, dy_2\, dz_2 \times \psi'^*(\mathbf{r}_1)\phi'^*(\mathbf{r}_2)\psi(\mathbf{r}_1)\phi(\mathbf{r}_1)$$

As a final example, let V_1 be a subspace of L^2, consisting of functions $\psi(\mathbf{r})$, and let V_2 be the two-dimensional space of 2×1 matrices with constant entries. Let $\psi_j(\mathbf{r})$ ($j = 1, 2, \ldots, d_1$) be an ortho-normal basis of V_1 and

$$\phi_1 = \begin{bmatrix} 1 \\ 0 \end{bmatrix}, \qquad \phi_2 = \begin{bmatrix} 0 \\ 1 \end{bmatrix}$$

an ortho-normal basis of V_2 (it being assumed that the inner product of two elements

$$\begin{bmatrix} a \\ b \end{bmatrix} \text{ and } \begin{bmatrix} a' \\ b' \end{bmatrix}$$

of V_2 is $(a^*a' + b^*b')$). Then $V_1 \otimes V_2$ is the set of two-component quantities

$$\begin{bmatrix} \psi(\mathbf{r}) \\ \psi'(\mathbf{r}) \end{bmatrix}$$

($\psi(\mathbf{r}), \psi'(\mathbf{r}) \in V$), with inner product defined by

$$\left(\begin{bmatrix} \psi(\mathbf{r}) \\ \psi'(\mathbf{r}) \end{bmatrix}, \begin{bmatrix} \eta(\mathbf{r}) \\ \eta'(\mathbf{r}) \end{bmatrix}\right) = \int\int\int \{\psi^*(\mathbf{r})\eta(\mathbf{r}) + \psi'^*(\mathbf{r})\eta'(\mathbf{r})\}\, dx\, dy\, dz$$

for all $\psi(\mathbf{r}), \psi'(\mathbf{r}), \eta(\mathbf{r}), \eta'(\mathbf{r}) \in V_1$.

It is shown in Chapter 5, Section 4, that if V_1 and V_2 are both subspaces of L^2 consisting of functions of the *same* variable $\mathbf{r}$, then it may *not* be possible to identify $V_1 \otimes V_2$ with the set of linear combinations of ordinary products of members of V_1 and V_2 and at the same time retain the inner product of L^2 as the inner product of $V_1 \otimes V_2$.

8 Quaternions

The concept of a quaternion can be obtained by generalizing that of a complex number.

To each complex number $z = x + iy$ there can be assigned an element $\psi(z) = xe_0 + ye_1$ of a two-dimensional real vector space whose basis elements are e_0 and e_1, this assignment being one-to-one. If a multiplication operation is *defined* in this space so that

$$e_0e_0 = e_0, \qquad e_1e_1 = -e_0, \qquad e_0e_1 = e_1e_0 = e_1, \tag{B.36}$$

and so that

$$(xe_0+ye_1)(x'e_0+y'e_1)=xx'e_0e_0+xy'e_0e_1+yx'e_1e_0+yy'e_1e_1, \tag{B.37}$$

then

$$(xe_0+ye_1)(x'e_0+y'e_1)=(xx'-yy')e_0+(xy'+y'x)e_1.$$

However, if $z=x+iy$ and $z'=x'+iy'$, then $zz'=(xx'-yy')+i(xy'+yx')$, so that

$$\psi(z)\psi(z')=\psi(zz').$$

Thus ψ provides an isomorphic mapping of the set of complex numbers onto this two-dimensional real vector space. The operation of conjugation in this space may be *defined* by

$$\psi(z)^*=xe_0-ye_1$$

so that $\psi(z)^*=\psi(z^*)$. Then $\psi(z)^*\psi(z)=|z|^2e_0$.

In the same way a *quaternion* q may be regarded as an element in a four-dimensional real vector space with basis elements e_0, e_1, e_2 and e_3. That is

$$q=\sum_{r=0}^{3} q_r e_r \tag{B.38}$$

with q_0, q_1, q_2 and q_3 all real numbers. The generalization of Equations (B.36) is the requirement that

$$e_0e_r=e_re_0=e_r \quad (r=0,1,2,3) \tag{B.39}$$

and

$$\left.\begin{array}{c} e_re_r=-e_0, \quad (r=0,1,2,3) \\ e_1e_2=-e_2e_1=e_3, \qquad e_2e_3=-e_3e_2=e_1, \qquad e_3e_1=-e_1e_3=e_2. \end{array}\right\} \tag{B.40}$$

These latter relations (Equations (B.40)) may be written more concisely as:

$$e_re_s=-e_0\delta_{rs}+\sum_{t=1}^{3}\varepsilon_{rst}e_t, \quad r,s=1,2,3, \tag{B.41}$$

where ε_{rst} is the antisymmetric quantity defined in (10.20). Similarly, the generalization of Equation (B.37) is

$$qq'=\sum_{r,s=0}^{3} q_rq'_se_re_s. \tag{B.42}$$

where $q' = \sum_{s=0}^{3} q'_s e_s$. It should be noted that in general

$$qq' \neq q'q.$$

that is, multiplication of quaternions is ***not commutative***. The conjugate of the quaternion q of Equation (B.38) is defined to be

$$q^* = q_0 e_0 - \sum_{r=1}^{3} q_r e_r, \tag{B.43}$$

and the norm $N(q)$ of q may be defined as the non-negative real number such that

$$N(q)^2 = \sum_{r=0}^{3} q_r^2$$

so that

$$q^* q = qq^* = N(q)^2 e_0.$$

(Just as e_1 corresponds to the purely imaginary number of unit norm i for the case of complex numbers, for the quaternions e_1, e_2 and e_3 can be thought of as corresponding to three different "imaginary" numbers of unit norm.)

A concrete realization of the basis elements e_0, e_1, e_2 and e_3 provided by the 2×2 matrices

$$\mathbf{e}_0 = \mathbf{1}_2, \qquad \mathbf{e}_r = i\boldsymbol{\sigma}_r \quad (r = 1, 2, 3),$$

where $\boldsymbol{\sigma}_1, \boldsymbol{\sigma}_2, \boldsymbol{\sigma}_3$ are the Pauli spin matrices of Equations (3.24), which satisfy the relations in Equations (B.39) and (B.40). For any quaternion $q = \sum_{r=0}^{3} q_r e_r$ (with q_0, q_1, q_2, q_3 all real) let $\mathbf{q} = \phi(q)$ be the 2×2 matrix defined by

$$\mathbf{q} = \boldsymbol{\phi}(q) = q_0 \mathbf{1}_2 + i \sum_{r=1}^{3} q_r \boldsymbol{\sigma}_r. \tag{B.44}$$

Then for any two quaternions q and q'

$$\boldsymbol{\phi}(q)\boldsymbol{\phi}(q') = \boldsymbol{\phi}(qq'),$$

so that $\boldsymbol{\phi}$ provides an isomorphic mapping of the set of quaternions onto this set of 2×2 matrices (Equation (B.44)).

It is possible to construct matrices whose elements are quaternions instead of real or complex numbers. The product of two such matrices may be defined in terms of products of matrix elements exactly as in Equation (A.1), but the product of quaternionic matrix elements is given by Equation (B.42). As multiplication of quaternions is not commutative, it is *not true* that $\operatorname{tr}\mathbf{AB} = \operatorname{tr}\mathbf{BA}$ if $\mathbf{A}$ and $\mathbf{B}$ have quaternionic matrix elements.

Appendix C

Proofs of Certain Theorems on Group Representations

In order to present the main ideas of group representations as concisely as possible, a number of the longer proofs of the theorems of Chapters 4 and 5 have been relegated to this appendix.

1 Proofs of Theorems I and IV of Chapter 4, Section 3

Theorem I If $\mathcal{G}$ is a *finite* group or a *compact* Lie group, then *every* representation of $\mathcal{G}$ is *equivalent* to a *unitary* representation.

Proof The proof will be given explicitly for a finite group of order g. It immediately generalizes to a compact Lie group if every sum of the form $(1/g)\sum_{T\in\mathcal{G}} f(T)$ is replaced by the corresponding invariant integral

$$\int_{\mathcal{G}} f(T)\,dT$$

(see Chapter 3, Section 5).

Let $\boldsymbol{\Gamma}$ be a d-dimensional representation of $\mathcal{G}$. Define the $d\times d$ matrix $\mathbf{H}$ by

$$\mathbf{H}=(1/g)\sum_{T\in\mathcal{G}} \boldsymbol{\Gamma}(T)^{\dagger}\boldsymbol{\Gamma}(T). \tag{C.1}$$

(As $\{\boldsymbol{\Gamma}^{\dagger}(T)\boldsymbol{\Gamma}(T)\}_{jk}$ is a continuous function of T for every $j, k =$

1, 2, ..., d, for a compact group the integrals

$$\int_{\mathcal{G}} \{\mathbf{\Gamma}^{\dagger}(\mathrm{T})\mathbf{\Gamma}(\mathrm{T})\}_{jk}\, d\mathrm{T}$$

are all *finite*. There is *no* guarantee that this will be so for a non-compact Lie group. Indeed, an instancc is given in the Example I of Chapter 4, Section 3 for which such an integral is not finite.)

Then, for any $\mathrm{T}' \in \mathcal{G}$

$$\begin{aligned} \mathbf{\Gamma}(\mathrm{T}')^{\dagger}\mathbf{H}\mathbf{\Gamma}(\mathrm{T}') &= (1/g) \sum_{\mathrm{T}\in\mathcal{G}} \{\mathbf{\Gamma}(\mathrm{T})\mathbf{\Gamma}(\mathrm{T}')\}^{\dagger}\{\mathbf{\Gamma}(\mathrm{T})\mathbf{\Gamma}(\mathrm{T}')\} \\ &= (1/g) \sum_{\mathrm{T}\in\mathcal{G}} \mathbf{\Gamma}(\mathrm{TT}')^{\dagger}\mathbf{\Gamma}(\mathrm{TT}') \\ &= (1/g) \sum_{\mathrm{T}\in\mathcal{G}} \mathbf{\Gamma}(\mathrm{T})^{\dagger}\mathbf{\Gamma}(\mathrm{T}) = \mathbf{H} \end{aligned} \quad \text{(C.2)}$$

by virtue of the right-invariance of the sum. (It is at this stage that the right-invariance property in Equation (3.22) is required for a compact Lie group.)

It follows from Equation (C.1) that $\mathbf{H}$ is Hermitian (i.e. $\mathbf{H}^{\dagger} = \mathbf{H}$), so (see Appendix A, Section 2) there exist a $d \times d$ unitary matrix $\mathbf{U}$ such that $\mathbf{H}' = \mathbf{U}^{-1}\mathbf{H}\mathbf{U}$ is diagonal,. Also if $\mathbf{x}$ is a $d \times 1$ matrix with complex elements then

$$\mathbf{x}^{\dagger}\mathbf{H}\mathbf{x} = (1/g) \sum_{j=1}^{d} \sum_{\mathrm{T}\in\mathcal{G}} |(\mathbf{\Gamma}(\mathrm{T})\mathbf{x})_{j1}|^2,$$

which is real and positive for any such matrix $\mathbf{x}$ whose components are not all zero. With $\mathbf{y} = \mathbf{U}^{-1}\mathbf{x}$, $\mathbf{y}^{\dagger}\mathbf{H}'\mathbf{y} = \mathbf{x}^{\dagger}\mathbf{H}\mathbf{x}$, so $\mathbf{y}^{\dagger}\mathbf{H}'\mathbf{y}$ is also real and positive for any $d \times 1$ matrix $\mathbf{y}$ whose components are not all zero. However, as $\mathbf{H}'$ is diagonal, $\mathbf{y}^{\dagger}\mathbf{H}'\mathbf{y} = \sum_{j=1}^{d} \mathrm{H}'_{jj}\, |y_{j1}|^2$, so all the diagonal elements of $\mathbf{H}'$ must be real and positive. It is then possible to define its "square root" $\mathbf{h}$ by $\mathrm{h}_{jk} = 0$ if $j \neq k$ and $\mathrm{h}_{jj} = (\mathrm{H}'_{jj})^{1/2}$, so that $\mathbf{hh} = \mathbf{H}'$.

Now define the $d \times d$ matrix $\mathbf{S}$ by

$$\mathbf{S} = \mathbf{U}\mathbf{h}^{-1}\mathbf{U}^{-1}$$

so that $\mathbf{S}^{-1} = \mathbf{U}\mathbf{h}\mathbf{U}^{-1}$. As $\mathbf{U}$ is unitary and $\mathbf{h}^{-1}$ is Hermitian, $\mathbf{S}^{\dagger} = \mathbf{S}$. Consequently

$$\mathbf{H} = (\mathbf{S}^{\dagger})^{-1}\mathbf{S}^{-1}, \quad \text{(C:3)}$$

for

$$\mathbf{H} = \mathbf{U}\mathbf{H}'\mathbf{U}^{-1} = \mathbf{U}\mathbf{h}\mathbf{h}\mathbf{U}^{-1} = \mathbf{U}\mathbf{h}\mathbf{U}^{-1}\mathbf{U}\mathbf{h}\mathbf{U}^{-1} = \mathbf{S}^{-1}\mathbf{S}^{-1} = (\mathbf{S}^{\dagger})^{-1}\mathbf{S}^{-1}.$$

Finally, define the representation $\boldsymbol{\Gamma}'$ of $\mathcal{G}$ by

$$\boldsymbol{\Gamma}'(T) = \mathbf{S}^{-1}\boldsymbol{\Gamma}(T)\mathbf{S} \tag{C.4}$$

for all $T \in \mathcal{G}$. Then from Equations (C.2), (C.3) and (C.4)

$$(\mathbf{S}\boldsymbol{\Gamma}'(T)\mathbf{S}^{-1})^{\dagger}(\mathbf{S}^{\dagger})^{-1}\mathbf{S}^{-1}(\mathbf{S}\boldsymbol{\Gamma}'(T)\mathbf{S}^{-1}) = (\mathbf{S}^{\dagger})^{-1}\mathbf{S}^{-1},$$

so that

$$\boldsymbol{\Gamma}'(T)^{\dagger}\boldsymbol{\Gamma}'(T) = \mathbf{1}.$$

Thus $\boldsymbol{\Gamma}'$ is a *unitary* representation that is equivalent to $\boldsymbol{\Gamma}$.

Theorem IV If $\boldsymbol{\Gamma}$ and $\boldsymbol{\Gamma}'$ are two equivalent representations of a group $\mathcal{G}$ related by a similarity transformation

$$\boldsymbol{\Gamma}'(T) = \mathbf{S}^{-1}\boldsymbol{\Gamma}(T)\mathbf{S} \tag{C.5}$$

for all $T \in \mathcal{G}$, and if $\boldsymbol{\Gamma}$ is a unitary representation and $\mathbf{S}$ is a unitary matrix, then $\boldsymbol{\Gamma}'$ is also a unitary representation. Conversely, if $\boldsymbol{\Gamma}$ and $\boldsymbol{\Gamma}'$ are equivalent representations that are both unitary, then the matrix $\mathbf{S}$ in the similarity transformation relating them can always be chosen to be unitary.

Proof If $\boldsymbol{\Gamma}(T)$ and $\mathbf{S}$ are unitary matrices, then Equation (C.5) implies that

$$\boldsymbol{\Gamma}'(T)^{\dagger} = \mathbf{S}^{\dagger}\boldsymbol{\Gamma}(T)^{\dagger}(\mathbf{S}^{-1})^{\dagger} = \mathbf{S}^{-1}\boldsymbol{\Gamma}(T)^{-1}\mathbf{S} = \boldsymbol{\Gamma}'(T)^{-1},$$

so that $\boldsymbol{\Gamma}'(T)$ is also unitary.

Now consider the converse proposition, where $\boldsymbol{\Gamma}$ and $\boldsymbol{\Gamma}'$ are two unitary representations related by the similarity transformation in Equation (C.5) involving a matrix $\mathbf{S}$. It will be shown that in Equation (C.5) $\mathbf{S}$ can be replaced by a unitary matrix $\mathbf{U}$.

From Equation (C.5)

$$\mathbf{S}\boldsymbol{\Gamma}'(T) = \boldsymbol{\Gamma}(T)\mathbf{S}, \tag{C.6}$$

so that $\boldsymbol{\Gamma}'(T)^{\dagger}\mathbf{S}^{\dagger} = \mathbf{S}^{\dagger}\boldsymbol{\Gamma}(T)^{\dagger}$ and, as $\boldsymbol{\Gamma}$ and $\boldsymbol{\Gamma}'$ are unitary representations, $\boldsymbol{\Gamma}'(T^{-1})\mathbf{S}^{\dagger} = \mathbf{S}^{\dagger}\boldsymbol{\Gamma}(T^{-1})$ for all $T \in \mathcal{G}$. Since every element of $\mathcal{G}$ has a corresponding inverse, this can be rewritten as $\boldsymbol{\Gamma}'(T)\mathbf{S}^{\dagger} = \mathbf{S}^{\dagger}\boldsymbol{\Gamma}(T)$ for all $T \in \mathcal{G}$, which when combined with Equation (C.6) gives

$$\boldsymbol{\Gamma}(T)\mathbf{S}\mathbf{S}^{\dagger} = \mathbf{S}\mathbf{S}^{\dagger}\boldsymbol{\Gamma}(T) \tag{C.7}$$

for all $T \in \mathcal{G}$.

As $\mathbf{S}\mathbf{S}^{\dagger}$ is a Hermitian matrix there exists (see Appendix A, Section

2) a unitary matrix $\mathbf{V}$ such that

$$\mathbf{D}=\mathbf{V}^{-1}(\mathbf{S}\mathbf{S}^{\dagger})\mathbf{V} \tag{C.8}$$

is diagonal. A similar argument to that employed in the immediately preceding proof shows that the diagonal elements of $\mathbf{D}$ are real and positive, so that its "square root" $\mathbf{d}$ may be defined by $d_{jk}=0$ if $j\neq k$ and $d_{jj}=(D_{jj})^{1/2}$, and consequently

$$\mathbf{dd}=\mathbf{D}. \tag{C.9}$$

Moreover, from Equations (C.7) and (C.8), for all $T\in\mathcal{G}$

$$\{\mathbf{V}^{-1}\boldsymbol{\Gamma}(T)\mathbf{V}\}\mathbf{D}=\mathbf{D}\{\mathbf{V}^{-1}\boldsymbol{\Gamma}(T)\mathbf{V}\}.$$

Consideration of the (j, k) elements of this equation gives

$$(\mathbf{V}^{-1}\boldsymbol{\Gamma}(T)\mathbf{V})_{jk}\{D_{kk}-D_{jj}\}=0.$$

Thus if $D_{kk}\neq D_{jj}$ then $(\mathbf{V}^{-1}\boldsymbol{\Gamma}(T)\mathbf{V})_{jk}=0$. Consequently

$$(\mathbf{V}^{-1}\boldsymbol{\Gamma}(T)\mathbf{V})_{jk}\{(D_{kk})^{1/2}-(D_{jj})^{1/2}\}=0,$$

and hence

$$\{\mathbf{V}^{-1}\boldsymbol{\Gamma}(T)\mathbf{V}\}\mathbf{d}=\mathbf{d}\{\mathbf{V}^{-1}\boldsymbol{\Gamma}(T)\mathbf{V}\} \tag{C.10}$$

for all $T\in\mathcal{G}$.

The matrix $\mathbf{U}$ defined by

$$\mathbf{U}=\mathbf{V}\mathbf{d}^{-1}\mathbf{V}^{-1}\mathbf{S}$$

has the required properties. Firstly, it is unitary, as

$$\mathbf{U}\mathbf{U}^{\dagger}=(\mathbf{V}\mathbf{d}^{-1}\mathbf{V}^{-1}\mathbf{S})(\mathbf{S}^{\dagger}\mathbf{V}\mathbf{d}^{-1}\mathbf{V}^{-1})$$

(as $\mathbf{V}$ is unitary and $\mathbf{d}^{-1}$ is diagonal with real positive diagonal elements)

$$\begin{aligned}&=\mathbf{V}\mathbf{d}^{-1}\mathbf{D}\mathbf{d}^{-1}\mathbf{V}^{-1}\quad\text{(by Equation (C.8))}\\&=\mathbf{V}\mathbf{V}^{-1}\ \text{(by Equation (C.9))}\\&=\mathbf{1}.\end{aligned}$$

Secondly, $\mathbf{U}$ relates $\boldsymbol{\Gamma}$ and $\boldsymbol{\Gamma}'$ in a similarity transformation, for

$$\begin{aligned}\mathbf{U}^{-1}\boldsymbol{\Gamma}(T)\mathbf{U}&=(\mathbf{S}^{-1}\mathbf{V}\mathbf{d}\mathbf{V}^{-1})\boldsymbol{\Gamma}(T)(\mathbf{V}\mathbf{d}^{-1}\mathbf{V}\mathbf{S})\\&=(\mathbf{S}^{-1}\mathbf{V})\{\mathbf{d}(\mathbf{V}^{-1}\boldsymbol{\Gamma}(T)\mathbf{V}\}(\mathbf{d}^{-1}\mathbf{V}\mathbf{S})\\&=(\mathbf{S}^{-1}\mathbf{V})\{(\mathbf{V}^{-1}\boldsymbol{\Gamma}(T)\mathbf{V})\mathbf{d}\}(\mathbf{d}^{-1}\mathbf{V}\mathbf{S})\quad\text{(by Equation (C.10))}\\&=\mathbf{S}^{-1}\boldsymbol{\Gamma}(T)\mathbf{S}\\&=\boldsymbol{\Gamma}'(T).\end{aligned}$$

2 Proof of lemma required for Theorem I of Chapter 4, Section 4

Lemma If Γ is a reducible *unitary* representation of a group $\mathcal{G}$, then Γ is *completely reducible.*

Proof Suppose that the representation Γ has dimension d. Let $\psi_1, \psi_2, \ldots, \psi_d$ be an ortho-normal basis for the carrier space V of Γ. Then as Γ is unitary the operators $\Phi(T)$ defined by Equation (4.1) are unitary operators for each $T \in \mathcal{G}$.

As Γ is reducible it is equivalent to a representation Γ' that has the form

$$\Gamma'(T) = \begin{bmatrix} \Gamma'_{11}(T) & \Gamma'_{12}(T) \\ \mathbf{0} & \Gamma'_{22}(T) \end{bmatrix} \tag{C.11}$$

for all $T \in \mathcal{G}$, where the zero matrix $\mathbf{0}$ has dimensions $s_2 \times s_1$, with $s_1 + s_2 = d$, $s_1 \geq 1$, $s_2 \geq 1$, Then, by Equations (4.7) and (C.11), V has a s_1-dimensional invariant subspace S_1 such that if $\phi_1 \in S_1$ then $\Phi(T)\phi_1 \in S_1$ for all $T \in \mathcal{G}$.

Let $\psi''_1, \psi''_2, \ldots, \psi''_{s_1}$ be an ortho-normal basis for the subspace S_1, and let $\psi''_1, \psi''_2, \ldots, \psi''_{s_1}, \psi''_{s_1+1}, \ldots, \psi''_d$ be an ortho-normal basis for the carrier space V. Then $\psi''_1, \psi''_2, \ldots, \psi''_d$ form a basis for a *unitary* representation Γ'' that is equivalent to Γ and Γ', and as S_1 is an invariant subspace $\Gamma''(T)_{mn} = 0$ for $m = s_1+1, s_1+2, \ldots, d$ and $n = 1, 2, \ldots, s_1$. Thus Γ'' is unitary *and* has the same form as Γ', i.e. for all $T \in \mathcal{G}$

$$\Gamma''(T) = \begin{bmatrix} \Gamma''_{11}(T) & \Gamma''_{12}(T) \\ \mathbf{0} & \Gamma''_{22}(T) \end{bmatrix} \tag{C.12}$$

where the zero matrix $\mathbf{0}$ again has dimensions $s_2 \times s_1$.

Let S_2 be the s_2-dimensional subspace of V with basis $\psi''_{s_1+1}, \psi''_{s_1+2}, \ldots, \psi''_d$. It will now be shown that S_2 is also an invariant subspace. First it should be noted that $(\phi_1, \phi_2) = 0$ if $\phi_1 \in S_1$ and $\phi_2 \in S_2$. Moreover, if $(\phi_1, \phi_2) = 0$ for *all* $\phi_1 \in S_1$, then $\phi_2 \in S_2$. Consequently for any $\phi_1 \in S_1$ and $\phi_2 \in S_2$, as $\Phi(T)$ is unitary and S_1 is an invariant subspace,

$$(\phi_1, \Phi(T)\phi_2) = (\Phi(T)^{-1}\phi_1, \phi_2) = (\Phi(T^{-1})\phi_1, \phi_2) = 0,$$

so that $\Phi(T)\phi_2 \in S_2$ for all $T \in \mathcal{G}$.

However, for $n = s_1+1, s_1+2, \ldots, d$, Equations (4.1) and (C.12) imply that

$$\Phi(T)\psi''_n = \sum_{m=1}^{s_1} \Gamma''_{12}(T)_{m,n-s_1}\psi''_m + \sum_{m=s_1+1}^{s_2} \Gamma''_{22}(T)_{m-s_1,n-s_1}\psi''_m,$$

so $\boldsymbol{\Gamma}''_{12}(T) = \mathbf{0}$ for all $T \in \mathcal{G}$. Thus $\boldsymbol{\Gamma}''$ has the form

$$\boldsymbol{\Gamma}''(T) = \left[\begin{array}{c|c} \boldsymbol{\Gamma}''_{11}(T) & \mathbf{0} \\ \hline \mathbf{0} & \boldsymbol{\Gamma}''_{22}(T) \end{array}\right]$$

Clearly the representations $\boldsymbol{\Gamma}''_{11}$ and $\boldsymbol{\Gamma}''_{22}$ are also unitary.

If $\boldsymbol{\Gamma}''_{11}$ and $\boldsymbol{\Gamma}''_{22}$ are not both irreducible the argument can be repeated for the submatrices until the form of Equation (4.12) is reached.

3 Proofs of Theorems I and IV of Chapter 4, Section 5

Theorem I Let $\boldsymbol{\Gamma}$ and $\boldsymbol{\Gamma}'$ be two *irreducible* representations of a group $\mathcal{G}$, of dimensions d and d' respectively, and suppose that there exists a $d \times d'$ matrix $\mathbf{A}$ such that

$$\boldsymbol{\Gamma}(T)\mathbf{A} = \mathbf{A}\boldsymbol{\Gamma}'(T) \tag{C.13}$$

for all $T \in \mathcal{G}$. Then either $\mathbf{A} = \mathbf{0}$, or $d = d'$ and $\det \mathbf{A} \neq 0$.

Proof Let $\psi_1, \psi_2, \ldots, \psi_d$ be a basis for the carrier space V of $\boldsymbol{\Gamma}$. Let $\phi_k = \sum_{j=1}^{d} A_{jk}\psi_j$, $k = 1, 2, \ldots, d'$, and let S be the subspace of V generated by $\phi_1, \phi_2, \ldots, \phi_{d'}$. Then, for any $T \in \mathcal{G}$, by Equations (4.1) and (4.2),

$$\begin{aligned}\Phi(T)\phi_k &= \sum_{j=1}^{d'} \sum_{l=1}^{d} A_{jk}\Gamma(T)_{lj}\psi_l = \sum_{l=1}^{d} (\boldsymbol{\Gamma}(T)\mathbf{A})_{lk}\psi_l \\ &= \sum_{l=1}^{d} (\mathbf{A}\boldsymbol{\Gamma}'(T))_{lk}\psi_l, \text{ by Equation (C.7)}, \\ &= \sum_{l=1}^{d} \sum_{j=1}^{d'} A_{lj}\Gamma'(T)_{jk}\psi_l = \sum_{j=1}^{d'} \Gamma'(T)_{jk}\phi_j,\end{aligned}$$

so that $\Phi(T)\phi_k$ is also a member of S for $k = 1, 2, \ldots, d'$, and hence S is an invariant subspace of V. As $\boldsymbol{\Gamma}$ is irreducible, V has no invariant subspaces of smaller dimension than itself. Thus either $\phi_k = 0$ for $k = 1, 2, \ldots, d'$ (which implies $\mathbf{A} = \mathbf{0}$), or else S coincides with V. In this latter situation there are two possibilities, either $\phi_1, \phi_2, \ldots, \phi_{d'}$ are linearly independent, in which case necessarily $d = d'$ and

$\det \mathbf{A} \neq 0$, or this set is not linearly independent. This last circumstance could only occur if $d < d'$. However, as Equation (C.13) can be rewritten as

$$\tilde{\mathbf{\Gamma}}'(\mathrm{T})^{-1}\tilde{\mathbf{A}} = \tilde{\mathbf{A}}\tilde{\mathbf{\Gamma}}(\mathrm{T})^{-1},$$

the same reasoning can be applied with $\mathbf{\Gamma}$ and $\mathbf{\Gamma}'$ replaced by $\tilde{\mathbf{\Gamma}}'^{-1}$ and $\tilde{\mathbf{\Gamma}}^{-1}$ respectively, for it is easily shown that these two latter representations are also irreducible and have dimensions d' and d respectively. Thus $\mathbf{A} = \mathbf{0}$, or $d = d'$ and $\det \mathbf{A} \neq 0$, or $d' < d$. Clearly the only compatible possibilities are $\mathbf{A} = \mathbf{0}$, or $d = d'$ with $\det \mathbf{A} \neq 0$.

It is worthwhile stating and proving Theorem IV of Chapter 4, Section 5, in a slightly more general form:

Theorem IV Let $\mathbf{\Gamma}^p$ and $\mathbf{\Gamma}^q$ be two *unitary irreducible* representations of a finite group $\mathcal{G}$ of order g. Then if $\mathbf{\Gamma}^p$ and $\mathbf{\Gamma}^q$ are not equivalent

$$(1/g) \sum_{\mathrm{T} \in \mathcal{G}} \Gamma^p(\mathrm{T})^*_{jk} \Gamma^q(\mathrm{T})_{st} = 0,$$

whereas if $\mathbf{\Gamma}^p$ and $\mathbf{\Gamma}^q$ are equivalent, and the similarity transformation relating them is

$$\mathbf{\Gamma}^p(\mathrm{T}) = \mathbf{S}^{-1}\mathbf{\Gamma}^q(\mathrm{T})\mathbf{S}$$

for all $\mathrm{T} \in \mathcal{G}$, then

$$(1/g) \sum_{\mathrm{T} \in \mathcal{G}} \Gamma^p(\mathrm{T})^*_{jk} \Gamma^q(\mathrm{T})_{st} = S_{sj}(\mathbf{S}^{-1})_{kt}/d_p.$$

If $\mathcal{G}$ is a compact Lie group similar results apply, the summations above being replaced by invariant integrals.

Proof The proof will be given explicitly for the case in which $\mathcal{G}$ is a finite group of order g. It immediately generalizes to a compact Lie group if every sum of the form $(1/g)\sum_{\mathrm{T} \in \mathcal{G}} f(\mathrm{T})$ is replaced by the corresponding invariant integral

$$\int_{\mathcal{G}} f(\mathrm{T})\, d\mathrm{T}$$

(see Chapter 3, Section 5).

Let d_q be the dimension of $\mathbf{\Gamma}^q$ and let t and k be any fixed integers from the sets $1, 2, \ldots, d_q$ and $1, 2, \ldots, d_p$ respectively. Let $\mathbf{B}$ be the $d_q \times d_p$ matrix whose (t, k) element is 1, all other elements being 0.

Let

$$\mathbf{A} = (1/g) \sum_{T \in \mathcal{G}} \mathbf{\Gamma}^q(T)\mathbf{B}\mathbf{\Gamma}^p(T^{-1}),$$

so that

$$\mathbf{\Gamma}^q(T')\mathbf{A} = \mathbf{A}\mathbf{\Gamma}^p(T') \tag{C.14}$$

for every $T' \in \mathcal{G}$. (This follows as

$$\mathbf{\Gamma}^q(T')\mathbf{A}\mathbf{\Gamma}^p(T'^{-1}) = (1/g) \sum_{T \in \mathcal{G}} \mathbf{\Gamma}^q(T'T)\mathbf{B}\mathbf{\Gamma}^p((T'T)^{-1}) = \mathbf{A}$$

by the Rearrangement Theorem of Chapter 2, Section 1.)

Suppose first that $\mathbf{\Gamma}^p$ is not equivalent to $\mathbf{\Gamma}^q$. Then the first of the two lemmas of Schur applied to Equation (C.14) implies $\mathbf{A} = \mathbf{0}$, so

$$(1/g) \sum_{T \in \mathcal{G}} \mathbf{\Gamma}^q(T)\mathbf{B}\mathbf{\Gamma}^p(T^{-1}) = \mathbf{0}. \tag{C.15}$$

As $\mathbf{\Gamma}^p$ is assumed unitary, $\mathbf{\Gamma}^p(T^{-1}) = \tilde{\mathbf{\Gamma}}^p(T)^*$, so consideration of the (s, j) elements of Equation (C.15) with the particular choice of $\mathbf{B}$ made earlier gives

$$(1/g) \sum_{T \in \mathcal{G}} \Gamma^q(T)_{st} \Gamma^p(T)^*_{jk} = 0.$$

Now suppose that $\mathbf{\Gamma}^p$ and $\mathbf{\Gamma}^q$ are equivalent. Then Equation (C.14) becomes

$$\mathbf{\Gamma}^p(T')(\mathbf{S}^{-1}\mathbf{A}) = (\mathbf{S}^{-1}\mathbf{A})\mathbf{\Gamma}^p(T')$$

for all $T' \in \mathcal{G}$. The second of Schur's lemmas then shows that $\mathbf{S}^{-1}\mathbf{A} = a\mathbf{1}$, where a is some complex number, so that $\mathbf{A} = a\mathbf{S}$. However the definition of $\mathbf{A}$ can be rewritten as

$$\mathbf{A}\mathbf{S}^{-1} = (1/g) \sum_{T \in \mathcal{G}} \mathbf{\Gamma}^q(T')\mathbf{B}\mathbf{S}^{-1}\mathbf{\Gamma}^q(T')^{-1}.$$

From the invariance property of the trace of a matrix under a similarity transformation (see Appendix A) it follows that $\operatorname{tr}(\mathbf{A}\mathbf{S}^{-1}) = \operatorname{tr}(\mathbf{B}\mathbf{S}^{-1})$ and hence $ad_p = (\mathbf{S}^{-1})_{kt}$. Thus

$$(1/g) \sum_{T \in \mathcal{G}} \mathbf{\Gamma}^q(T)\mathbf{B}\tilde{\mathbf{\Gamma}}^p(T)^* = \{(\mathbf{S}^{-1})_{kt}/d_p\}\mathbf{S}.$$

Consideration of the (s, j) elements then gives

$$(1/g) \sum_{T \in \mathcal{G}} \Gamma^q(T)_{st} \Gamma^p(T)^*_{jk} = S_{sj}(\mathbf{S}^{-1})_{kt}/d_p.$$

The orthogonality theorem quoted in Chapter 4, Section 5, is merely a special case in which $\mathbf{\Gamma}^p$ and $\mathbf{\Gamma}^q$ are not equivalent for $p \neq q$, while for $p = q$, $\mathbf{S} = 1$, so that $S_{sj} = \delta_{sj}$ and $(\mathbf{S}^{-1})_{kt} = \delta_{kt}$.

4 Proofs of Theorems IV, VII, VIII and IX of Chapter 4, Section 6

Theorem IV If $\mathscr{G}$ is a finite group or a compact Lie group then a sufficient condition for two representations to be equivalent is provided by the equality of their character systems.

Proof Let $\mathbf{\Gamma}$ and $\mathbf{\Gamma}'$ be two representations of $\mathscr{G}$ and let $\chi(T)$ and $\chi'(T)$ be their characters for the group element T.

Suppose first that both these representations are *irreducible*. Suppose also that they have identical character systems but are not equivalent. Then, applying Theorem III of Chapter 4, Section 6, to a finite group $\mathscr{G}$ with $\chi^p = \chi$ and $\chi^q = \chi'$,

$$(1/g) \sum_{T \in \mathscr{G}} \chi'(T)\chi(T)^* = 0,$$

whereas with $\chi^p = \chi^q = \chi$

$$(1/g) \sum_{T \in \mathscr{G}} \chi(T)\chi(T)^* = 1.$$

(The obvious generalizations apply for compact Lie groups.) Clearly with $\chi(T) = \chi'(T)$ for all $T \in \mathscr{G}$ these equations are inconsistent, so the initial assumptions must be incompatible. Thus, if two irreducible representations of a finite group or a compact Lie group have identical character systems, they are equivalent.

Now suppose that at least one of the representations $\mathbf{\Gamma}$ and $\mathbf{\Gamma}'$ is *reducible* and that they have identical character systems. Then, by Theorem I of Chapter 4, Section 4, a reducible representation is completely reducible. Suppose that $\mathbf{\Gamma}$ is equivalent to the direct sum in which the irreducible representation $\mathbf{\Gamma}^1$ appears n_1 times, the irreducible representation $\mathbf{\Gamma}^2$ appears n_2 times and so on. Then

$$\chi(T) = \sum_q n_q \chi^q(T)$$

for all $T \in \mathscr{G}$. Similarly, suppose that $\mathbf{\Gamma}'$ is equivalent to the direct sum in which $\mathbf{\Gamma}^1$ appears n_1' times, $\mathbf{\Gamma}^2$ appears n_2' times and so on, so that

$$\chi'(T) = \sum_q n_q' \chi^q(T)$$

for all $T \in \mathcal{G}$. Then, as $\chi'(T) = \chi(T)$,

$$\sum_q (n'_q - n_q)\chi^q(T) = 0$$

for all $T \in \mathcal{G}$. Thus for a finite group $\mathcal{G}$ and any irreducible representation Γ^p of $\mathcal{G}$,

$$\sum_q (n'_q - n_q)(1/g) \sum_{T \in \mathcal{G}} \chi^q(T)\chi^p(T)^* = 0,$$

so that, from Theorem III of Chapter 4, Section 6,

$$n'_q = n_q$$

for all q. (An essentially similar argument also applies for the case in which $\mathcal{G}$ is a compact Lie group.) Thus the two direct sums contain the same irreducible representations (at least up to equivalence) and must therefore be equivalent. Consequently Γ and Γ' must be equivalent.

The proofs of Theorems VII and IX of Chapter 4, Section 6 require the concept of the "regular representation" Γ^{reg} of a finite group $\mathcal{G}$. Let $\mathcal{G}$ be of order g and suppose that $T_1(=E), T_2, \ldots, T_g$ are the elements of $\mathcal{G}$. Then for any $T_s \in \mathcal{G}$ define the $g \times g$ matrix $\Gamma^{reg}(T_s)$ by

$$\Gamma^{reg}(T_s)_{kl} = \begin{cases} 1, & \text{if } T_sT_l = T_k, \\ 0, & \text{if } T_sT_l \neq T_k, \end{cases}$$

which implies

$$T_sT_l = \sum_{k=1}^{g} \Gamma^{reg}(T_s)_{kl}T_k \tag{C.16}$$

for $s, l = 1, 2, \ldots, g$. This set of matrices form a g-dimensional representation of $\mathcal{G}$, as, from Equation (C.16), for any T_s, T_t of $\mathcal{G}$,

$$(T_sT_t)T_l = \sum_{k=1}^{g} \Gamma^{reg}(T_sT_t)_{kl}T_k,$$

while

$$(T_sT_t)T_l = T_s(T_tT_l) = T_s \sum_{m=1}^{g} \Gamma^{reg}(T_t)_{ml}T_m$$

$$= \sum_{k,m=1}^{g} \Gamma^{reg}(T_s)_{km}\Gamma^{reg}(T_t)_{ml}T_k,$$

so that, on equating coefficients of T_k on the right-hand side,

$$\Gamma^{reg}(T_sT_t)_{kl} = \sum_{m=1}^{g} \Gamma^{reg}(T_s)_{km}\Gamma^{reg}(T_t)_{ml}$$

Moreover, as $T_s T_l = T_l$ if and only if $T_s = E$,

$$\Gamma^{reg}(T_s)_{ll} = \begin{cases} 1, & \text{if} \quad T_s = E, \\ 0, & \text{if} \quad T_s \neq E, \end{cases}$$

so that the character $\chi^{reg}(T_s)$ is given by

$$\chi^{reg}(T_s) = \begin{cases} g, & \text{if} \quad T_s = E, \\ 0, & \text{if} \quad T_s \neq E. \end{cases} \tag{C.17}$$

Theorem VII For a finite group $\mathscr{G}$, the sum of the squares of the dimensions of the inequivalent irreducible representations is equal to the order of $\mathscr{G}$.

Proof Suppose that $\mathscr{G}$ has order g and Γ^{reg} is its regular representation. Let Γ^p be any irreducible representation of $\mathscr{G}$ and let d_p be its dimension. Then, by Theorem V of Chapter 4, Section 6, the number of times n_p that Γ^p (or a representation equivalent to it) appears in Γ^{reg} is given by

$$n_p = (1/g) \sum_{T \in \mathscr{G}} \chi^{reg}(T) \chi^p(T)^*,$$

which, on using Equation (C.17), gives $n_p = d_p$. As the dimension of Γ^{reg} is equal to the sum of the dimensions of all the irreducible representations that appear in its direct sum decomposition, this implies that

$$g = \sum_p n_p d_p = \sum_p (d_p)^2.$$

The proof of Theorem IX of Chapter 4, Section 6, requires the idea of "class multiplication". For the rest of this section the convention will be adopted that two sets of elements of a group $\mathscr{G}$ are to be considered equal if and only if each element appears as often in one set as in the other set. With this convention it is obvious that

$$X\mathscr{C}X^{-1} = \mathscr{C}$$

where $\mathscr{C}$ is any class of $\mathscr{G}$ (see Chapter 2, Section 2), $X \in \mathscr{G}$, and $X\mathscr{C}X^{-1}$ is defined to be the set of elements XTX^{-1}, where $T \in \mathscr{C}$. Conversely, if $X\mathscr{C}X^{-1} = \mathscr{C}$ for all $X \in \mathscr{G}$, then $\mathscr{C}$ consists entirely of complete classes. (This follows on considering the situation when all complete classes are removed from both sides of the equation, for if $\mathscr{R}$ is the remainder then $X\mathscr{R}X^{-1} = \mathscr{R}$ for all $X \in \mathscr{G}$. Thus, if $T \in \mathscr{R}$ then $XTX^{-1} \in \mathscr{R}$ for all $X \in \mathscr{G}$, so that $\mathscr{R}$ contains the complete class of T, giving a contradiction unless $\mathscr{R}$ is the empty set.)

The product $\mathscr{C}_i\mathscr{C}_j$ of two classes $\mathscr{C}_i$ and $\mathscr{C}_j$ is defined to be the set of elements of $\mathscr{G}$ formed by multiplying each of the elements of $\mathscr{C}_i$ by each of the elements of $\mathscr{C}_j$. Thus, if $\mathscr{C}_i$ has N_i elements and $\mathscr{C}_j$ has N_j elements, then $\mathscr{C}_i\mathscr{C}_j$ must have N_iN_j elements, some of which may be equal. As $\mathscr{C}_i\mathscr{C}_j = X(\mathscr{C}_i\mathscr{C}_j)X^{-1}$ for all $X \in \mathscr{C}$, it follows that $\mathscr{C}_i\mathscr{C}_j$ consists entirely of complete classes. Thus one may write for a finite group $\mathscr{G}$

$$\mathscr{C}_i\mathscr{C}_j = \sum_k c_{ijk}\mathscr{C}_k,$$

where c_{ijk} is a non-negative integer specifying how often the class $\mathscr{C}_k$ appears in the product $\mathscr{C}_i\mathscr{C}_j$.

The coefficients c_{ijk} have four useful properties:

(*a*) $c_{ijk} = c_{jik}$, which is a consequence of the equality $\mathscr{C}_i\mathscr{C}_j = \mathscr{C}_j\mathscr{C}_i$, which itself can be proved as follows. Let TT′ be a typical element of $\mathscr{C}_i\mathscr{C}_j$, where $T \in \mathscr{C}_i$ and $T' \in \mathscr{C}_j$. As $TT' = (TT'T^{-1})T$ and $TT'T^{-1} \in \mathscr{C}_j$ then $TT' \in \mathscr{C}_j\mathscr{C}_i$. Thus $\mathscr{C}_i\mathscr{C}_j$ is contained in $\mathscr{C}_j\mathscr{C}_i$ and the set equality follows on repeating the argument with i and j interchanged.

(*b*) If $\mathscr{C}_1 = \{E\}$ then $c_{1jk} = c_{j1k} = \delta_{jk}$.

(*c*) Corresponding to each class $\mathscr{C}_i$ there exists a class $\mathscr{C}_{i'}$ consisting of the inverses of the elements of $\mathscr{C}_i$. $\mathscr{C}_i$ and $\mathscr{C}_{i'}$ contain the same number of elements N_i and may in certain cases be identical. Then obviously $\mathscr{C}_i\mathscr{C}_j$ contains $\mathscr{C}_1$ ($=\{E\}$) N_i times if $j = i'$ but does not contain $\mathscr{C}_1$ if $j \neq i'$, so that

$$c_{ij1} = N_i\delta_{ji'}. \tag{C.18}$$

(*d*) If Γ^p is an irreducible representation of dimension d_p of $\mathscr{G}$, then

$$N_iN_j\chi^p(\mathscr{C}_i)\chi^p(\mathscr{C}_j) = d_p \sum_k c_{ijk}N_k\chi^p(\mathscr{C}_k). \tag{C.19}$$

This result is sometimes used in the construction of character tables. Its proof is as follows. Let

$$\mathbf{C}_i = \sum_{T \in \mathscr{C}_i} \Gamma^p(T), \tag{C.20}$$

so that

$$\mathbf{C}_i\mathbf{C}_j = \sum_k c_{ijk}\mathbf{C}_k,$$

As $\mathrm{XTX}^{-1} \in \mathscr{C}_i$ for every $\mathrm{T} \in \mathscr{C}_i$ and $\mathrm{X} \in \mathscr{G}$, then $\mathbf{\Gamma}^{\mathrm{p}}(\mathrm{X})\mathbf{C}_i = \mathbf{C}_i\mathbf{\Gamma}^{\mathrm{p}}(X)$, so $\mathbf{C}_i = \gamma_i \mathbf{1}$ by the second of Schur's lemmas. This implies that

$$\gamma_i \gamma_j = \sum_k c_{ijk} \gamma_k \tag{C.21}$$

and $\operatorname{tr} \mathbf{C}_i = \gamma_i d_{\mathrm{p}}$. However, from Equation (C.20), $\operatorname{tr} \mathbf{C}_i = N_i \chi^{\mathrm{p}}(\mathscr{C}_i)$, so that $\gamma_i = N_i \chi^{\mathrm{p}}(\mathscr{C}_i)/d_{\mathrm{p}}$. Substituting into Equation (C.21) gives Equation (C.19) immediately.

Theorem IX If $\chi^{\mathrm{p}}(\mathscr{C}_j)$ is the character of the class $\mathscr{C}_j$ of a finite group $\mathscr{G}$ for the irreducible representation $\mathbf{\Gamma}^{\mathrm{p}}$ of $\mathscr{G}$ then

$$\sum_{\mathrm{p}} \chi^{\mathrm{p}}(\mathscr{C}_j)^* \chi^{\mathrm{p}}(\mathscr{C}_k) N_j = g\delta_{jk},$$

where the sum is over all the inequivalent irreducible representations of $\mathscr{G}$, g is the order of $\mathscr{G}$, and N_j is the number of elements in the class $\mathscr{C}_j$.

Proof As noted in the proof of Theorem VII above, the number of times n_{p} that $\mathbf{\Gamma}^{\mathrm{p}}$ (or a representation equivalent to it) appears in the regular representation $\mathbf{\Gamma}^{\mathrm{reg}}$ is given by $n_{\mathrm{p}} = d_{\mathrm{p}}$, so that $\chi^{\mathrm{reg}}(\mathrm{T}) = \sum_{\mathrm{p}} d_{\mathrm{p}} \chi^{\mathrm{p}}(\mathrm{T})$ for any $\mathrm{T} \in \mathscr{G}$. Hence, by Equation (C.17),

$$\sum_{\mathrm{p}} d_{\mathrm{p}} \chi^{\mathrm{p}}(\mathrm{T}) = \begin{cases} g, & \text{if} \quad \mathrm{T} = \mathrm{E}, \\ 0, & \text{if} \quad \mathrm{T} \neq \mathrm{E}, \end{cases} \tag{C.22}$$

Summing both sides of Equation (C.19) over inequivalent irreducible representations of $\mathscr{G}$ and using Equations (C.18) and (C.22),

$$\sum_{\mathrm{p}} N_i N_j \chi^{\mathrm{p}}(\mathscr{C}_i) \chi^{\mathrm{p}}(\mathscr{C}_j) = g N_i \delta_{ji'}, \tag{C.23}$$

where $\mathscr{C}_{i'}$ is the class containing the inverses of $\mathscr{C}_i$. However, as $\mathbf{\Gamma}^{\mathrm{p}}$ can be chosen to be unitary, $\chi^{\mathrm{p}}(\mathscr{C}_j) = \chi^{\mathrm{p}}(\mathscr{C}_{j'})^*$, and as $N_j = N_{j'}$, Equation (C.23) becomes

$$\sum_{\mathrm{p}} N_{j'} \chi^{\mathrm{p}}(\mathscr{C}_i) \chi^{\mathrm{p}}(\mathscr{C}_{j'})^* = g\delta_{ji'}.$$

The theorem follows on putting $k = j'$ as $\delta_{k'i'} = \delta_{ki}$.

Theorem VIII For a finite group $\mathscr{G}$ the number of inequivalent irreducible representations is equal to the number of classes of $\mathscr{G}$.

Proof Suppose that there are r classes and s inequivalent irreducible representations. The first orthogonality theorem for characters may be written for the case $p=q$ in the form

$$\sum_{j=1}^{r} N_j\chi^p(\mathscr{C}_j)^*\chi^p(\mathscr{C}_j)=g,$$

so that

$$\sum_{p=1}^{s}\sum_{j=1}^{r} N_j\chi^p(\mathscr{C}_j)^*\chi^p(\mathscr{C}_j)=gs.$$

Similarly, summing the equality of the second orthogonality theorem for characters over all classes $\mathscr{C}_j$ and $\mathscr{C}_k$ gives

$$\sum_{p=1}^{s}\sum_{j=1}^{r} N_j\chi^p(\mathscr{C}_j)^*\chi^p(\mathscr{C}_j)=gr.$$

Hence $s=r$.

5 Proofs of theorems of Chapter 5, Section 1

Theorem I Any function $\phi(\mathbf{r})$ of L^2 can be written as a linear combination of basis functions of the unitary irreducible representations of a group $\mathscr{G}$ of coordinate transformations in $\mathbb{R}^3$. That is

$$\phi(\mathbf{r})=\sum_p\sum_{j=1}^{d_p} a_j^p\phi_j^p(\mathbf{r}),$$

where $\phi_j^p(\mathbf{r})$ is a normalized basis function transforming as the jth row of the d_p-dimensional unitary irreducible representation Γ^p of $\mathscr{G}$, a_j^p are a set of complex numbers, and the sum over p is over all the inequivalent unitary irreducible representations of $\mathscr{G}$.

Proof Suppose first that $\mathscr{G}$ is a finite group. Construct the functions $P(T)\phi(\mathbf{r})$ for all $T\in\mathscr{G}$. Suppose this set contains d linearly independent functions. Let these be $\theta_k(\mathbf{r})$, $k=1,2,\ldots,d$. As they are members of L^2, the Schmidt orthogonalization process (see Appendix B, Section 2) may be applied to them to give an ortho-normal set $\psi_n(\mathbf{r})$, $n=1,2,\ldots,d$. Then $P(T)\phi(\mathbf{r})$ is a linear combination of $\psi_1(\mathbf{r}),\psi_2(\mathbf{r}),\ldots,\psi_d(\mathbf{r})$ for every $T\in\mathscr{G}$ and, conversely, every $\psi_n(\mathbf{r})$, $n=1,2,\ldots,d$ is a linear combination of the functions $P(T)\phi(\mathbf{r})$. Then for any $T'\in\mathscr{G}$, as $P(T')P(T)\phi(\mathbf{r})=P(T'T)\phi(\mathbf{r})$, it follows that $P(T)\psi_n(\mathbf{r})$ is a linear combination of $\psi_1(\mathbf{r}),\psi_2(\mathbf{r}),\ldots,\psi_d(\mathbf{r})$ for each $n=1,2,\ldots,d$, so one can write

$$P(T)\psi_n(\mathbf{r})=\sum_{m=1}^{d}\Gamma(T)_{mn}\psi_m(\mathbf{r}),\qquad n=1,2,\ldots,d,$$

where the $\Gamma(T)_{mn}$ are a set of coefficients that depend on T, m and n.

These equations are exactly of the form of Equation (1.27). The argument following Equation (1.27) can therefore be repeated to show that the matrices $\mathbf{\Gamma}(T)$ form a d-dimensional representation of $\mathscr{G}$ and that $\psi_1(\mathbf{r}), \psi_2(\mathbf{r}), \ldots, \psi_d(\mathbf{r})$ are a set of basis functions for this representation. As $\phi(\mathbf{r}) = P(E)\phi(\mathbf{r})$, $\phi(\mathbf{r})$ is a linear combination of $\psi_1(\mathbf{r}), \psi_2(\mathbf{r}), \ldots, \psi_d(\mathbf{r})$. If $\mathbf{\Gamma}$ is irreducible, then the theorem is proved. If $\mathbf{\Gamma}$ is reducible, let $\mathbf{S}$ be a $d \times d$ unitary matrix such that $\mathbf{\Gamma}'(T) = \mathbf{S}^{-1}\mathbf{\Gamma}(T)\mathbf{S}$ is a direct sum of unitary irreducible representations. Then, as shown in Theorem II of Chapter 4, Section 2, the functions $\psi'_n(\mathbf{r}) = \sum_{m=1}^{d} S_{mn}\psi_m(\mathbf{r})$ are basis functions of $\mathbf{\Gamma}'$. Moreover, as noted in Chapter 4, Section 4, the $\psi'_n(\mathbf{r})$ will be basis functions of the various irreducible representations that appear in the direct sum decomposition, and they form an ortho-normal set. Then, as $\phi(\mathbf{r})$ is a linear combination of $\psi'_1(\mathbf{r}), \psi'_2(\mathbf{r}) \ldots, \psi'_d(\mathbf{r})$, the theorem is proved for this case as well.

Now suppose that $\mathscr{G}$ is the group of all rotations in $\mathbb{R}^3$. It will be shown in Chapter 12, Section 4, that the basis functions of $\mathscr{G}$ are of the form $Y_l^m(\theta, \phi)R_l(r)$, where $Y_l^m(\theta, \phi)$ are spherical harmonics (θ, ϕ and r being the spherical polar coordinates) and $R_l(r)$ is any function of r such that the product is normalizable. As any function of θ and ϕ can be expanded as a linear combination of the spherical harmonics, the theorem must be true for this group.

If $\mathscr{G}$ is the Euclidean group of $\mathbb{R}^3$, then as noted in Example II of Chapter 4, Section 3, its only *unitary* irreducible representations are given by those of the group of rotations. Consequently, the sets of basis functions belonging to the *unitary* irreducible representations of the Euclidean group in $\mathbb{R}^3$ and of the group of all rotations in $\mathbb{R}^3$ coincide, so the theorem is true for the Euclidean group in $\mathbb{R}^3$ as well.

Finally, any other group of transformations in $\mathbb{R}^3$ is either a subgroup of the group of rotations, in which case its basis functions are spherical harmonics, or it is a subgroup of the Euclidean group containing a non-countable number of translations, in which case its *unitary* irreducible representations are given by those of its rotational subgroup, so the relevant basis functions are again spherical harmonics. Thus for every such group the theorem is true.

Theorem II The projection operators $\mathscr{P}^p_{mn}$ have the following properties.

(a) For any two functions $\phi(\mathbf{r})$ and $\psi(\mathbf{r})$ of L^2

$$(\mathscr{P}^p_{mn}\,\phi, \psi) = (\phi, \mathscr{P}^p_{nm}\,\psi).$$

In particular,

$$(\mathcal{P}^{p}_{nn}\phi, \psi) = (\phi, \mathcal{P}^{p}_{nn}\psi),$$

so that $\mathcal{P}^{p}_{nn}$ is a self-adjoint operator.

(*b*) For any two projection operators $\mathcal{P}^{p}_{mn}$ and $\mathcal{P}^{q}_{jk}$

$$\mathcal{P}^{p}_{mn}\mathcal{P}^{q}_{jk} = \delta_{pq}\delta_{nj}\mathcal{P}^{q}_{mk}.$$

In particular,

$$(\mathcal{P}^{p}_{nn})^2 = \mathcal{P}^{p}_{nn}.$$

(*c*) If $\psi^{q}_{1}(\mathbf{r}), \psi^{q}_{2}(\mathbf{r}), \ldots$ are basis functions transforming as the unitary irreducible representation $\boldsymbol{\Gamma}^{q}$ of $\mathcal{G}$ then

$$\mathcal{P}^{p}_{mn}\psi^{q}_{j}(\mathbf{r}) = \delta_{pq}\delta_{nj}\psi^{p}_{m}(\mathbf{r}).$$

(*d*) For any function $\phi(\mathbf{r})$ of L^2

$$\mathcal{P}^{p}_{nn}\phi(\mathbf{r}) = a^{p}_{n}\phi^{p}_{n}(\mathbf{r}),$$

where a^{p}_{n} and $\phi^{p}_{n}(\mathbf{r})$ are the coefficients and basis functions of the expansion of $\phi(\mathbf{r})$ (Equation (5.1)) that relate to the nth row of $\boldsymbol{\Gamma}^{p}$.

Proof Each statement will be proved for the case in which $\mathcal{G}$ is of finite order, the generalization to the case in which $\mathcal{G}$ is a compact Lie group being obvious.

(*a*) For any functions $\phi(\mathbf{r})$ and $\psi(\mathbf{r})$ of L^2, from Equations (1.20) and (5.2),

$$\begin{aligned}(\mathcal{P}^{p}_{mn}\phi, \psi) &= (d_p/g) \sum_{T\in\mathcal{G}} \Gamma^{p}(T)_{mn}(P(T)\phi, \psi)\\ &= (d_p/g) \sum_{T\in\mathcal{G}} \Gamma^{p}(T)_{mn}(\phi, P(T^{-1})\psi).\end{aligned}$$

As the set of inverse elements of $\mathcal{G}$ is merely a rearrangement of the set of elements of $\mathcal{G}$,

$$\begin{aligned}(\mathcal{P}^{p}_{mn}\phi, \psi) &= (d_p/g) \sum_{T'\in\mathcal{G}} \Gamma^{p}(T'^{-1})_{mn}(\phi, P(T')\psi)\\ &= (d_p/g) \sum_{T'\in\mathcal{G}} \Gamma^{p}(T')^{*}_{nm}(\phi, P(T')\psi)\end{aligned}$$

(as $\boldsymbol{\Gamma}^{p}(T')$ is a unitary matrix)

$$= (\phi, \mathcal{P}^{p}_{nm}\psi).$$

(*b*) From Equations (1.22) and (5.2)

$$\mathcal{P}^{p}_{mn}\mathcal{P}^{q}_{jk} = (d_p d_q/g^2) \sum_{T,T'\in\mathcal{G}} \Gamma^p(T)^*_{mn}\Gamma^q(T')^*_{jk}P(TT')$$

$$= (d_p d_q/g^2) \sum_{T,T''\in\mathcal{G}} \Gamma^p(T)^*_{mn}\Gamma^q(T^{-1}T'')^*_{jk}P(T'')$$

(on using the Rearrangement Theorem of Chapter 2, Section 1)

$$= (d_p d_q/g^2) \sum_{T,T''\in\mathcal{G}} \sum_{l=1}^{d_q} \Gamma^p(T)^*_{mn}\Gamma^q(T^{-1})^*_{jl}\Gamma^q(T'')^*_{lk}P(T'')$$

$$= \delta_{pq}\delta_{nj}\mathcal{P}^{q}_{mk}$$

(on using the unitary property $\Gamma^q(T^{-1})^*_{jl} = \Gamma^q(T)_{lj}$ and the orthogonality theorem for matrix representations, Theorem IV of Chapter 4, Section 5).

(*c*) From Equations (1.26) and (5.2)

$$\mathcal{P}^{p}_{mn}\psi^q_j(\mathbf{r}) = (d_p/g) \sum_{T\in\mathcal{G}} \sum_{k=1}^{d_q} \Gamma^p(T)^*_{mn}\Gamma^q(T)_{kj}\psi^q_j(\mathbf{r})$$

$$= \sum_{k=1}^{d_q} \delta_{pq}\delta_{mk}\delta_{nj}\psi^q_j(\mathbf{r})$$

(on using the orthogonality theorem for matrix representations)

$$= \delta_{pq}\delta_{nj}\psi^p_m(\mathbf{r}).$$

(*d*) This follows immediately on applying the result of part (*c*) to Equation (5.1).

6 Proof of the Wigner–Eckart Theorem, Theorem II of Chapter 5, Section 3

Theorem II Let $\mathcal{G}$ be a group of coordinate transformations that is either a finite group or a compact Lie group. Let Γ^p, Γ^q and Γ^r be unitary irreducible representations of $\mathcal{G}$ of dimensions d_p, d_q and d_r respectively, and suppose that $\phi^p_j(\mathbf{r})$, $j = 1, 2, \ldots, d_p$ and $\psi^r_l(\mathbf{r})$, $l = 1, 2, \ldots, d_r$ are sets of basis functions for Γ^p and Γ^r respectively. Finally, let Q^q_k, $k = 1, 2, \ldots, d_q$, be a set of irreducible tensor operators of Γ^q. Then

$$(\psi^r_l, Q^q_k\phi^p_j) = \sum_{\alpha=1}^{n^r_{pq}} \begin{pmatrix} p & q \\ j & k \end{pmatrix}\begin{vmatrix} r,\alpha \\ l \end{pmatrix}^* (r\,|Q^q|p)_\alpha \tag{C.24}$$

for all $j=1,2,\ldots,d_p$, $k=1,2,\ldots,d_q$, and $r=1,2,\ldots,d_r$, where $(r\,|Q^q|\,p)_\alpha$ are a set of n^r_{pq} "reduced matrix elements" that are independent of j, k and l.

Proof As P(T) is a unitary operator, from Equation (1.20)

$$\begin{aligned}(\psi^r_l, Q^q_k\phi^p_j) &= (P(T)\psi^r_l, P(T)\{Q^q_k\phi^p_j\}) \\ &= (P(T)\psi^r_l, \{P(T)Q^q_kP(T)^{-1}\}\{P(T)\phi^p_j\}),\end{aligned}$$

so that, by Equations (1.26) and (5.34),

$$(\psi^r_l, Q^q_k\phi^p_j) = \sum_{s=1}^{d_p}\sum_{t=1}^{d_q}\sum_{u=1}^{d_r} \Gamma^p(T)_{sj}\Gamma^q(T)_{tk}\Gamma^r(T)^*_{ul}(\psi^r_u, Q^q_t\phi^p_s).$$

Suppose first that $\mathscr{G}$ is a finite group. Replacing each T by T^{-1}, summing both sides of the above equation over all $T\in\mathscr{G}$ and invoking the unitary property of Γ^p, Γ^q and Γ^r gives

$$g(\psi^r_l, Q^q_k\phi^p_j) = \sum_{s=1}^{d_p}\sum_{t=1}^{d_q}\sum_{u=1}^{d_r}\sum_{T\in\mathscr{G}} \Gamma^p(T)^*_{js}\Gamma^q(T)^*_{kt}\Gamma^r(T)_{lu}(\psi^r_u, Q^q_t\phi^p_s),$$

Thus, from Equation (5.27),

$$(\psi^r_l, Q^q_k\phi^p_j) = (1/d_r)\sum_{s=1}^{d_p}\sum_{t=1}^{d_q}\sum_{u=1}^{d_r}\sum_{\alpha=1}^{n^r_{pq}} \left(\begin{matrix} p & q \\ j & k \end{matrix}\middle|\begin{matrix} r, & \alpha \\ l & \end{matrix}\right)^* \left(\begin{matrix} p & q \\ s & t \end{matrix}\middle|\begin{matrix} r, & \alpha \\ u & \end{matrix}\right)(\psi^r_u, Q^q_t\phi^p_s). \quad \text{(C.25)}$$

For a compact Lie group the same result is obtained from Equation (5.29) after summation has been replaced by invariant integration. With the reduced matrix elements defined by

$$(r\,|Q^q|\,p)_\alpha = (1/d_r)\sum_{s=1}^{d_p}\sum_{t=1}^{d_q}\sum_{u=1}^{d_r} \left(\begin{matrix} p & q \\ s & t \end{matrix}\middle|\begin{matrix} r, & \alpha \\ u & \end{matrix}\right)(\psi^r_u, Q^q_t\phi^p_s), \quad \text{(C.26)}$$

Equation (C.25) immediately gives Equation (C.24).

7 Proof of Theorem II of Chapter 5, Section 5

Theorem. If $\mathscr{G}_1\otimes\mathscr{G}_2$ is a finite group or a compact linear Lie group and Γ_1 and Γ_2 are irreducible representations of $\mathscr{G}_1$ and $\mathscr{G}_2$ respectively, then the representation Γ defined by Equation (5.44) is an *irreducible* representation of $\mathscr{G}_1\otimes\mathscr{G}_2$. Moreover, *every* irreducible representation of $\mathscr{G}_1\otimes\mathscr{G}_2$ is equivalent to a representation constructed in this way.

Proof From Equations (5.44) and (A.8) the characters χ of Γ are given in terms of the characters χ_1 and χ_2 of Γ_1 and Γ_2 by

$$\chi((T_1, T_2)) = \chi_1(T_1)\chi_2(T_2). \tag{C.27}$$

Suppose first that $\mathcal{G}_1$ and $\mathcal{G}_2$ are both finite groups of order g_1 and g_2 respectively. The irreducible nature of Γ then follows from Theorem VI of Chapter 4, Section 6, for

$$\begin{aligned}(1/g_1g_2) \sum_{(T_1,T_2)\in\mathcal{G}_1\otimes\mathcal{G}_2} |\chi((T_1, T_2))|^2 \\ &= \{(1/g_1) \sum_{T_1\in\mathcal{G}_1} |\chi_1(T_1)|^2\}\{(1/g_2) \sum_{T_2\in\mathcal{G}_2} |\chi_2(T_2)|^2\} \\ &= 1.\end{aligned}$$

When $\mathcal{G}_1\otimes\mathcal{G}_2$ is a compact linear Lie group the same line of argument can be applied, with the summation over group elements being replaced by the appropriate invariant integral.

It remains to show that every irreducible representation of $\mathcal{G}_1\otimes\mathcal{G}_2$ is equivalent to a representation constructed in this way. First it should be noted that Theorems I and IV of Chapter 4, Section 6, applied to Equation (C.27) show that the representations $\Gamma_1\otimes\Gamma_2$ and $\Gamma_1'\otimes\Gamma_2'$ of $\mathcal{G}_1\otimes\mathcal{G}_2$ are equivalent if and only if Γ_1 and Γ_1' are equivalent representations of $\mathcal{G}_1$ and Γ_2 and Γ_2' are equivalent representations of $\mathcal{G}_2$. When $\mathcal{G}_1$ and $\mathcal{G}_2$ are both finite the remainder of the demonstration is simple, for the sum of the squares of the dimensions of the inequivalent irreducible representations of $\mathcal{G}_1\otimes\mathcal{G}_2$ constructed as in Equation (5.44) is equal to $\{\sum_p (d_1^p)^2\}\{\sum_q (d_2^q)^2\}$, where the sums over p and q are over all inequivalent irreducible representations Γ_1^p and Γ_2^q of $\mathcal{G}_1$ and $\mathcal{G}_2$ respectively, d_1^p and d_2^q being the dimensions of Γ_1^p and Γ_2^q. However, Theorem VII of Chapter 4, Section 6, shows that $\sum_p (d_1^p)^2 = g_1$ and $\sum_q (d_2^q)^2 = g_2$. As g_1g_2 is the order of $\mathcal{G}_1\otimes\mathcal{G}_2$, a further application of this theorem implies that the set of inequivalent irreducible representations given by Equation (5.44) completely exhausts the set of all inequivalent irreducible representations of $\mathcal{G}_1\otimes\mathcal{G}_2$. This line of argument cannot be extended to the case in which $\mathcal{G}_1\otimes\mathcal{G}_2$ is a compact linear Lie group. In that case the proof requires a part of the theorem of Peter and Weyl (1927), which has not been described in this book. An outline may be found on pages 215 and 216 of the book by Miller (1972).

8 Proofs of the theorems of Chapter 5, Section 7

Theorem I Let $\mathcal{S}$ be a subgroup of order s of a group $\mathcal{G}$ of order g and let $T_1, T_2, \ldots$ be a set of $M = g/s$ coset representatives for the decomposition of $\mathscr{G}$ into right cosets with respect to $\mathscr{S}$. Let $\boldsymbol{\Delta}$ be a d-dimensional unitary representation of $\mathscr{S}$. Then the set of $Md \times Md$ matrices $\boldsymbol{\Gamma}(\mathrm{T})$ defined for all $\mathrm{T} \in \mathscr{G}$ by

$$\Gamma(\mathrm{T})_{kt,jr} = \begin{cases} \Delta(\mathrm{T}_k\mathrm{T}\mathrm{T}_j^{-1})_{tr}, & \text{if} \quad \mathrm{T}_k\mathrm{T}\mathrm{T}_j^{-1} \in \mathscr{S}, \\ 0, & \text{if} \quad \mathrm{T}_k\mathrm{T}\mathrm{T}_j^{-1} \notin \mathscr{S}, \end{cases} \tag{C.28}$$

provide an Md-dimensional unitary representation of $\mathscr{G}$. If $\psi(\mathrm{S})$ are the characters of the representation $\boldsymbol{\Delta}$ of $\mathscr{S}$, then the characters $\chi(\mathrm{T})$ of the representation $\boldsymbol{\Gamma}$ of $\mathscr{S}$ are given by

$$\chi(\mathrm{T}) = \sum_j \psi(\mathrm{T}_j\mathrm{T}\mathrm{T}_j^{-1}), \tag{C.29}$$

where the sum is over all the coset representatives T_j such that $\mathrm{T}_j\mathrm{T}\mathrm{T}_j^{-1} \in \mathscr{S}$.

Proof Let T and T′ be any two elements of $\mathscr{G}$. Then

$$(\boldsymbol{\Gamma}(\mathrm{T})\boldsymbol{\Gamma}(\mathrm{T}'))_{kt,jr} = \sum_{l=1}^{M} \sum_{u=1}^{d} \Gamma(\mathrm{T})_{kt,lu}\Gamma(\mathrm{T}')_{lu,jr}$$

where in the sum over l only those coset representatives T_l are included for which $\mathrm{T}_k\mathrm{T}\mathrm{T}_l^{-1} \in \mathscr{S}$ ***and*** $\mathrm{T}_l\mathrm{T}'\mathrm{T}_j^{-1} \in \mathscr{S}$. However, $\mathrm{T}_k\mathrm{T}\mathrm{T}_l^{-1} \in \mathscr{S}$ if and only if $\mathrm{T}_k\mathrm{T} \in \mathscr{S}\mathrm{T}_l$, so this sum contains at most only one term. Let $\mathrm{T}_{l'}$ be ***the*** coset representative (from the chosen set $\mathrm{T}_1, \mathrm{T}_2, \ldots$) such that $\mathrm{T}_k\mathrm{T} \in \mathscr{S}\mathrm{T}_{l'}$. As $\mathrm{T}_k(\mathrm{T}\mathrm{T}')\mathrm{T}_j^{-1} = (\mathrm{T}_k\mathrm{T}\mathrm{T}_{l'}^{-1})(\mathrm{T}_{l'}\mathrm{T}'\mathrm{T}_j^{-1})$, it follows that $\mathrm{T}_{l'}\mathrm{T}'\mathrm{T}_j^{-1} \in \mathscr{S}$ if and only if $\mathrm{T}_k(\mathrm{T}\mathrm{T}')\mathrm{T}_j^{-1} \in \mathscr{S}$, and when this is so

$$\sum_{u=1}^{d} \Delta(\mathrm{T}_k\mathrm{T}\mathrm{T}_{l'}^{-1})_{tu}\Delta(\mathrm{T}_{l'}\mathrm{T}'\mathrm{T}_j^{-1})_{ur} = \Delta(\mathrm{T}_k(\mathrm{T}\mathrm{T}')\mathrm{T}_j^{-1})_{tr}.$$

Thus the above sum is

$$\begin{cases} \Delta(\mathrm{T}_k(\mathrm{T}\mathrm{T}')\mathrm{T}_j^{-1})_{tr}, & \text{if} \quad \mathrm{T}_k(\mathrm{T}\mathrm{T}')\mathrm{T}_j^{-1} \in \mathscr{S} \\ 0, & \text{if} \quad \mathrm{T}_k(\mathrm{T}\mathrm{T}')\mathrm{T}_j^{-1} \notin \mathscr{S} \end{cases}$$
$$= \Gamma(\mathrm{T}\mathrm{T}')_{kt,jr},$$

so that the matrices $\boldsymbol{\Gamma}(\mathrm{T})$ do provide a representation of $\mathscr{G}$.

The representation $\boldsymbol{\Gamma}$ of $\mathscr{G}$ is unitary, as, for any $\mathrm{T} \in \mathscr{G}$.

$$(\tilde{\boldsymbol{\Gamma}}(\mathrm{T}^{-1}))^*_{kt,jr} = \Gamma(\mathrm{T}^{-1})^*_{jr,kt}$$
$$= \begin{cases} \Delta(\mathrm{T}_j\mathrm{T}^{-1}\mathrm{T}_k^{-1})^*_{rt}, & \text{if} \quad \mathrm{T}_j\mathrm{T}^{-1}\mathrm{T}_k^{-1} \in \mathscr{S}, \\ 0, & \text{if} \quad \mathrm{T}_j\mathrm{T}^{-1}\mathrm{T}_k^{-1} \notin \mathscr{S}, \end{cases}$$
$$= \Gamma(\mathrm{T})_{kt,jr},$$

as $\mathrm{T}_j\mathrm{T}^{-1}\mathrm{T}_k^{-1} = (\mathrm{T}_k\mathrm{T}\mathrm{T}_j^{-1})^{-1}$ and $\boldsymbol{\Delta}$ is unitary.

Finally, the formula (Equation (C.29)) for the characters follows immediately from Equations (C.28), as for any $T \in \mathscr{G}$

$$\chi(T) = \sum_{j=1}^{M} \sum_{r=1}^{d} \Gamma(T)_{jr,jr}$$
$$= \sum_{j} \sum_{r=1}^{d} \Delta(T_j T T_j^{-1})_{rr} = \sum_{j} \psi(T_j T T_j^{-1}),$$

where the sum over j is over all coset representatives T_j such that $T_j T T_j^{-1} \in \mathscr{S}$.

Theorem II Let $\Gamma^{q,p}$ be the unitary representation of the semi-direct product group $\mathscr{G}(=\mathscr{A} \circledS \mathscr{B})$ defined by Equations (5.55). Then

(*a*) $\Gamma^{q,p}$ is an *irreducible* representation of $\mathscr{G}$; and
(*b*) the *complete* set of unitary irreducible representations of $\mathscr{G}$ may be determined (up to equivalence) by choosing one q in each orbit and then constructing $\Gamma^{q,p}$ for each $\Gamma^{p}_{\mathscr{B}(q)}$ of $\mathscr{B}(q)$.

Proof

(*a*) From Equation (5.56)

$$(1/g) \sum_{T \in \mathscr{G}} |\chi^{q,p}(T)|^2 = (1/ab) \sum_{A \in \mathscr{A}} \sum_{B \in \mathscr{B}} \sum_{j} |\chi_{\mathscr{A}}^{B_j(q)}(A) \chi^{p}_{\mathscr{B}(q)}(B_j B B_j^{-1})|^2,$$

where the sum over j is over all coset representatives B_j such that $B_j B B_j^{-1} \in \mathscr{B}(q)$. But this sum can be rewritten as

$$\sum_{j} \sum_{k} \chi_{\mathscr{A}}^{B_j(q)}(A) \chi_{\mathscr{A}}^{B_k(q)}(A)^* \chi^{p}_{\mathscr{B}(q)}(B_j B B_j^{-1}) \chi^{p}_{\mathscr{B}(q)}(B_k B B_k^{-1})^*.$$

However, the characters $\chi_{\mathscr{A}}^{B_j(q)}$ and $\chi_{\mathscr{A}}^{B_k(q)}$ correspond to inequivalent irreducible representations of $\mathscr{A}$ if $j \neq k$ (as for one-dimensional representations equivalence implies identity of character systems, and by Equations (5.51) and (5.52) $\chi_{\mathscr{A}}^{B_j(q)}(A)$ and $\chi_{\mathscr{A}}^{B_k(q)}(A)$ give identical character systems if and only if $B_k^{-1} B_j \in \mathscr{B}(q)$, that is, if and only if $B_j = B_k$). Consequently, by Theorem III of Chapter 4, Section 6,

$$(1/a) \sum_{A \in \mathscr{A}} \chi_{\mathscr{A}}^{B_j(q)}(A) \chi_{\mathscr{A}}^{B_k(q)}(A)^* = \delta_{jk},$$

so that

$$(1/g) \sum_{T \in \mathscr{G}} |\chi^{q,p}(T)|^2 = (1/b) \sum_{B \in \mathscr{B}} \sum_{j} |\chi^{p}_{\mathscr{B}(q)}(B_j B B_j^{-1})|^2.$$

For each j the mapping $B \rightarrow B_j^{-1}BB_j$ maps $\mathscr{B}(q)$ one-to-one onto a subset of $\mathscr{B}$. Thus

$$(1/g) \sum_{T \in \mathscr{G}} |\chi^{q,p}(T)|^2 = (M(q)/b) \sum_{B \in \mathscr{B}(q)} |\chi^{p}_{\mathscr{B}(q)}(B)|^2$$
$$= \{M(q)b(q)\}/b = 1$$

on using Theorem VI of Chapter 4, Section 6, a further application of which shows that $\Gamma^{q,p}$ of $\mathscr{G}$ is irreducible.

(b) It will first be shown that $\Gamma^{q,p}$ and $\Gamma^{q',p'}$ are equivalent representations of $\mathscr{G}$ only if q' is in the orbit of q and the representation $\Gamma^{p}_{\mathscr{B}(q)}$ of $\mathscr{B}(q)$ is equivalent to the representation $\Gamma^{p'}_{\mathscr{B}(q')}$ of the isomorphic group $\mathscr{B}(q')$. By the Theorem I of Chapter 4, Section 6, and Equation (5.56), if $\Gamma^{q,p}$ and $\Gamma^{q',p'}$ are equivalent then for all $A \in \mathscr{A}$ and all $B \in \mathscr{B}$,

$$\sum_j \chi_{\mathscr{A}}^{B_j(q)}(A)\chi^{p}_{\mathscr{B}(q)}(B_jBB_j^{-1}) = \sum_k \chi_{\mathscr{A}}^{B'_k(q')}(A)\chi^{p'}_{\mathscr{B}(q')}(B'_kBB_k'^{-1}), \tag{C.30}$$

where the sum over j is over all coset representatives B_j of the decomposition of $\mathscr{B}$ into right cosets with respect to $\mathscr{B}(q)$ such that $B_jBB_j^{-1} \in \mathscr{B}(q)$, while the sum over k involves the representatives B'_k of the decomposition with respect to $\mathscr{B}(q')$ such that $B'_kBB_k'^{-1} \in \mathscr{B}(q')$.

First consider Equation (C.30) for the special case $B = E$. Then as $B_jEB_j^{-1} = E \in \mathscr{B}(q)$ for all B_j and $B'_kEB_k'^{-1} = E \in \mathscr{B}(q')$ for all B'_k, Equation (C.30) becomes

$$\sum_{j=1}^{M(q)} \chi_{\mathscr{A}}^{B_j(q)}(A) = \sum_{k=1}^{M(q')} \chi_{\mathscr{A}}^{B'_k(q')}(A) \tag{C.31}$$

for all $A \in \mathscr{A}$. Applying Theorems IV and V of Chapter 4, Section 6, to $\mathscr{A}$ alone then shows that the irreducible representations of $\mathscr{A}$ on the right-hand side of Equation (C.31) must at most be merely a rearrangement of those on the left-hand side of Equation (C.31). As $B'_1(q') = q'$, there must exist a coset representative B_l (from the set $B_1, B_2, \ldots$) such that $q' = B_l(q)$, so q' must be in the orbit of q.

Now putting $A = E$ in Equation (C.30) gives

$$\sum_j \chi^{p}_{\mathscr{B}(q)}(B_jBB_j^{-1}) = \sum_k \chi^{p'}_{\mathscr{B}(B_l(q))}(B'_kBB_k'^{-1}) \tag{C.32}$$

for all $B \in \mathscr{B}$. However, as mentioned in note (g) of Chapter 5, Section 7, the mapping $\phi_l(B) = B_lBB_l^{-1}$ gives a one-to-one mapping of $\mathscr{B}(B_l(q))$ into $\mathscr{B}(q)$. As ϕ_l is an automorphism of $\mathscr{B}$, the coset representatives for the decomposition of $\mathcal{B}$ into right cosets with respect to $\mathcal{B}(\mathrm{B}_l(q))$ may be taken to be $\mathrm{B}_l^{-1}\mathrm{B}_j\mathrm{B}_l$, $j = 1, 2, \ldots, M(q)$. Thus the right-hand side of Equation (C.32) becomes

$$\sum_j \chi^{\mathrm{p}'}_{\mathcal{B}(\mathrm{B}_l(q))}((\mathrm{B}_l^{-1}\mathrm{B}_j\mathrm{B}_l)\mathrm{B}(\mathrm{B}_l^{-1}\mathrm{B}_j\mathrm{B}_l)^{-1}),$$

where the sum is over B_j such that $(\mathrm{B}_l^{-1}\mathrm{B}_j\mathrm{B}_l)\mathrm{B}(\mathrm{B}_l^{-1}\mathrm{B}_j\mathrm{B}_l)^{-1} \in \mathcal{B}(\mathrm{B}_l(q))$, that is, such that $(\mathrm{B}_j\mathrm{B}_l)\mathrm{B}(\mathrm{B}_j\mathrm{B}_l)^{-1} \in \mathcal{B}(q)$, so that the right-hand side of Equation (C.32) is

$$\sum_j \chi^{\mathrm{p}'}_{\mathcal{B}(q)}((\mathrm{B}_j\mathrm{B}_l)\mathrm{B}(\mathrm{B}_j\mathrm{B}_l)^{-1}).$$

As $\mathcal{B}(q)(\mathrm{B}_j\mathrm{B}_l)$ is distinct from $\mathcal{B}(q)(\mathrm{B}_k\mathrm{B}_l)$ if and only if $\mathcal{B}(q)\mathrm{B}_j$ is distinct from $\mathcal{B}(q)\mathrm{B}_k$, the set $\mathrm{B}_j\mathrm{B}_l$, $j = 1, 2, \ldots, M(q)$, is an alternative set of coset representatives for the decomposition of $\mathcal{B}$ into right cosets with respect to $\mathcal{B}(q)$. Consequently Equation (C.32) becomes

$$\sum_j \chi^{\mathrm{p}}_{\mathcal{B}(q)}(\mathrm{B}_j\mathrm{B}\mathrm{B}_j^{-1}) = \sum_j \chi^{\mathrm{p}'}_{\mathcal{B}(q)}(\mathrm{B}_j\mathrm{B}\mathrm{B}_j^{-1})$$

for all $\mathrm{B} \in \mathcal{B}$, from which it follows that Γ^{p} and $\Gamma^{\mathrm{p}'}$ must be equivalent.

It remains to show that by choosing one q in each orbit and constructing $\Gamma^{q,\mathrm{p}}$ for each inequivalent irreducible representation $\Gamma^{\mathrm{p}}_{\mathcal{B}(q)}$ of $\mathcal{B}(q)$, the set of *all* irreducible representations of $\mathcal{G}$ is obtained (up to equivalence). According to Theorem VII of Chapter 4, Section 6, to prove this it is sufficient to show that the sum of the squares of the dimensions of the irreducible representations so obtained, S, is equal to the order $g(=ab)$ of $\mathcal{G}$. But the part of S for a particular q is $\sum_{\mathrm{p}} \{M(q)d_{\mathrm{p}}\}^2 = M(q)^2 \sum_{\mathrm{p}} (d_{\mathrm{p}})^2 = M(q)^2 b(q) = bM(q)$. As each of the a irreducible representations of $\mathcal{A}$ lies in one and only one orbit, the sum of $M(q)$ over different orbits is precisely a, so that indeed $S = ab$.

Theorem III With the notation of Theorem I of this section, let V_Δ be a carrier space for the representation Δ of $\mathcal{S}$ and let $\Phi_\Delta(\mathrm{S})$ be a set of operators defined for each $\mathrm{S} \in \mathcal{S}$ to act in V_Δ in such a way that

$$\Phi_\Delta(\mathrm{S})\psi_n = \sum_{m=1}^{d} \Delta(\mathrm{S})_{mn}\psi_m, \tag{C.33}$$

where $\psi_1, \psi_2, \ldots, \psi_d$ are a basis for V_Δ. Let V be the vector space of all mappings $\phi(\mathcal{S}\mathrm{T}_j)$ into V_Δ of right cosets $\mathcal{S}\mathrm{T}_1, \mathcal{S}\mathrm{T}_2, \ldots$ of $\mathcal{G}$ with respect to $\mathscr{S}$. For each $T \in \mathscr{G}$, define the operator $\Phi(T)$ by

$$\Phi(T)\phi(\mathscr{S}T_k) = \Phi_\Delta(T_k T T_j^{-1})\phi(\mathscr{S}T_j), \tag{C.34}$$

where ϕ is any member of V and T_j is the coset representative such that $T_k T \in \mathscr{S}T_j$. Then

(*a*) for any $T, T' \in \mathscr{G}$

$$\Phi(TT') = \Phi(T)\Phi(T'),$$

and

(*b*) there exists a basis ϕ_{kt} of V ($k = 1, 2, \ldots, M$; $t = 1, 2, \ldots, d$) such that for all $T \in \mathscr{G}$ and $j = 1, 2, \ldots, M$, $r = 1, 2, \ldots, d$.

$$\Phi(T)\phi_{jr} = \sum_{k=1}^{M} \sum_{t=1}^{d} \Gamma(T)_{kt,jr}\phi_{kt}, \tag{C.35}$$

where Γ is the induced representation of $\mathscr{G}$ defined by Equations (5.49). Thus V is a carrier space for the induced representation Γ of $\mathscr{G}$ and the $\Phi(T)$ are the corresponding operators acting in V.

Proof A "mapping" ϕ of the space of right cosets of $\mathscr{G}$ with respect to $\mathscr{S}$ into V_Δ is simply an assignment to each coset $\mathscr{S}T_j$ of some element $\phi(\mathscr{S}T_j)$ of V_Δ. If ϕ and ϕ' are any two such mappings and a and a' are any complex numbers, then $(a\phi + a'\phi')$ can be defined as the mapping in which $\mathscr{S}T_j$ is mapped into the element $\{a\phi(\mathscr{S}T_j) + a'\phi'(\mathscr{S}T_j)\}$ of V_Δ. The "zero mapping" may be defined as the mapping that assigns the zero element of V_Δ to every coset $\mathscr{S}T_j$. It is then easily verified that the set of such mappings do indeed form a complex vector space (see Appendix B, Section 1). It must be emphasized that V is completely different from V_Δ.

Equation (C.34) is well defined because $\phi(\mathscr{S}T_j) \in V_\Delta$ and Φ_Δ is implicitly defined by Equation (C.33) to act on every element of V_Δ. It should be noted that Equation (C.33) is of the form of Equation (4.1), so that

$$\Phi_\Delta(SS') = \Phi_\Delta(S)\Phi_\Delta(S') \tag{C.36}$$

for all $S, S' \in \mathscr{S}$.

(*a*) Let T and T′ be any two members of $\mathscr{S}$ and let $\mathscr{S}T_k$ be any right coset of $\mathscr{G}$ with respect to $\mathscr{S}$. Then, if $T_k T' \in \mathscr{S}T_l$, from Equation (C.34)

$$\Phi(T')\phi(\mathscr{S}T_k) = \Phi_\Delta(T_k T' T_l^{-1})\phi(\mathscr{S}T_l).$$

where the right-hand side is a member of V_Δ whose "value" depends on $\mathscr{S}T_k$. Accordingly, write

$$\theta(\mathscr{S}T_k) = \Phi_\Delta(T_k T' T_l^{-1})\phi(\mathscr{S}T_l).$$

Then

$$\begin{aligned}\Phi(\mathrm{T})\Phi(\mathrm{T}')\phi(\mathcal{S}\mathrm{T}_k) &= \Phi(\mathrm{T})\theta(\mathcal{S}\mathrm{T}_k)\\ &= \Phi_\Delta(\mathrm{T}_k\mathrm{T}\mathrm{T}_j^{-1})\theta(\mathcal{S}\mathrm{T}_j) \text{ (where } \mathrm{T}_k\mathrm{T}\in\mathcal{S}\mathrm{T}_j)\\ &= \Phi_\Delta(\mathrm{T}_k\mathrm{T}\mathrm{T}_j^{-1})\Phi_\Delta(\mathrm{T}_j\mathrm{T}'\mathrm{T}_l^{-1})\phi(\mathcal{S}\mathrm{T}_l)\\ &= \Phi_\Delta(\mathrm{T}_k\mathrm{T}\mathrm{T}'\mathrm{T}_l^{-1})\phi(\mathcal{S}\mathrm{T}_l) \text{ (by Equation (C.36))}\\ &= \Phi(\mathrm{T}\mathrm{T}')\phi(\mathcal{S}\mathrm{T}_k).\end{aligned}$$

(*b*) Define the elements ϕ_{kt} ($k = 1, 2, \ldots, M$; $t = 1, 2, \ldots, d$) of V to be the set of mappings such that

$$\phi_{kt}(\mathcal{S}\mathrm{T}_j) = \delta_{kj}\psi_t, \qquad \text{(C.37)}$$

where $\psi_1, \psi_2, \ldots, \psi_d$ provide the basis for V_Δ. This is obviously a mapping of the space of right cosets into V_Δ. Moreover, *any* mapping ϕ of the space of right cosets into V_Δ can be written as a linear combination of the ϕ_{kt}. (For suppose that $\phi(\mathcal{S}\mathrm{T}_j) = \sum_{t=1}^{d} a_t(\mathcal{S}\mathrm{T}_j)\psi_t$, where the complex numbers $a_t(\mathcal{S}\mathrm{T}_j)$ which depend on $\mathcal{S}\mathrm{T}_j$ completely specify ϕ. Then

$$\phi = \sum_{k=1}^{M}\sum_{t=1}^{d} a_t(\mathcal{S}\mathrm{T}_k)\phi_{kt},$$

for, by Equation (C.37),

$$\begin{aligned}\phi(\mathcal{S}\mathrm{T}_j) &= \sum_{k=1}^{M}\sum_{t=1}^{d} a_t(\mathcal{S}\mathrm{T}_k)\phi_{kt}(\mathcal{S}\mathrm{T}_j)\\ &= \sum_{k=1}^{M}\sum_{t=1}^{d} a_t(\mathcal{S}\mathrm{T}_k)\delta_{jk}\psi_t = \sum_{t=1}^{d} a_t(\mathcal{S}\mathrm{T}_j)\psi_t,\end{aligned}$$

as required.) This demonstrates that V has dimension Md, for the elements ϕ_{kt} are obviously linearly independent.

For any $\mathrm{T}\in\mathcal{G}$ and any coset $\mathcal{S}\mathrm{T}_p$, from Equation (C.34)

$$\begin{aligned}\Phi(\mathrm{T})\phi_{jr}(\mathcal{S}\mathrm{T}_p) &= \Phi_\Delta(\mathrm{T}_p\mathrm{T}\mathrm{T}_q^{-1})\phi_{jr}(\mathcal{S}\mathrm{T}_q) && \text{(where } \mathrm{T}_p\mathrm{T}\in\mathcal{S}\mathrm{T}_q)\\ &= \Phi_\Delta(\mathrm{T}_p\mathrm{T}\mathrm{T}_q^{-1})\delta_{jq}\psi_r && \text{(by Equation (C.37))}\\ &= \begin{cases}\sum_{t=1}^{d}\Delta(\mathrm{T}_p\mathrm{T}\mathrm{T}_j^{-1})_{tr}\psi_t, & \text{if } \mathrm{T}_p\mathrm{T}\in\mathcal{S}\mathrm{T}_j,\\ 0, & \text{if } \mathrm{T}_p\mathrm{T}\notin\mathcal{S}\mathrm{T}_j,\end{cases}\\ &= \sum_{t=1}^{d}\Gamma(\mathrm{T})_{pt,jr}\psi_t && \text{(by Equation (5.49))}\\ &= \sum_{k=1}^{M}\sum_{t=1}^{d}\Gamma(\mathrm{T})_{kt,jr}\phi_{kt}(\mathcal{S}\mathrm{T}_p) && \text{(by Equation (C.37))},\end{aligned}$$

thereby verifying Equation (C.35).

9 Proofs of Theorems II, III and IV of Chapter 5, Section 8

Theorem II Suppose that the unitary irreducible representations $\mathbf{\Gamma}$ and $\mathbf{\Gamma}^*$ of a group $\mathscr{G}$ that is finite or is a compact Lie group are equivalent, the similarity transformation relating them being

$$\mathbf{\Gamma}^*(T) = \mathbf{Z}^{-1}\mathbf{\Gamma}(T)\mathbf{Z} \tag{C.38}$$

for all $T \in \mathscr{G}$, $\mathbf{Z}$ being a $d \times d$ non-singular matrix. Then

$$\mathbf{Z}^*\mathbf{Z} = c_z\mathbf{1},$$

where c_z is a real non-zero number. Moreover, $\mathbf{\Gamma}$ is potentially real if $c_z > 0$ and is pseudo-real if $c_z < 0$. In particular if $\mathbf{Z}$ is chosen to be unitary then $(c_z)^2 = 1$, $\mathbf{\Gamma}$ being potentially real if $c_z = 1$ and pseudo-real if $c_z = -1$.

Proof Taking the complex conjugate of both sides of Equation (C.38) and substituting back in Equation (C.38) gives $\mathbf{\Gamma}^*(T) = (\mathbf{Z}^*\mathbf{Z})^{-1}\mathbf{\Gamma}^*(T)(\mathbf{Z}^*\mathbf{Z})$. The second of Schur's Lemmas (see Chapter 4, Section 5) then implies $\mathbf{Z}^*\mathbf{Z} = c_z\mathbf{1}$, where c_z is some complex number. As $\det(\mathbf{Z}^*\mathbf{Z}) = |\det \mathbf{Z}|^2 = (c_z)^d$ and $\det \mathbf{Z} \neq 0$, c_z must be non-zero. It will first be shown that c_z must also be real.

Suppose that $\mathbf{Z}'$ is another $d \times d$ non-singular matrix relating $\mathbf{\Gamma}$ and $\mathbf{\Gamma}^*$, that is, $\mathbf{\Gamma}^*(T) = (\mathbf{Z}')^{-1}\mathbf{\Gamma}(T)\mathbf{Z}'$ for all $T \in \mathscr{G}$, as in Equation (C.38). Then $\mathbf{Z}'$ must be a scalar multiple of $\mathbf{Z}$, for combining the two similarity transformations gives $\mathbf{\Gamma}(T) = (\mathbf{Z}'\mathbf{Z}^{-1})^{-1}\mathbf{\Gamma}(T)(\mathbf{Z}'\mathbf{Z}^{-1})$ for all $T \in \mathscr{G}$, which by a further application of Schur's Lemma gives $\mathbf{Z}'\mathbf{Z}^{-1} = b\mathbf{1}$ and hence $\mathbf{Z}' = b\mathbf{Z}$, where b is some complex number. It follows that $(\mathbf{Z}')^*\mathbf{Z}' = |b|^2\mathbf{Z}^*\mathbf{Z} = |b|^2c_z\mathbf{1}$. However, $\mathbf{Z}'$ may be chosen to be unitary, in which case this implies $\mathbf{Z}' = |b|^2 c_z\tilde{\mathbf{Z}}'$. Taking the transpose and substituting back gives $\{|b|^2c_z\}^2 = 1$, so $|b|^2 c_z = \pm 1$. As $|b|^2$ is real, c_z must also be real.

Now suppose that $\mathbf{\Gamma}$ is equivalent to a real representation $\mathbf{\Gamma}'$. Then there exists a non-singular $d \times d$ matrix $\mathbf{S}$ such that $\mathbf{\Gamma}'(T) = \mathbf{S}^{-1}\mathbf{\Gamma}(T)\mathbf{S}$ for all $T \in \mathscr{G}$. Thus $\mathbf{\Gamma}'(T) = \mathbf{S}^{*-1}\mathbf{\Gamma}^*(T)\mathbf{S}^*$ and hence $\mathbf{\Gamma}^*(T) = (\mathbf{S}\mathbf{S}^{*-1})^{-1}\mathbf{\Gamma}(T)(\mathbf{S}\mathbf{S}^{*-1})$. As this is of the form of Equation (C.38), $\mathbf{Z}$ must be a scalar multiple of $\mathbf{S}\mathbf{S}^{*-1}$. Let $\mathbf{Z} = a\mathbf{S}\mathbf{S}^{*-1}$. Then $\mathbf{Z}^*\mathbf{Z} = |a|^2\mathbf{1}$, so $c_z = |a|^2 > 0$.

Conversely, suppose $c_z > 0$ and let $\mathbf{S} = (c_z^{-1/2}\mathbf{Z} + \kappa\mathbf{1})$, where $c_z^{1/2}$ is the positive square root of c_z and κ is a complex number of modulus unity chosen to make $\mathbf{S}$ is non-singular. Then $\mathbf{Z}\mathbf{S}^* = \kappa^* c^{1/2}\mathbf{S}$. With $\mathbf{\Gamma}'(T)$ defined by $\mathbf{\Gamma}'(T) = \mathbf{S}^{-1}\mathbf{\Gamma}(T)\mathbf{S}$,

$$\begin{aligned}\mathbf{\Gamma}'(T)^* &= \mathbf{S}^{*-1}\mathbf{\Gamma}^*(T)\mathbf{S}^* = (\mathbf{Z}\mathbf{S}^*)^{-1}\mathbf{\Gamma}(T)(\mathbf{Z}\mathbf{S}^*)^{-1} \text{ (by Equation (C.38))}\\ &= (\kappa^* c^{1/2}\mathbf{S})^{-1}\mathbf{\Gamma}(T)(\kappa^* c^{1/2}\mathbf{S}) = \mathbf{\Gamma}'(T),\end{aligned}$$

so that the representation $\mathbf{\Gamma}'$ (to which $\mathbf{\Gamma}$ is equivalent) is real. Thus $\mathbf{\Gamma}$ is equivalent to a real representation if and only if $c_z > 0$.

If $\mathbf{Z}$ is chosen to be unitary, Equation (C.38) implies $\mathbf{Z} = c_z\tilde{\mathbf{Z}}$. Taking the transpose and substituting back gives $(c_z)^2 = 1$, so $c_z = \pm 1$.

Theorem III If $\mathscr{G}$ is a finite group of order g and $\chi(T)$ is the character of $T \in \mathscr{G}$ in the irreducible representation $\mathbf{\Gamma}$, then

$$(1/g) \sum_{T\in\mathscr{G}} \chi(T^2) = \begin{cases} 1, & \text{if } \mathbf{\Gamma} \text{ is potentially real,} \\ 0, & \text{if } \mathbf{\Gamma} \text{ is essentially complex,} \\ -1, & \text{if } \mathbf{\Gamma} \text{ is pseudo-real.} \end{cases}$$

Similarly, if $\mathscr{G}$ is a compact Lie group,

$$\int_{\mathscr{G}} \chi(T^2)\, dT = \begin{cases} 1, & \text{if } \mathbf{\Gamma} \text{ is potentially real,} \\ 0, & \text{if } \mathbf{\Gamma} \text{ is essentially complex,} \\ -1, & \text{if } \mathbf{\Gamma} \text{ is pseudo-real.} \end{cases}$$

Proof The proof will be given explicitly for a finite group. It generalizes in the obvious way for a compact Lie group.

Suppose first that $\mathbf{\Gamma}$ is essentially complex. Then from the orthogonality theorem for matrix representations, Theorem IV of Chapter 4, Section 5, with $\mathbf{\Gamma}^p = \mathbf{\Gamma}^*$ and $\mathbf{\Gamma}^q = \mathbf{\Gamma}$,

$$(1/g) \sum_{T\in\mathscr{G}} \Gamma(T)_{jk}\Gamma(T)_{st} = 0$$

Putting $k = s$ and $j = t$, and then summing over j and k gives

$$(1/g) \sum_{T\in\mathscr{G}} \chi(T^2) = 0.$$

Now suppose that $\mathbf{\Gamma}$ is equivalent to $\mathbf{\Gamma}^*$, the similarity transformation being Equation (C.38). Then from the more general form of the orthogonality theorem for matrix representations as given in Theorem IV of Section 3 of this appendix, with $\mathbf{\Gamma}^p = \mathbf{\Gamma}^*, \mathbf{\Gamma}^q = \mathbf{\Gamma}, \mathbf{S} = \mathbf{Z}$ and $d_p = d$,

$$(1/g) \sum_{T\in\mathscr{G}} \Gamma(T)_{jk}\Gamma(T)_{st} = Z_{sj}(\mathbf{Z}^{-1})_{kt}/d. \tag{C.39}$$

The matrix $\mathbf{Z}$ may be taken to be unitary, so that $(\mathbf{Z}^{-1})_{kt} = Z^*_{tk}$. On putting $k = s$ and $j = t$ and then summing over j and k, Equation (C.39) becomes

$$(1/g) \sum_{T\in\mathscr{G}} \chi(T^2) = (1/d) \sum_{j=1}^{d} (\mathbf{Z}^*\mathbf{Z})_{jj}.$$

As $\mathbf{Z}^*\mathbf{Z}=\mathbf{1}$ when $\mathbf{\Gamma}$ is potentially real and $\mathbf{Z}^*\mathbf{Z}=-\mathbf{1}$ when $\mathbf{\Gamma}$ is pseudo-real, the quoted result follows immediately.

Theorem IV If $\mathbf{\Gamma}$ is a representation that is real and reducible, then any pseudo-real irreducible representation that appears in its reduction can only occur an even number of times. Moreover, if an essentially complex irreducible representation $\mathbf{\Gamma}^p$ is contained in the reduction of $\mathbf{\Gamma}$, then its complex conjugate representation $\mathbf{\Gamma}^{p*}$ occurs the same number of times as $\mathbf{\Gamma}^p$.

Proof Suppose that $\mathbf{\Gamma}'$ is the completely reduced form of $\mathbf{\Gamma}$, with $\mathbf{\Gamma}'(T)=\mathbf{S}^{-1}\mathbf{\Gamma}(T)\mathbf{S}$ for all $T\in\mathcal{G}$ and $\mathbf{\Gamma}'=\sum_p\oplus n_p\mathbf{\Gamma}^p$, n_p being the number of times that the irreducible representation $\mathbf{\Gamma}^p$ appears in $\mathbf{\Gamma}'$. Then as $\mathbf{\Gamma}$ is real

$$\mathbf{\Gamma}'(T)^*=\mathbf{U}^{-1}\mathbf{\Gamma}'(T)\mathbf{U}, \tag{C.40}$$

where $\mathbf{U}=\mathbf{S}^{-1}\mathbf{S}^*$, so that

$$\mathbf{U}^*\mathbf{U}=\mathbf{1}. \tag{C.41}$$

Equation (C.40) has the form

$$\mathbf{U}\begin{bmatrix}\mathbf{\Gamma}^{1*}(T) & \mathbf{0} & \cdots & \mathbf{0} & \cdots\\ \mathbf{0} & \mathbf{\Gamma}^{1*}(T) & \cdots & \mathbf{0} & \cdots\\ \vdots & \vdots & \ddots & \vdots & \\ \mathbf{0} & \mathbf{0} & \cdots & \mathbf{\Gamma}^{2*}(T) & \cdots\\ \vdots & \vdots & & & \ddots\end{bmatrix}=\begin{bmatrix}\mathbf{\Gamma}^{1}(T) & \mathbf{0} & \cdots & \mathbf{0} & \cdots\\ \mathbf{0} & \mathbf{\Gamma}^{1}(T) & \cdots & \mathbf{0} & \cdots\\ \vdots & \vdots & \ddots & \vdots & \\ \mathbf{0} & \mathbf{0} & \cdots & \mathbf{\Gamma}^{2}(T) & \cdots\\ \vdots & \vdots & & & \ddots\end{bmatrix}\mathbf{U} \tag{C.42}$$

for all $T\in\mathcal{G}$. Now suppose that $\mathbf{\Gamma}^q$ is a pseudo-real unitary irreducible representation of dimension d_q appearing in $\mathbf{\Gamma}'$. As $\mathbf{\Gamma}^{q*}$ is not equivalent to $\mathbf{\Gamma}^p$ for any $p\neq q$, Schur's Lemma applied to Equation (C.42) shows that $\mathbf{U}$ has the form

$$\mathbf{U}=\begin{bmatrix}\cdots & \mathbf{0} & \cdots\\ \mathbf{0} & \mathbf{U}_q & \mathbf{0}\\ \cdots & \mathbf{0} & \cdots\end{bmatrix}$$

where $\mathbf{U}_q$ is an $n_qd_q\times n_qd_q$ matrix corresponding to $\mathbf{\Gamma}^q$ and the submatrices not indicated explicitly may be non-zero. From Equation (C.41)

$$\mathbf{U}_q^*\mathbf{U}_q=\mathbf{1}_{n_qd_q}. \tag{C.43}$$

Moreover,

$$\mathbf{U}_q\begin{bmatrix}\mathbf{\Gamma}^{q*}(T) & \mathbf{0} & \cdots\\ \mathbf{0} & \mathbf{\Gamma}^{q*}(T) & \cdots\\ \vdots & \vdots & \end{bmatrix}=\begin{bmatrix}\mathbf{\Gamma}^{q}(T) & \mathbf{0} & \cdots\\ \mathbf{0} & \mathbf{\Gamma}^{q}(T) & \cdots\\ \vdots & \vdots & \end{bmatrix}\mathbf{U}_q \tag{C.44}$$

for all $T \in \mathcal{G}$. Let $\mathbf{Z}_q$ be a $d_q \times d_q$ non-singular matrix such that $\Gamma^{q*}(T) = \mathbf{Z}_q^{-1}\Gamma^q(T)\mathbf{Z}_q$ for all $T \in \mathcal{G}$. Then Equation (C.44) becomes

$$\mathbf{U}_q \begin{bmatrix} \mathbf{Z}_q^{-1} & \mathbf{0} & \cdot\cdot \\ \mathbf{0} & \mathbf{Z}_q^{-1} & \cdot\cdot \\ \vdots & \vdots & \end{bmatrix} \begin{bmatrix} \Gamma^q(T) & \mathbf{0} & \cdot\cdot \\ \mathbf{0} & \Gamma^q(T) & \cdot\cdot \\ \vdots & \vdots & \end{bmatrix} = \begin{bmatrix} \Gamma^q(T) & \mathbf{0} & \cdot\cdot \\ \mathbf{0} & \Gamma^q(T) & \cdot\cdot \\ \vdots & \vdots & \end{bmatrix} \begin{bmatrix} \mathbf{Z}_q^{-1} & \mathbf{0} & \cdot\cdot \\ \mathbf{0} & \mathbf{Z}_q^{-1} & \cdot\cdot \\ \vdots & \vdots & \end{bmatrix} \mathbf{U}_q,$$

and a further application of Schur's Lemma gives

$$\mathbf{U}_q \begin{bmatrix} \mathbf{Z}_q^{-1} & \mathbf{0} & \cdot\cdot \\ \mathbf{0} & \mathbf{Z}_q^{-1} & \cdot\cdot \\ \vdots & \vdots & \end{bmatrix} = \begin{bmatrix} \tau_{11}\mathbf{1}_{n_q} & \tau_{12}\mathbf{1}_{n_q} & \cdot\cdot \\ \tau_{21}\mathbf{1}_{n_q} & \tau_{22}\mathbf{1}_{n_q} & \cdot\cdot \\ \vdots & \vdots & \end{bmatrix},$$

so that

$$\mathbf{U}_q = \begin{bmatrix} \tau_{11}\mathbf{Z}_q & \tau_{12}\mathbf{Z}_q & \cdot\cdot \\ \tau_{12}\mathbf{Z}_q & \tau_{22}\mathbf{Z}_q & \cdot\cdot \\ \vdots & \vdots & \end{bmatrix} = \boldsymbol{\tau} \otimes \mathbf{Z}_q,$$

where $\boldsymbol{\tau}$ is an $n_q \times n_q$ non-singular matrix. Thus, by Equation (A.9), $\mathbf{U}_q^*\mathbf{U}_q = (\boldsymbol{\tau}^*\boldsymbol{\tau}) \otimes \mathbf{Z}_q^*\mathbf{Z}_q$. But $\mathbf{Z}_q^*\mathbf{Z}_q = c_z \mathbf{1}_{d_q}$ with $c_z < 0$. Thus, from Equation (C.43), $\boldsymbol{\tau}^*\boldsymbol{\tau} = (1/c_z)\mathbf{1}_{n_q}$ and hence $|\det \boldsymbol{\tau}|^2 = (1/c_z)^{n_q}$. As $|\det \boldsymbol{\tau}|^2 > 0$ and $c_z < 0$, n_q must be even.

Now suppose that Γ^q is an essentially complex irreducible representation. The number of times n_q that Γ^q and the number of times n_{q^*} that Γ^{q^*} appear in Γ are given by

$$\left. \begin{aligned} n_q &= (1/g) \sum_{T \in \mathcal{G}} \chi(T)\chi^q(T)^* \\ n_{q^*} &= (1/g) \sum_{T \in \mathcal{G}} \chi(T)\chi^{q^*}(T)^* \end{aligned} \right\},$$

χ, χ^q and χ^{q^*} denoting the characters of Γ, Γ^q and Γ^{q^*} respectively, it being assumed that $\mathcal{G}$ is a finite group of order g. As n_q, n_{q^*} and $\chi(T)$ are real, and as $\chi^{q^*}(T) = \chi^q(T)^*$ for all $T \in \mathcal{G}$, it follows immediately on taking the complex conjugate of the first equation that $n_q = n_{q^*}$. This proof generalizes in the obvious way to compact Lie groups.

Appendix D

Character Tables of Point Groups

1 The single crystallographic point groups

The 32 crystallographic point groups will be listed in roughly decreasing order of complexity. For each group the following details are given:

(*a*) ***The group elements.*** The notation for rotations is as in Chapter 1, Section 2(*a*), C_{nj} denoting a proper rotation through $2\pi/n$ in the right-hand screw sense about the axis *Oj* and I denoting the spatial inversion operator. All the axes involved are indicated in Figures D.1

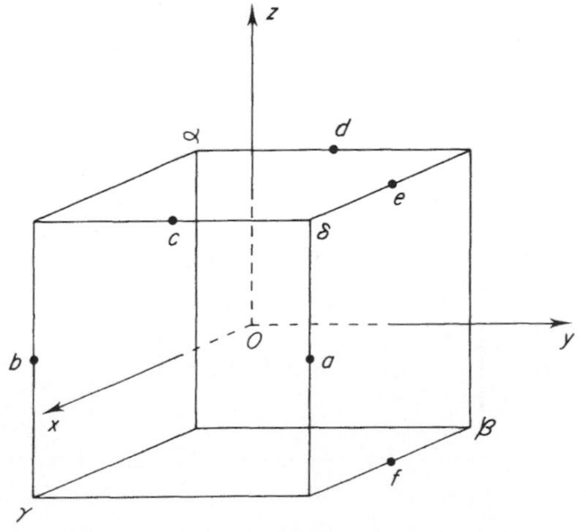

Figure D.1 The axes *Oa*, *Ob*, *Oc*, *Od*, *Oe*, *Of*, *Oα*, *Oβ*, *Oγ* and *Oδ*.

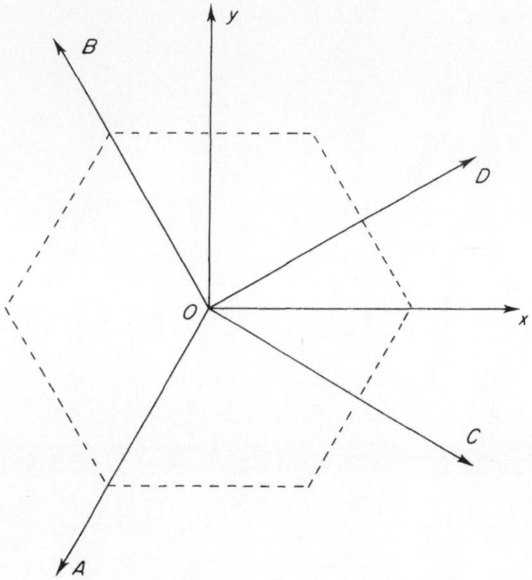

Figure D.2 The axes *OA*, *OB*, *OC* and *OD*. (All these axes lie in the plane *Oxy*.)

and D.2. The matrices **R**(T) for every relevant proper rotation are specified in Table D.1. The rotations are listed in classes.

(*b*) *The character table*. Several alternative systems of labelling are given, the first column merely giving an arbitrary listing. In the labelling of the second column one-dimensional representations are

$$\mathbf{R}(\mathrm{E}) = \begin{bmatrix} 1 & 0 & 0 \\ 0 & 1 & 0 \\ 0 & 0 & 1 \end{bmatrix}, \qquad \mathbf{R}(\mathrm{C}_{3\alpha}) = \begin{bmatrix} 0 & 1 & 0 \\ 0 & 0 & -1 \\ -1 & 0 & 0 \end{bmatrix},$$

$$\mathbf{R}(\mathrm{C}_{3\beta}) = \begin{bmatrix} 0 & -1 & 0 \\ 0 & 0 & -1 \\ 1 & 0 & 0 \end{bmatrix}, \qquad \mathbf{R}(\mathrm{C}_{3\gamma}) = \begin{bmatrix} 0 & -1 & 0 \\ 0 & 0 & 1 \\ -1 & 0 & 0 \end{bmatrix},$$

$$\mathbf{R}(\mathrm{C}_{3\delta}) = \begin{bmatrix} 0 & 1 & 0 \\ 0 & 0 & 1 \\ 1 & 0 & 0 \end{bmatrix}, \qquad \mathbf{R}(\mathrm{C}_{3\alpha}^{-1}) = \begin{bmatrix} 0 & 0 & -1 \\ 1 & 0 & 0 \\ 0 & -1 & 0 \end{bmatrix},$$

$$\mathbf{R}(\mathrm{C}_{3\beta}^{-1}) = \begin{bmatrix} 0 & 0 & 1 \\ -1 & 0 & 0 \\ 0 & -1 & 0 \end{bmatrix}, \qquad \mathbf{R}(\mathrm{C}_{3\gamma}^{-1}) = \begin{bmatrix} 0 & 0 & -1 \\ -1 & 0 & 0 \\ 0 & 1 & 0 \end{bmatrix},$$

Table D.1 The matrices **R**(T) for the proper rotations T appearing in various crystallographic point groups.

$$\mathbf{R}(C_{3\delta}^{-1})=\begin{bmatrix}0&0&1\\1&0&0\\0&1&0\end{bmatrix},\qquad \mathbf{R}(C_{2x})=\begin{bmatrix}1&0&0\\0&-1&0\\0&0&-1\end{bmatrix},$$

$$\mathbf{R}(C_{2y})=\begin{bmatrix}-1&0&0\\0&1&0\\0&0&-1\end{bmatrix},\qquad \mathbf{R}(C_{2z})=\begin{bmatrix}-1&0&0\\0&-1&0\\0&0&1\end{bmatrix},$$

$$\mathbf{R}(C_{4x})=\begin{bmatrix}1&0&0\\0&0&1\\0&-1&0\end{bmatrix},\qquad \mathbf{R}(C_{4y})=\begin{bmatrix}0&0&-1\\0&1&0\\1&0&0\end{bmatrix},$$

$$\mathbf{R}(C_{4z})=\begin{bmatrix}0&1&0\\-1&0&0\\0&0&1\end{bmatrix},\qquad \mathbf{R}(C_{4x}^{-1})=\begin{bmatrix}1&0&0\\0&0&-1\\0&1&0\end{bmatrix},$$

$$\mathbf{R}(C_{4y}^{-1})=\begin{bmatrix}0&0&1\\0&1&0\\-1&0&0\end{bmatrix},\qquad \mathbf{R}(C_{4z}^{-1})=\begin{bmatrix}0&-1&0\\1&0&0\\0&0&1\end{bmatrix},$$

$$\mathbf{R}(C_{2a})=\begin{bmatrix}0&1&0\\1&0&0\\0&0&-1\end{bmatrix},\qquad \mathbf{R}(C_{2b})=\begin{bmatrix}0&-1&0\\-1&0&0\\0&0&-1\end{bmatrix},$$

$$\mathbf{R}(C_{2c})=\begin{bmatrix}0&0&1\\0&-1&0\\1&0&0\end{bmatrix},\qquad \mathbf{R}(C_{2d})=\begin{bmatrix}0&0&-1\\0&-1&0\\-1&0&0\end{bmatrix},$$

$$\mathbf{R}(C_{2e})=\begin{bmatrix}-1&0&0\\0&0&1\\0&1&0\end{bmatrix},\qquad \mathbf{R}(C_{2f})=\begin{bmatrix}-1&0&0\\0&0&-1\\0&-1&0\end{bmatrix},$$

$$\mathbf{R}(C_{3z})=\begin{bmatrix}-\frac{1}{2}&\frac{1}{2}\sqrt{3}&0\\-\frac{1}{2}\sqrt{3}&-\frac{1}{2}&0\\0&0&1\end{bmatrix},\qquad \mathbf{R}(C_{3z}^{-1})=\begin{bmatrix}-\frac{1}{2}&-\frac{1}{2}\sqrt{3}&0\\\frac{1}{2}\sqrt{3}&-\frac{1}{2}&0\\0&0&1\end{bmatrix},$$

$$\mathbf{R}(C_{6z})=\begin{bmatrix}\frac{1}{2}&\frac{1}{2}\sqrt{3}&0\\-\frac{1}{2}\sqrt{3}&\frac{1}{2}&0\\0&0&1\end{bmatrix},\qquad \mathbf{R}(C_{6z}^{-1})=\begin{bmatrix}\frac{1}{2}&-\frac{1}{2}\sqrt{3}&0\\\frac{1}{2}\sqrt{3}&\frac{1}{2}&0\\0&0&1\end{bmatrix},$$

$$\mathbf{R}(C_{2A})=\begin{bmatrix}-\frac{1}{2}&\frac{1}{2}\sqrt{3}&0\\\frac{1}{2}\sqrt{3}&\frac{1}{2}&0\\0&0&-1\end{bmatrix},\qquad \mathbf{R}(C_{2B})=\begin{bmatrix}-\frac{1}{2}&-\frac{1}{2}\sqrt{3}&0\\-\frac{1}{2}\sqrt{3}&\frac{1}{2}&0\\0&0&-1\end{bmatrix},$$

$$\mathbf{R}(C_{2C})=\begin{bmatrix}\frac{1}{2}&-\frac{1}{2}\sqrt{3}&0\\-\frac{1}{2}\sqrt{3}&-\frac{1}{2}&0\\0&0&-1\end{bmatrix},\qquad \mathbf{R}(C_{2D})=\begin{bmatrix}\frac{1}{2}&\frac{1}{2}\sqrt{3}&0\\\frac{1}{2}\sqrt{3}&-\frac{1}{2}&0\\0&0&-1\end{bmatrix}.$$

Table D.1 (*continued*)

denoted by A or B, two-dimensional irreducible representations by E and three-dimensional irreducible representations by T, in all cases with subscripts and/or superscripts attached. (The subscripts g and *u* (standing for *gerade* and *ungerade*) indicated representations that are even and odd under I respectively.) However, pairs of one-dimensional complex conjugate representations are bracketed together and labelled as a two-dimensional representation, as they correspond to degenerate eigenvalues (see Chapter 6, Section 5(*a*) and Chapter 7, Section 3(*f*)).

For a point group that is isomorphic to a group $\mathscr{G}_0(\mathbf{k})$ (see Chapter 9, Section 2(*a*)) for O_h^1, O_h^5 and O_h^9, the third column gives the labelling convention of Bouckaert *et al.* (1936). In such cases the corresponding **k**-vector is as defined in Tables 9.1, 9.2 or 9.3. As described in Chapter 9, Section 2(*b*), it is possible for two or more **k**-vectors in different stars to have point groups $\mathscr{G}_0(\mathbf{k})$ that are isomorphic. The group elements for each such $\mathscr{G}_0(\mathbf{k})$ belonging to O_h^1, O_h^5 and O_h^9 are specified when this occurs.

(*c*) ***Matrices for the irreducible representations of dimension greater than one.*** (Of course these are only unique up to a similarity transformation.) For one-dimensional representations the characters themselves are the matrix elements.

The notation employed for the point groups is that of Schönfliess (1923). More information on these groups may be found in the book of Koster *et al.* (1964), which is wholly devoted to this subject, and in the articles of Altmann (1962, 1963).

(1) O_h:

(*a*) Classes [for $\mathscr{G}_0(\mathbf{k})$ of Γ, H and R]:

$$\mathscr{C}_1 = E;\ \mathscr{C}_2 = C_{3\alpha}, C_{3\beta}, C_{3\gamma}, C_{3\delta}, C_{3\alpha}^{-1}, C_{3\beta}^{-1}, C_{3\gamma}^{-1}, C_{3\delta}^{-1};$$

$$\mathscr{C}_3 = C_{2x}, C_{2y}, C_{2z};\ \mathscr{C}_4 = C_{4x}, C_{4y}, C_{4z}, C_{4x}^{-1}, C_{4y}^{-1}, C_{4z}^{-1};$$

$$\mathscr{C}_5 = C_{2a}, C_{2b}, C_{2c}, C_{2d}, C_{2e}, C_{2f};\ \mathscr{C}_6 = I;$$

$$\mathscr{C}_7 = IC_{3\alpha}, IC_{3\beta}; IC_{3\gamma}, IC_{3\delta}, IC_{3\alpha}^{-1}, IC_{3\beta}^{-1}, IC_{3\gamma}^{-1}, IC_{3\delta}^{-1};$$

$$\mathscr{C}_8 = IC_{2x}, IC_{2y}, IC_{2z};\ \mathscr{C}_9 = IC_{4x}, IC_{4y}, IC_{4z}, IC_{4x}^{-1}, IC_{4y}^{-1}, IC_{4z}^{-1};$$

$$\mathscr{C}_{10} = IC_{2a}, IC_{2b}, IC_{2c}, IC_{2d}, IC_{2e}, IC_{2f}.$$

(*b*) The character table is given in Table D.2.

(*c*) Matrices for irreducible representations of dimension greater

			$\mathscr{C}_1$	$\mathscr{C}_2$	$\mathscr{C}_3$	$\mathscr{C}_4$	$\mathscr{C}_5$	$\mathscr{C}_6$	$\mathscr{C}_7$	$\mathscr{C}_8$	$\mathscr{C}_9$	$\mathscr{C}_{10}$
Γ^1	A_{1g}	$\Gamma_1 H_1 R_1$	1	1	1	1	1	1	1	1	1	1
Γ^2	A_{2g}	$\Gamma_2 H_2 R_2$	1	1	1	−1	−1	1	1	1	−1	−1
Γ^3	E_g	$\Gamma_{12} H_{12} R_{12}$	2	−1	2	0	0	2	−1	2	0	0
Γ^4	T_{1g}	$\Gamma_{15'} H_{15'} R_{15'}$	3	0	−1	1	−1	3	0	−1	1	−1
Γ^5	T_{2g}	$\Gamma_{25'} H_{25'} R_{25'}$	3	0	−1	−1	1	3	0	−1	−1	1
Γ^6	A_{1u}	$\Gamma_{1'} H_{1'} R_{1'}$	1	1	1	1	1	−1	−1	−1	−1	−1
Γ^7	A_{2u}	$\Gamma_{2'} H_{2'} R_{2'}$	1	1	1	−1	−1	−1	−1	−1	1	1
Γ^8	E_u	$\Gamma_{12'} H_{12'} R_{12'}$	2	−1	2	0	0	−2	1	−2	0	0
Γ^9	T_{1u}	$\Gamma_{15} H_{15} R_{15}$	3	0	−1	1	−1	−3	0	1	−1	1
Γ^{10}	T_{2u}	$\Gamma_{25} H_{25} R_{25}$	3	0	−1	−1	1	−3	0	1	1	−1

Table D.2 Character table for O_h.

than one, for any proper rotation C_{nj} of O_h:

$$\mathbf{\Gamma}^3(C_{nj}) = \mathbf{\Gamma}^8(C_{nj}) = \mathbf{\Gamma}''(C_{nj});\ \mathbf{\Gamma}^3(IC_{nj}) = -\mathbf{\Gamma}^8(IC_{nj}) = \mathbf{\Gamma}''(C_{nj});$$

$$\mathbf{\Gamma}^4(C_{nj}) = \mathbf{\Gamma}^9(C_{nj}) = \mathbf{R}(C_{nj});$$

$$\mathbf{\Gamma}^4(IC_{nj}) = -\mathbf{\Gamma}^9(IC_{nj}) = \mathbf{R}(C_{nj});\ \mathbf{\Gamma}^5(C_{nj}) = \mathbf{\Gamma}^{10}(C_{nj}) = \mathbf{\Gamma}'(C_{nj});$$

$$\mathbf{\Gamma}^5(IC_{nj}) = -\mathbf{\Gamma}^{10}(IC_{nj}) = \mathbf{\Gamma}'(C_{nj});$$

where the matrices $\mathbf{\Gamma}'(C_{nj})$, $\mathbf{\Gamma}''(C_{nj})$ and $\mathbf{R}(C_{nj})$ are given in Tables D.3, D.4 and D.1 respectively.

(2) D_{6h}:

(*a*) Classes:

$$\mathscr{C}_1 = E;\ \mathscr{C}_2 = C_{6z}, C_{6z}^{-1};\ \mathscr{C}_3 = C_{3z}, C_{3z}^{-1};\ \mathscr{C}_4 = C_{2z};\ \mathscr{C}_5 = C_{2x}, C_{2A}, C_{2B};$$

$$\mathscr{C}_6 = C_{2y}, C_{2C}, C_{2D};\ \mathscr{C}_7 = I;\ \mathscr{C}_8 = IC_{6z}, IC_{6z}^{-1};\ \mathscr{C}_9 = IC_{3z}, IC_{3z}^{-1};\ \mathscr{C}_{10} = IC_{2z};$$

$$\mathscr{C}_{11} = IC_{2x}, IC_{2A}, IC_{2B};\ \mathscr{C}_{12} = IC_{2y}, IC_{2C}, IC_{2D}.$$

(*b*) The character table is given in Table D.5.

(*c*) Matrices for irreducible representations of dimension greater than one, for any proper rotation C_{nj} of D_{6h}:

$$\mathbf{\Gamma}^5(C_{nj}) = \mathbf{\Gamma}^{11}(C_{nj}) = \mathbf{D}'(C_{nj});\ \mathbf{\Gamma}^5(IC_{nj}) = -\mathbf{\Gamma}^{11}(IC_{nj}) = \mathbf{D}'(C_{nj});$$

$$\mathbf{\Gamma}^6(C_{nj}) = \mathbf{\Gamma}^{12}(C_{nj}) = \mathbf{D}''(C_{nj});\ \mathbf{\Gamma}^6(IC_{nj}) = -\mathbf{\Gamma}^{12}(IC_{nj}) = \mathbf{D}''(C_{nj});$$

where the matrices $\mathbf{D}'(C_{nj})$ and $\mathbf{D}''(C_{nj})$ are given in Tables D.6 and D.7 respectively.

$$\mathbf{\Gamma}'(\mathrm{E})=\begin{bmatrix}1&0&0\\0&1&0\\0&0&1\end{bmatrix},\qquad \mathbf{\Gamma}'(\mathrm{C}_{3\alpha})=\begin{bmatrix}0&-1&0\\0&0&1\\-1&0&0\end{bmatrix},$$

$$\mathbf{\Gamma}'(\mathrm{C}_{3\beta})=\begin{bmatrix}0&1&0\\0&0&-1\\-1&0&0\end{bmatrix},\qquad \mathbf{\Gamma}'(\mathrm{C}_{3\gamma})=\begin{bmatrix}0&-1&0\\0&0&-1\\1&0&0\end{bmatrix},$$

$$\mathbf{\Gamma}'(\mathrm{C}_{3\delta})=\begin{bmatrix}0&1&0\\0&0&1\\1&0&0\end{bmatrix},\qquad \mathbf{\Gamma}'(\mathrm{C}_{3\alpha}^{-1})=\begin{bmatrix}0&0&-1\\-1&0&0\\0&1&0\end{bmatrix},$$

$$\mathbf{\Gamma}'(\mathrm{C}_{3\beta}^{-1})=\begin{bmatrix}0&0&-1\\1&0&0\\0&-1&0\end{bmatrix},\qquad \mathbf{\Gamma}'(\mathrm{C}_{3\gamma}^{-1})=\begin{bmatrix}0&0&1\\-1&0&0\\0&-1&0\end{bmatrix},$$

$$\mathbf{\Gamma}'(\mathrm{C}_{3\delta}^{-1})=\begin{bmatrix}0&0&1\\1&0&0\\0&1&0\end{bmatrix},\qquad \mathbf{\Gamma}'(\mathrm{C}_{2x})=\begin{bmatrix}-1&0&0\\0&1&0\\0&0&-1\end{bmatrix},$$

$$\mathbf{\Gamma}'(\mathrm{C}_{2y})=\begin{bmatrix}-1&0&0\\0&-1&0\\0&0&1\end{bmatrix},\qquad \mathbf{\Gamma}'(\mathrm{C}_{2z})=\begin{bmatrix}1&0&0\\0&-1&0\\0&0&-1\end{bmatrix},$$

$$\mathbf{\Gamma}'(\mathrm{C}_{4x})=\begin{bmatrix}0&0&1\\0&-1&0\\-1&0&0\end{bmatrix},\qquad \mathbf{\Gamma}'(\mathrm{C}_{4y})=\begin{bmatrix}0&-1&0\\1&0&0\\0&0&-1\end{bmatrix},$$

$$\mathbf{\Gamma}'(\mathrm{C}_{4z})=\begin{bmatrix}-1&0&0\\0&0&-1\\0&1&0\end{bmatrix},\qquad \mathbf{\Gamma}'(\mathrm{C}_{4x}^{-1})=\begin{bmatrix}0&0&-1\\0&-1&0\\1&0&0\end{bmatrix},$$

$$\mathbf{\Gamma}'(\mathrm{C}_{4y}^{-1})=\begin{bmatrix}0&1&0\\-1&0&0\\0&0&-1\end{bmatrix},\qquad \mathbf{\Gamma}'(\mathrm{C}_{4z}^{-1})=\begin{bmatrix}-1&0&0\\0&0&1\\0&-1&0\end{bmatrix},$$

$$\mathbf{\Gamma}'(\mathrm{C}_{2a})=\begin{bmatrix}1&0&0\\0&0&-1\\0&-1&0\end{bmatrix},\qquad \mathbf{\Gamma}'(\mathrm{C}_{2b})=\begin{bmatrix}1&0&0\\0&0&1\\0&1&0\end{bmatrix},$$

$$\mathbf{\Gamma}'(\mathrm{C}_{2c})=\begin{bmatrix}0&-1&0\\-1&0&0\\0&0&1\end{bmatrix},\qquad \mathbf{\Gamma}'(\mathrm{C}_{2d})=\begin{bmatrix}0&1&0\\1&0&0\\0&0&1\end{bmatrix},$$

$$\mathbf{\Gamma}'(\mathrm{C}_{2e})=\begin{bmatrix}0&0&-1\\0&1&0\\-1&0&0\end{bmatrix},\qquad \mathbf{\Gamma}'(\mathrm{C}_{2f})=\begin{bmatrix}0&0&1\\0&1&0\\1&0&0\end{bmatrix}.$$

Table D.3 The matrices $\mathbf{\Gamma}'$ for O_h, T_d, O, T_h and T.

$$\mathbf{\Gamma}''(\mathrm{E}) = \mathbf{\Gamma}''(\mathrm{C}_{2x}) = \mathbf{\Gamma}''(\mathrm{C}_{2y}) = \mathbf{\Gamma}''(\mathrm{C}_{2z}) = \begin{bmatrix} 1 & 0 \\ 0 & 1 \end{bmatrix},$$

$$\mathbf{\Gamma}''(\mathrm{C}_{3\alpha}) = \mathbf{\Gamma}''(\mathrm{C}_{3\beta}) = \mathbf{\Gamma}''(\mathrm{C}_{3\gamma}) = \mathbf{\Gamma}''(\mathrm{C}_{3\delta}) = \begin{bmatrix} -\frac{1}{2} & -\frac{1}{2}\sqrt{3} \\ \frac{1}{2}\sqrt{3} & -\frac{1}{2} \end{bmatrix},$$

$$\mathbf{\Gamma}''(\mathrm{C}_{3\alpha}^{-1}) = \mathbf{\Gamma}''(\mathrm{C}_{3\beta}^{-1}) = \mathbf{\Gamma}''(\mathrm{C}_{3\gamma}^{-1}) = \mathbf{\Gamma}''(\mathrm{C}_{3\delta}^{-1}) = \begin{bmatrix} -\frac{1}{2} & \frac{1}{2}\sqrt{3} \\ -\frac{1}{2}\sqrt{3} & -\frac{1}{2} \end{bmatrix},$$

$$\mathbf{\Gamma}''(\mathrm{C}_{4x}) = \mathbf{\Gamma}''(\mathrm{C}_{4x}^{-1}) = \mathbf{\Gamma}''(\mathrm{C}_{2e}) = \mathbf{\Gamma}''(\mathrm{C}_{2f}) = \begin{bmatrix} \frac{1}{2} & -\frac{1}{2}\sqrt{3} \\ -\frac{1}{2}\sqrt{3} & -\frac{1}{2} \end{bmatrix},$$

$$\mathbf{\Gamma}''(\mathrm{C}_{4y}) = \mathbf{\Gamma}''(\mathrm{C}_{4y}^{-1}) = \mathbf{\Gamma}''(\mathrm{C}_{2c}) = \mathbf{\Gamma}''(\mathrm{C}_{2d}) = \begin{bmatrix} \frac{1}{2} & \frac{1}{2}\sqrt{3} \\ \frac{1}{2}\sqrt{3} & -\frac{1}{2} \end{bmatrix},$$

$$\mathbf{\Gamma}''(\mathrm{C}_{4z}) = \mathbf{\Gamma}''(\mathrm{C}_{4z}^{-1}) = \mathbf{\Gamma}''(\mathrm{C}_{2a}) = \mathbf{\Gamma}''(\mathrm{C}_{2b}) = \begin{bmatrix} -1 & 0 \\ 0 & 1 \end{bmatrix}.$$

Table D.4 The matrices $\mathbf{\Gamma}''$ for O_h, T_d, O, D_{3d} and C_{3v}.

		$\mathscr{C}_1$	$\mathscr{C}_2$	$\mathscr{C}_3$	$\mathscr{C}_4$	$\mathscr{C}_5$	$\mathscr{C}_6$	$\mathscr{C}_7$	$\mathscr{C}_8$	$\mathscr{C}_9$	$\mathscr{C}_{10}$	$\mathscr{C}_{11}$	$\mathscr{C}_{12}$
$\mathbf{\Gamma}^1$	A_{1g}	1	1	1	1	1	1	1	1	1	1	1	1
$\mathbf{\Gamma}^2$	A_{2g}	1	1	1	1	−1	−1	1	1	1	1	−1	−1
$\mathbf{\Gamma}^3$	B_{1g}	1	−1	1	−1	1	−1	1	−1	1	−1	1	−1
$\mathbf{\Gamma}^4$	B_{2g}	1	−1	1	−1	−1	1	1	−1	1	−1	−1	1
$\mathbf{\Gamma}^5$	E_{2g}	2	−1	−1	2	0	0	2	−1	−1	2	0	0
$\mathbf{\Gamma}^6$	E_{1g}	2	1	−1	−2	0	0	2	1	−1	−2	0	0
$\mathbf{\Gamma}^7$	A_{1u}	1	1	1	1	1	1	−1	−1	−1	−1	−1	−1
$\mathbf{\Gamma}^8$	A_{2u}	1	1	1	1	−1	−1	−1	−1	−1	−1	1	1
$\mathbf{\Gamma}^9$	B_{1u}	1	−1	1	−1	1	−1	−1	1	−1	1	−1	1
$\mathbf{\Gamma}^{10}$	B_{2u}	1	−1	1	−1	−1	1	−1	1	−1	1	1	−1
$\mathbf{\Gamma}^{11}$	E_{2u}	2	−1	−1	2	0	0	−2	1	1	−2	0	0
$\mathbf{\Gamma}^{12}$	E_{1u}	2	1	−1	−2	0	0	−2	−1	1	2	0	0

Table D.5 Character table for D_{6h}.

$$\mathbf{D}'(\mathrm{E}) = \mathbf{D}'(\mathrm{C}_{2z}) = \begin{bmatrix} 1 & 0 \\ 0 & 1 \end{bmatrix}, \qquad \mathbf{D}'(\mathrm{C}_{6z}) = \mathbf{D}'(\mathrm{C}_{3z}^{-1}) = \begin{bmatrix} -\frac{1}{2} & -\frac{1}{2}\sqrt{3} \\ \frac{1}{2}\sqrt{3} & -\frac{1}{2} \end{bmatrix},$$

$$\mathbf{D}'(\mathrm{C}_{6z}^{-1}) = \mathbf{D}'(\mathrm{C}_{3z}) = \begin{bmatrix} -\frac{1}{2} & \frac{1}{2}\sqrt{3} \\ -\frac{1}{2}\sqrt{3} & -\frac{1}{2} \end{bmatrix}, \qquad \mathbf{D}'(\mathrm{C}_{2x}) = \mathbf{D}'(\mathrm{C}_{2y}) = \begin{bmatrix} -1 & 0 \\ 0 & 1 \end{bmatrix},$$

$$\mathbf{D}'(\mathrm{C}_{2A}) = \mathbf{D}'(\mathrm{C}_{2C}) = \begin{bmatrix} \frac{1}{2} & -\frac{1}{2}\sqrt{3} \\ -\frac{1}{2}\sqrt{3} & -\frac{1}{2} \end{bmatrix}, \qquad \mathbf{D}'(\mathrm{C}_{2B}) = \mathbf{D}'(\mathrm{C}_{2D}) = \begin{bmatrix} \frac{1}{2} & \frac{1}{2}\sqrt{3} \\ \frac{1}{2}\sqrt{3} & -\frac{1}{2} \end{bmatrix}.$$

Table D.6 The matrices $\mathbf{D}'$ for D_{6h}, D_{3h}, C_{6v} and D_6.

$$\mathbf{D}''(\mathrm{E}) = \begin{bmatrix} 1 & 0 \\ 0 & 1 \end{bmatrix}, \qquad \mathbf{D}''(\mathrm{C}_{6z}) = \begin{bmatrix} \frac{1}{2} & \frac{1}{2}\sqrt{3} \\ -\frac{1}{2}\sqrt{3} & \frac{1}{2} \end{bmatrix},$$

$$\mathbf{D}''(\mathrm{C}_{6z}^{-1}) = \begin{bmatrix} \frac{1}{2} & -\frac{1}{2}\sqrt{3} \\ \frac{1}{2}\sqrt{3} & \frac{1}{2} \end{bmatrix}, \qquad \mathbf{D}''(\mathrm{C}_{3z}) = \begin{bmatrix} -\frac{1}{2} & \frac{1}{2}\sqrt{3} \\ -\frac{1}{2}\sqrt{3} & -\frac{1}{2} \end{bmatrix},$$

$$\mathbf{D}''(\mathrm{C}_{3z}^{-1}) = \begin{bmatrix} -\frac{1}{2} & -\frac{1}{2}\sqrt{3} \\ \frac{1}{2}\sqrt{3} & -\frac{1}{2} \end{bmatrix}, \qquad \mathbf{D}''(\mathrm{C}_{2z}) = \begin{bmatrix} -1 & 0 \\ 0 & -1 \end{bmatrix},$$

$$\mathbf{D}''(\mathrm{C}_{2x}) = \begin{bmatrix} 1 & 0 \\ 0 & -1 \end{bmatrix}, \qquad \mathbf{D}''(\mathrm{C}_{2A}) = \begin{bmatrix} -\frac{1}{2} & \frac{1}{2}\sqrt{3} \\ \frac{1}{2}\sqrt{3} & \frac{1}{2} \end{bmatrix},$$

$$\mathbf{D}''(\mathrm{C}_{2B}) = \begin{bmatrix} -\frac{1}{2} & -\frac{1}{2}\sqrt{3} \\ -\frac{1}{2}\sqrt{3} & \frac{1}{2} \end{bmatrix}, \qquad \mathbf{D}''(\mathrm{C}_{2y}) = \begin{bmatrix} -1 & 0 \\ 0 & 1 \end{bmatrix},$$

$$\mathbf{D}''(\mathrm{C}_{2C}) = \begin{bmatrix} \frac{1}{2} & -\frac{1}{2}\sqrt{3} \\ -\frac{1}{2}\sqrt{3} & -\frac{1}{2} \end{bmatrix}, \qquad \mathbf{D}''(\mathrm{C}_{2D}) = \begin{bmatrix} \frac{1}{2} & \frac{1}{2}\sqrt{3} \\ \frac{1}{2}\sqrt{3} & -\frac{1}{2} \end{bmatrix}.$$

Table D.7 The matrices $\mathbf{D}''$ for D_{6h}, C_{6v}, D_6 and D_3.

(3) T_d:

(*a*) Classes [for $\mathcal{G}_0(\mathbf{k})$ of P]:

$$\mathscr{C}_1 = \mathrm{E},\ \mathscr{C}_2 = \mathrm{C}_{3\alpha}, \mathrm{C}_{3\beta}, \mathrm{C}_{3\gamma}, \mathrm{C}_{3\delta}, \mathrm{C}_{3\alpha}^{-1}, \mathrm{C}_{3\beta}^{-1}, \mathrm{C}_{3\gamma}^{-1}, \mathrm{C}_{3\delta}^{-1};$$

$$\mathscr{C}_3 = \mathrm{C}_{2x}, \mathrm{C}_{2y}, \mathrm{C}_{2z};\ \mathscr{C}_4 = \mathrm{IC}_{4x}, \mathrm{IC}_{4y}, \mathrm{IC}_{4z}, \mathrm{IC}_{4x}^{-1}, \mathrm{IC}_{4y}^{-1}, \mathrm{IC}_{4z}^{-1};$$

$$\mathscr{C}_5 = \mathrm{IC}_{2a}, \mathrm{IC}_{2b}, \mathrm{IC}_{2c}, \mathrm{IC}_{2d}, \mathrm{IC}_{2e}, \mathrm{IC}_{2f}.$$

(*b*) The character table is given in Table D.8.

(*c*) Matrices for irreducible representations of dimension greater than one, for proper rotations C_{nj} of T_d:

$$\mathbf{\Gamma}^3(\mathrm{C}_{nj}) = \mathbf{\Gamma}''(\mathrm{C}_{nj});\ \mathbf{\Gamma}^4(\mathrm{C}_{nj}) = \mathbf{R}(\mathrm{C}_{nj});\ \mathbf{\Gamma}^5(\mathrm{C}_{nj}) = \mathbf{\Gamma}'(\mathrm{C}_{nj});$$

and for the improper rotations IC_{nj} of T_d:

$$\mathbf{\Gamma}^3(\mathrm{IC}_{nj}) = -\mathbf{\Gamma}''(\mathrm{C}_{nj});\ \mathbf{\Gamma}^4(\mathrm{IC}_{nj}) = -\mathbf{R}(\mathrm{C}_{nj});\ \mathbf{\Gamma}^5(\mathrm{IC}_{nj}) = -\mathbf{\Gamma}'(\mathrm{C}_{nj});$$

where the matrices $\mathbf{\Gamma}'(C_{nj})$, $\mathbf{\Gamma}''(C_{nj})$ and $\mathbf{R}(C_{nj})$ are given in Tables D.3, D.4 and D.1 respectively.

			$\mathscr{C}_1$	$\mathscr{C}_2$	$\mathscr{C}_3$	$\mathscr{C}_4$	$\mathscr{C}_5$
$\mathbf{\Gamma}^1$	A_1	P_1	1	1	1	1	1
$\mathbf{\Gamma}^2$	A_2	P_2	1	1	1	−1	−1
$\mathbf{\Gamma}^3$	E	P_3	2	−1	2	0	0
$\mathbf{\Gamma}^4$	T_2	P_4	3	0	−1	−1	1
$\mathbf{\Gamma}^5$	T_1	P_5	3	0	−1	1	−1

Table D.8 Character table for T_d and 0.

(4) O:

(*a*) Classes:

$$\mathscr{C}_1 = \mathrm{E};\ \mathscr{C}_2 = \mathrm{C}_{3\alpha}, \mathrm{C}_{3\beta}, \mathrm{C}_{3\gamma}, \mathrm{C}_{3\delta}, \mathrm{C}_{3\alpha}^{-1}, \mathrm{C}_{3\beta}^{-1}, \mathrm{C}_{3\gamma}^{-1}, \mathrm{C}_{3\delta}^{-1};$$

$$\mathscr{C}_3 = \mathrm{C}_{2x}, \mathrm{C}_{2y}, \mathrm{C}_{2z};\ \mathscr{C}_4 = \mathrm{C}_{4x}, \mathrm{C}_{4y}, \mathrm{C}_{4z}, \mathrm{C}_{4x}^{-1}, \mathrm{C}_{4y}^{-1}, \mathrm{C}_{4z}^{-1};$$

$$\mathscr{C}_5 = \mathrm{C}_{2a}, \mathrm{C}_{2b}, \mathrm{C}_{2c}, \mathrm{C}_{2d}, \mathrm{C}_{2e}, \mathrm{C}_{2f}.$$

(*b*) The character table is given in Table D.8.

(*c*) Matrices for irreducible representations of dimension greater than one, for all rotations C_{nj} of O:

$$\mathbf{\Gamma}^3(\mathrm{C}_{nj}) = \mathbf{\Gamma}''(\mathrm{C}_{nj});\ \mathbf{\Gamma}^4(\mathrm{C}_{nj}) = \mathbf{\Gamma}'(\mathrm{C}_{nj});\ \mathbf{\Gamma}^5(\mathrm{C}_{nj}) = \mathbf{R}(\mathrm{C}_{nj});$$

where the matrices $\mathbf{\Gamma}'(\mathrm{C}_{nj})$, $\mathbf{\Gamma}''(\mathrm{C}_{nj})$ and $\mathbf{R}(\mathrm{C}_{nj})$ are given in Tables D.3, D.4 and D.1 respectively.

(5) T_h:

(*a*) Classes:

$$\mathscr{C}_1 = \mathrm{E};\ \mathscr{C}_2 = \mathrm{C}_{3\alpha}, \mathrm{C}_{3\beta}, \mathrm{C}_{3\gamma}, \mathrm{C}_{3\delta};\ \mathscr{C}_3 = \mathrm{C}_{3\alpha}^{-1}, \mathrm{C}_{3\beta}^{-1}, \mathrm{C}_{3\gamma}^{-1}, \mathrm{C}_{3\delta}^{-1};$$

$$\mathscr{C}_4 = \mathrm{C}_{2x}, \mathrm{C}_{2y}, \mathrm{C}_{2z};\ \mathscr{C}_5 = \mathrm{I};\ \mathscr{C}_6 = \mathrm{IC}_{3\alpha}, \mathrm{IC}_{3\beta}, \mathrm{IC}_{3\gamma}, \mathrm{IC}_{3\delta};$$

$$\mathscr{C}_7 = \mathrm{IC}_{3\alpha}^{-1}, \mathrm{IC}_{3\beta}^{-1}, \mathrm{IC}_{3\gamma}^{-1}, \mathrm{IC}_{3\delta}^{-1},\ \mathscr{C}_8 = \mathrm{IC}_{2x}, \mathrm{IC}_{2y}, \mathrm{IC}_{2z}.$$

(*b*) The character table is given in Table D.9.

(*c*) Matrices for irreducible representations of dimension greater than one, for any proper rotation C_{nj} of T_h:

$$\mathbf{\Gamma}^4(\mathrm{C}_{nj}) = \mathbf{\Gamma}^8(\mathrm{C}_{nj}) = \mathbf{\Gamma}'(\mathrm{C}_{nj});\ \mathbf{\Gamma}^4(\mathrm{IC}_{nj}) = -\mathbf{\Gamma}^8(\mathrm{IC}_{nj}) = \mathbf{\Gamma}'(\mathrm{C}_{nj});$$

where the matrices $\mathbf{\Gamma}'(\mathrm{C}_{nj})$ are given in Table D.3.

		$\mathscr{C}_1$	$\mathscr{C}_2$	$\mathscr{C}_3$	$\mathscr{C}_4$	$\mathscr{C}_5$	$\mathscr{C}_6$	$\mathscr{C}_7$	$\mathscr{C}_8$
$\mathbf{\Gamma}^1$	A_g	1	1	1	1	1	1	1	1
$\mathbf{\Gamma}^2$	E_g	1	ϕ	ϕ^2	1	1	ϕ	ϕ^2	1
$\mathbf{\Gamma}^3$		1	ϕ^2	ϕ	1	1	ϕ^2	ϕ	1
$\mathbf{\Gamma}^4$	T_g	3	0	0	−1	3	0	0	−1
$\mathbf{\Gamma}^5$	A_u	1	1	1	1	−1	−1	−1	−1
$\mathbf{\Gamma}^6$	E_u	1	ϕ	ϕ^2	1	−1	$-\phi$	$-\phi^2$	−1
$\mathbf{\Gamma}^7$		1	ϕ^2	ϕ	1	−1	$-\phi^2$	$-\phi$	−1
$\mathbf{\Gamma}^8$	T_u	3	0	0	−1	−3	0	0	1

Table D.9 Character table for T_h ($\phi = \exp(\frac{2}{3}\pi i)$).

			$\mathscr{C}_1$	$\mathscr{C}_2$	$\mathscr{C}_3$	$\mathscr{C}_4$	$\mathscr{C}_5$	$\mathscr{C}_6$	$\mathscr{C}_7$	$\mathscr{C}_8$	$\mathscr{C}_9$	$\mathscr{C}_{10}$
Γ^1	A_{1g}	X_1, M_1	1	1	1	1	1	1	1	1	1	1
Γ^2	B_{1g}	X_2, M_2	1	1	1	−1	−1	1	1	1	−1	−1
Γ^3	B_{2g}	X_3, M_3	1	−1	1	−1	1	1	−1	1	−1	1
Γ^4	A_{2g}	X_4, M_4	1	−1	1	1	−1	1	−1	1	1	−1
Γ^5	E_g	X_5, M_5	2	0	−2	0	0	2	0	−2	0	0
Γ^6	A_{1u}	$X_{1'}, M_{1'}$	1	1	1	1	1	−1	−1	−1	−1	−1
Γ^7	B_{1u}	$X_{2'}, M_{2'}$	1	1	1	−1	−1	−1	−1	−1	1	1
Γ^8	B_{2u}	$X_{3'}, M_{3'}$	1	−1	1	−1	1	−1	1	−1	1	−1
Γ^9	A_{2u}	$X_{4'}, M_{4'}$	1	−1	1	1	−1	−1	1	−1	−1	1
Γ^{10}	E_u	$X_{5'}, M_{5'}$	2	0	−2	0	0	−2	0	2	0	0

Table D.10 Character table for D_{4h}.

(6) D_{4h}:

(*a*) Classes [for $\mathscr{G}_0(\mathbf{k})$ of X]:

$$\mathscr{C}_1 = E;\ \mathscr{C}_2 = C_{2x}, C_{2y};\ \mathscr{C}_3 = C_{2z};\ \mathscr{C}_4 = C_{4z}, C_{4z}^{-1};$$

$$\mathscr{C}_5 = C_{2a}, C_{2b};\ \mathscr{C}_6 = I;\ \mathscr{C}_7 = IC_{2x}, IC_{2y};\ \mathscr{C}_8 = IC_{2z};\ \mathscr{C}_9 = IC_{4z}, IC_{4z}^{-1};$$

$$\mathscr{C}_{10} = IC_{2a}, IC_{2b}.$$

Classes [for $\mathscr{G}_0(\mathbf{k})$ of M]:

$$\mathscr{C}_1 = E,\ \mathscr{C}_2 = C_{2y}, C_{2z};\ \mathscr{C}_3 = C_{2x};\ \mathscr{C}_4 = C_{4x}, C_{4x}^{-1};\ \mathscr{C}_5 = C_{2e}, C_{2f};\ \mathscr{C}_6 = I;$$

$$\mathscr{C}_7 = IC_{2y}, IC_{2z};\ \mathscr{C}_8 = IC_{2x};\ \mathscr{C}_9 = IC_{4x}, IC_{4x}^{-1};\ \mathscr{C}_{10} = IC_{2e}, IC_{2f}.$$

(*b*) The character table is given in Table D.10.

(*c*) Matrices for irreducible representations of dimension greater than one, for any proper rotation C_{nj} of D_{4h}:

$$\mathbf{\Gamma}^5(C_{nj}) = \mathbf{\Gamma}^{10}(C_{nj}) = \mathbf{D}(C_{nj});\ \mathbf{\Gamma}^5(IC_{nj}) = -\mathbf{\Gamma}^{10}(IC_{nj}) = \mathbf{D}(C_{nj});$$

where for $\mathscr{G}_0(\mathbf{k})$ of M the matrices $\mathbf{D}(C_{nj})$ are given in Table D.11, while for $\mathscr{G}_0(\mathbf{k})$ of X they are given in Table D.12.

$$\mathbf{D}(E) = \begin{bmatrix} 1 & 0 \\ 0 & 1 \end{bmatrix},\quad \mathbf{D}(C_{2x}) = \begin{bmatrix} -1 & 0 \\ 0 & -1 \end{bmatrix},\quad \mathbf{D}(C_{2y}) = \begin{bmatrix} 1 & 0 \\ 0 & -1 \end{bmatrix},$$

$$\mathbf{D}(C_{2z}) = \begin{bmatrix} -1 & 0 \\ 0 & 1 \end{bmatrix},\quad \mathbf{D}(C_{4x}) = \begin{bmatrix} 0 & 1 \\ -1 & 0 \end{bmatrix},\quad \mathbf{D}(C_{4x}^{-1}) = \begin{bmatrix} 0 & -1 \\ 1 & 0 \end{bmatrix},$$

$$\mathbf{D}(C_{2e}) = \begin{bmatrix} 0 & 1 \\ 1 & 0 \end{bmatrix},\quad \mathbf{D}(C_{2f}) = \begin{bmatrix} 0 & -1 \\ -1 & 0 \end{bmatrix}.$$

Table D.11 The matrices **D** for D_{4h} for $\mathscr{G}_0(\mathbf{k})$ of M.

$$\mathbf{D}(E)=\begin{bmatrix}1&0\\0&1\end{bmatrix},\quad \mathbf{D}(C_{2x})=\begin{bmatrix}1&0\\0&-1\end{bmatrix},\quad \mathbf{D}(C_{2y})=\begin{bmatrix}-1&0\\0&1\end{bmatrix},$$

$$\mathbf{D}(C_{2z})=\begin{bmatrix}-1&0\\0&-1\end{bmatrix},\quad \mathbf{D}(C_{4z})=\begin{bmatrix}0&1\\-1&0\end{bmatrix},\quad \mathbf{D}(C_{4z}^{-1})=\begin{bmatrix}0&-1\\1&0\end{bmatrix},$$

$$\mathbf{D}(C_{2a})=\begin{bmatrix}0&1\\1&0\end{bmatrix},\quad \mathbf{D}(C_{2b})=\begin{bmatrix}0&-1\\-1&0\end{bmatrix}.$$

Table D.12 The matrices **D** for D_{4h} for $\mathscr{G}_0(\mathbf{k})$ of X.

(7) D_{3h}:

(*a*) Classes

$$\mathscr{C}_1=E;\ \mathscr{C}_2=C_{3z},\ C_{3z}^{-1};\ \mathscr{C}_3=C_{2x},\ C_{2A},\ C_{2B};\ \mathscr{C}_4=IC_{2z};\ \mathscr{C}_5=IC_{6z},\ IC_{6z}^{-1};$$

$$\mathscr{C}_6=IC_{2y},\ IC_{2C},\ IC_{2D}.$$

(*b*) The character table is given in Table D.13.

(*c*) Matrices for irreducible representations of dimension greater than one, for any proper rotation C_{nj} of D_{3h}:

$$\boldsymbol{\Gamma}^3(C_{nj})=\boldsymbol{\Gamma}^6(C_{nj})=\mathbf{D}'(C_{nj});$$

and for any improper rotation IC_{nj} of D_{3h}:

$$\boldsymbol{\Gamma}^3(IC_{nj})=-\boldsymbol{\Gamma}^6(IC_{nj})=\mathbf{D}'(C_{nj});$$

where the matrices $\mathbf{D}'(C_{nj})$ are given in Table D.6.

(8) D_{3d}:

(*a*) Classes [for $\mathscr{G}_0(\mathbf{k})$ of L]:

$$\mathscr{C}_1=E;\ \mathscr{C}_2=C_{3\delta},\ C_{3\delta}^{-1};\ \mathscr{C}_3=C_{2b},\ C_{2d},\ C_{2f};\ \mathscr{C}_4=I;\ \mathscr{C}_5=IC_{3\delta},\ IC_{3\delta}^{-1};$$

$$\mathscr{C}_6=IC_{2b},\ IC_{2d},\ IC_{2f}.$$

(*b*) The character table is given in Table D.13.

	D_{3h}	D_{3d}	C_{6v}	D_6		$\mathscr{C}_1$	$\mathscr{C}_2$	$\mathscr{C}_3$	$\mathscr{C}_4$	$\mathscr{C}_5$	$\mathscr{C}_6$
Γ^1	A_1'	A_{1g}	A_1	A_1	L_1	1	1	1	1	1	1
Γ^2	A_2'	A_{2g}	A_2	A_2	L_2	1	1	−1	1	1	−1
Γ^3	E'	E_g	E_2	E_2	L_3	2	−1	0	2	−1	0
Γ^4	A_1''	A_{1u}	B_1	B_1	$L_{1'}$	1	1	1	−1	−1	−1
Γ^5	A_2''	A_{2u}	B_2	B_2	$L_{2'}$	1	1	−1	−1	−1	1
Γ^6	E''	E_u	E_1	E_1	$L_{3'}$	2	−1	0	−2	1	0

Table D.13 Character table for D_{3h}, D_{3d}, C_{6v} and D_6.

(*c*) Matrices for irreducible representations of dimension greater than one, for any proper rotation C_{nj} of D_{3d}:

$$\mathbf{\Gamma}^3(C_{nj}) = \mathbf{\Gamma}^6(C_{nj}) = \mathbf{\Gamma}''(C_{nj}); \mathbf{\Gamma}^3(IC_{nj}) = -\mathbf{\Gamma}^6(IC_{nj}) = \mathbf{\Gamma}''(C_{nj});$$

where the matrices $\mathbf{\Gamma}''(C_{nj})$ are given in Table D.4.

(9) C_{6v}:

(*a*) Classes:

$$\mathscr{C}_1 = E;\ \mathscr{C}_2 = C_{3z},\ C_{3z}^{-1};\ \mathscr{C}_3 = IC_{2x},\ IC_{2A},\ IC_{2B};\ \mathscr{C}_4 = C_{2z};$$
$$\mathscr{C}_5 = C_{6z}, C_{6z}^{-1};\ \mathscr{C}_6 = IC_{2y}, IC_{2C}, IC_{2D}.$$

(*b*) The character table is given in Table D.13.

(*c*) Matrices for irreducible representations of dimension greater than one, for any proper rotation C_{nj} of C_{6v}:

$$\mathbf{\Gamma}^6(C_{nj}) = \mathbf{D}''(C_{nj}); \mathbf{\Gamma}^3(C_{nj}) = \mathbf{D}'(C_{nj});$$

and for any improper rotation IC_{nj} of C_{6v}:

$$\mathbf{\Gamma}^6(IC_{nj}) = \mathbf{D}''(C_{nj}); \mathbf{\Gamma}^3(IC_{nj}) = \mathbf{D}'(C_{nj});$$

where the matrices $\mathbf{D}'(C_{nj})$ and $\mathbf{D}''(C_{nj})$ are given in Tables D.6 and D.7 respectively.

(10) C_{6h}:

(*a*) Classes:

$$\mathscr{C}_1 = E; \mathscr{C}_2 = C_{6z}; \mathscr{C}_3 = C_{3z}; \mathscr{C}_4 = C_{2z}; \mathscr{C}_5 = C_{3z}^{-1}; \mathscr{C}_6 = C_{6z}^{-1}; \mathscr{C}_7 = I;$$
$$\mathscr{C}_8 = IC_{6z}; \mathscr{C}_9 = IC_{3z}; \mathscr{C}_{10} = IC_{2z}; \mathscr{C}_{11} = IC_{3z}^{-1}; \mathscr{C}_{12} = IC_{6z}^{-1}.$$

(*b*) The character table is given in Table D.14.

(11) D_6:

(*a*) Classes:

$$\mathscr{C}_1 = E; \mathscr{C}_2 = C_{3z}, C_{3z}^{-1}; \mathscr{C}_3 = C_{2x}, C_{2A}, C_{2B}; \mathscr{C}_4 = C_{2z}; \mathscr{C}_5 = C_{6z}, C_{6z}^{-1};$$
$$\mathscr{C}_6 = C_{2y}, C_{2C}, C_{2D}.$$

(*b*) The character table is given in Table D.13.

(*c*) Matrices for irreducible representations of dimension greater than one, for any rotation C_{nj} of D_6:

$$\mathbf{\Gamma}^6(C_{nj}) = \mathbf{D}''(C_{nj}); \mathbf{\Gamma}^3(C_{nj}) = \mathbf{D}'(C_{nj});$$

where the matrices $\mathbf{D}'(C_{nj})$ and $\mathbf{D}''(C_{nj})$ are given in Tables D.6 and D.7 respectively.

		$\mathscr{C}_1$	$\mathscr{C}_2$	$\mathscr{C}_3$	$\mathscr{C}_4$	$\mathscr{C}_5$	$\mathscr{C}_6$	$\mathscr{C}_7$	$\mathscr{C}_8$	$\mathscr{C}_9$	$\mathscr{C}_{10}$	$\mathscr{C}_{11}$	$\mathscr{C}_{12}$
Γ^1	A_g	1	1	1	1	1	1	1	1	1	1	1	1
Γ^2	B_g	1	−1	1	−1	1	−1	1	−1	1	−1	1	−1
Γ^3	E'_g	1	ω	ω^2	−1	$-\omega$	$-\omega^2$	1	ω	ω^2	−1	$-\omega$	$-\omega^2$
Γ^4		1	$-\omega^2$	$-\omega$	−1	ω^2	ω	1	$-\omega^2$	$-\omega$	−1	ω^2	ω
Γ^5	E''_g	1	ω^2	$-\omega$	1	ω^2	$-\omega$	1	ω^2	$-\omega$	1	ω^2	$-\omega$
Γ^6		1	$-\omega$	ω^2	1	$-\omega$	ω^2	1	$-\omega$	ω^2	1	$-\omega$	ω^2
Γ^7	A_u	1	1	1	1	1	1	−1	−1	−1	−1	−1	−1
Γ^8	B_u	1	−1	1	−1	1	−1	−1	1	−1	1	−1	1
Γ^9	E'_u	1	ω	ω^2	−1	$-\omega$	$-\omega^2$	−1	$-\omega$	$-\omega^2$	1	ω	ω^2
Γ^{10}		1	$-\omega^2$	$-\omega$	−1	ω^2	ω	−1	ω^2	ω	1	$-\omega^2$	$-\omega$
Γ^{11}	E''_u	1	ω^2	$-\omega$	1	ω^2	$-\omega$	−1	$-\omega^2$	ω	−1	$-\omega^2$	ω
Γ^{12}		1	$-\omega$	ω^2	1	$-\omega$	ω^2	−1	ω	$-\omega^2$	−1	ω	$-\omega^2$

Table D.14 Character table for C_{6h} ($\omega = \exp(\frac{1}{3}\pi i)$).

(12) T:

(*a*) Classes:

$$\mathscr{C}_1 = E;\ \mathscr{C}_2 = C_{3\alpha}, C_{3\beta}, C_{3\gamma}, C_{3\delta};\ \mathscr{C}_3 = C_{3\alpha}^{-1}, C_{3\beta}^{-1}, C_{3\gamma}^{-1}, C_{3\delta}^{-1};$$

$$\mathscr{C}_4 = C_{2x}, C_{2y}, C_{2z}.$$

(*b*) The character table is given in Table D.15.

(*c*) Matrices for irreducible representation of dimension greater than one, for any rotation C_{nj} of T:

$$\Gamma^4(C_{nj}) = \Gamma'(C_{nj});$$

where the matrices $\Gamma'(C_{nj})$ are given in Table D.3.

(13) D_{2h} or V_h:

(*a*) Classes [for $\mathscr{G}_0(\mathbf{k})$ of N]:

$$\mathscr{C}_1 = E;\ \mathscr{C}_2 = C_{2x};\ \mathscr{C}_3 = C_{2e};\ \mathscr{C}_4 = C_{2f};\ \mathscr{C}_5 = I;\ \mathscr{C}_6 = IC_{2x};$$

$$\mathscr{C}_7 = IC_{2e};\ \mathscr{C}_8 = IC_{2f}.$$

(*b*) The character table is given in Table D.16.

		$\mathscr{C}_1$	$\mathscr{C}_2$	$\mathscr{C}_3$	$\mathscr{C}_4$
Γ^1	A	1	1	1	1
Γ^2	E	1	ϕ	ϕ^2	1
Γ^3		1	ϕ^2	ϕ	1
Γ^4	T	3	0	0	−1

Table D.15 Character table for T ($\phi = \exp(\frac{2}{3}\pi i)$).

			$\mathscr{C}_1$	$\mathscr{C}_2$	$\mathscr{C}_3$	$\mathscr{C}_4$	$\mathscr{C}_5$	$\mathscr{C}_6$	$\mathscr{C}_7$	$\mathscr{C}_8$
Γ^1	A_{1g}	N_1	1	1	1	1	1	1	1	1
Γ^2	B_{1g}	N_2	1	−1	1	−1	1	−1	1	−1
Γ^3	B_{2g}	N_3	1	−1	−1	1	1	−1	−1	1
Γ^4	B_{3g}	N_4	1	1	−1	−1	1	1	−1	−1
Γ^5	A_{1u}	$N_{2'}$	1	1	1	1	−1	−1	−1	−1
Γ^6	B_{1u}	$N_{1'}$	1	−1	1	−1	−1	1	−1	1
Γ^7	B_{2u}	$N_{4'}$	1	−1	−1	1	−1	1	1	−1
Γ^8	B_{3u}	$N_{3'}$	1	1	−1	−1	−1	−1	1	1

Table D.16 Character table for D_{2h}.

(14) C_{4v}:

(*a*) Classes [for $\mathscr{G}_0(\mathbf{k})$ of Δ]:

$$\mathscr{C}_1 = E;\ \mathscr{C}_2 = C_{2z};\ \mathscr{C}_3 = C_{4z},\ C_{4z}^{-1};\ \mathscr{C}_4 = IC_{2x},\ IC_{2y};\ \mathscr{C}_5 = IC_{2a},\ IC_{2b}.$$

Classes [for $\mathscr{G}_0(\mathbf{k})$ of T]:

$$\mathscr{C}_1 = E;\ \mathscr{C}_2 = C_{2x};\ \mathscr{C}_3 = C_{4x},\ C_{4x}^{-1};\ \mathscr{C}_4 = IC_{2y},\ IC_{2z};\ \mathscr{C}_5 = IC_{2e},\ IC_{2f}.$$

(*b*) The character table is given in Table D.17.

(*c*) Matrices for irreducible representation of dimension greater than one, for any proper rotation C_{nj} of C_{4v}:

$$\Gamma^5(C_{nj}) = \mathbf{D}(C_{nj});$$

and for any improper rotation IC_{nj} of C_{4v}:

$$\Gamma^5(IC_{nj}) = -\mathbf{D}(C_{nj});$$

where for $\mathscr{G}_0(\mathbf{k})$ of Δ the matrices $\mathbf{D}(C_{nj})$ are given in Table D.12, while for $\mathscr{G}_0(\mathbf{k})$ of T they are given in Table D.11.

				$\mathscr{C}_1$	$\mathscr{C}_2$	$\mathscr{C}_3$	$\mathscr{C}_4$	$\mathscr{C}_5$
Γ^1	A_1	$\Delta_1, T_1,$	W_1	1	1	1	1	1
Γ^2	B_1	$\Delta_2, T_2,$	$W_{2'}$	1	1	−1	1	−1
Γ^3	A_2	$\Delta_{1'}, T_{1'},$	W_2	1	1	1	−1	−1
Γ^4	B_2	$\Delta_{2'}, T_{2'},$	$W_{1'}$	1	1	−1	−1	1
Γ^5	E	$\Delta_5, T_5,$	W_3	2	−2	0	0	0

Table D.17 Character table for C_{4v}, D_4 and D_{2d}.

(15) D_4:

(*a*) Classes:

$$\mathscr{C}_1 = E;\ \mathscr{C}_2 = C_{2y};\ \mathscr{C}_3 = C_{4y},\ C_{4y}^{-1},\ \mathscr{C}_4 = C_{2x},\ C_{2z};\ \mathscr{C}_5 = C_{2c},\ C_{2d}.$$

(*b*) The character table is given in Table D.17.

(*c*) Matrices for irreducible representation of dimension greater than one, for any rotation C_{nj} of D_4:

$$\boldsymbol{\Gamma}^5(C_{nj}) = \mathbf{S}(C_{nj});$$

where the matrices $\mathbf{S}(C_{nj})$ are given in Table D.18.

(16) D_{2d} or V_d:

(*a*) Classes [for $\mathscr{G}_0(\mathbf{k})$ of W]:

$$\mathscr{C}_1 = E;\ \mathscr{C}_2 = C_{2y};\ \mathscr{C}_3 = IC_{4y},\ IC_{4y}^{-1};\ \mathscr{C}_4 = IC_{2x},\ IC_{2z};\ \mathscr{C}_5 = C_{2c},\ C_{2d}.$$

(*b*) The character table is given in Table D.17.

(*c*) Matrices for irreducible representation of dimension greater than one, for any proper rotation C_{nj} of D_{2d}:

$$\boldsymbol{\Gamma}^5(C_{nj}) = \mathbf{S}(C_{nj});$$

and for any improper rotation IC_{nj} of D_{2d}:

$$\boldsymbol{\Gamma}^5(IC_{nj}) = -\mathbf{S}(C_{nj});$$

where the matrices $\mathbf{S}(C_{nj})$ are given in Table D.18.

(17) C_{4h}:

(*a*) Classes:

$$\mathscr{C}_1 = E;\ \mathscr{C}_2 = C_{4z};\ \mathscr{C}_3 = C_{2z};\ \mathscr{C}_4 = C_{4z}^{-1};\ \mathscr{C}_5 = I;\ \mathscr{C}_6 = IC_{4z};$$
$$\mathscr{C}_7 = IC_{2z};\ \mathscr{C}_8 = IC_{4z}^{-1}.$$

(*b*) The character table is given in Table D.19.

$$\mathbf{S}(E) = \begin{bmatrix} 1 & 0 \\ 0 & 1 \end{bmatrix}, \quad \mathbf{S}(C_{2y}) = \begin{bmatrix} -1 & 0 \\ 0 & -1 \end{bmatrix}, \quad \mathbf{S}(C_{2c}) = \begin{bmatrix} 0 & 1 \\ 1 & 0 \end{bmatrix},$$

$$\mathbf{S}(C_{2d}) = \begin{bmatrix} 0 & -1 \\ -1 & 0 \end{bmatrix}, \quad \mathbf{S}(C_{4y}) = \begin{bmatrix} 0 & -1 \\ 1 & 0 \end{bmatrix}, \quad \mathbf{S}(C_{4y}^{-1}) = \begin{bmatrix} 0 & 1 \\ -1 & 0 \end{bmatrix},$$

$$\mathbf{S}(C_{2x}) = \begin{bmatrix} 1 & 0 \\ 0 & -1 \end{bmatrix}, \quad \mathbf{S}(C_{2z}) = \begin{bmatrix} -1 & 0 \\ 0 & 1 \end{bmatrix}.$$

Table D.18 The matrices **S** for D_4 and D_{2d}.

		$\mathscr{C}_1$	$\mathscr{C}_2$	$\mathscr{C}_3$	$\mathscr{C}_4$	$\mathscr{C}_5$	$\mathscr{C}_6$	$\mathscr{C}_7$	$\mathscr{C}_8$
Γ^1	A_g	1	1	1	1	1	1	1	1
Γ^2	B_g	1	−1	1	−1	1	−1	1	−1
Γ^3	E_g	1	i	−1	$-i$	1	i	−1	$-i$
Γ^4		1	$-i$	−1	i	1	$-i$	−1	i
Γ^5	A_u	1	1	1	1	−1	−1	−1	−1
Γ^6	B_u	1	−1	1	−1	−1	1	−1	1
Γ^7	E_u	1	i	−1	$-i$	−1	$-i$	1	i
Γ^8		1	$-i$	−1	i	−1	i	1	$-i$

Table D.19 Character table for C_{4h}.

(18) C_{3h}:

(*a*) Classes:

$$\mathscr{C}_1 = E;\ \mathscr{C}_2 = IC_{6z};\ \mathscr{C}_3 = C_{3z};\ \mathscr{C}_4 = IC_{2z};\ \mathscr{C}_5 = C_{3z};\ \mathscr{C}_6 = IC_{6z}^{-1}.$$

(*b*) The character table is given in Table D.20.

	C_{3h}	C_{3i}	C_6	$\mathscr{C}_1$	$\mathscr{C}_2$	$\mathscr{C}_3$	$\mathscr{C}_4$	$\mathscr{C}_5$	$\mathscr{C}_6$
Γ^1	A′	A_g	A	1	1	1	1	1	1
Γ^2	A″	A_u	B	1	−1	1	−1	1	−1
Γ^3	E″	E_u	E′	1	ω	ω^2	−1	$-\omega$	$-\omega^2$
Γ^4				1	$-\omega^2$	$-\omega$	−1	ω^2	ω
Γ^5	E′	E_g	E″	1	ω^2	$-\omega$	1	ω^2	$-\omega$
Γ^6				1	$-\omega$	ω^2	1	$-\omega$	ω^2

Table D.20 Character table for C_{3h}, C_{3i} and C_6 ($\omega = \exp(\frac{1}{3}\pi i)$).

(19) C_{3v}:

(*a*) Classes [for $\mathscr{G}_0(\mathbf{k})$ of Λ]:

$$\mathscr{C}_1 = E;\ \mathscr{C}_2 = C_{3\delta}, C_{3\delta}^{-1};\ \mathscr{C}_3 = IC_{2b}, IC_{2d}, IC_{2f}.$$

Classes [for $\mathscr{G}_0(\mathbf{k})$ of F]:

$$\mathscr{C}_1 = E;\ \mathscr{C}_2 = C_{3\alpha}, C_{3\alpha}^{-1};\ \mathscr{C}_3 = IC_{2b}, IC_{2c}, IC_{2e}.$$

(*b*) The character table is given in Table D.21.

			$\mathscr{C}_1$	$\mathscr{C}_2$	$\mathscr{C}_3$
Γ^1	A_1	F_1, Λ_1	1	1	1
Γ^2	A_2	F_2, Λ_2	1	1	−1
Γ^3	E	F_3, Λ_3	2	−1	0

Table D.21 Character table for C_{3v} and D_3.

(*c*) Matrices for irreducible representation of dimension greater than one, for any proper rotation C_{nj} of C_{3v}:

$$\boldsymbol{\Gamma}^3(C_{nj}) = \boldsymbol{\Gamma}''(C_{nj});$$

and for any improper rotation IC_{nj} of C_{3v}:

$$\boldsymbol{\Gamma}^3(IC_{nj}) = \boldsymbol{\Gamma}''(C_{nj});$$

where the matrices $\boldsymbol{\Gamma}''(C_{nj})$ are given in Table D.4.

(20) D_3:

(*a*) Classes:

$$\mathscr{C}_1 = E;\ \mathscr{C}_2 = C_{3z}, C_{3z}^{-1};\ \mathscr{C}_3 = C_{2x}, C_{2A}, C_{2B}.$$

(*b*) The character table is given in Table D.21.
(*c*) Matrices for irreducible representation of dimension greater than one, for any rotation C_{nj} of D_3:

$$\boldsymbol{\Gamma}^3(C_{nj}) = \mathbf{D}'(C_{nj});$$

where the matrices $\mathbf{D}'(C_{nj})$ are given in Table D.6.

(21) C_{3i} or S_6:

(*a*) Classes:

$$\mathscr{C}_1 = E;\ \mathscr{C}_2 = IC_{3z}^{-1};\ \mathscr{C}_3 = C_{3z};\ \mathscr{C}_4 = I;\ \mathscr{C}_5 = C_{3z}^{-1};\ \mathscr{C}_6 = IC_{3z}.$$

(*b*) The character table is given in Table D.20.

(22) C_6:

(*a*) Classes:

$$\mathscr{C}_1 = E;\ \mathscr{C}_2 = C_{6z};\ \mathscr{C}_3 = C_{3z};\ \mathscr{C}_4 = C_{2z};\ \mathscr{C}_5 = C_{3z}^{-1},\ \mathscr{C}_6 = C_{6z}^{-1}.$$

(*b*) The character table is given in Table D.20.

(23) C_{2v}:

(*a*) Classes [for $\mathscr{G}_0(\mathbf{k})$ for Σ]:

$$\mathscr{C}_1 = E;\ \mathscr{C}_2 = C_{2e};\ \mathscr{C}_3 = IC_{2x};\ \mathscr{C}_4 = IC_{2f}.$$

Classes [for $\mathscr{G}_0(\mathbf{k})$ for D]:

$$\mathscr{C}_1 = E;\ \mathscr{C}_2 = C_{2x};\ \mathscr{C}_3 = IC_{2e};\ \mathscr{C}_4 = IC_{2f}.$$

Classes [for $\mathscr{G}_0(\mathbf{k})$ for S]:

$$\mathscr{C}_1 = E;\ \mathscr{C}_2 = C_{2a};\ \mathscr{C}_3 = IC_{2z};\ \mathscr{C}_4 = IC_{2b}.$$

	C_{2v}	C_{2h}	D_2						$\mathscr{C}_1$	$\mathscr{C}_2$	$\mathscr{C}_3$	$\mathscr{C}_4$
Γ^1	A_1	A_g	A_1	Σ_1	S_1	Z_1	G_1	D_1	1	1	1	1
Γ^2	A_2	A_u	B_1	Σ_2	S_2	Z_2	G_2	D_2	1	1	−1	−1
Γ^3	B_1	B_u	B_2	Σ_3	S_3	Z_3	G_3	D_3	1	−1	−1	1
Γ^4	B_2	B_g	B_3	Σ_4	S_4	Z_4	G_4	D_4	1	−1	1	−1

Table D.22 Character table for C_{2v}, C_{2h} and D_2.

Classes [for $\mathscr{G}_0(\mathbf{k})$ for Z]:

$$\mathscr{C}_1 = E;\ \mathscr{C}_2 = C_{2y};\ \mathscr{C}_3 = IC_{2z};\ \mathscr{C}_4 = IC_{2x}.$$

Classes [for $\mathscr{G}_0(\mathbf{k})$ for G]:

$$\mathscr{C}_1 = E;\ \mathscr{C}_2 = C_{2f};\ \mathscr{C}_3 = IC_{2x};\ \mathscr{C}_4 = IC_{2e}.$$

(*b*) The character table is given in Table D.22.

(24) C_{2h}:

(*a*) Classes:

$$\mathscr{C}_1 = E;\ \mathscr{C}_2 = C_{2z};\ \mathscr{C}_3 = I;\ \mathscr{C}_4 = IC_{2z}.$$

(*b*) The character table is given in Table D.22.

(25) D_2 or V:

(*a*) Classes:

$$\mathscr{C}_1 = E;\ \mathscr{C}_2 = C_{2x};\ \mathscr{C}_3 = C_{2y};\ \mathscr{C}_4 = C_{2z}.$$

(*b*) The character table is given in Table D.22.

(26) C_4:

(*a*) Classes:

$$\mathscr{C}_1 = E;\ \mathscr{C}_2 = C_{4z};\ \mathscr{C}_3 = C_{2z};\ \mathscr{C}_4 = C_{4z}^{-1}.$$

(*b*) The character table is given in Table D.23.

(27) S_4:

(*a*) Classes:

$$\mathscr{C}_1 = E;\ \mathscr{C}_2 = IC_{4y};\ \mathscr{C}_3 = C_{2y};\ \mathscr{C}_4 = IC_{4y}^{-1}.$$

(*b*) The character table is given in Table D.23.

		$\mathscr{C}_1$	$\mathscr{C}_2$	$\mathscr{C}_3$	$\mathscr{C}_4$
Γ^1	A	1	1	1	1
Γ^2	B	1	−1	1	−1
Γ^3	E {	1	i	−1	$-i$
Γ^4		1	$-i$	−1	i

Table D.23 Character table for C_4 and S_4.

(28) C_3:

(*a*) Classes:

$$\mathscr{C}_1 = E;\ \mathscr{C}_2 = C_{3z};\ \mathscr{C}_3 = C_{3z}^{-1}.$$

(*b*) The character table is given in Table D.24.

		$\mathscr{C}_1$	$\mathscr{C}_2$	$\mathscr{C}_3$
Γ^1	A	1	1	1
Γ^2	E {	1	ϕ	ϕ^2
Γ^3		1	ϕ^2	ϕ

Table D.24 Character table for C_3 ($\phi = \exp(\frac{2}{3}\pi i)$).

(29) C_s or C_{1h}:

(*a*) Classes:

$$\mathscr{C}_1 = E;\ \mathscr{C}_2 = IC_{2z}.$$

(*b*) The character table is given in Table D.25.

	C_s	C_2	C_i		$\mathscr{C}_1$	$\mathscr{C}_2$
Γ^1	A′	A	A_g	Q_1	1	1
Γ^2	A″	B	A_u	Q_2	1	−1

Table D.25 Character table for C_s, C_2 and C_i.

(30) C_2:

(*a*) Classes [for $\mathscr{G}_0(\mathbf{k})$ of Q]:

$$\mathscr{C}_1 = E;\ \mathscr{C}_2 = C_{2d}.$$

(*b*) The character table is given in Table D.25.

(31) C_i or S_2:

(*a*) Classes:

$$\mathscr{C}_1 = E; \mathscr{C}_2 = I.$$

(*b*) The character table is given in Table D.25

(32) C_1:

(*a*) This group consists of E alone.
(*b*) $\chi(E) = 1$.

2 The double crystallographic point groups

The following is a list of the double crystallographic point groups, arranged in the same order as the single groups of the previous section. For each group the following details are given:

(*a*) *The group elements*. These are arranged in classes. For brevity the following conventions have been used;

(i) The generalized rotations $[C_{nj} \mid \mathbf{0}]$ and $[\bar{C}_{nj} \mid \mathbf{0}]$, defined in Chapter 6, Section 4(*b*), are written simply as C_{nj} and $\bar{C}_{nj}$, with a similar convention for improper rotations. The product of two such generalized rotations is defined in Equations (6.32) and (6.33), and, as noted there, is *not* necessarily the same as the product of the corresponding elements of the single group. The notation for the corresponding elements of the single groups is specified in the previous section. The matrices $\mathbf{u}(C_{nj})$ $(=\mathbf{u}(\mathbf{R}_p(C_{nj}))$ are listed in Table D.26.
(ii) If the class $\mathscr{C}_j$ consists entirely of a set of "unbarred" elements and the class $\mathscr{C}_j$ consists of the corresponding "barred" elements, then $\mathscr{C}_j$ is said to be equal to $\bar{\mathscr{C}}_i$. For example, if $\mathscr{C}_2 = C_{3\alpha}, C_{3\alpha}^{-1}$, and $\mathscr{C}_5 = \bar{C}_{3\alpha}, \bar{C}_{3\alpha}^{-1}$, then $\mathscr{C}_5 = \bar{\mathscr{C}}_2$.

(*b*) *The character systems of the "extra" representations* (as defined in Chapter 6, Section 4(*e*)). If $\chi(\bar{\mathscr{C}}_i)$ denotes the character of the class $\bar{\mathscr{C}}_i$ in an "extra" irreducible representation then $\chi(\bar{\mathscr{C}}_i) = -\chi(\mathscr{C}_i)$. Accordingly the tables have been shortened by omitting the characters of such classes $\bar{\mathscr{C}}_i$.

For a point group that is isomorphic to a group $\mathscr{G}_0^D(\mathbf{k})$ for O_h^1, O_h^5 and O_h^9, the first column gives the labelling convention of Elliott (1954). In such cases the corresponding $\mathbf{k}$-vector is as defined in Tables 9.1, 9.2 or 9.3.

$$\mathbf{u}(\mathrm{E}) = \begin{bmatrix} 1 & 0 \\ 0 & 1 \end{bmatrix}, \qquad \mathbf{u}(\mathrm{C}_{3\alpha}) = \frac{1}{2}\begin{bmatrix} 1+i & -1-i \\ 1-i & 1-i \end{bmatrix},$$

$$\mathbf{u}(\mathrm{C}_{3\beta}) = \frac{1}{2}\begin{bmatrix} 1-i & 1-i \\ 1-i & 1+i \end{bmatrix}, \qquad \mathbf{u}(\mathrm{C}_{3\gamma}) = \frac{1}{2}\begin{bmatrix} 1-i & -1+i \\ 1+i & 1+i \end{bmatrix},$$

$$\mathbf{u}(\mathrm{C}_{3\delta}) = \frac{1}{2}\begin{bmatrix} 1+i & 1+i \\ -1+i & 1-i \end{bmatrix}, \qquad \mathbf{u}(\mathrm{C}_{3\alpha}^{-1}) = \frac{1}{2}\begin{bmatrix} 1-i & 1+i \\ -1+i & 1+i \end{bmatrix},$$

$$\mathbf{u}(\mathrm{C}_{3\beta}^{-1}) = \frac{1}{2}\begin{bmatrix} 1+i & -1+i \\ 1+i & 1-i \end{bmatrix}, \qquad \mathbf{u}(\mathrm{C}_{3\gamma}^{-1}) = \frac{1}{2}\begin{bmatrix} 1+i & 1-i \\ -1-i & 1-i \end{bmatrix},$$

$$\mathbf{u}(\mathrm{C}_{3\delta}^{-1}) = \frac{1}{2}\begin{bmatrix} 1-i & -1-i \\ 1-i & 1+i \end{bmatrix}, \qquad \mathbf{u}(\mathrm{C}_{2x}) = \begin{bmatrix} 0 & i \\ i & 0 \end{bmatrix},$$

$$\mathbf{u}(\mathrm{C}_{2y}) = \begin{bmatrix} 0 & 1 \\ -1 & 0 \end{bmatrix}, \qquad \mathbf{u}(\mathrm{C}_{2z}) = \begin{bmatrix} 1 & 0 \\ 0 & -1 \end{bmatrix},$$

$$\mathbf{u}(\mathrm{C}_{4x}) = \frac{1}{\sqrt{2}}\begin{bmatrix} 1 & i \\ i & 1 \end{bmatrix}, \qquad \mathbf{u}(\mathrm{C}_{4y}) = \frac{1}{\sqrt{2}}\begin{bmatrix} 1 & 1 \\ -1 & 1 \end{bmatrix},$$

$$\mathbf{u}(\mathrm{C}_{4z}) = \frac{1}{\sqrt{2}}\begin{bmatrix} 1+i & 0 \\ 0 & 1-i \end{bmatrix}, \qquad \mathbf{u}(\mathrm{C}_{4x}^{-1}) = \frac{1}{\sqrt{2}}\begin{bmatrix} 1 & -i \\ -i & 1 \end{bmatrix},$$

$$\mathbf{u}(\mathrm{C}_{4y}^{-1}) = \frac{1}{\sqrt{2}}\begin{bmatrix} 1 & -1 \\ 1 & 1 \end{bmatrix}, \qquad \mathbf{u}(\mathrm{C}_{4z}^{-1}) = \frac{1}{\sqrt{2}}\begin{bmatrix} 1-i & 0 \\ 0 & 1+i \end{bmatrix},$$

$$\mathbf{u}(\mathrm{C}_{2a}) = \frac{1}{\sqrt{2}}\begin{bmatrix} 0 & 1+i \\ -1+i & 0 \end{bmatrix}, \qquad \mathbf{u}(\mathrm{C}_{2b}) = \frac{1}{\sqrt{2}}\begin{bmatrix} 0 & 1-i \\ -1-i & 0 \end{bmatrix},$$

$$\mathbf{u}(\mathrm{C}_{2c}) = \frac{1}{\sqrt{2}}\begin{bmatrix} i & i \\ i & -i \end{bmatrix}, \qquad \mathbf{u}(\mathrm{C}_{2d}) = \frac{1}{\sqrt{2}}\begin{bmatrix} i & -i \\ -i & -i \end{bmatrix},$$

$$\mathbf{u}(\mathrm{C}_{2e}) = \frac{1}{\sqrt{2}}\begin{bmatrix} i & 1 \\ -1 & -i \end{bmatrix}, \qquad \mathbf{u}(\mathrm{C}_{2f}) = \frac{1}{\sqrt{2}}\begin{bmatrix} i & -1 \\ 1 & -i \end{bmatrix},$$

$$\mathbf{u}(\mathrm{C}_{3z}) = \frac{1}{2}\begin{bmatrix} 1+i\sqrt{3} & 0 \\ 0 & 1-i\sqrt{3} \end{bmatrix}, \qquad \mathbf{u}(\mathrm{C}_{3z}^{-1}) = \frac{1}{2}\begin{bmatrix} 1-i\sqrt{3} & 0 \\ 0 & 1+i\sqrt{3} \end{bmatrix},$$

$$\mathbf{u}(\mathrm{C}_{6z}) = \frac{1}{2}\begin{bmatrix} \sqrt{3}+i & 0 \\ 0 & \sqrt{3}-i \end{bmatrix}, \qquad \mathbf{u}(\mathrm{C}_{6z}^{-1}) = \frac{1}{2}\begin{bmatrix} \sqrt{3}-i & 0 \\ 0 & \sqrt{3}+i \end{bmatrix},$$

$$\mathbf{u}(\mathrm{C}_{2A}) = \frac{1}{2}\begin{bmatrix} 0 & \sqrt{3}+i \\ -\sqrt{3}+i & 0 \end{bmatrix}, \qquad \mathbf{u}(\mathrm{C}_{2B}) = \frac{1}{2}\begin{bmatrix} 0 & \sqrt{3}-i \\ -\sqrt{3}-i & 0 \end{bmatrix},$$

$$\mathbf{u}(\mathrm{C}_{2C}) = \frac{1}{2}\begin{bmatrix} 0 & 1-i\sqrt{3} \\ -1-i\sqrt{3} & 0 \end{bmatrix}, \qquad \mathbf{u}(\mathrm{C}_{2D}) = \frac{1}{2}\begin{bmatrix} 0 & 1+i\sqrt{3} \\ -1+i\sqrt{3} & 0 \end{bmatrix}.$$

Table D.26 The matrices $\mathbf{u}(\mathrm{C}_{nj})$ $(=\mathbf{u}(\mathbf{R}_p(\mathrm{C}_{nj})))$ for the proper rotations of the crystallographic point groups. (See last paragraph of Chapter 6, Section 4(*b*), for note on conventions.)

Explicit matrices for the "extra" irreducible representations of O_h^D, T_d^D, O^D, D_{4h}^D, D_{3d}^D, C_{4v}^D, C_{3v}^D and C_{2v}^D have been given by Onodera and Okazaki (1966).

(1) O_h^D:

(*a*) Classes [for $\mathscr{G}_0^D(\mathbf{k})$ of Γ, H and R]:

$$\mathscr{C}_1 = E;\ \mathscr{C}_2 = C_{3\alpha}, C_{3\beta}, C_{3\gamma}, C_{3\delta}, C_{3\alpha}^{-1}, C_{3\beta}^{-1}, C_{3\gamma}^{-1}, C_{3\delta}^{-1};$$

$$\mathscr{C}_3 = C_{2x}, C_{2y}, C_{2z}, \bar{C}_{2x}, \bar{C}_{2y}, \bar{C}_{2z};$$

$$\mathscr{C}_4 = C_{4x}, C_{4y}, C_{4z}, C_{4x}^{-1}, C_{4y}^{-1}, C_{4z}^{-1};$$

$$\mathscr{C}_5 = C_{2a}, C_{2b}, C_{2c}, C_{2d}, C_{2e}, C_{2f}, \bar{C}_{2a}, \bar{C}_{2b}, \bar{C}_{2c}, \bar{C}_{2d}, \bar{C}_{2e}, \bar{C}_{2f};$$

$$\mathscr{C}_6 = I;\ \mathscr{C}_7 = IC_{3\alpha}, IC_{3\beta}, IC_{3\gamma}, IC_{3\delta}, IC_{3\alpha}^{-1}, IC_{3\beta}^{-1}, IC_{3\gamma}^{-1}, IC_{3\delta}^{-1};$$

$$\mathscr{C}_8 = IC_{2x}, IC_{2y}, IC_{2z}, \overline{IC}_{2x}, \overline{IC}_{2y}, \overline{IC}_{2z};$$

$$\mathscr{C}_9 = IC_{4x}, IC_{4y}, IC_{4z}, IC_{4x}^{-1}, IC_{4y}^{-1}, IC_{4z}^{-1};$$

$$\mathscr{C}_{10} = IC_{2a}, IC_{2b}, IC_{2c}, IC_{2d}, IC_{2e}, IC_{2f}, \overline{IC}_{2a}, \overline{IC}_{2b}, \overline{IC}_{2c}, \overline{IC}_{2d}, \overline{IC}_{2e}, \overline{IC}_{2f};$$

$$\mathscr{C}_{11} = \bar{\mathscr{C}}_1;\ \mathscr{C}_{12} = \bar{\mathscr{C}}_2;\ \mathscr{C}_{13} = \bar{\mathscr{C}}_4;\ \mathscr{C}_{14} = \bar{\mathscr{C}}_6;\ \mathscr{C}_{15} = \bar{\mathscr{C}}_7;\ \mathscr{C}_{16} = \bar{\mathscr{C}}_9.$$

(*b*) The character table is given in Table D.27.

	$\mathscr{C}_1$	$\mathscr{C}_2$	$\mathscr{C}_3$	$\mathscr{C}_4$	$\mathscr{C}_5$	$\mathscr{C}_6$	$\mathscr{C}_7$	$\mathscr{C}_8$	$\mathscr{C}_9$	$\mathscr{C}_{10}$
$\Gamma_6^+ H_6^+ R_6^+$	2	1	0	$\sqrt{2}$	0	2	1	0	$\sqrt{2}$	0
$\Gamma_7^+ H_7^+ R_7^+$	2	1	0	$-\sqrt{2}$	0	2	1	0	$-\sqrt{2}$	0
$\Gamma_8^+ H_8^+ R_8^+$	4	-1	0	0	0	4	-1	0	0	0
$\Gamma_6^- H_6^- R_6^-$	2	1	0	$\sqrt{2}$	0	-2	-1	0	$-\sqrt{2}$	0
$\Gamma_7^- H_7^- R_7^-$	2	1	0	$-\sqrt{2}$	0	-2	-1	0	$\sqrt{2}$	0
$\Gamma_8^- H_8^- R_8^-$	4	-1	0	0	0	-4	1	0	0	0

Table D.27 Characters of "extra" irreducible representations of O_h^D.

(2) D_{6h}^D:

(*a*) Classes:

$$\mathscr{C}_1 = E;\ \mathscr{C}_2 = C_{6z}, C_{6z}^{-1};\ \mathscr{C}_3 = C_{3z}, C_{3z}^{-1};\ \mathscr{C}_4 = C_{2z}, \bar{C}_{2z};$$

$$\mathscr{C}_5 = C_{2x}, C_{2A}, C_{2B}, \bar{C}_{2x}, \bar{C}_{2A}, \bar{C}_{2B};\ \mathscr{C}_6 = C_{2y}, C_{2C}, C_{2D}, \bar{C}_{2y}, \bar{C}_{2C}, \bar{C}_{2D};$$

$$\mathscr{C}_7 = I;\ \mathscr{C}_8 = IC_{6z}, IC_{6z}^{-1};\ \mathscr{C}_9 = IC_{3z}, IC_{3z}^{-1};\ \mathscr{C}_{10} = IC_{2z}, \overline{IC}_{2z};$$

$$\mathscr{C}_{11} = IC_{2x}, IC_{2A}, IC_{2B}, \overline{IC}_{2x}, \overline{IC}_{2A}, \overline{IC}_{2B};$$

$$\mathscr{C}_{12} = IC_{2y}, IC_{2C}, IC_{2D}, \overline{IC}_{2y}, \overline{IC}_{2C}, \overline{IC}_{2D};$$

$$\mathscr{C}_{13} = \bar{\mathscr{C}}_1;\ \mathscr{C}_{14} = \bar{\mathscr{C}}_2;\ \mathscr{C}_{15} = \bar{\mathscr{C}}_3;\ \mathscr{C}_{16} = \bar{\mathscr{C}}_7;\ \mathscr{C}_{17} = \bar{\mathscr{C}}_8;\ \mathscr{C}_{18} = \bar{\mathscr{C}}_9.$$

(*b*) The character table is given in Table D.28.

$\mathscr{C}_1$	$\mathscr{C}_2$	$\mathscr{C}_3$	$\mathscr{C}_4$	$\mathscr{C}_5$	$\mathscr{C}_6$	$\mathscr{C}_7$	$\mathscr{C}_8$	$\mathscr{C}_9$	$\mathscr{C}_{10}$	$\mathscr{C}_{11}$	$\mathscr{C}_{12}$
2	$\sqrt{3}$	1	0	0	0	2	$\sqrt{3}$	1	0	0	0
2	$-\sqrt{3}$	1	0	0	0	2	$-\sqrt{3}$	1	0	0	0
2	0	-2	0	0	0	2	0	-2	0	0	0
2	$\sqrt{3}$	1	0	0	0	-2	$-\sqrt{3}$	-1	0	0	0
2	$-\sqrt{3}$	1	0	0	0	-2	$\sqrt{3}$	-1	0	0	0
2	0	-2	0	0	0	-2	0	2	0	0	0

Table D.28 Characters of "extra" irreducible representations of D_{6h}^D.

(3) T_d^D:

(*a*) Classes [for $\mathscr{G}_0^D(\mathbf{k})$ of P]:

$$\mathscr{C}_1 = E;\ \mathscr{C}_2 = C_{3\alpha}, C_{3\beta}, C_{3\gamma}, C_{3\delta}, C_{3\alpha}^{-1}, C_{3\beta}^{-1}, C_{3\gamma}^{-1}, C_{3\delta}^{-1};$$

$$\mathscr{C}_3 = C_{2x}, C_{2y}, C_{2z}, \bar{C}_{2x}, \bar{C}_{2y}, \bar{C}_{2z};$$

$$\mathscr{C}_4 = IC_{4x}, IC_{4y}, IC_{4z}, IC_{4x}^{-1}, IC_{4y}^{-1}, IC_{4z}^{-1};$$

$$\mathscr{C}_5 = IC_{2a}, IC_{2b}, IC_{2c}, IC_{2d}, IC_{2e}, IC_{2f}, \overline{IC}_{2a}, \overline{IC}_{2b}, \overline{IC}_{2c}, \overline{IC}_{2d}, \overline{IC}_{2e}, \overline{IC}_{2f};$$

$$\mathscr{C}_6 = \bar{\mathscr{C}}_1;\ \mathscr{C}_7 = \bar{\mathscr{C}}_2;\ \mathscr{C}_8 = \bar{\mathscr{C}}_4.$$

(*b*) The character table is given in Table D.29.

	$\mathscr{C}_1$	$\mathscr{C}_2$	$\mathscr{C}_3$	$\mathscr{C}_4$	$\mathscr{C}_5$
P_6	2	1	0	$\sqrt{2}$	0
P_7	2	1	0	$-\sqrt{2}$	0
P_8	4	-1	0	0	0

Table D.29 Characters of "extra" irreducible representations of T_d^D and O^D.

(4) O^D:

(*a*) Classes:

$$\mathscr{C}_1 = E;\ \mathscr{C}_2 = C_{3\alpha}, C_{3\beta}, C_{3\gamma}, C_{3\delta}, C_{3\alpha}^{-1}, C_{3\beta}^{-1}, C_{3\gamma}^{-1}, C_{3\delta}^{-1};$$

$$\mathscr{C}_3 = C_{2x}, C_{2y}, C_{2z}, \bar{C}_{2x}, \bar{C}_{2y}, \bar{C}_{2z};\ \mathscr{C}_4 = C_{4x}, C_{4y}, C_{4z}, C_{4x}^{-1}, C_{4y}^{-1}, C_{4z}^{-1};$$

$$\mathscr{C}_5 = C_{2a}, C_{2b}, C_{2c}, C_{2d}, C_{2e}, C_{2f}, \bar{C}_{2a}, \bar{C}_{2b}, \bar{C}_{2c}, \bar{C}_{2d}, \bar{C}_{2e}, \bar{C}_{2f};$$

$$\mathscr{C}_6 = \bar{\mathscr{C}}_1;\ \mathscr{C}_7 = \bar{\mathscr{C}}_2;\ \mathscr{C}_8 = \bar{\mathscr{C}}_4.$$

(*b*) The character table is given in Table D.29.

(5) T_h^D:

(*a*) Classes:

$$\mathscr{C}_1 = E;\ \mathscr{C}_2 = C_{3\alpha}, C_{3\beta}, C_{3\gamma}, C_{3\delta};\ \mathscr{C}_3 = C_{3\alpha}^{-1}, C_{3\beta}^{-1}, C_{3\gamma}^{-1}, C_{3\delta}^{-1};$$

$$\mathscr{C}_4 = C_{2x}, C_{2y}, C_{2z}, \bar{C}_{2x}, \bar{C}_{2y}, \bar{C}_{2z};\ \mathscr{C}_5 = I;\ \mathscr{C}_6 = IC_{3\alpha}, IC_{3\beta}, IC_{3\gamma}, IC_{3\delta};$$

$$\mathscr{C}_7 = IC_{3\alpha}^{-1}, IC_{3\beta}^{-1}, IC_{3\gamma}^{-1}, IC_{3\delta}^{-1};\ \mathscr{C}_8 = IC_{2x}, IC_{2y}, IC_{2z}, \overline{IC}_{2x}, \overline{IC}_{2y}, \overline{IC}_{2z};$$

$$\mathscr{C}_9 = \bar{\mathscr{C}}_1;\ \mathscr{C}_{10} = \bar{\mathscr{C}}_2;\ \mathscr{C}_{11} = \bar{\mathscr{C}}_3;\ \mathscr{C}_{12} = \bar{\mathscr{C}}_5;\ \mathscr{C}_{13} = \bar{\mathscr{C}}_6;\ \mathscr{C}_{14} = \bar{\mathscr{C}}_7.$$

(*b*) The character table is given in Table D.30.

$\mathscr{C}_1$	$\mathscr{C}_2$	$\mathscr{C}_3$	$\mathscr{C}_4$	$\mathscr{C}_5$	$\mathscr{C}_6$	$\mathscr{C}_7$	$\mathscr{C}_8$
2	1	1	0	2	1	1	0
2	ω	ω^2	0	2	ω	ω^2	0
2	$-\omega^2$	$-\omega$	0	2	$-\omega^2$	$-\omega$	0
2	1	1	0	-2	-1	-1	0
2	ω	ω^2	0	-2	$-\omega$	$-\omega^2$	0
2	$-\omega^2$	$-\omega$	0	-2	ω^2	ω	0

Table D.30 Characters of "extra" irreducible representations of T_h^D ($\omega = \exp(\frac{1}{3}\pi i)$).

(6) D_{4h}^D:

(*a*) Classes [for $\mathscr{G}_0^D(\mathbf{k})$ of X]:

$$\mathscr{C}_1 = E;\ \mathscr{C}_2 = C_{2x}, C_{2y}, \bar{C}_{2x}, \bar{C}_{2y};\ \mathscr{C}_3 = C_{2z}, \bar{C}_{2z};\ \mathscr{C}_4 = C_{4z}, C_{4z}^{-1};$$

$$\mathscr{C}_5 = C_{2a}, C_{2b}, \bar{C}_{2a}, \bar{C}_{2b};\ \mathscr{C}_6 = I;\ \mathscr{C}_7 = IC_{2x}, IC_{2y}, \overline{IC}_{2x}, \overline{IC}_{2y};$$

$$\mathscr{C}_8 = IC_{2z}, \overline{IC}_{2z};\ \mathscr{C}_9 = IC_{4z}, IC_{4z}^{-1};\ \mathscr{C}_{10} = IC_{2a}, IC_{2b}, \overline{IC}_{2a}, \overline{IC}_{2b};$$

$$\mathscr{C}_{11} = \bar{\mathscr{C}}_1;\ \mathscr{C}_{12} = \bar{\mathscr{C}}_4;\ \mathscr{C}_{13} = \bar{\mathscr{C}}_6;\ \mathscr{C}_{14} = \bar{\mathscr{C}}_9.$$

Classes [for $\mathscr{G}_0^D(\mathbf{k})$ of M]:

$$\mathscr{C}_1 = E;\ \mathscr{C}_2 = C_{2y}, C_{2z}, \bar{C}_{2y}, \bar{C}_{2z};\ \mathscr{C}_3 = C_{2x}, \bar{C}_{2x};\ \mathscr{C}_4 = C_{4x}, C_{4x}^{-1};$$

$$\mathscr{C}_5 = C_{2e}, C_{2f}, \bar{C}_{2e}, \bar{C}_{2f};\ \mathscr{C}_6 = I;\ \mathscr{C}_7 = IC_{2y}, IC_{2z}, \overline{IC}_{2y}, \overline{IC}_{2z};$$

$$\mathscr{C}_8 = IC_{2x}, \overline{IC}_{2x};\ \mathscr{C}_9 = IC_{4x}, IC_{4x}^{-1};\ \mathscr{C}_{10} = IC_{2e}, IC_{2f}, \overline{IC}_{2e}, \overline{IC}_{2f};$$

$$\mathscr{C}_{11} = \bar{\mathscr{C}}_1;\ \mathscr{C}_{12} = \bar{\mathscr{C}}_4;\ \mathscr{C}_{13} = \bar{\mathscr{C}}_6;\ \mathscr{C}_{14} = \bar{\mathscr{C}}_9.$$

(*b*) The character table is given in Table D.31.

	$\mathscr{C}_1$	$\mathscr{C}_2$	$\mathscr{C}_3$	$\mathscr{C}_4$	$\mathscr{C}_5$	$\mathscr{C}_6$	$\mathscr{C}_7$	$\mathscr{C}_8$	$\mathscr{C}_9$	$\mathscr{C}_{10}$
X_6^+, M_6^+	2	0	0	$\sqrt{2}$	0	2	0	0	$\sqrt{2}$	0
X_7^+, M_7^+	2	0	0	$-\sqrt{2}$	0	2	0	0	$-\sqrt{2}$	0
X_6^-, M_6^-	2	0	0	$\sqrt{2}$	0	-2	0	0	$-\sqrt{2}$	0
X_7^-, M_7^-	2	0	0	$-\sqrt{2}$	0	-2	0	0	$\sqrt{2}$	0

Table D.31 Characters of "extra" irreducible representations of D_{4h}^D.

(7) D_{3h}^D:

(*a*) Classes:

$$\mathscr{C}_1 = E;\ \mathscr{C}_2 = C_{3z}, C_{3z}^{-1};\ \mathscr{C}_3 = C_{2x}, C_{2A}, C_{2B}, \bar{C}_{2x}, \bar{C}_{2A}, \bar{C}_{2B};$$

$$\mathscr{C}_4 = IC_{2z}, \overline{IC}_{2z};\ \mathscr{C}_5 = IC_{6z}, IC_{6z}^{-1};$$

$$\mathscr{C}_6 = IC_{2y}, IC_{2C}, IC_{2D}, \overline{IC}_{2y}, \overline{IC}_{2C}, \overline{IC}_{2D};\ \mathscr{C}_7 = \bar{\mathscr{C}}_1;\ \mathscr{C}_8 = \bar{\mathscr{C}}_2;\ \mathscr{C}_9 = \bar{\mathscr{C}}_5.$$

(*b*) The character table is given in Table D.32.

$\mathscr{C}_1$	$\mathscr{C}_2$	$\mathscr{C}_3$	$\mathscr{C}_4$	$\mathscr{C}_5$	$\mathscr{C}_6$
2	1	0	0	$\sqrt{3}$	0
2	1	0	0	$-\sqrt{3}$	0
2	-2	0	0	0	0

Table D.32 Characters of "extra" irreducible representations of D_{3h}^D, C_{6v}^D and D_6^D.

(8) D_{3d}^D:

(*a*) Classes [for $\mathscr{G}_0^D(\mathbf{k})$ of L]:

$$\mathscr{C}_1 = E;\ \mathscr{C}_2 = C_{3\delta}, C_{3\delta}^{-1};\ \mathscr{C}_3 = C_{2b}, C_{2d}, C_{2f};\ \mathscr{C}_4 = I;$$

$$\mathscr{C}_5 = IC_{3\delta}, IC_{3\delta}^{-1};\ \mathscr{C}_6 = IC_{2b}, IC_{2d}, IC_{2f};$$

$$\mathscr{C}_7 = \bar{\mathscr{C}}_1;\ \mathscr{C}_8 = \bar{\mathscr{C}}_2;\ \mathscr{C}_9 = \bar{\mathscr{C}}_3;\ \mathscr{C}_{10} = \bar{\mathscr{C}}_4;\ \mathscr{C}_{11} = \bar{\mathscr{C}}_5;\ \mathscr{C}_{12} = \bar{\mathscr{C}}_6.$$

(*b*) The character table is given in Table D.33.

	$\mathscr{C}_1$	$\mathscr{C}_2$	$\mathscr{C}_3$	$\mathscr{C}_4$	$\mathscr{C}_5$	$\mathscr{C}_6$
L_4^+	1	-1	i	1	-1	i
L_5^+	1	-1	$-i$	1	-1	$-i$
L_6^+	2	1	0	2	1	0
L_4^-	1	-1	i	-1	1	$-i$
L_5^-	1	-1	$-i$	-1	1	i
L_6^-	2	1	0	-2	-1	0

Table D.33 Characters of "extra" irreducible representations of D_{3d}^D.

(9) C^D_{6v}:

(*a*) Classes:

$$\mathscr{C}_1 = E;\ \mathscr{C}_2 = C_{3z}, C_{3z}^{-1};\ \mathscr{C}_3 = IC_{2x}, IC_{2A}, IC_{2B}, \overline{IC}_{2x}, \overline{IC}_{2A}, \overline{IC}_{2B};$$
$$\mathscr{C}_4 = C_{2z}, \bar{C}_{2z};\ \mathscr{C}_5 = C_{6z}, C_{6z}^{-1};\ \mathscr{C}_6 = IC_{2y}, IC_{2C}, IC_{2D}, \overline{IC}_{2y}, \overline{IC}_{2C}, \overline{IC}_{2D};$$
$$\mathscr{C}_7 = \bar{\mathscr{C}}_1;\ \mathscr{C}_8 = \bar{\mathscr{C}}_2;\ \mathscr{C}_9 = \bar{\mathscr{C}}_5.$$

(*b*) The character table is given in Table D.32.

(10) C^D_{6h}:

(*a*) Classes:

$$\mathscr{C}_1 = E;\ \mathscr{C}_2 = C_{6z};\ \mathscr{C}_3 = C_{3z};\ \mathscr{C}_4 = C_{2z};\ \mathscr{C}_5 = C_{3z}^{-1};\ \mathscr{C}_6 = C_{6z}^{-1};\ \mathscr{C}_7 = I;$$
$$\mathscr{C}_8 = IC_{6z};\ \mathscr{C}_9 = IC_{3z};\ \mathscr{C}_{10} = IC_{2z};\ \mathscr{C}_{11} = IC_{3z}^{-1};$$
$$\mathscr{C}_{12} = IC_{6z}^{-1};\ \mathscr{C}_{(12+n)} = \bar{\mathscr{C}}_n,\ n = 1, 2, \ldots, 12.$$

(*b*) The character table is given in Table D.34.

$\mathscr{C}_1$	$\mathscr{C}_2$	$\mathscr{C}_3$	$\mathscr{C}_4$	$\mathscr{C}_5$	$\mathscr{C}_6$	$\mathscr{C}_7$	$\mathscr{C}_8$	$\mathscr{C}_9$	$\mathscr{C}_{10}$	$\mathscr{C}_{11}$	$\mathscr{C}_{12}$
1	θ	θ^2	θ^3	$-\theta^4$	$-\theta^5$	1	θ	θ^2	θ^3	$-\theta^4$	$-\theta^5$
1	$-\theta$	θ^2	$-\theta^3$	$-\theta^4$	θ^5	1	$-\theta$	θ^2	$-\theta^3$	$-\theta^4$	θ^5
1	θ^5	$-\theta^4$	θ^3	θ^2	$-\theta$	1	θ^5	$-\theta^4$	θ^3	θ^2	$-\theta$
1	$-\theta^5$	$-\theta^4$	$-\theta^3$	θ^2	θ	1	$-\theta^5$	$-\theta^4$	$-\theta^3$	θ^2	θ
1	i	-1	$-i$	1	i	1	i	-1	$-i$	1	i
1	$-i$	-1	i	1	$-i$	1	$-i$	-1	i	1	$-i$
1	θ	θ^2	θ^3	$-\theta^4$	$-\theta^5$	-1	$-\theta$	$-\theta^2$	$-\theta^3$	θ^4	θ^5
1	$-\theta$	θ^2	$-\theta^3$	$-\theta^4$	θ^5	-1	θ	$-\theta^2$	θ^3	θ^4	$-\theta^5$
1	θ^5	$-\theta^4$	θ^3	θ^2	$-\theta$	-1	$-\theta^5$	θ^4	$-\theta^3$	$-\theta^2$	θ
1	$-\theta^5$	$-\theta^4$	$-\theta^3$	θ^2	θ	-1	θ^5	θ^4	θ^3	$-\theta^2$	$-\theta$
1	i	-1	$-i$	1	i	-1	$-i$	1	i	-1	$-i$
1	$-i$	-1	i	1	$-i$	-1	i	1	$-i$	-1	i

Table D.34 Characters of "extra" irreducible representations of C^D_{6h} ($\theta = \exp(\frac{1}{6}\pi i)$).

(11) D^D_6:

(*a*) Classes:

$$\mathscr{C}_1 = E;\ \mathscr{C}_2 = C_{3z}, C_{3z}^{-1};\ \mathscr{C}_3 = C_{2x}, C_{2A}, C_{2B}, \bar{C}_{2x}, \bar{C}_{2A}, \bar{C}_{2B};$$
$$\mathscr{C}_4 = C_{2z}, \bar{C}_{2z};\ \mathscr{C}_5 = C_{6z}, C_{6z}^{-1};\ \mathscr{C}_6 = C_{2y}, C_{2C}, C_{2D}, \bar{C}_{2y}, \bar{C}_{2C}, \bar{C}_{2D};$$
$$\mathscr{C}_7 = \bar{\mathscr{C}}_1;\ \mathscr{C}_8 = \bar{\mathscr{C}}_2;\ \mathscr{C}_9 = \bar{\mathscr{C}}_5.$$

(*b*) The character table is given in Table D.32.

(12) T^D:

(*a*) Classes:

$$\mathscr{C}_1 = E;\ \mathscr{C}_2 = C_{3\alpha}, C_{3\beta}, C_{3\gamma}, C_{3\delta};\ \mathscr{C}_3 = C_{3\alpha}^{-1}, C_{3\beta}^{-1}, C_{3\gamma}^{-1}, C_{3\delta}^{-1};$$

$$\mathscr{C}_4 = C_{2x}, C_{2y}, C_{2z}, \bar{C}_{2x}, \bar{C}_{2y}, \bar{C}_{2z};$$

$$\mathscr{C}_5 = \bar{\mathscr{C}}_1;\ \mathscr{C}_6 = \bar{\mathscr{C}}_2;\ \mathscr{C}_7 = \bar{\mathscr{C}}_3.$$

(*b*) The character table is given in Table D.35.

$\mathscr{C}_1$	$\mathscr{C}_2$	$\mathscr{C}_3$	$\mathscr{C}_4$
2	1	1	0
2	ω	ω^2	0
2	$-\omega^2$	$-\omega$	0

Table D.35 Characters of "extra" irreducible representations of T^D ($\omega = \exp(\frac{1}{3}\pi i)$).

(13) D_{2h}^D or V_h^D:

(*a*) Classes [for $\mathscr{G}_0^D(\mathbf{k})$ of N]:

$$\mathscr{C}_1 = E;\ \mathscr{C}_2 = C_{2x}, \bar{C}_{2x};\ \mathscr{C}_3 = C_{2e}, \bar{C}_{2e};\ \mathscr{C}_4 = C_{2f}, \bar{C}_{2f};\ \mathscr{C}_5 = I;$$

$$\mathscr{C}_6 = IC_{2x}, \overline{IC}_{2x};\ \mathscr{C}_7 = IC_{2e}, \overline{IC}_{2e};\ \mathscr{C}_8 = IC_{2f}, \overline{IC}_{2f};\ \mathscr{C}_9 = \bar{\mathscr{C}}_1;\ \mathscr{C}_{10} = \bar{\mathscr{C}}_5.$$

(*b*) The character table is given in Table D.36.

	$\mathscr{C}_1$	$\mathscr{C}_2$	$\mathscr{C}_3$	$\mathscr{C}_4$	$\mathscr{C}_5$	$\mathscr{C}_6$	$\mathscr{C}_7$	$\mathscr{C}_8$
N_5	2	0	0	0	2	0	0	0
$N_{5'}$	2	0	0	0	-2	0	0	0

Table D.36 Characters of "extra" irreducible representations of D_{2h}^D.

(14) C_{4v}^D:

(*a*) Classes [for $\mathscr{G}_0^D(\mathbf{k})$ of Δ]:

$$\mathscr{C}_1 = E;\ \mathscr{C}_2 = C_{2z}, \bar{C}_{2z};\ \mathscr{C}_3 = C_{4z}, C_{4z}^{-1};\ \mathscr{C}_4 = IC_{2x}, IC_{2y}, \overline{IC}_{2x}, \overline{IC}_{2y};$$

$$\mathscr{C}_5 = IC_{2a}, IC_{2b}, \overline{IC}_{2a}, \overline{IC}_{2b};\ \mathscr{C}_6 = \bar{\mathscr{C}}_1;\ \mathscr{C}_7 = \bar{\mathscr{C}}_3.$$

Classes [for $\mathscr{G}_0^D(\mathbf{k})$ of T]:

$$\mathscr{C}_1 = E;\ \mathscr{C}_2 = C_{2x}, \bar{C}_{2x};\ \mathscr{C}_3 = C_{4x}, C_{4x}^{-1};$$

$$\mathscr{C}_4 = IC_{2y}, IC_{2z}, \overline{IC}_{2y}, \overline{IC}_{2z};\ \mathscr{C}_5 = IC_{2e}, IC_{2f}, \overline{IC}_{2e}, \overline{IC}_{2f};\ \mathscr{C}_6 = \bar{\mathscr{C}}_1;\ \mathscr{C}_7 = \bar{\mathscr{C}}_3.$$

(*b*) The character table is given in Table D.37.

	$\mathscr{C}_1$	$\mathscr{C}_2$	$\mathscr{C}_3$	$\mathscr{C}_4$	$\mathscr{C}_5$
Δ_6, T_6, W_6	2	0	$\sqrt{2}$	0	0
Δ_7, T_7, W_7	2	0	$-\sqrt{2}$	0	0

Table D.37 Characters of "extra" irreducible representations of C^D_{4v}, D^D_4 and D^D_{2d}.

(15) D^D_4:

(*a*) Classes:

$$\mathscr{C}_1 = E;\ \mathscr{C}_2 = C_{2y}, \bar{C}_{2y};\ \mathscr{C}_3 = C_{4y}, C_{4y}^{-1};\ \mathscr{C}_4 = C_{2x}, C_{2z}, \bar{C}_{2x}, \bar{C}_{2z};$$
$$\mathscr{C}_5 = C_{2c}, C_{2d}, \bar{C}_{2c}, \bar{C}_{2d};\ \mathscr{C}_6 = \bar{\mathscr{C}}_1;\ \mathscr{C}_7 = \bar{\mathscr{C}}_3.$$

(*b*) The character table is given in Table D.37.

(16) D^D_{2d} or V^D_d:

(*a*) Classes [for $\mathscr{G}^D_0(\mathbf{k})$ of W]:

$$\mathscr{C}_1 = E;\ \mathscr{C}_2 = C_{2y}, \bar{C}_{2y};\ \mathscr{C}_3 = IC_{4y}, IC_{4y}^{-1};\ \mathscr{C}_4 = IC_{2x}, IC_{2z}, \overline{IC}_{2x}, \overline{IC}_{2z};$$
$$\mathscr{C}_5 = C_{2c}, C_{2d}, \bar{C}_{2c}, \bar{C}_{2d};\ \mathscr{C}_6 = \bar{\mathscr{C}}_1;\ \mathscr{C}_7 = \bar{\mathscr{C}}_3.$$

(*b*) The character table is given in Table D.37.

(17) C^D_{4h}:

(*a*) Classes:

$$\mathscr{C}_1 = E;\ \mathscr{C}_2 = C_{4z};\ \mathscr{C}_3 = C_{2z};\ \mathscr{C}_4 = C_{4z}^{-1};\ \mathscr{C}_5 = I;\ \mathscr{C}_6 = IC_{4z};$$
$$\mathscr{C}_7 = IC_{2z};\ \mathscr{C}_8 = IC_{4z}^{-1};\ \mathscr{C}_{(8+n)} = \bar{\mathscr{C}}_n,\ n = 1, 2, \ldots, 8.$$

(*b*) The character table is given in Table D.38.

$\mathscr{C}_1$	$\mathscr{C}_2$	$\mathscr{C}_3$	$\mathscr{C}_4$	$\mathscr{C}_5$	$\mathscr{C}_6$	$\mathscr{C}_7$	$\mathscr{C}_8$
1	ψ	i	$-\psi^3$	1	ψ	i	$-\psi^3$
1	$-\psi$	i	ψ^3	1	$-\psi$	i	ψ^3
1	ψ^3	$-i$	$-\psi$	1	ψ^3	$-i$	$-\psi$
1	$-\psi^3$	$-i$	ψ	1	$-\psi^3$	$-i$	ψ
1	ψ	i	$-\psi^3$	-1	$-\psi$	$-i$	ψ^3
1	$-\psi$	i	ψ^3	-1	ψ	$-i$	$-\psi^3$
1	ψ^3	$-i$	$-\psi$	-1	$-\psi^3$	i	ψ
1	$-\psi^3$	$-i$	ψ	-1	ψ^3	i	$-\psi$

Table D.38 Characters of "extra" irreducible representations of C^D_{4h} ($\psi = \exp(\frac{1}{4}\pi i)$).

(18) C_{3h}^{D}:

(*a*) Classes:

$$\mathscr{C}_1 = E;\ \mathscr{C}_2 = IC_{6z};\ \mathscr{C}_3 = C_{3z};\ \mathscr{C}_4 = IC_{2z};\ \mathscr{C}_5 = IC_{3z}^{-1};$$
$$\mathscr{C}_6 = IC_{6z}^{-1};\ \mathscr{C}_{(6+n)} = \bar{\mathscr{C}}_n,\ n = 1, 2, \ldots, 6.$$

(*b*) The character table is given in Table D.39.

$\mathscr{C}_1$	$\mathscr{C}_2$	$\mathscr{C}_3$	$\mathscr{C}_4$	$\mathscr{C}_5$	$\mathscr{C}_6$
1	θ	θ^2	θ^3	$-\theta^4$	$-\theta^5$
1	$-\theta$	θ^2	$-\theta^3$	$-\theta^4$	θ^5
1	θ^5	$-\theta^4$	θ^3	θ^2	$-\theta$
1	$-\theta^5$	$-\theta^4$	$-\theta^3$	θ^2	θ
1	i	-1	$-i$	1	i
1	$-i$	-1	i	1	$-i$

Table D.39 Characters of "extra" irreducible representations of C_{3h}^{D}, C_{3i}^{D} and C_6^{D} ($\theta = \exp(\frac{1}{6}\pi i)$).

(19) C_{3v}^{D}:

(*a*) Classes [for $\mathscr{G}_0^D(\mathbf{k})$ of Λ]:

$$\mathscr{C}_1 = E;\ \mathscr{C}_2 = C_{3\delta}, C_{3\delta}^{-1};\ \mathscr{C}_3 = IC_{2b}, IC_{2d}, IC_{2f};$$
$$\mathscr{C}_4 = \bar{\mathscr{C}}_1;\ \mathscr{C}_5 = \bar{\mathscr{C}}_2;\ \mathscr{C}_6 = \bar{\mathscr{C}}_3.$$

Classes [for $\mathscr{G}_3^D(\mathbf{k})$ of F]:

$$\mathscr{C}_1 = E;\ \mathscr{C}_2 = C_{3\alpha}, C_{3\alpha}^{-1};\ \mathscr{C}_3 = IC_{2b}, IC_{2c}, IC_{2e};\ \mathscr{C}_4 = \bar{\mathscr{C}}_1;\ \mathscr{C}_5 = \bar{\mathscr{C}}_2;\ \mathscr{C}_6 = \bar{\mathscr{C}}_3.$$

(*b*) The character table is given in Table D.40. The notations in the first column are those of Parmenter (1955).

	$\mathscr{C}_1$	$\mathscr{C}_2$	$\mathscr{C}_3$
Λ_4, F_4	1	-1	i
Λ_5, F_5	1	-1	$-i$
Λ_6, F_6	2	1	0

Table D.40 Characters of "extra" irreducible representations of C_{3v}^{D} and D_3^{D}.

(20) D_3^{D}:

(*a*) Classes:

$$\mathscr{C}_1 = E;\ \mathscr{C}_2 = C_{3z}, C_{3z}^{-1};\ \mathscr{C}_3 = C_{2x}, C_{2A}, C_{2B};\ \mathscr{C}_4 = \bar{\mathscr{C}}_1;\ \mathscr{C}_5 = \bar{\mathscr{C}}_2;\ \mathscr{C}_6 = \bar{\mathscr{C}}_3.$$

(*b*) The character table is given in Table D.40.

(21) C_{3i}^D or S_6^D:

(*a*) Classes:

$$\mathscr{C}_1 = E;\ \mathscr{C}_2 = IC_{3z}^{-1};\ \mathscr{C}_3 = C_{3z};\ \mathscr{C}_4 = I;\ \mathscr{C}_5 = C_{3z}^{-1};\ \mathscr{C}_6 = IC_{3z};$$
$$\mathscr{C}_{(6+n)} = \bar{\mathscr{C}}_n,\ n = 1, 2, \ldots, 6.$$

(*b*) The character table is given in Table D.39.

(22) C_6^D:

(*a*) Classes:

$$\mathscr{C}_1 = E;\ \mathscr{C}_2 = C_{6z};\ \mathscr{C}_3 = C_{3z};\ \mathscr{C}_4 = C_{2z};\ \mathscr{C}_5 = C_{3z}^{-1};\ \mathscr{C}_6 = C_{6z}^{-1};$$
$$\mathscr{C}_{(6+n)} = \bar{\mathscr{C}}_n,\ n = 1, 2, \ldots, 6.$$

(*b*) The character table is given in Table D.39.

(23) C_{2v}^D:

(*a*) Classes [for $\mathscr{G}_0^D(\mathbf{k})$ of Σ]:

$$\mathscr{C}_1 = E;\ \mathscr{C}_2 = C_{2e}, \bar{C}_{2e};\ \mathscr{C}_3 = IC_{2x}, \overline{IC}_{2x};\ \mathscr{C}_4 = IC_{2f}, \overline{IC}_{2f};\ \mathscr{C}_5 = \bar{\mathscr{C}}_1.$$

Classes [for $\mathscr{G}_0^D(\mathbf{k})$ of D]:

$$\mathscr{C}_1 = E;\ \mathscr{C}_2 = C_{2x}, \bar{C}_{2x};\ \mathscr{C}_3 = IC_{2e}, \overline{IC}_{2e};\ \mathscr{C}_4 = IC_{2f}, \overline{IC}_{2f};\ \mathscr{C}_5 = \bar{\mathscr{C}}_1.$$

Classes [for $\mathscr{G}_0^D(\mathbf{k})$ of S]:

$$\mathscr{C}_1 = E;\ \mathscr{C}_2 = C_{2a}, \bar{C}_{2a};\ \mathscr{C}_3 = IC_{2z}, \overline{IC}_{2z};\ \mathscr{C}_4 = IC_{2b}, \overline{IC}_{2b};\ \mathscr{C}_5 = \bar{\mathscr{C}}_1.$$

Classes [for $\mathscr{G}_0^D(\mathbf{k})$ of Z]:

$$\mathscr{C}_1 = E;\ \mathscr{C}_2 = C_{2y}, \bar{C}_{2y};\ \mathscr{C}_3 = IC_{2z}, \overline{IC}_{2z};\ \mathscr{C}_4 = IC_{2x}, \overline{IC}_{2x};\ \mathscr{C}_5 = \bar{\mathscr{C}}_1.$$

Classes [for $\mathscr{G}_0^D(\mathbf{k})$ of G]:

$$\mathscr{C}_1 = E;\ \mathscr{C}_2 = C_{2f}, \bar{C}_{2f};\ \mathscr{C}_3 = IC_{2x}, \overline{IC}_{2x};\ \mathscr{C}_4 = IC_{2e}, \overline{IC}_{2e};\ \mathscr{C}_5 = \bar{\mathscr{C}}_1.$$

(*b*) The character table is given in Table D.41.

$\mathscr{C}_1$	$\mathscr{C}_2$	$\mathscr{C}_3$	$\mathscr{C}_4$
2	0	0	0

Table D.41 Characters of the "extra" irreducible representation of C_{2v}^D and D_2^D.

(24) C_{2h}^D:

(*a*) Classes:

$$\mathscr{C}_1 = E;\ \mathscr{C}_2 = C_{2z};\ \mathscr{C}_3 = I;\ \mathscr{C}_4 = IC_{2z};\ \mathscr{C}_{(4+n)} = \bar{\mathscr{C}}_n,\ n = 1, 2, 3, 4.$$

(*b*) The character table is given in Table D.42.

$\mathscr{C}_1$	$\mathscr{C}_2$	$\mathscr{C}_3$	$\mathscr{C}_4$
1	i	1	i
1	$-i$	1	$-i$
1	i	-1	$-i$
1	$-i$	-1	i

Table D.42 Characters of "extra" irreducible representations of C_{2h}^D.

(25) D_2^D or V^D:

(*a*) Classes:

$$\mathscr{C}_1 = E;\ \mathscr{C}_2 = C_{2x}, \bar{C}_{2x};\ \mathscr{C}_3 = C_{2y}, \bar{C}_{2y};\ \mathscr{C}_4 = C_{2z}, \bar{C}_{2z};\ \mathscr{C}_5 = \bar{\mathscr{C}}_1.$$

(*b*) The character table is given in Table D.41.

(26) C_4^D:

(*a*) Classes:

$$\mathscr{C}_1 = E;\ \mathscr{C}_2 = C_{4z};\ \mathscr{C}_3 = C_{2z};\ \mathscr{C}_4 = C_{4z}^{-1};\ \mathscr{C}_{(4+n)} = \bar{\mathscr{C}}_n,\ n = 1, 2, 3, 4.$$

(*b*) The character table is given in Table D.43.

$\mathscr{C}_1$	$\mathscr{C}_2$	$\mathscr{C}_3$	$\mathscr{C}_4$
1	ψ	i	$-\psi^3$
1	$-\psi$	i	ψ^3
1	ψ^3	$-i$	$-\psi$
1	$-\psi^3$	$-i$	ψ

Table D.43 Characters of "extra" irreducible representations of C_4^D and S_4^D ($\psi = \exp(\frac{1}{4}\pi i)$).

(27) S_4^D:

(*a*) Classes:

$$\mathscr{C}_1 = E;\ \mathscr{C}_2 = IC_{4y};\ \mathscr{C}_3 = C_{2y};\ \mathscr{C}_4 = IC_{4y}^{-1};\ \mathscr{C}_{(4+n)} = \bar{\mathscr{C}}_n,\ n = 1, 2, 3, 4.$$

(*b*) The character table is given in Table D.43.

(28) C_3^D:

(*a*) Classes:

$$\mathscr{C}_1 = E;\ \mathscr{C}_2 = C_{3z};\ \mathscr{C}_3 = C_{3z}^{-1};\ \mathscr{C}_{(3+n)} = \bar{\mathscr{C}}_n,\ n = 1, 2, 3.$$

(*b*) The character table is given in Table D.44.

$\mathscr{C}_1$	$\mathscr{C}_2$	$\mathscr{C}_3$
1	ω	ω^2
1	$-\omega^2$	$-\omega$
1	-1	1

Table D.44 Characters of "extra" irreducible representations of C_3^D ($\omega = \exp(\frac{1}{3}\pi i)$).

(29) C_s^D or C_{1h}^D:

(*a*) Classes:

$$\mathscr{C}_1 = E;\ \mathscr{C}_2 = IC_{2z};\ \mathscr{C}_3 = \bar{\mathscr{C}}_1;\ \mathscr{C}_4 = \bar{\mathscr{C}}_2.$$

(*b*) The character table is given in Table D.45.

$\mathscr{C}_1$	$\mathscr{C}_2$
1	i
1	$-i$

Table D.45 Characters of "extra" irreducible representations of C_s^D, C_2^D and C_i^D.

(30) C_2^D:

(*a*) Classes [for $\mathscr{G}_0^D(\mathbf{k})$ of Q]:

$$\mathscr{C}_1 = E;\ \mathscr{C}_2 = C_{2d};\ \mathscr{C}_3 = \bar{\mathscr{C}}_1;\ \mathscr{C}_4 = \bar{\mathscr{C}}_2.$$

(*b*) The character table is given in Table D.45.

(31) C_i^D:

(*a*) Classes:

$$\mathscr{C}_1 = E;\ \mathscr{C}_2 = I;\ \mathscr{C}_3 = \bar{\mathscr{C}}_1;\ \mathscr{C}_4 = \bar{\mathscr{C}}_2.$$

(*b*) The character table is given in Table D.45.

(32) C_1^D:

(*a*) Classes:

$$\mathscr{C}_1 = E;\ \mathscr{C}_2 = \bar{\mathscr{C}}_1.$$

(*b*) $\chi(\mathscr{C}_1) = 1.$

3 The linear infinite point groups $D_{\infty h}$ and $C_{\infty h}$

The elements and classes of $D_{\infty h}$ and $C_{\infty h}$ are described in detail in Example VI of Chapter 2, Section 7. As $D_{\infty h}$ and $C_{\infty h}$ are compact Lie groups they each possess a countable infinity of inequivalent irreducible representations (see Theorem X of Chapter 4, Section 6). (Because of the semi-direct product structure of $D_{\infty h}$ and $C_{\infty h}$, these representations may be induced (see Chapter 5, Section 7) from the Abelian invariant subgroup $\mathscr{G}'_1$, which, being isomorphic to U(1), has the irreducible representations specified in Chapter 4, Section 7(*b*).) The character tables are given in Tables D.46 and D.47. (As $\chi^{\mathrm{p}}(C_\pi) = \lim_{\theta\to\pi}\chi^{\mathrm{p}}(C_\theta)$ and $\chi^{\mathrm{p}}(IC_\pi) = \lim_{\theta\to\pi}\chi^{\mathrm{p}}(IC_\theta)$, the values of $\chi^{\mathrm{p}}(C_\pi)$ and $\chi^{\mathrm{p}}(IC_\pi)$ are not listed separately.) The labelling notation is similar to that described for finite point groups in Section 1.

	E	C_θ, $C_{-\theta}$	$C_{2\Phi}$	I	IC_θ, $IC_{-\theta}$	$IC_{2\Phi}$
A_{1g}	1	1	1	1	1	1
A_{1u}	1	1	1	−1	−1	−1
A_{2g}	1	1	−1	1	1	−1
A_{2u}	1	1	−1	−1	−1	1
E_{ng}	2	$2\cos(n\theta)$	0	2	$2\cos(n\theta)$	0
E_{nu}	2	$2\cos(n\theta)$	0	−2	$-2\cos(n\theta)$	0

Table D.46 Character table for $D_{\infty h}$. (In this table n takes each of the values $1, 2, 3, \ldots$.)

	E	C_θ, $C_{-\theta}$	$IC_{2\Phi}$
A_1	1	1	1
A_2	1	1	−1
E_n	2	$2\cos(n\theta)$	0

Table D.47 Character table for $C_{\infty h}$. (In this table n takes each of the values $1, 2, 3, \ldots$.)

References

Abers, E. and Lee, B. (1973). *Phys. Rep.* **9C,** 2–141.

Abrams, G. S., Briggs, D., Chinowsky, W., Friedberg, C. E., Goldhaber, G., Hollebeek, R. J., Kadyk, J. A., Litke, A., Lulu, B., Pierre, F., Sadoulet, B., Trilling, G. H., Whitaker, J. S., Wiss, J., Zipse, J. E., Augustin, J.-E., Boyarski, A. M., Breidenbach, M., Bulos, F., Feldman, G. J., Fischer, G. E., Fryberger, D., Hanson, G., Jean-Marie, B., Larsen, R. R., Luth, V., Lynch, H. L., Lyon, D., Morehouse, C. C., Paterson, J. M., Perl, M. L., Richter, B., Rapidis, P., Schwitters, R. F., Tanenbaum, W. and Vannucci, F. (1974). *Phys. Rev. Letts.* **33,** 1453–1455.

Adams, J. F. (1969). "Lectures on Lie Groups". W. A. Benjamin, New York.

Adler, S. L. (1969). *Phys. Rev.* **177,** 2426–2438.

Ado, I. D. (1947). *Uspeki Mat. Nauk* (*N.S.*) **2, No. 6(22),** 159–173. [Translation (1962): *Amer. Math. Soc. Trans. Series* 1, **9,** 308–327.]

Agrawala, V. K. (1980). *J. Math. Phys.* **21,** 1562–1565.

Aguiler-Benitez, M., Crawford, R. L., Frosch, R., Gopal, G. P., Hendrick, R. E., Kelly, R. L., Losty, M. J., Montanet, L., Porter, F. C., Rittenberg, A., Roos, M., Roper, L. D., Shimada, T., Shrock, R. E., Trippe, T. G., Walck, Ch., Wohl, C. G., Yost, G. P. and Armstrong, B. (1982). *Phys. Letts.* **111B,** i–xxi and 1–294.

Aitchison, I. J. R. (1982). "An Informal Introduction to Gauge Theories". Cambridge U.P., Cambridge, England.

Aitchison, I. J. R. and Hey, A. J. G. (1982). "Gauge Theories in Particle Physics". Adam Hilger, Bristol.

Akhiezer, N. I. and Glazman, I. M. (1961). "Theory of Linear Operators in Hilbert Space", Vol. I. Frederick Ungar, New York.

Altmann, S. L. (1957). *Proc. Camb. Phil. Soc.* **53,** 343–367.

Altmann, S. L. (1962). *Phil. Trans.* **A255,** 216–240.

Altmann, S. L. (1963). *Rev. Mod. Phys.* **35,** 641–645.

Altmann, S. L. and Cracknell, A. P. (1965). *Rev. Mod. Phys.* **37,** 19–32.

Amati, D., Bacry, H., Nuyts, J. and Prentki, J. (1964). *Nuovo Cimento* **34,** 1732–1750.

Anderson, H. L., Fermi, E., Martin, R. and Nagle, D. E. (1953). *Phys. Rev.* **91,** 155–168.

Antoine, J.-P., Speiser, D. and Oakes, R. J. (1966). *Phys. Rev.* **141,** 1542–1553.

Arnison, G., Astbury, A., Aubert, B., Bacci, C., Bauer, G., Bezaguet, A., Bock, R., Bowcock, T. J. V., Calvetti, M., Carroll, T., Catz, P., Cennini, P., Centro, S., Ceradini, F., Cittolin, S., Cline, D., Cochet, C., Colas, J., Corden, M., Dallman, D., DcBeer, M., Della Negra, M., Demoulin, M., Denegri, D., DiCiaccio, A., DiBitonto, D., Dobrzynski, L., Dowell, J. D., Edwards, M., Eggert, K., Eisenhandler, E., Ellis, N., Erhard, P., Faissner, H., Fontaine, G., Frey, R., Fruhwirth, R., Garvey, J., Geer, S., Ghesquiere, C., Ghez, P., Giboni, K. L., Gibson, W. R., Giraud-Héraud, Y., Givernaud, A., Gonidec, A., Grayer, G., Gutierrez, P., Hansl-Kozanecka, T., Haynes, W. J. Hertzberger, L. O., Hodges, C., Hoffman, D., Hoffmann, H., Holthuizen, D. J., Homer, R. J., Honma, A., Jank, W., Jorat, G., Kalmus, P. I. P., Karimaki, V., Keeler, R., Kenyon, I., Kernan, A., Kinnunen, R., Kowalski, H., Kozanecki, W., Kryn, D., Lacava, F., Laugier, J.-P., Lees, J.-P., Lehmann, H., Leuchs, K., Lévêque, A., Linglin, D., Locci, E., Loret, M., Malosse, J.-J., Markiewicz, T., Maurin, G., McMahon, T., Mendiburu, J.-P., Minard, M.-N., Moricca, M., Muirhead, H., Muller, F., Nandi, A. K., Naumann, L., Norton, A., Orkin-Lecourtois, A., Paoluzi, G., Petrucci, G., Piano Mortari, G., Pimiä, M., Placci, A., Radermacher, E., Ransdell, J., Reithler, H., Revol, J.-P., Rich, J., Rijssenbeek, M., Roberts, C., Rohlf, J., Rossi, P., Rubbia, C., Sadoulet, B., Sajot, G., Salvi, G., Salvini, G., Sass, J., Saudraix, J., Savoy-Navarro, A., Schinzel, D., Scott, W., Shah, T. P., Spiro, M., Strauss, J., Sumorok, K., Szoncso, F., Smith, D., Tao, C. Thompson, G., Timmer, J., Tscheslog, E., Tuominiemi, J., van der Meer, S., Vialle, J.-P., Vrana, J., Vuillemin, V., Wahl, H. D., Watkins, P , Wilson, J., Xie, Y. G., Yvert, M. and Zurfluh, E. (1983a). *Phys. Letts.* **122B,** 103–116.

Arnison, G., Astbury, A., Aubert, B., Bacci, C., Bauer, G., Bézaguet, A., Böck, R., Bowcock, T. J. V., Calvetti, M., Catz, P., Cennini, P., Centro, S., Ceradini, F., Cittolin, S., Cline, D., Cochet, C., Colas, J., Corden, M., Dallman, D., Dau, D., DeBeer, M., Della Negra, M., Demoulin, M., Denegri, D., DiCiaccio, A., DiBitonto, D., Dobrzynski, L., Dowell, J. D., Eggert, K., Eisenhandler, E., Ellis, N., Erhard, P., Faissner, H., Fincke, M., Fontaine, G., Frey, R., Frühwirth, R., Garvey, J., Geer, S., Ghesquière, C., Ghez, P., Giboni, K., Gibson, W. R., Giraud-Héraud, Y., Givernaud, A., Gonidec, A., Grayer, G., Hansl-Kozanecka, T., Haynes, W. J., Hertzberger, L. O., Hodges, C., Hoffman, D., Hoffmann, H., Holthuizen, D. J., Homer, R. J., Honma, A., Jank, W., Jorat, G., Kalmus, P. I. P., Karimäki, V., Keeler, R., Kenyon, I., Kernan, A., Kinnunen, R., Kozanecki, W., Kryn, D., Lacava, F., Laugier, J.-P., Lees, J.-P., Lehmann, H., Leuchs, R., Lévêque, A., Linglin, D., Locci, E., McMahon, T., Malosse, J.-J., Markiewicz, T., Maurin, G., Mendiburu, J.-P., Minard, M.-N., Mohammadi, M., Moricca, M., Morgan, K., Muirhead, H., Muller, F., Nandi, A. K., Naumann, L., Norton, A., Orkin-Lecourtois, A., Paoluzi, L., Pauss, F., Piano Mortari, G., Pietarinen, E., Pimiä, M., Placci, A., Porte, J. P., Radermacher, E., Ransdell, J., Reithler, H., Revol, J.-P., Rich, J., Rijssenbeek, M., Roberts, C., Rohlf, J., Rossi, P., Rubbia, C., Sadoulet, B., Sajot, G., Salvi, G., Salvini, G., Sass, J., Saudraix, J., Savoy-Navarro, A., Schinzel, D., Scott, W., Shah, T. P., Spiro, M., Strauss, J., Streets, J., Sumorok, K., Szoncso, F., Smith, D., Tao, C., Thompson, G., Timmer, J., Tscheslog, E., Tuominiemi, J., van Eijk, B.,

Vialle, J.-P., Vrana, J., Vuillemin, V., Wahl, H. D., Watkins, P., Wilson, J., Wulz, C., Xie, G. Y., Yvert, M. and Zurfluh, E. (1983b). *Phys. Letts.* **126B,** 398–410.

Aronson, E. B., Malkin, I. A. and Man'ko, V. I. (1974). *Sov. J. Particles and Nuclei.* **5,** 47–67.

Aubert, J. J., Becker, U., Biggs, P. J., Burger, J., Chen, M., Everhart, G., Goldhagen, P., Leong, J., McCorriston, T., Rhoades, T. G., Rohde, M., Ting, S. C. C., Wu, Sau Lan and Lee, Y. Y. (1974). *Phys. Rev. Letts.* **33,** 1404–1406.

Augustin, J. E., Boyarski, A. M., Breidenbach, M., Bulos, F., Dakin, J. T., Feldman, G. J., Fischer, G. E., Fryberger, D., Hanson, G., Jean-Marie, B., Larsen, R. R., Luth, V., Lynch, H. L., Lyon, D., Morehouse, C. C., Paterson, J. M., Perl, M. L., Richter, B., Rapidis, P., Schwitters, R. F., Tanenbaum, W. M., Vannucci, F., Abrams, G. S., Briggs, D., Chinowsky, W., Friedberg, C. E., Goldhaber, G., Hollebeek, R. J., Kadyk, J. A., Lulu, B., Pierre, F., Trilling, G. H., Whitaker, J. S., Wiss, J. and Zipse, J. E. (1974). *Phys. Rev. Letts.* **33,** 1406–1408.

Bailin, D. (1982). "Weak Interactions". Adam Hilger, Bristol.

Baird, G. E. and Biedenharn, L. C. (1963). *J. Math. Phys.* **4,** 1449–1466.

Baker, H. F. (1905). *Proc. London Math. Soc.* (2) **3,** 24–47.

Balkanski, M. and Nusimovici, M. (1964). *Phys. Status Solidi* **5,** 635–647.

Bargmann, V. (1936). *Z. Phys.* **99,** 576–582.

Bargmann, V. (1947). *Ann. Math.* **48,** 568–640.

Bargmann, V. (1954). *Ann. Math.* **59,** 1–46.

Bargmann, V. and Wigner, E. P. (1948). *Proc. Nat. Acad. Sci.* **34,** 211–223.

Barnes, V. E., Connolly, P. L., Crennell, D. J., Culwick, B. B., Delaney, W. C., Fowler, W. B., Hagerty, P. E., Hart, E. L., Horwitz, N., Hough, P. V. C., Jensen, J. E., Kopp, J. K., Lai, K. W., Leitner, J., Lloyd, J. L., London, G. W., Morris, T. W., Oren, Y., Palmer, R. B., Prodell, A. G., Radojicic, D., Rahm, D. C., Richardson, C. R., Samios, N. P., Sanford, J. R., Shutt, R. P., Smith, J. R., Stonehill, D. L., Strang, R. C., Thorndike, A. M., Webster, M. S., Willis, W. J. and Yamamoto, S. S. (1964). *Phys. Rev. Letts.* **12,** 204–206.

Barut, A. O. and Raczka, R. (1977). "Theory of Group Representations and Applications". Polish Scientific Publishers, Warsaw.

Bateman, H. (1932). "Partial Differential Equations of Mathematical Physics". Cambridge U.P., Cambridge, England.

Behrends, R. E. (1966). *In* "Lectures in Theoretical Physics" (Ed. W. E. Britten), Vol. VIIIB, 363–373. Gordon and Breach, New York.

Behrends, R. E. (1968). *In* "Group Theory and its Applications" (Ed. E. M. Loebl), Vol. I, 541–627. Academic Press, Orlando, New York and London.

Behrends, R. E., Dreitlein, J., Fronsdal C. and Lee, W. (1962). *Rev. Mod. Phys.* **34,** 1–40.

Bell, D. G. (1954). *Rev. Mod. Phys.* **26,** 311–320.

Bell, J. S. and Jackiw, R. (1969). *Nuovo Cimento* **60A,** 47–61.

Berenson, R. and Birman, J. L. (1975). *J. Math. Phys.* **16,** 227–235.

Berenson, R., Itzkan, I. and Birman J. L. (1975). *J. Math. Phys.* **16,** 236–242.

Berestetskii, V. B. (1965). *Soviet Phys. Uspekhi* **8,** 147–176.

Bernstein, J. (1974). *Rev. Mod. Phys.* **46,** 7–48.

Bertrand, J. (1966). *Ann. Inst. Henri Poincare,* **5,** 235–256.

Bethe, H. A. (1929). *Ann. Physik* **3,** 133–208.

Bethe, H. A. and de Hoffmann, F. (1955). "Mesons and Fields. Vol. II: Mesons". Row, Peterson and Co., Evanston.
Bhabha, H. J. (1945). *Rev. Mod. Phys.* **17,** 200–216.
Biedenharn, L. C. and Louck, J. D. (1979a). "Angular Momentum in Quantum Mechanics". Addison-Wesley, Reading, Massachusetts.
Biedenharn, L. C. and Louck, J. D. (1979b). "The Racah-Wigner Algebra in Quantum Theory". Addison-Wesley, Reading, Massachusetts.
Bilenky, S. M. (1982). "Introduction to the Physics of Electroweak Interactions". Pergamon Press, Oxford.
Birman, J. L. (1962). *Phys. Rev.* **127,** 1093–1106.
Birman, J. L. (1963). *Phys. Rev.* **131,** 1489–1496.
Birman, J. L. (1966). *Phys. Rev.* **150,** 771–782.
Bjorken, J. D. and Drell, S. D. (1964). "Relativistic Quantum Mechanics". McGraw-Hill, New York.
Bjorken, J. D. and Drell, S. D. (1965). "Relativistic Quantum Fields". McGraw-Hill, New York.
Bjorken, J. D. and Glashow, S. L. (1964). *Phys. Letts.* **11,** 255–257.
Bloch, F. (1928). *Z. Phys.* **52,** 555–600.
Boerner, H. (1963). "Representations of Groups". North Holland, Amsterdam.
Bogoliubov, N. N. and Shirkov, D. V. (1959). "Introduction to the Theory of Quantised Fields". Interscience, New York and London.
Born, M. and Huang, K. (1954). "Dynamical Theory of Crystal Lattices". Oxford U.P., London and New York.
Born, M. and Oppenheimer, J. R. (1927). *Ann. Physik.* **84,** 457–484.
Born, M. and von Karman, T. (1912). *Phys. Z.* **13,** 297–309.
Bouckaert, L. P., Smoluchowski, R. and Wigner, E. P. (1936). *Phys. Rev.* **50,** 58–67.
Bourbaki, N. (1975). "Groupes et Algebres de Lie", Chapters VII and VIII. Hermann, Paris.
Bradley, C. J. (1966). *J. Math. Phys.* **7,** 1145–1152.
Bradley, C. J. and Cracknell, A. P. (1972). "The Mathematical Theory of Symmetry in Solids: Representation Theory for Point Groups and Space Groups". Oxford U.P., Oxford.
Brennich, R. H. (1970). *Ann. Inst. Henri Poincare* **13,** 137–161.
Brescansin, L. M., Padial, N. T. and Shukla, M. M. (1976). *Nuovo Cimento* **34B,** 103–112.
Brickell, F. and Clark, R. S. (1970). "Differentiable Manifolds". Van Nostrand Reinhold, New York.
Brout, R. and Englert, F. (1964). *Phys. Rev. Letts.* **13,** 321–323.
Burns, G. and Glazer, A. M. (1978). "Introduction to Space Groups for Solid State Scientists". Academic Press, Orlando, New York and London.
Butler, P. H. (1981). "Point Group Symmetry Applications, Methods and Tables". Plenum, New York.
Callaway, J. (1958). "Electron Energy Bands in Solids". *In* "Solid State Physics, Advances in Research and Applications", Vol. 7 (Eds. F. Seitz and D. Turnbull), 100–212. Academic Press, Orlando, New York and London.
Callaway, J. (1964). "Energy Band Theory". Academic Press, Orlando, New York and London.
Campbell, J. E. (1897a). *Proc. London Math. Soc.* (1) **28,** 381–390.
Campbell, J. E. (1897b). *Proc. London Math. Soc.* (1) **29,** 14–32.
Carmeli, M. and Malin, S. (1976). "Representations of the Rotation and

Lorentz Groups — An Introduction". Marcel Dekker, New York.
Carruthers, P. (1966). "Unitary Symmetry of Strong Interactions". Wiley-Interscience, New York.
Cartan, E. (1894). "Sur la Structure des Groupes de Transformations Finis et Continus". Thesis, Paris. [Reprint (1952): *In* "Oeuvres Completes", Partie I, Vol. I, 137–287. Vuibert, Paris.]
Cartan, E. (1913). *Bull. Soc. Math. Franc.* **41,** 53–96. [Reprint (1952): *In* "Oeuvres Completes", Partie I, Vol. I, 355–398. Vuibert, Paris.]
Cartan, E. (1914). *Ann. l'Ecole Norm. Sup. 3-eme serie* **31,** 263–355.
Cartan, E. (1929). *J. Math. Pure Appl.* **8,** 1–33.
Casimir, H. (1931). *Proc. R. Acad. Amstd.* **34,** 844–846.
Charap, J. M., Jones, R. B. and Williams, P. G. (1967). *Repts. Progr. Phys.* **30,** 227–283.
Chelikowsky, J. R. and Cohen, M. L. (1976). *Phys. Rev.* **B14,** 556–582.
Chevalley, C. (1946). "Theory of Lie Groups", Vol. 1. Princeton U.P., Princeton.
Chevalley, C. (1955). *Tohoku Math. J.* (2) **I,** 14–66.
Chinowsky, W. (1977). *In* "Annual Review of Nuclear Science" (Eds. E. Segre, J. R. Grover and H. P. Noyes), Vol. 27, 393–464. Annual Reviews, Palo Alto, California.
Close, F. E. (1979). "An Introduction to Quarks and Partons". Academic Press, London, Orlando and New York.
Cochrane, W. (1973). "The Dynamics of Atoms in Crystals". Edward Arnold, London.
Cohen, E. R. (1974). "Tables of the Clebsch–Gordan Coefficients". North American Rockwell Science Center, Thousand Oaks, California.
Coleman, A. J. (1968). "Induced and Subduced Representations". *In* "Group Theory and its Applications" (Ed. E. M. Loebl), 57–118. Academic Press, Orlando, New York and London.
Coleman, S. and Glashow, S. L. (1961). *Phys. Rev. Letts.* **6,** 423–425.
Coleman, S. and Mandula, J. (1967). *Phys. Rev.* **159,** 1251–1256.
Condon, E. U. and Odabasi, H. (1980). "Atomic Structure". Cambridge U.P., Cambridge, England.
Condon, E. U. and Shortley, G. H. (1935). "The Theory of Atomic Spectra". Cambridge U.P., Cambridge, England.
Copson, E. T. (1935). "Theory of Functions of a Complex Variable". Oxford U.P., Oxford.
Cordero, P. and Ghirardi, G. C. (1972). *Forts. d. Phys.* **20,** 105–133.
Cornwell, J. F. (1964). *Phys. Konders. Materie* **1,** 161–165.
Cornwell, J. F. (1969). "Group Theory and Electronic Energy Bands in Solids". North Holland, Amsterdam.
Cornwell, J. F. (1970). *Phys. Status Solidi* **37,** 225–235.
Cornwell, J. F. (1971a). *Repts. Math. Phys.* **2,** 153–163.
Cornwell, J. F. (1971b). *Repts. Math. Phys.* **2,** 229–238.
Cornwell, J. F. (1971c). *Repts. Math. Phys.* **2,** 239–261.
Cornwell, J. F. (1971d). *Repts. Math. Phys.* **2,** 289–309.
Cornwell, J. F. (1971e). *Phys. Status Solidi* **43(b),** 763–767.
Cornwell, J. F. (1972a). *Phys. Status Solidi* **52(b),** 275–283.
Cornwell, J. F. (1972b). *Repts. Math. Phys.* **3,** 91–107.
Cornwell, J. F. and Ekins, J. M. (1975). *J. Math. Phys.* **16,** 394–399.
Cracknell, A. P. and Davies, B. L. (1979). "Kronecker Product Tables. Volume 3:

Wave Vector Selection Rules for Irreducible Representations of Triclinic, Monoclinic, Tetragonal, Trigonal, and Hexagonal Space Groups". Plenum, New York.

Cracknell, A. P., Davies, B. L., Miller, S. C. and Love, W. F. (1979). "Kronecker Product Tables. Volume 1: General Introduction and Tables of Irreducible Representations of Space Groups", Plenum, New York.

Croom, F. H. (1978). "Basic Concepts of Algebraic Topology". Springer-Verlag, New York, Heidelberg and Berlin.

Dashen, R. F. and Sharp, D. H. (1964). *Phys. Rev.* **B133,** 1585–1588.

Davies, B. L. and Cracknell, A. P. (1979). "Kronecker Product Tables. Volume 2: Wave Vector Selection Rules and Reductions of Kronecker Products for Irreducible Representations of Orthorhombic and Cubic Space Groups". Plenum, New York.

Davies, B. L. and Cracknell, A. P. (1980). "Kronecker Product Tables. Volume 4: Symmetrised Powers of Irreducible Representations of Space Groups". Plenum, New York.

de Angelis, B. A., Newnham, R. E. and White, R. B. (1972). *Amer. Mineral.* **57,** 255–268.

de Franceschi, G. and Maiani, L. (1965). *Fortschr. Phys.* **13,** 279–384.

de Swart, J. J. (1963). *Rev. Mod. Phys.* **35,** 916–939.

de Swart, J. J. (1965). *Rev. Mod. Phys.* **37,** 326.

de Swart, J. J. (1974). *Proc. Third International Colloquium on Group Theoretical Methods in Physics*, Vol. 2, 378–387. Centre National de la Recherche Scientifique, Marseille.

Dirac, P. A. M. (1928). *Proc. Roy. Soc.* (*London*). **A117,** 610–624.

Dirl, R. (1979a). *J. Math. Phys.* **20,** 659–663.

Dirl, R. (1979b). *J. Math. Phys.* **20,** 664–670.

Dirl, R. (1979c). *J. Math. Phys.* **20,** 671–678.

Dirl, R. (1979d). *J. Math. Phys.* **20,** 679–684.

Donnay, J. D. H. and Nowacki, W. (1954). "Crystal Data". Geological Society of America, Memoir 60, New York.

Dresselhaus, G. (1955). *Phys. Rev.* **100,** 580–586.

Dynkin, E. B. (1947). *Uspeki. Mat. Nauk.* (*N.S.*) **2, No. 4(20),** 59–127. [Translation (1950): *Amer. Math. Soc. Trans. Series 1,* **9,** 328–469.]

Dynkin, E. B. (1952). *Mat. Sbornik.* **30(72)**, 349–462. [Translation (1957): *Amer. Math. Soc. Trans. Series* 2, **6,** 111–244.]

Dynkin, E. B. and Oniscik, A. L. (1955). *Uspeki. Mat. Nauk.* (*N.S.*) **10, No. 4(66),** 3–74. [Translation (1962): *Amer. Math. Soc. Trans. Series 2,* **21,** 119–192.]

Dyson, F. (1966). "Symmetry Groups in Nuclear and Particle Physics". W. A. Benjamin, New York.

Edmonds, A. R. (1957). "Angular Momentum in Quantum Mechanics". Princeton U.P., Princeton.

Ekins, J. M. and Cornwell, J. F. (1974). *Repts. Math. Phys.* **5,** 17–49.

Ekins, J. M. and Cornwell, J. F. (1975). *Repts. Math. Phys.* **5,** 167–203.

Elliott, R. J. (1954). *Phys. Rev.* **96,** 280–287.

Elliott, R. J. and Loudon, R. (1960). *J. Phys. Chem. Solids* **15,** 146–151.

Ellis, J. (1981). *In* "Gauge Theories and Experiments at High Energies" (Eds. K. C. Bowler and D. G. Sutherland), 201–302. Scottish Universities Summer Schools in Physics Publications, Edinburgh.

Emmerson, J. McL. (1972). "Symmetries in Particle Physics". Oxford U.P., Oxford.

Englefield, M. J. and King, R. C. (1980). *J. Phys.* **A13,** 2297–2317.

Faddeev, L. D. and Slavnov, A. A. (1980). "Gauge Fields: Introduction to Quantum Theory". Benjamin/Cummins Publ. Co., Reading, Massachusetts.

Fermi, E. (1934a). *Nuovo Cimento* **11,** 1–19.

Fermi, E. (1934b). *Z. Phys.* **88,** 161–177.

Fischler, M. (1980). "Young-Tableaux Methods for Kronecker Products of Representations of the Classical Groups". Fermi Laboratory Report FERMILAB-PUB-80/49-THY Batavia, Illinois.

Flamm, D. and Schöberl, F. (1982). "Introduction to the Quark Model of Elementary Particles. I: Quantum Numbers, Gauge Theories, and Hadron Spectroscopy". Gordon and Breach, New York and London.

Fleming, W. (1977). "Functions of Several Variables", second edition. Springer-Verlag, New York, Heidelberg and Berlin.

Fletcher, G. C. (1971). "The Electron Band Theory of Solids". North Holland, Amsterdam.

Fock, V. (1935). *Z. Phys.* **98,** 145–154.

Fonda, L. and Ghirardi, G. C. (1970). "Symmetry Principles in Quantum Physics". Marcel Dekker, New York.

Frazer, W. R. (1966). "Elementary Particles". Prentice-Hall, Englewood Cliffs, New Jersey.

Freudenthal, H. (1964). *Adv. in Maths.* **1,** 143–190.

Freudenthal, H. and de Vries, H. (1969). "Linear Lie Groups". Academic Press, Orlando, New York and London.

Fritzsch, H., Gell-Mann, M. and Leutwyler, H. (1973). *Phys. Letts.* **47B,** 365–368.

Frobenius, G. and Schur, I. (1906). *Sitzgsber. Preuss. Akad. Wiss.*, 186–208.

Gaillard, M. K., Lee, B. W. and Rosner, J. (1975). *Rev. Mod. Phys.* **47,** 277–310.

Gantmacher, F. R. (1939a). *Rec. Math.* (*Mat. Sbornik*) *N.S.* **5(47),** 101–144.

Gantmacher, F. R. (1939b). *Rec. Math.* (*Mat. Sbornik*) *N.S.* **5(47),** 217–250.

Gantmacher, F. R. (1959). "The Theory of Matrices", Vol. 1. Chelsea Publishing, New York.

Gatto, R. (1964). *Nuovo Cimento Suppl.* **4,** 414–464.

Gel'fand, I. M. and Shapiro, A. Ya (1956). *Amer. Math. Soc. Trans. Series* 2, **2,** 207–316. [Original (1952): *Uspeki Mat. Nauk* (*N.S.*) **7, No. 1 (47),** 3–117.]

Gel'fand, I. M. Minlos, R. A. and Shapiro, Z. Ya. (1963). "Representations of the Rotation and Lorentz Groups and their Applications". Pergamon Press, London.

Gell-Mann, M. (1953). *Phys. Rev.* **92,** 833–834.

Gell-Mann, M. (1961). California Institute of Technology Report CTSL-20, unpublished.

Gell-Mann, M. (1962). *Phys. Rev.* **125,** 1067–1084.

Gell-Mann, M. (1964). *Phys. Letts.* **8,** 214–215.

Gell-Mann, M. and Ne'eman, Y. (1964). "The Eightfold Way". W. A. Benjamin, New York.

Gell-Mann, M. and Pais, A. (1955). *In* "Proceedings of the 1954 Glasgow Conference on Nuclear and Meson Physics" (Eds. E. H. Bellamy and R. G. Moorhouse), 342–352. Pergamon Press, London and New York.

Gell-Mann, M., Ramond, P. and Slansky, R. (1978). *Rev. Mod. Phys.* **50,** 721–744.

Georgi, H. and Glashow, S. L. (1972). *Phys. Rev.* **D6,** 429–431.

Georgi, H. and Glashow, S. L. (1974). *Phys. Rev. Letts.* **32,** 438–441.
Gerstein, I. S. (1965). *J. Math. Phys.* **6,** 469–473.
Gerstein, I. S. and Whippman, M. L. (1965). *Phys. Rev.* **137,** B1522–1529.
Ghirardi, G. C. (1973). *Forts. d. Phys.* **21,** 653–666.
Girardi, G., Sciarrino, A. and Sorba, P. (1982). *J. Phys.* **A15,** 1119–1129.
Girardi, G., Sciarrino, A. and Sorba, P. (1983). *J. Phys.* **A16,** 2609–2614.
Glashow, S. L. (1980). *Rev. Mod. Phys.* **52,** 539–543.
Glashow, S. L., Iliopoulos, J. and Maiani, L. (1970). *Phys. Rev.* **D2,** 1285–1292.
Glück, M., Gur, Y. and Zak, J. (1967). *J. Math. Phys.* **8,** 787–790.
Goldstein, H. (1950). "Classical Mechanics". Addison-Wesley, Reading, Massachusetts.
Goldstone, J. (1961). *Nuovo Cimento* **19,** 154–164.
Goldstone, J., Salam, A. and Weinberg, S. (1962). *Phys. Rev.* **127,** 965–970.
Good, R. H. (1955). *Rev. Mod. Phys.* **27,** 187–211.
Goto, M. and Grosshans, F. D. (1978). "Semi-simple Lie Algebras". Marcel Dekker, New York and Basel.
Gourdin, M. (1967). "Unitary Symmetries". North Holland, Amsterdam.
Greenberg, D. F. (1966). *Amer. J. Phys.* **34,** 1101–1109.
Grib, A. A., Damaskinskii, E. V. and Maksimov, V. M. (1970). *Uspeki. Fiz. Nauk.* **102,** 587–620. [Translation (1971): *Soviet Phys. Uspekhi* **13,** 798–815.]
Gross, D. and Wilczek, F. (1973). *Phys. Rev. Letts.* **30,** 1343–1346.
Gruber, B. (1968). *Ann. Inst. Henri Poincare* **A8,** 43–51.
Gruber, B. (1970a). *J. Math. Phys.* **11,** 1783–1790.
Gruber, B. (1970b). *J. Math. Phys.* **11,** 3077–3090.
Gruber, B. and O'Raifeartaigh, L. (1964). *J. Math. Phys.* **5,** 1796–1804.
Gruber, B., Han, B. H. and Saldanha, J. A. (1971). *J. Math. Phys.* **12,** 821–827.
Gunaydin, M. (1973). *Phys. Rev.* **D7,** 1924–1927.
Gunaydin, M. and Gursey, F. (1973), *J. Math. Phys.* **14,** 1651–1667.
Gunn, J. C. (1962). *In* "Quantum Theory" (Ed. D. R. Bates), Vol. III, 195–285. Academic Press, Orlando, New York and London.
Guralnik, G. S., Hagen, C. R. and Kibble, T. W. B. (1968). *In* "Advances in Particle Physics", Vol. II, 567–708. Interscience, New York and London.
Haake, E. M., Moffat, J. W. and Savaria, P. (1976). *J. Math. Phys.* **17,** 2041–2066.
Haar, A. (1933). *Ann. Math.* **34,** 147–169.
Halmos, P. R. (1950). "Measure Theory". D. Van Nostrand, New York.
Halpern, F. R. (1968). "Special Relativity and Quantum Mechanics". Prentice-Hall, Englewood Cliffs, New Jersey.
Hamermesh, M. (1962). "Group Theory and its Application to Physical Problems". Addison-Wesley, Reading, Massachusetts.
Harish-Chandra (1951). *Trans. Amer. Math. Soc.* **75,** 185–243.
Harris, E. G. (1972). "A Pedestrian Approach to Quantum Field Theory". Wiley-Interscience, New York and London.
Harris, D. C. and Bertolucci, M. D. (1978). "Symmetry and Spectroscopy. An Introduction to Vibrational and Electronic Spectroscopy". Oxford U.P., Oxford.
Hasert, F. J., Kabe, S., Krenz, W., von Krogh, J., Lanske, D., Morfin, J., Schultze, K., Weerts, H., Bertrand-Coremans, G. H., Sacton, J., van Doninck, W., Vilain, P., Camerini, U., Cundy, D. C., Baldi, R., Danil-

chenko, I., Fry, W. F., Haidt, D., Natali, S., Mussett, P., Osculati, B., Palmer, R., Pattison, J. B. M., Perkins, D. H. Pullia, A., Rousset, A., Venus, W., Wachsmuth, H., Brisson, V., Degrange, B., Haguenauer, M., Kluberg, L., Nguyen-Khac, U., Petiau, P., Belotti, E., Bonetti, S., Cavalli, D., Conta, C., Fiorini, E., Rollier, M., Aubert, B., Blum, D., Chounet, L. M., Heusse, P., Legarrigue, A., Lutz, A. M., Orkin-Lecourtois, A., Vialle, J. P., Bullock, F. W., Esten, M. J., Jones, T. W., McKenzie, J., Michette, A. G., Myatt, G. and Scott, W. G. (1973). *Phys. Letts.* **B46,** 138–140.

Hausdorff, F. (1906). *Leipz. Ber.* **58,** 19–48.

Hausner, M. and Schwartz, J. T. (1968). "Lie Groups. Lie Algebras". Thomas Nelson and Sons, London.

Hegerfeldt, G. C. and Hennig, J. (1968). *Forts. der Phys.* **16,** 491–543.

Helgason, S. (1962). 'Differential Geometry and Symmetric Spaces". Academic Press, Orlando, New York and London.

Helgason, S. (1978). "Differential Geometry, Lie Groups, and Symmetric Spaces". Academic Press, Orlando, New York and London.

Henley, E. M. and Thirring, W. (1962). "Elementary Quantum Field Theory". McGraw-Hill, New York and London.

Henry, N. F. M. and Lonsdale, K. (1965). "International Tables for X-ray Crystallography", Vol. I. International Union of Crystallography, Kynoch Press, Birmingham, England.

Herring, C. (1937a). *Phys. Rev.* **52,** 361–365.

Herring, C. (1937b). *Phys. Rev.* **52,** 365–373.

Herring, C. (1942). *J. Franklin Inst.* **233,** 525–543.

Hewitt, E. and Ross, K. A. (1963). "Abstract Harmonic Analysis", Vol. I. Springer-Verlag, Berlin.

Higgs, P. W. (1964a). *Phys. letts.* **12,** 132–133.

Higgs, P. W. (1964b). *Phys. Rev. Letts.* **13,** 508–509.

Higgs, P. W. (1966). *Phys. Rev.* **145,** 1156–1163.

Higgs, P. W. (1974). *In* "Phenomenology of Particles at High Energies" (Eds. R. L. Crawford and R. Jennings), 529–552. Academic Press, Orlando, New York and London.

Howarth, D. J. and Jones, H. (1952). *Proc. Phys. Soc.* **A65,** 355–368.

Hughes, I. S. (1972). "Elementary Particles". Penguin Books, Harmondsworth.

Humphreys, J. E. (1972). "Introduction to Lie Algebras and Representation Theory". Springer-Verlag, New York, Heidelberg and Berlin.

Ince, E. (1956). "Ordinary Differential Equations". Dover, New York.

Inönü, E. and Wigner, E. P. (1952). *Nuovo Cimento.* **9,** 705–718.

Itzykson, C. and Nauenberg, M. (1966). *Rev. Mod. Phys.* **38,** 95–120.

Itzykson, C. and Zuber, J-B. (1980). "Quantum Field Theory". McGraw-Hill, New York.

Jacobson, N. (1962). "Lie Algebras". Interscience, New York.

Jacobson, N. (1971). "Exceptional Lie Algebras". Marcel Dekker, New York.

Jauch, J. M. and Hill, E. L. (1940). *Phys. Rev.* **57,** 641–645.

Joshi, S. K. and Rajagopal, A. K. (1968). *In* "Solid State Physics, Advances in Research and Applications" (Eds. F. Seitz, D. Turnbull and H. Ehrenreich), Vol. 22, 159–312. Academic Press, Orlando, New York and London.

Kemmer, N. (1938). *Proc. Camb. Phil. Soc.* **34,** 354–364.

Kibble, T. W. B. (1967). *Phys. Rev.* **155,** 1554–1561.

Killing, W. (1888). *Math. Ann.* **31,** 252–290.

Killing, W. (1889a). *Math. Ann.* **33,** 1–48.
Killing, W. (1889b). *Math. Ann.* **34,** 57–122.
Killing, W. (1890). *Math. Ann.* **36,** 161–189.
Kim, J. E., Langacker, P., Levine, M. and Williams, H. H. (1981). *Rev. Mod. Phys.* **53,** 211–252.
Kitz, A. (1965). *Physica Status Solidi* **8,** 813–829.
Kleima, D., Holman, W. J. III and Biedenharn, L. C. (1968). *In* "Group Theory and its Applications" (Ed. E. M. Loebl), 1–56. Academic Press, Orlando, New York and London.
Klimyk, A. U. (1968a). *Amer. Math. Soc. Trans. Series 2,* **76,** 63–73. [Original (1966): *Ukrain. Mat. Z.* **18, No. 5,** 19–27.]
Klimyk, A. U. (1968b). *Amer. Math. Soc. Trans. Series 2,* **76,** 75–88. [Original (1967): *Ukrain. Mat. Z.* **19, No. 2,** 11–22.]
Klimyk, A. U. (1971). *Teor. and Mat. Fiz.* (USSR) **8,** 55–60.
Klimyk, A. U. (1972). *Teor. and Mat. Fiz.* (USSR) **13,** 327–342. [Translation (1972): *Theor. and Math. Phys.* **13,** 1171–1182.]
Klimyk, A. U. (1975). *Repts. Math. Phys.* **7,** 153–166.
Kobayashi, S. and Nomizu, K. (1963). "Foundations of Differential Geometry", Vol. I. J. Wiley and Sons, New York and London.
Kokkedee, J. J. J. (1969). "The Quark Model". W. A. Benjamin, New York.
Konopleva, N. P. and Popov, V. N. (1981). "Gauge Fields". Harwood Academic Publishers, London.
Konuma, M., Shima, K. and Wada, M. (1963). *Progr. Theor. Phys., Suppl.* **28,** 1–131.
Koster, G. F. (1957). *In* "Solid State Physics, Advances in Research and Applications" (Eds. F. Seitz and D. Turnbull), Vol. 5, 173–256. Academic Press, Orlando, New York and London.
Koster, G. F. (1958). *Phys. Rev.* **109,** 227–231.
Koster, G. F., Dimmock, J. O., Wheeler, R. E. and Statz, H. (1964). "Properties of the Thirty-two Point Groups". M.I.T. Press, Cambridge, Massachusetts.
Kovalev, O. V. (1965). "Irreducible Representations of the Space Groups". Gordon and Breach, New York. [Translation of book originally published by the Academy of Science USSR Press, Kiev. (1961).]
Krajcik, R. A. and Nieto, M. M. (1974). *Phys. Rev.* **D10,** 4049–4063.
Krajcik, R. A. and Nieto, M. M. (1975a). *Phys. Rev.* **D11,** 1442–1458.
Krajcik, R. A. and Nieto, M. M. (1975b). *Phys. Rev.* **D11,** 1459–1471.
Krajcik, R. A. and Nieto, M. M. (1976a). *Phys. Rev.* **D13,** 924–941.
Krajcik, R. A. and Nieto, M. M. (1976b). *Phys. Rev.* **D14,** 418–436.
Krajcik, R. A. and Nieto, M. M. (1977a). *Phys. Rev.* **D15,** 433–444.
Krajcik, R. A. and Nieto, M. M. (1977b). *Phys. Rev.* **D15,** 445–452.
Krajcik, R. A. and Nieto, M. M. (1977c). *Amer. J. Phys.* **45,** 818–822.
Krämer, M. (1978). *Repts. Math. Phys.* **13,** 295–304.
Kudryavtseva, N. V. (1967). *Fiz. Tverd. Tela* **9,** 2364–2368. [Translation (1968): *Soviet Physics, Solid State* **9,** 1850–1853.]
Lax, M. (1965). *Phys. Rev.* **138,** A793–802.
Lax, M. and Hopfield, J. J. (1961). *Phys. Rev.* **124,** 115–123.
Leader, E. and Predazzi, E. (1982). "An Introduction to Gauge Theories and the New Physics". Cambridge U.P., Cambridge, England.
Lee, T. D. and Yang, C. N. (1956). *Phys. Rev.* **104,** 254–258.
Lee, T. D. (1982). "Particle Physics and an Introduction to Field Theory". Harwood Academic Publishers, New York.

Leites Lopes, J. (1981). "Gauge Field Theories—An Introduction". Pergamon Press, Oxford.
Lenz, W. (1924). *Z. Phys.* **24,** 197–207.
Lichtenberg, D. B. (1978). "Unitary Symmetry and Elementary Particles". Academic Press, Orlando, New York and London.
Lipsman, R. L. (1974). "Group Representations", Lecture Notes in Mathematics, Vol. 388. Springer-Verlag, Berlin.
Litvin, D. B. and Zak, J. (1968). *J. Math. Phys.* **9,** 212–221.
Lomont, J. S. (1959). "Applications of Finite Groups". Academic Press, Orlando, New York and London.
London, G. W. (1964). *Fortschr. Phys.* **12,** 643–666.
Longo, M. J. (1973). "Fundamentals of Elementary Particle Physics". McGraw-Hill Kogakusha, Tokyo.
Loomis, L. H. (1953). "An Introduction to Abstract Harmonic Analysis". D. Van Nostrand, Princeton.
Luehrmann, A. W. (1968). *Adv. Phys.* **17,** 1–77.
Luttinger, J. M. (1960). *Phys. Rev.* **119,** 1153–1163.
McGlinn, W. D. (1964). *Phys. Rev. Letts.* **12,** 467–469.
McIntosh, H. V. (1963). *J. Mol. Spectry.* **10,** 51–74.
McKay, W., Patera, J. and Sharp, R. T. (1976). *J. Math. Phys.* **17,** 1371–1375.
McKay, W. G. and Patera J. (1981). "Tables of Dimensions, Indices, and Branching Rules for Representations of Simple Lie Algebras". Marcel Dekker, Basel.
Mackey, G. W. (1955). "Theory of Group Representations". Notes by Drs. Fell and Lowdenslager, Dept. of Math., University of Chicago.
Mackey, G. W. (1963a). "Mathematical Problems in Relativistic Physics" (Ed. I. E. Segal), 113–130. Amer. Math. Soc., Providence, Rhode Island.
Mackey, G. W. (1963b). *Bull. Amer. Math. Soc.* **69,** 628–686.
Mackey, G. W. (1968). "Induced Representation of Groups and Quantum Mechanics". W. A. Benjamin, New York.
Mackey, G. W. (1976). "The Theory of Unitary Group Representations". University of Chicago Press, Chicago and London.
Majorana, E. (1937). *Nuovo Cimento* **14,** 171–184.
Malenka B. J. and Primakoff, H. (1957). *Phys. Rev.* **105,** 338–343.
Mal'tsev, A. I. (1943). *Dokl. Akad. Nauk. S.S.S.R.* **40,** 108–110.
Mal'tsev, A. I. (1945). *Rec. Math.* (*Mat. Sbornik*) *N.S.* **16 (58),** 163–190.
Mal'tsev, A. I. (1946). *Rec. Math.* (*Mat. Sbornik*) *N.S.* **19 (61),** 523–524.
Mandl, F. (1959). "Introduction to Quantum Field Theory". Interscience, New York and London.
Maradudin, A. A., Montroll, E. W. and Weiss, G. H. (1963). "Theory of Lattice Dynamics in the Harmonic Approximation". Academic Press, Orlando, New York and London.
Marciano, W. and Pagels, H. (1978). *Phys. Repts.* **36,** 137–276.
Martin, R. M. (1969). *Phys. Rev.* **186,** 871–884.
Mathews, P. T. (1967). *In* "High Energy Physics" (Ed. E. H. Burhop), Academic Press, Orlando, New York and London.
Mehta, M. L. (1966). *J. Math. Phys.* **7,** 1824–1832.
Mehta, M. L. and Srivastava, P. K. (1966). *J. Math. Phys.* **7,** 1833–1835.
Miller, S. C. and Love, W. H. (1967). "Irreducible Representations of Space Groups". Pruett Press, Boulder, Colorado.
Miller, W. (1964). *Commun. Pure Appl. Math.* **17,** 527–540.

Miller, W. (1972). "Symmetry Groups and their Applications". Academic Press, Orlando, New York and London.

Mura, T. (1982). "Quantum Chromodynamics—Short Distance Regime". World Publishing, Singapore.

Naimark, M. A. (1957). *Amer. Math. Soc. Trans. Series 2*, **6,** 379–458. [Original (1954): *Uspeki Mat. Nauk* (*N.S.*) **9, No. 4(62),** 19–93.]

Naimark, M. A. (1964). "Linear Representations of the Lorentz Group". Pergamon Press, London.

Nakano, T. and Nishijima, K. (1953). *Progr. Theor. Phys.* **10,** 581–582.

Ne'eman, Y. (1961). *Nucl. Phys.* **26,** 222–229.

Ne'eman, Y. (1965). *In* "Progress in Elementary Particle and Cosmic Ray Physics" (Eds. J. G. Wilson and S. A. Wouthysen), 69–118. North Holland, Amsterdam.

Newton, T. D. (1965). *In* "Theory of Groups in Classical and Quantum Physics". (Ed. T. Kahan), Vol. I, 207–255. Oliver and Boyd, Edinburgh and London.

Niederer, U. H. and O'Raifeartaigh, L. (1974a). *Forts. der Physik* **22,** 111–129.

Niederer, U. H. and O'Raifeartaigh, L. (1974b). *Forts. der Physik* **22,** 131–157.

Nishijima, K. (1954). *Progr. Theor. Phys.* **12,** 107–108.

Novosadov, B. K., Saulevich, L. K., Sviridov, D. T. and Smirnov, Yu. F. (1969). *Dokl. Akad. Nauk. S.S.S.R.* **184,** 82–84. [Translation (1969): *Soviet Phys. Doklady* **14,** 50–52.]

Okubo, S. (1962). *Prog. Theor. Phys.* **27,** 949–966.

Okubo, S. (1977). *J. Math. Phys.* **18,** 2382–2394.

Okubo, S. (1982). *J. Math. Phys.* **23,** 8–20.

Okun, L. B. (1981). "Leptons and Quarks". North Holland, Amsterdam.

Olbrychski, K. (1963). *Phys. Status Solidi* **3,** 1868–1875.

Omnes, R. (1970). "Introduction to Particle Physics". Wiley-Interscience, London and New York.

Onodera, Y. and Okazaki, M. (1966). *Proc. Phys. Soc. Japan* **21,** 2400–2408.

Opechowski, W. (1940). *Physica* **7,** 552–562.

O'Raifeartaigh, L. (1965). *Phys. Rev. Letts.* **14,** 575–577.

O'Raifeartaigh, L. (1968). *In* "Group Theory and its Applications" (Ed. E. M. Loebl), Vol. I, 469–540. Academic Press, Orlando, New York and London.

O'Raifeartaigh, L. (1979). *Rep. Progr. Phys.* **42,** 159–223.

Pais, V. (1952). *Phys. Rev.* **86,** 663–672.

Parmenter, R. H. (1955). *Phys. Rev.* **100,** 573–579.

Patera, J. and Sankoff, D. (1973). "Tables of Branching Rules for Representations of Simple Lie Algebras". L'Université de Montreal, Montreal.

Patera, J., Sharp, R. T. and Winternitz, P. (1976). *J. Math. Phys.* **17,** 1972–1979.

Patera, J., Sharp, R. T. and Winternitz, P. (1977). *J. Math. Phys.* **18,** 1519.

Pauli, W. (1926). *Z. Phys.* **36,** 336–363.

Pauli, W. (1958). "Theory of Relativity". Pergamon Press, London.

Peter, F. and Weyl, H. (1927). *Math. Ann.* **97,** 737–755.

Phillips, J. C. (1956). *Phys. Rev.* **104,** 1263–1277.

Pilkuhn, H. M. (1979). "Relativistic Particle Physics". Springer-Verlag, New York, Heidelberg and Berlin.

Pincherle, L. (1960). *Rep. Progr. Phys.* **23,** 355–394.

Pincherle, L. (1971). "Electronic Energy Bands in Solids". McDonald, London.

Pontrjagin, L. (1946). "Topological Groups". Princeton U.P., Princeton.

Price, J. F. (1977). "Lie Groups and Compact Groups". Cambridge U.P., Cambridge, England.

Rabi, V., Campbell, G. and Wali, K. C. (1975). *J. Math. Phys.* **16,** 2494–2506.

Racah, G. (1937). *Nuovo Cimento* **14,** 322–328.

Racah, G. (1942). *Phys. Rev.* **62,** 438–462.

Racah, G. (1950). *Rend. Lincei* **8,** 108–112.

Racah, G. (1951). *Ergebnisse der Exacten Naturwissenschaften,* **37,** 28–84.

Raghavacharyulu, I. V. V. (1961). *Can. J. Phys.* **39,** 830–840.

Raghavacharyulu, I. V. V. and Bhavanacharyulu, I. (1962). *Can. J. Phys.* **40,** 1490–1495.

Ramond, P. (1976). "Introduction to Exceptional Lie Groups and Algebras". California Institute of Technology Report CALT 68 577, Pasadena.

Ramond, P. (1981). "Field Theory. A Modern Primer". Benjamin/Cummings, Reading, Massachusetts.

Rashba, E. I. (1959). *Fiz. Tvend. Tela* **1,** 407–421. [Translation (1959): *Soviet Phys. Solid State* **1,** 368–380.]

Riesz, F. and Sz.-Nagy, B. (1956). "Functional Analysis". Blackie and Son, London and Glasgow.

Reitz, J. R. (1955). *In* "Solid State Physics, Advances in Research and Applications", Vol. 1 (Eds. F. Seitz and D. Turnbull), 1–95. Academic Press, Orlando, New York and London.

Roman, P. (1969). "Introduction to Quantum Field Theory". J. Wiley and Sons, New York and London.

Rose, M. E. (1957). "Elementary Theory of Angular Momentum". J. Wiley and Sons, New York.

Ross, G. G. (1981a). *In* "Gauge Theories and Experiments at High Energies" (Eds. K. C. Bowler and D. G. Sutherland), 1–100. Scottish Universities Summer Schools in Physics Publications, Edinburgh.

Ross, G. G. (1981b). *Rep. Progr. Phys.* **44,** 655–718.

Rotman, J. J. (1965). "The Theory of Groups". Allyn and Bacon, Boston.

Rudra, P. (1965). *J. Math. Phys.* **6,** 1278–1282.

Runge, C. (1923). "Vector Analysis". Methuen and Co., London.

Ryder, L. H. (1975). "Elementary Particles and Their Symmetries". Gordon and Breach, New York and London.

Sagle, A. A. and Walde, R. E. (1973). "Introduction to Lie Groups and Lie Algebras". Academic Press, Orlando, New York and London.

Sakata, S. (1956). *Progr. Theor. Phys.* **16,** 686–688.

Sakurai, J. J. (1962). *Phys. Rev. Letts.* **9,** 472–475.

Salam, A. (1968). *In* "Proceedings of the Eighth Nobel Symposium on Elementary Particle Theory, Relativistic Groups, and Analyticity" (Ed. N. Svartholm), 367–377. Almquist and Wikell, Stockholm.

Salam, A. (1980). *Rev. Mod. Phys.* **52,** 525–538.

Salam, A. and Strathdee, J., (1974). *Nucl. Phys.* **B76,** 477–482.

Samelson, H. (1969). "Notes on Lie Algebras". Van Nostrand Reinhold, New York.

Schiff, L. I. (1968). "Quantum Mechanics". McGraw-Hill, New York and London.

Schlosser, H. (1962). *J. Phys. Chem. Solids* **23,** 963–969.

Schönfliess, A. (1923). "Theorie der Kristallstruktur". Bomtraeger, Berlin.

Schur, I. (1904). *J. Reine Angew. Math.* **127,** 20–50.

Schur, I. (1907). *J. Reine Angew. Math.* **132,** 85–137.
Schweber, S. S. (1961). "An Introduction to Relativistic Quantum Field Theory". Harper and Row, New York.
Schwinger, J. (1952). "On Angular Momentum". U.S. Atomic Energy Commission, NYO-3071.
Shaw, R. (1955). "The Problem of Particle Types and Other Contributions to the Theory of Elementary Particles". Ph.D. Thesis, Cambridge University, unpublished.
Sheka, V. I. (1960). *Fiz. Tvend. Tela,* **2,** 1211–1219. [Translation (1960): *Soviet Phys. Solid State* **2,** 1096–1104.]
Shephard, G. C. (1966). "Vector Spaces of Finite Dimension". Oliver and Boyd, Edinburgh and London.
Shubnikov, A. V. and Koptsik, V. A. (1974). "Symmetry in Science and Art". Plenum, New York.
Simmons, G. F. (1963). "Topology and Modern Analysis". McGraw-Hill, New York and London.
Simms, D. J. (1973). *Z. Naturf.* **28a,** 538–540.
Sirota, A. I. and Solodovnikov, A. S. (1963). *Russian Math. Surveys.* **18 (3),** 85–140. [Original (1963): *Uspeki Mat. Nauk.* **18,** 87–140.]
Slansky, R. (1981). *Phys. Repts.* **79,** 1–128.
Slechta, J. (1966). *Cesk. Casopis Fys.* **A16,** 1–5.
Smorodinsky, Ya. A. (1965). *Soviet Phys. Uspekhi* **7,** 637–655.
Solodovnikov, A. S. (1959). *Dokl. Akad. Nauk S.S.S.R.* **129,** 272–275.
Speiser, D. R. (1964). *In* "Group Theoretical Concepts and Methods in Elementary Particle Physics" (Ed. F. Gursey), 201–276. Gordon and Breach, New York and London.
Stein, E. M. (1965). *In* "High Energy Physics and Elementary Particles" (Ed. A. Salam), 563–584. International Atomic Energy, Vienna.
Steinberg, A. (1961). *Bull. Amer. Math. Soc.* **67,** 406–407.
Streitwolf, H. W. (1971). "Group Theory in Solid State Physics". McDonald, London.
Tarjanne, P. (1964). *Phys. Rev.* **136,** B1532–1534.
Taylor, J. C. (1976). "Gauge Theories of Weak Interactions". Cambridge U.P., Cambridge, England.
Teleman, E. and Glodeanu, A. (1967). *Rev. Roumaine Sci. Tech. Electrotech. Energet.* **12,** 551–560.
't Hooft, G. (1971a) *Nucl. Phys.* **B33,** 173–199.
't Hooft, G. (1971b). *Nucl. Phys.* **B35,** 167–188.
Tits, J. (1953). *Bull. Acad. Roy. Belg. Sci.* **39,** 309–329.
Tits, J. (1959). *Pub. Math. Inst. Hauts Etudes Sci.* **2,** 13–60.
Toyama, H. (1948). *Proc. Japan Acad.* **24,** 13–16.
Trilling, G. H. (1981). *Phys. Repts.* **75,** 57–124.
van den Broek, P. M. (1979). *J. Math. Phys.* **20,** 2028–2035.
van den Broek, P. M. and Cornwell, J. F. (1978). *Phys. Stat. Sol.* (*b*) **90,** 211–224.
van Hove, L. (1953). *Phys. Rev.* **89,** 1189–1193.
Varadarajan, V. S. (1974). "Lie Groups, Lie Algebras, and their Representations". Prentice-Hall, Englewood Cliffs, New Jersey.
Velo, G. and Wightman, A. S. (1978). "Invariant Wave Equations". Springer-Verlag, Berlin.
Voisin, J. (1965a). *J. Math. Phys.* **6,** 1519–1529.

Voisin, J. (1965b). *J. Math. Phys.* **6,** 1822–1832.
Voisin, J. (1966). *J. Math. Phys.* **7,** 2235–2237.
von der Lage, F. C. and Bethe, H. A. (1947). *Phys. Rev.* **71,** 612–622.
Warner, F. W. (1971). "Foundations of Differentiable Manifolds and Lie Groups". Scott, Foreman and Co., Glenview, Illinois and London.
Warner, G. (1972). "Harmonic Analysis on Semi-Simple Lie Groups", Vol. I. Springer-Verlag, Berlin, Heidelberg and New York.
Weinberg, S. (1967). *Phys. Rev. Letts.* **19,** 1264–1266.
Weinberg, S. (1971). *Phys. Rev. Letts.* **27,** 1688–1691.
Weinberg, S. (1973). *Phys. Rev.* **D8,** 4482–4498.
Weinberg, S. (1974). *Rev. Mod. Phys.* **46,** 255–278.
Weinberg, S. (1980). *Rev. Mod. Phys.* **52,** 515–523.
Weyl, H. (1925). *Math. Zeits.* **23,** 271–309.
Weyl, H. (1926a). *Math. Zeits.* **24,** 328–376.
Weyl, H. (1926b). *Math. Zeits.* **24,** 377–395.
Weyl, H. (1939). "The Classical Groups". Princeton U.P., Princeton.
Wigner, E. P. (1939). *Ann. Math.* **40,** 149–204.
Wigner, E. P. (1959). "Group Theory and its Application to the Quantum Mechanics of Atomic Spectra". Academic Press, Orlando, New York and London.
Wigner, E. P. (1964). *In* "Group Theoretical Methods and Concepts in Elementary Particle Physics" (Ed. F. Gursey), 37–80. Gordon and Breach, New York and London.
Williams, W. S. C. (1971). "An Introduction to Elementary Particles". Academic Press, Orlando, New York and London.
Wilson, E. B., Decius, J. C. and Cross, P. C. (1955). "Molecular Vibrations". McGraw-Hill, New York and London.
Wondratschek, H. and Neubüser, J. (1967). *Acta. Cryst.* **23,** 349–352.
Wood, J. H. (1962). *Phys. Rev.* **126,** 517–527.
Wulfman, C. E. (1971). *In* "Group Theory and its Applications" (Ed. E. M. Loebl), Vol. II, 145–197. Academic Press, Orlando, New York and London.
Wybourne, B. G. (1970). "Symmetry Principles and Atomic Spectroscopy". J. Wiley and Sons, New York.
Wyckoff, R. W. G. (1963). "Crystal Structures 1". Interscience, New York.
Wyckoff, R. W. G. (1964). "Crystal Structures 2". Interscience, New York.
Wyckoff, R. W. G. (1965). "Crystal Structures 3". Interscience, New York.
Yang, C. N. and Mills, R. L. (1954). *Phys. Rev.* **96,** 191–195.
Young, A. (1928). *Proc. London Math. Soc.* **28,** 255–292.
Zak, J. (1960). *J. Math. Phys.* **1,** 165–171.
Zak, J. (1962). *J. Math. Phys.* **3,** 1278–1279.
Zee, A. (1982). "Unity of Forces in the Universe". World Publishing, Singapore.
Ziman, J. M. (1960). "Electrons and Phonons". Oxford U.P., London and New York.
Zweig, G. (1964). *CERN Repts.* TH 401 and TH 402.

Subject Index

Printed and bound by CPI Group (UK) Ltd, Croydon, CR0 4YY
08/10/2024
01042187-0001